Pergamon Series in Analytical Chemistry
Volume 5

General Editors: R. Belcher[†] (Chairman), D. Betteridge & L. Meites

Applied Complexometry

Related Pergamon Titles of Interest

BOOKS

Earlier Volumes in the Pergamon Series in Analytical Chemistry

Volume 1
MEITES: An Introduction to Chemical Equilibrium and Kinetics

Volume 2
PATAKI & ZAPP: Basic Analytical Chemistry

Volume 3
YINON & ZITRIN: The Analysis of Explosives

Volume 4
JEFFERY & HUTCHISON: Chemical Methods of Rock Analysis, 3rd Edition

PŘIBIL: Analytical Applications of EDTA and Related Compounds

JOURNALS

Journal of Pharmaceutical and Biomedical Analysis
Talanta

Full details of all the above publications/free specimen copy of any Pergamon journal available on request from your nearest Pergamon office.

RONALD BELCHER

Professor Emeritus, Birmingham University 1909-1982

Applied
Complexometry

by
RUDOLF PŘIBIL
Czechoslovak Academy of Sciences

Translated by
Rudolf Přibil and Madeleine Štulíková

Edited by
ROBERT A. CHALMERS
University of Aberdeen

PERGAMON PRESS
OXFORD · NEW YORK · TORONTO · SYDNEY · PARIS · FRANKFURT

U.K.	Pergamon Press Ltd., Headington Hill Hall, Oxford OX3 0BW, England
U.S.A.	Pergamon Press Inc., Maxwell House, Fairview Park, Elmsford, New York 10523, U.S.A.
CANADA	Pergamon Press Canada Ltd., Suite 104, 150 Consumers Rd., Willowdale, Ontario M2J 1P9, Canada
AUSTRALIA	Pergamon Press (Aust.) Pty. Ltd., P.O. Box 544, Potts Point, N.S.W. 2011, Australia
FRANCE	Pergamon Press SARL, 24 rue des Ecoles, 75240 Paris, Cedex 05, France
FEDERAL REPUBLIC OF GERMANY	Pergamon Press GmbH, 6242 Kronberg-Taunus, Hammerweg 6, Federal Republic of Germany

This edition 1982
Translated by Rudolf Přibil and
Madeleine Štulíková from the Czech
Komplexometrie, published by SNTL,
Prague, 1977

Library of Congress Cataloging in Publication Data
Přibil, Rudolf.
Applied complexometry.
(Pergamon series in analytical chemistry;
v. 5)
Translation of : Komplexometrie.
Includes index.
1. Complexometric titration. I. Title.
II. Series.
QD111.P713 1982   545′.2   82-16517

British Library Cataloguing in Publication Data
Přibil, Rudolf
Applied complexometry. —(Pergamon series in
analytical chemistry; 5)
1. Chemistry, Analytic 2. Complex compounds
I. Title II. Komplexometrie. *English*
543   QD77
ISBN 0-08-026277-5

In order to make this volume available as economically and as rapidly as possible the authors' typescripts have been reproduced in their original forms. This method unfortunately has its typographical limitations but it is hoped that they in no way distract the reader.

Printed in Great Britain by A. Wheaton & Co. Ltd., Exeter

To my wife

VLASTA

Preface

It is nearly forty years since Schwarzenbach surprised the analytical world by his papers on the titrimetric determination of calcium and magnesium with ethylenediaminetetra-acetic acid (EDTA), using first classical acid-base indicators and then murexide and Eriochrome Black T, and opened up a new field of titrimetric analysis, complexometric titration (complexometry, chelatometry), for further development.

The applicability of complexometry has been extended to the determination of more than 50 elements by the introduction of new indicators, masking agents and suitable separation methods for use in the fast analysis of multicomponent samples. More than 2000 papers have been published on the analysis of alloys, ores, minerals, slags, glass, fuels, pharmaceutical products, plating baths, fertilizers, soils, etc.

Complexometric titrations - at first sight so simple to perform - have been given a more thorough theoretical treatment than any other branch of analytical chemistry. The well known monographs of Schwarzenbach, Flaschka, and Ringbom, deal in all possible detail with the theory of complexometry. Only in the books by Schwarzenbach, Flaschka, Welcher, and West, do we find "recommended procedures" for single elements, chosen more or less arbitrarily. Practical applications are dealt with by giving references to the literature, which may not be available to all.

The analytical chemist very often meets the problem of how to apply EDTA and similar reagents for his routine work, and asks himself how to replace time-consuming older methods by more rapid volumetric methods. This monograph deals mostly with the applied procedures that have been developed in the last three decades. In this respect, complexometry developed so quickly that many methods described one day were replaced the next by more convenient ones, based, for example, on special procedures for opening out of the sample, which were designed expressly for complexometric purposes. Ion-exchange separation and solvent extraction would have lost much of their appeal and applicability without complexometric "finishes".

As already said, this book deals with complexometry from the practical point of view. The "theory" is reduced as far as possible to the

vii

most important facts, which are needed for its understanding. The
chapter dealing with complexometric (metallochromic) indicators
describes only those that are really used in practice.   Among almost
100 known indicators there are only a few which are really needed and
are also marketed.   In Chapter 6 the reader will find a short
description of the complexometric behaviour of the elements, together
with simple procedures and short hints on how to make the methods
more selective.

Practical applications are dependent on the composition of the sample.
In many cases the samples are relatively simple and the determination
of only one or two elements is required (plating baths, pigments,
etc.).   On the other hand, in the analysis of alloys or silicates,
a whole series of elements can be determined by EDTA titration of one
or more aliquots of a stock sample solution.   It was a hard task to
prepare this last part without unnecessary repetition of procedures
or excessive cross-reference to earlier sections.

We are aware that many chemists will be interested in only their own
field of work, in which they will be highly expert, and they may find
a favourite method omitted.   In spite of this we hope that the book
will be of some help to all who are interested in this unique
titrimetric method.   The Czech edition has been revised and updated.

This book was originally intended to form part of the well known
series of analytical monographs founded by Professor Ronald Belcher
and published by Pergamon Press.   Through various vicissitudes, its
English version was delayed, and Professor Belcher's death during the
very last days of the preparation of the text for press robbed him of
seeing the book finally appear in print.   We would therefore ask the
reader to regard the book as a tribute in memory of Professor Belcher,
who was a good friend of both of us.

                                          R. Přibil, Prague

                                          R.A. Chalmers, Aberdeen

Contents

CHAPTER 1

Volumetric Reagents in Complexometry

Of some 80 – 100 aminopolycarboxylic acids possessing at least one complex-forming grouping $-N(CH_2COOH)_2$, only a few can be used as volumetric reagents.

Let us start with the simplest compounds iminodiacetic acid (I: IDA) and nitrilotriacetic acid (II: NTA):

$$HN\begin{smallmatrix}CH_2COOH\\CH_2COOH\end{smallmatrix} \qquad HOOCCH_2\begin{smallmatrix}CH_2COOH\\CH_2COOH\end{smallmatrix}$$

I: IDA   II: NTA

Iminodiacetic acid is not of practical use in complexometry, but is very valuable for the synthesis of the most important metallochromic and metallofluorescent indicators.

Nitrilotriacetic acid forms less stable complexes than do the other reagents described here and is now rarely used as a volumetric reagent.

From coordination chemistry it is well known that the most stable chelate complexes are those where the central atom is bound to ligands to form five- or six-membered rings, and has its coordination sites fully occupied by the same ligand molecule. It is obvious that one iminodiacetate group cannot satisfy this requirement and that only compounds having two such groups are able to form 1:1 complexes in one step. Such a compound has been discovered in ethylenediaminetetra-acetic acid (III: EDTA)

$$^{\ominus}OOCCH_2 \diagdown \underset{HOOCCH_2}{\overset{\oplus}{NH}}-CH_2-CH_2-\overset{\oplus}{NH}\underset{CH_2COOH}{\overset{CH_2COO^{\ominus}}{\diagup}}$$

III: EDTA

With almost all metals, EDTA practically instantaneously forms slightly dissociated water-soluble complexes which are usually colourless. Coloured complexes are formed only with coloured cations. All the complexes have 1:1 composition; EDTA does not form binuclear complexes, M_2L, or complexes of the type ML_2.

Because of these exceptional properties, EDTA was predestined to be a most important analytical reagent. It still keeps its dominant position in complexometry, although in some cases other compounds have proved to be either a little more suitable or to differ in complex formation. These include:

(a) 2,3-propylenediaminetetra-acetic acid (methyl-EDTA) (IV: MEDTA):

IV: MEDTA

(b) diethylenetriaminepenta-acetic acid (V: DTPA)

V: DTPA

(c) triethylenetraminehexa-acetic acid (VI: TTHA)

VI: TTHA

(d) ethyleneglycol-bis(2-aminoethylether)tetra-acetic acid (VII: EGTA)

VII: EGTA

(e) 1,2-diaminocyclohexanetetra-acetic acid (VIII: DCTA)

VIII: DCTA (DCyTA, CDTA)

The complexometric properties of these compounds will be described on p. 4ff.

1.1   THE PROPERTIES OF EDTA

EDTA (ethylenediaminetetra-acetic acid, H_4Y, m.w. 292.1) is a white crystalline solid, slightly soluble in water, insoluble in organic solvents.   It was discovered in West Germany in 1935 and its synthesis is covered by a number of patents.

As a tetrabasic acid it forms four series of salts, of which the sodium salts are the most important.   Their solubility in water increases with increasing number of sodium atoms (see Table 1.1). The tri- and tetrasodium salts are strongly hydrolysed and are not usually prepared as the solids.   The disodium salt dihydrate, $Na_2H_2Y.2H_2O$, is marketed as an analytical reagent under various proprietary names.   The higher sodium salts are used only in strongly alkaline solution for technological purposes.

TABLE 1.1   Solubilities of EDTA and its sodium salts (g per 100 ml of water) at various temperatures.   (According to the leaflet "Sequestrol", Geigy Ltd., Manchester)

	22°C	40°C	80°C
H_4Y	0.2	0.2	0.5
NaH_3Y	1.4	1.4	2.1
Na_2H_2Y	10.8	13.7	23.6
Na_3HY	46.5	46.5	46.5
Na_4Y	60	59	61

Shortly after its discovery, great attention was paid to the preparation of the "double salts" $AMeY.xH_2O$, where Me = heavy metal and A = Na, K, Rb etc. [1 - 4].   The compounds were studied mostly from a preparative point of view and it seems that the use of EDTA as a complex-forming reagent escaped the notice of the authors.   A number of salts of the complex anion CoY^- were prepared by Schwarzenbach [5] and also some salts of the EDTA complex of vanadium [6].

Of the compounds prepared, the calcium-EDTA complex $Na_2CaY.2H_2O$ is used for injection as an antidote to lead poisoning, and iron complex FeY^- has been recommended for the treatment of iron chlorosis.

(a)   Acid-base properties of EDTA

The titration curve for neutralization with sodium hydroxide shows three buffer regions.   The first belongs to the dissociations

$$H_4Y \rightleftharpoons H_3Y^- + H^+$$
$$H_3Y^- \rightleftharpoons H_2Y^{2-} + H^+$$

which are practically simultaneous, and the other two belong to the formation of the anions HY^{3-} and Y^{4-}.   The stepwise dissociation of

EDTA can be written schematically as

$$H_4Y \xrightarrow[pK_1=2.07]{-H^+} H_3Y^- \xrightarrow[pK_2=2.75]{-H^+} H_2Y^{2-} \xrightarrow[pK_3=6.24]{-H^+} HY^{3-} \xrightarrow[pK_4=10.34]{-H^+} Y^{4-}$$

The distribution diagram for the various forms of EDTA, as a function of pH, is given in Fig. 1.1.

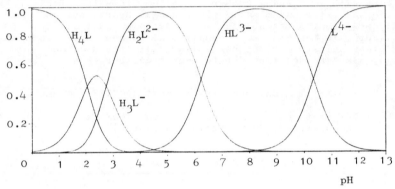

Fig.1.1   Distribution of forms of EDTA as a function of pH

To explain the low values of K_1 and K_2 Schwarzenbach and Ackermann [7] postulate that EDTA has a double betaine structure (see III on p. 1) and therefore the third and fourth protons, bound to the nitrogen atoms, dissociate at relatively higher pH.   Chapman [8] accounts for the infrared spectrum of a suspension of solid EDTA and its disodium salt in Nujol by assuming that the structure of the anion H_2Y^{2-} is

$$\left[H \begin{array}{c} ^{OOCCH_2} \\ \\ _{OOCCH_2} \end{array} N-CH_2-CH_2-N \begin{array}{c} ^{CH_2COO} \\ \\ _{CH_2COO} \end{array} H \right]^{2-}$$

The third proton dissociates from this anion at pH 6 and the fourth proton is transferred to a nitrogen atom and then dissociates at pH 10.   The existence of a double betaine he considers improbable because of mutual repulsion of both protons.

(b)   Complex-forming properties of EDTA

In qualitative analysis, the properties of EDTA are at first sight somewhat astonishing.   Most common reactions of cations totally fail in the presence of EDTA.   On the other hand, many reactions with organic and inorganic reagents become more selective.   In the presence of EDTA only titanium, beryllium, uranium, tin(IV), niobium and tantalum are precipitated with ammonia.   In alkaline solution hydrogen sulphide does not precipitate cobalt, nickel, zinc and manganese.   Similarly calcium does not react with oxalate, barium with sulphate etc.   EDTA and its derivatives are now commonly used as masking agents in gravimetry, colorimetry, chromatography, polarography, electrophoresis, and purification of several compounds.

1.2   THE STABILITY OF EDTA COMPLEXES

Although complexometric titrations are very simple, they cannot be developed properly without basic theoretical knowledge.   Analytical chemists are mainly interested in how stable these complexes are under various analytical conditions and which factors influence their stability.

All these theoretical problems were solved by Schwarzenbach [9] and in many details further developed by Ringbom [10].   We will try to describe as simply as possible the background theory and explain some basic terms.

(a)   The stability constants of EDTA complexes

The general reaction of a cation M with a ligand Y to form a complex MY (charges are omitted) is described by

$$M + Y \rightleftharpoons MY \tag{1.1}$$

and the stability constant of MY is identical with the equilibrium constant for the reaction:

$$K_{MY} = [MY]/[M][Y] \tag{1.2}$$

These concentration constants are related to the thermodynamic constants by means of appropriate activity coefficients, and have been measured by various methods under exactly known conditions of temperature, ionic strength etc.   The stability constants of EDTA complexes (MY) are given in Table 1.2.   For the analytical chemist they have similar significance to redox potentials and solubility products;   they are the working data for prediction of reaction performance in terms of thermodynamics, but say nothing about the rate of complex formation and dissociation.

(b)   Apparent stability constants

The stability of EDTA complexes is influenced by a number of factors such as temperature, ionic strength, acidity of the solution, presence of other complex forming anions etc.   Schwarzenbach was the first to pay attention to the quantitative treatment of all the main equilibria in EDTA solutions.   All these side-reactions cause considerable departure from the behaviour predicted from the thermodynamic stability constants.   Schwarzenbach coined the term "apparent constants" to describe the values of the stability constants as modified by the actual conditions used.   Reilley used the term "effective constants", and Kolthoff proposed the term "conditional constants".

The concentration of hydrogen ions has the most pronounced influence on the stability of the complexes.   During the titration of metal ion with EDTA the main equilibrium is given by equation (1.1), but we have to consider all the forms of EDTA present in the solution, the relative amounts of these being dependent on the hydrogen-ion concentration (see Fig. 1.1).   If we define, according to

Schwarzenbach, the apparent concentration of the ligand, $[Y']$, as the sum of the concentrations of all forms of EDTA with the exception of MY, then we get:

$$[Y'] = [Y^{4-}] + [HY^{3-}] + [H_2Y^{2-}] + [H_3Y^-] + [H_4Y] \qquad (1.3)$$

TABLE 1.2   Logarithm of the stability constants of EDTA complexes
Temperature 20°C, I = 0.1 (KNO_3)

M	log K_{MY}	M	log K_{MY}	Rare earth cations M	log K_{MY}
Ag^+	7.3	Mg^{2+}	8.7*	La^{3+}	11.50
Al^{3+}	16.1	Mn^{2+}	14.0	Ce^{3+}	15.98
Ba^{2+}	7.8*	Na^+	1.7*	Pr^{3+}	16.40
Bi^{3+}	27.9	Ni^{2+}	18.6	Nd^{3+}	16.61
Be^{2+}	9.3	Pb^{2+}	18.0	Sm^{3+}	17.14
Ca^{2+}	10.7	Ra^{2+}	7.1	Eu^{3+}	17.35
Cd^{2+}	16.5	Sc^{3+}	23.1	Gd^{3+}	17.37
Co^{2+}	16.3	Sn^{2+}	22.1	Tb^{3+}	17.93
Co^{3+}	36	Sr^{2+}	8.6	Dy^{3+}	18.30
Cr^{3+}	23	Th^{4+}	23.2	Ho^{3+}	18.74
Cu^{2+}	18.8	Ti^{3+}	21.3	Er^{3+}	18.85
Fe^{2+}	14.3*	TiO^{2+}	17.3	Tm^{3+}	19.32
Fe^{3+}	25.1*	V^{2+}	12.7*	Yb^{3+}	19.51
Ga^{3+}	20.3	V^{3+}	25.9*	Lu^{3+}	19.83
Hg^{3+}	21.8	VO^{2+}	18.8*		
In^{3+}	25.0	VO_2^+	18.1		
Li^+	1.8*	Y^{3+}	18.1		
		Zn^{2+}	16.5		

* 0.1M KCl

From the dissociation constants of EDTA (values on p. 7) with charges omitted for convenience:

$$K_{H_iY} = [H_{i-1}Y][H]/[H_iY] \qquad (1.4)$$

we can express $[Y']$ by

$$[Y'] = [Y](1 + \sum_{i=1}^{i} [H^+]^i \prod_{i=1}^{i} \frac{1}{K_{H_iY}}) \qquad (1.5)$$

From (1.5) it is obvious that the ratio $[Y']/[Y]$ has a constant value at a given pH and we call this the side-reaction coefficient $\alpha_{Y(H)}$:

$$\alpha_{Y(H)} = [Y']/[Y] \tag{1.6}$$

The apparent or conditional stability constant is then given by

$$K'_{MY} = K_{MY}/\alpha_{Y(H)} \tag{1.7}$$

or

$$\log K'_{MY} = \log K_{MY} - \log \alpha_{Y(H)} \tag{1.8}$$

The values of $\alpha_{Y(H)}$ can be calculated from equation (1.6) and are summarized in Table 1.3. Ringbom [10] makes extensive use of graphical presentation of α-values.

TABLE 1.3 Logarithms of $\alpha_{Y(H)}$ coefficients at various pH values. Dissociation constants of EDTA $(K_1 = 2.07, K_2 = 2.75, K_3 = 6.24, K_4 = 10.34)$

pH	$\log \alpha_{Y(H)}$	pH	$\log \alpha_{Y(H)}$	pH	$\log \alpha_{Y(H)}$	pH	$\log \alpha_{Y(H)}$
0	21.4	3	10.8	6	4.8	9	1.4
1	17.4	4	8.6	7	3.4	10	0.5
2	13.7	5	6.6	8	2.3	11	0.1

Both K_{MY} and $\alpha_{Y(H)}$ are important basic parameters of complexometry. Their significance can be illustrated by two examples. (1) Is it possible to determine zinc with EDTA at pH 5 in the presence of calcium? The logarithms of the apparent stability constants of the Zn and Ca complexes are calculated from the values in Tables 1.2 and 1.3:

$$\log K'_{ZnY} = 16.5 - 6.6 = 9.9;$$

$$\log K'_{CaY} = 10.7 - 6.6 = 4.1$$

The minimum apparent stability constant for reliable EDTA titration can easily be shown to be 10^7 (for 0.1M solutions). The titration of zinc is possible and calcium does not interfere unless it is present in very large excess.

(2) Similarly it is possible to determine tervalent iron in the presence of zinc at pH 2 because:

$$\log K'_{FeY} = 25.1 - 13.7 = 11.4;$$

$$\log K'_{ZnY} = 16.5 - 13.7 = 2.8$$

In these simple examples the concentration of the interfering metal is not considered.   The interference can also be treated quantitatively, which has great value for the evaluation of stepwise titrations of two or even three elements.   Details will be given on p. 13.

(c)   <u>Calculation of $\alpha_{Y(H)}$ values according to Reilley</u> [11]

The calculation of the $\alpha_{Y(H)}$ coefficient from equation (1.6) is somewhat elaborate, though once done it need not be repeated.   A very simple calculation has been proposed by Reilley, in which we need only the dissociation constants of EDTA.   At any pH one (or at most two) of the forms of EDTA will predominate in the complex-forming reaction (see Fig. 1.1, p. 4).   Then we can use the rule that $\log \alpha_{Y(H)}$ is obtained by adding together the log K values greater than the given pH and subtracting the pH times the number of dissociable protons still attached to the predominant form of EDTA.

(d)   <u>α-Coefficients of other side-reactions</u>

The hydrogen-ion concentration influences to some extent the composition of EDTA complexes even if the metal:ligand ratio remains 1:1.   In acidic medium protonated complexes MHY can be formed, which have stability constant K^H_{MHY}:

$$K^H_{MHY} = [MHY]/[MY][H] \qquad (1.9)$$

This reaction can also be characterized by a side-reaction coefficient $\alpha_{MY(H)}$:

$$\alpha_{MY(H)} = 1 + [H] K^H_{MHY} \qquad (1.10)$$

The complex MY might also be hydrolysed under certain conditions, forming a hydroxo-complex:

$$MY + OH^- = MY(OH) \qquad (1.11)$$

having a stability constant $K^{OH}_{MY(OH)}$:

$$K^{OH}_{MY(OH)} = [MY(OH)]/[MY][OH] \qquad (1.12)$$

This side-reaction is characterized by

$$\alpha_{MY(OH)} = 1 + K^{OH}_{MY(OH)}K_W/[H] \qquad (1.13)$$

where K_W is the ionic product of water.

Beside the hydrolysis of the complex ML, the cation itself might be hydrolysed, forming complexes $M(OH)$, $M(OH)_2$ ..... $M(OH)_j$, each having its own successive (or stepwise) stability constant $K^H_{M(OH)_j}$.   The value of $\alpha_{M(OH)}$ is then given by

$$\alpha_{M(OH)} = 1 + \sum_{j=1}^{j} [H]^j \prod_{j=1}^{j} K^H_{M(OH)_j} \tag{1.14}$$

If the values of all the side-reaction coefficients are known, then we can determine with sufficient accuracy the value of the apparent stability constant K'_{MY} according to the equation

$$K'_{MY} = [MY']/[M'][Y'] = K_{MY}\alpha_{MY}'/\alpha_{M(OH)}\alpha_{Y(H)} \tag{1.15}$$

The coefficient α_{MY}' includes both side-reactions (1.10) and (1.13) and has the value:

$$\alpha_{MY}' = \alpha_{MY(H)} + \alpha_{MY(OH)} - 1 \tag{1.16}$$

In practice α_{MY}' is determined by one of the coefficients on the right-hand side of equation (1.16), and depends on the acidity conditions.

The α-coefficients corresponding to these mixed complexes are of minor significance, since their only effect is to increase K'_{MY}. Formation of mixed complexes can therefore only be advantageous (provided the M:Y ratio remains constant). Usually the analytical conditions are chosen so that these side-reactions do not occur anyway, nor does the hydrolysis of the cation.

We can therefore agree with the opinion of Schwarzenbach [12] that it is not necessary to complicate the issue and we can use the basic coefficient $\alpha_{Y(H)}$. For analytical chemists the basic parameter is equation (1.8)

$$\log K'_{MY} = \log K_{MY} - \log \alpha_{Y(H)}$$

Much more important than these protonation or hydrolytic reactions is the influence of other strong complex-forming compounds. These will be considered in the chapter on masking (p. 58).

1.3   EDTA AS VOLUMETRIC REAGENT

For volumetric work the disodium salt $Na_2H_2Y.2H_2O$ is mostly used. The usual solutions, 0.01 - 0.05M, have a pH of about 5 (the H_2Y^{2-} anion is the prevailing form).

The large selection of very sensitive indicators permits the determination of metals over a very wide pH range, according to the stability of the individual complexes, e.g. from pH 0 (zirconium) to pH 13 (calcium). From Fig. 1.1 we can easily see in which form EDTA exists at various pH values. The reactions with metals proceed as follows:

at pH 4 - 5

$$H_2Y^{2-} + M^{n+} = MY^{(4-n)-} + 2H^+ \tag{1.17}$$

at pH 8 - 9

$$HY^{3-} + M^{n+} = MY^{(4-n)-} + H^+ \tag{1.18}$$

at pH 12

$$Y^{4-} + M^{n+} = MY^{(4-n)-} \tag{1.19}$$

In complex formation with EDTA, regardless of the charge on the metal ion, two hydrogen ions are always liberated. This is also valid for equations (1.18) and (1.19), because the EDTA is added in the form H_2Y^{2-} and for the formation of HY^{3-} and Y^{4-} one or two hydroxyl ions respectively will be consumed:

$$H_2Y^{2-} + OH^- = HY^{3-} + H_2O \tag{1.20}$$

$$H_2Y^{2-} + 2OH^- = Y^{4-} + 2H_2O \tag{1.21}$$

Similar reactions proceed during the titration in solutions of pH 0 - 1 where the form H_4L exists:

$$H_2Y^{2-} + 2H^+ = H_4Y \tag{1.22}$$

$$H_4Y + M^{n+} = MY^{(4-n)-} + 4H^+ \tag{1.23}$$

The sum of equations (1.22) and (1.23) gives equation (1.17) again. (In very acidic solutions two cationic forms of EDTA, H_5Y^+ and H_6Y^{2+}, are produced.)

The hydrogen ions liberated during the titration may influence the acidity of the titration solution. With strongly alkaline solutions this danger is very small; the concentration of hydroxide is always sufficient to keep the pH practically constant. During titrations in slightly acidic medium, however, it is necessary to keep the pH constant by addition of a suitable buffer or adjustment of pH during the titration. With media at pH 0 - 1 simple dilution with water will raise the pH (by 0.3 for 1:1 dilution).

Let us consider various types of EDTA titrations performed under optimal conditions with usual detection of the end-point (for details see Chapter 2).

(a)   Direct titration with EDTA

In this way we can determine a great number of metals, with the exception of those which react with EDTA too slowly. If an indicator is used, none of the metal ions present must block the indicator, and the indicator must react with the metal ion to be determined. Blocking the indicator means formation of a complex which reacts with EDTA very slowly or not at all. The cations which block indicators (usually cations of transition elements) include those of copper, nickel, cobalt, chromium, iron, aluminium and gallium. However, there are some indicators that react with certain of these elements: copper and nickel can be determined reliably by direct titration with EDTA if murexide is used as indicator; iron can be determined if Chromazurol S or salicylic acid is used as indicator. In some cases raising the temperature will accelerate the exchange reaction between EDTA and the metal-indicator complex at

the end-point.   Cobalt is usually determined at 60°C, even with
Xylenol Orange as indicator.   Similarly the titration of zirconium
is performed with nearly boiling solutions.

There are many cases of a cation not reacting with a particular
indicator or giving only an indistinct colour reaction.   A classical
example is Eriochrome Black T, which is not a good indicator for
absolutely pure solutions of calcium, mainly because of the very low
molar absorptivity of the calcium-indicator complex.   Traces of
magnesium considerably improve the colour-change at the end-point.

Higher temperature also improves the rate of reaction for several
metals which have aquo-complexes that react slowly with EDTA, e.g
zirconium and iron.   Aluminium and chromium(III) must be boiled with
EDTA for some time for quantitative complex formation.   Therefore
addition of excess of EDTA and back-titration is used for these
determinations.   On the other hand, DCTA reacts with aluminium very
quickly at room temperature.   This allows successive determination
of aluminium and chromium by titration with DCTA and EDTA
respectively [13].

(b) Back-titrations

These are based on titration of an excess of EDTA with a solution of
a suitable cation, usually (for convenience in calculation) of the
same molarity as the EDTA.   They can be used in all cases, but are
mainly applied to cations which cannot be determined directly for the
reasons given above.   Suppose we have in the solution a cation M^{2+},
which forms an EDTA complex MY_{2-}.   The excess of EDTA is titrated
with a solution of another cation Me^{2+} .   During the back-titration
the reaction

$$Me^{2+} + H_2Y^{2-} = MeY^{2-} + 2H^+ \qquad (1.24)$$

proceeds.   The first small excess of cation Me^{2+} reacts with the
indicator and shows the end-point.   This is valid only if the
apparent stability constant of the complex MeY^{2-} is lower than that
of the complex MY^{2-}, or when both constants are nearly equal.   When
the apparent stability constant of complex MeY^{2-} is sufficiently
greater than that of complex MY^{2-} a displacement reaction can proceed
after the end-point:

$$MY^{2-} + Me^{2+} = MeY^{2-} + M^{2+} \qquad (1.25)$$

At the end-point there are then three possibilities.

If both cations can react with the indicator it does not matter
whether reaction (1.25) proceeds or not.   The end-point is given by
a sudden change in the colour of the solution.

If the cation M does not react with the indicator and the reaction
(1.25) is fast, then the colour of the Me^{2+} indicator complex
disappears shortly after the end-point is reached.   On addition of
a small portion of Me^{2+} the colour appears and then disappears again.
We are overtitrating the solution and results for cation M must be
low.

However, most displacement reactions are sufficiently slow and the
Me^{2+}-indicator colour fades only after some minutes, so we can locate

the end-point without difficulty.   Some displacement reactions are
so slow that we can back-titrate the EDTA with a cation which forms
a more stable complex than does the cation present in the solution.
For example we can determine aluminium indirectly by back-titration
of EDTA with thorium nitrate even though the thorium-EDTA complex is
more stable than the aluminium-EDTA complex.   For the determination
of cations such as copper, nickel, etc., the method is very useful
and the surplus EDTA is generally back-titrated with lead or zinc
solution at pH 5 - 5.5 with Xylenol Orange as indicator.

In some titrations we observe that after the end-point the colour
change of the indicator becomes irreversible.   This is caused by the
cation blocking the indicator after displacing it from the complex.
According to the rate of the displacement reaction the irreversibil-
ity of the indicator appears after a shorter or longer time.   This
phenomenon causes some difficulties in so-called "pendulum"
titrations, where we locate the end-point by alternate additions of
small portions of EDTA and cation solution.

Back-titrations have great value when used in combination with masking agents.   For example the sum of nickel and copper can be
determined in ammoniacal solution by back-titration with calcium or
magnesium, using Thymolphthalexone as indicator.   After the addition
of thioglycollic acid (for masking Cu) we can titrate the EDTA
liberated and equivalent to the amount of copper.

With the use of two or three masking agents we can determine successively three or more elements in the same solution.

(c)   Substitution titrations

The two types of titrations just described suffice for almost all
determinations in practice.   We mention only briefly the other
methods of complexometric titration based on reactions of the type
(1.25).   If we add to a solution of "untitratable" cation a
sufficient amount of another metal-EDTA complex which is less stable,
then reaction (1.25) proceeds quantitatively.   In the early stages
of complexometry, when titrations were done only in alkaline medium,
mostly the Mg-EDTA complex was used.   It has been produced as the
potassium salt K_2MgY.   For example, in the determination of nickel,
which blocks many indicators, the Mg-EDTA complex was added in solid
form or as a solution:

$$MgY^{2-} + Ni^{2+} = NiY^{2-} + Mg^{2+} \tag{1.26}$$

The displaced magnesium was determined by direct titration with EDTA,
with Eriochrome Black T as indicator.   Addition of at least an
equivalent amount of Mg-EDTA complex was required to ensure a
distinct end-point.   More interesting are those titrations where the
"auxiliary complex" is added only in small amount.   If we add to a
solution of aluminium a small amount of Cu-EDTA then, on boiling, the
copper is displaced and reacts with PAN indicator, forming the violet
colour of the Cu-PAN complex.   Then we titrate the solution nearly
at boiling point with EDTA to the end-point colour change from violet-
red to yellow.

If a cation forms an EDTA complex that is too weak $(K < 10^7)$ for
accurate titration, we may use a substitution reaction with another
type of complex.   For example, silver can be determined by reaction

with tetracyanonickelate:

$$2Ag^+ + Ni(CN)_4^{2-} = 2Ag(CN)_2^- + Ni^{2+}$$

the nickel liberated being titrated with EDTA.

(d) Stepwise titrations

The thermodynamic constants of EDTA complexes (Table 1.2, p. 6) differ considerably, from 10^8 (Ba^{2+}) to 10^{36} (Co^{3+}).   There is a possibility of determining two, three and more elements.   A simple calculation shows that a cation can be determined if its apparent stability constant is at least 10^7.   Two elements can be determined successively if the ratio of their stability constants is at least $10^4 - 10^5$.   The concentration ratio of the cations must also be considered.

The quantitative treatment has been developed very simply by Ringbom [10].   Let us assume two titratable cations (M and N) to be present in the solution, with stability constants K_{MY} and K_{NY}.   In the titration of M the second cation N could interfere by the formation of complex NY.   This reaction can be considered as a normal side-reaction of anion Y, characterized by an α-coefficient $\alpha_{Y(N)}$:

$$\alpha_{Y(N)} = 1 + [N]K_{NY} \backsim [N]K_{NY} \qquad (1.27)$$

Then the apparent stability constant of the complex MY is given by the equation:

$$K'_{MY} = K_{MY}/\alpha_{Y(N)} = K_{MY}/[N]K_{NY} \qquad (1.28)$$

and in logarithmic form:

$$\log K'_{MY} = \Delta \log K + pN \qquad (1.29)$$

where $\Delta \log K = \log K_{MY} - \log K_{NY}$ and $pN = -\log [N]$.

When a metal M is titrated complexometrically in the presence of N, the apparent logarithmic constant $\log K_{MY}'$ is the difference between the logarithmic stability constants of the two complexes, plus pN, provided that $[N]K_{NY} > 1$ and also $\alpha_{Y(N)} > \alpha_{Y(H)}$ (because otherwise the protonation side-reaction would have the predominant effect on K'_{MY}). Hence N will cause interference if $\alpha_{Y(N)}$ is greater than the highest value of $\alpha_{Y(H)}$ permissible for successful titration of M.

As an example consider the determination of zinc in the presence of magnesium in concentrations (a) 0.1M and (b) 0.001M.   At which pH can we determine zinc without magnesium interference?   The stability constants of the Zn-EDTA and Mg-EDTA complexes are:

$$\log K_{ZnY} = 16.5 \text{ and } \log K_{MgY} = 8.7$$

The required apparent stability constants of the complex ZnY can be obtained from equation (1.29):

(a) $\log K'_{ZnY} = 16.5 - 8.7 + 1 = 8.8$

(b) $\log K'_{ZnY} = 16.5 - 8.7 + 3 = 10.8$

The values of $\alpha_{Y(Mg)}$ are obtained from equation (1.27):

$$\alpha_{Y(Mg)} = 10^{8.7} \times 10^{-1} = 10^{7.7}$$

$$\alpha_{Y(Mg)} = 10^{8.7} \times 10^{-3} = 10^{5.7}$$

The values for K'_{ZnY} are valid only if $\alpha_{Y(Mg)}$ is larger than $\alpha_{Y(H)}$. From Table 1.3 (p.7) we find the nearest values of $\log \alpha_{Y(H)}$ to $\log \alpha_{Y(Mg)}$ are 8.6 at pH 4, 6.6 at pH 5 and 4.8 at pH 6.   We find by interpolation the pH values at which magnesium does not interfere in titration of zinc.   These values are for (a) 4.5 and for (b) 5.5 as the upper pH limits for non-interference of magnesium.   In case (b) pH 5 - 5.5 is just optimal for the determination of zinc with Xylenol Orange as indicator.   In case (a) we have to titrate zinc at the somewhat lower pH 4.5.   A very suitable indicator for such a titration is Alizarin Complexone [14,15] for which the optimal pH is 4.2 - 4.5.

The effect of dilution is also easily seen.   Because the α coefficient is dependent on the concentration of cation N, its value drops on dilution and as a result the apparent stability constant of the complex MY increases, equation (1.29).

Let us come back to the case (a) where we have 0.1M magnesium solution.   If we pipette 10 ml of this solution and dilute it to 100 ml, then the concentration of magnesium decreases to 0.01M and $\alpha_{Y(Mg)}$ will have the value $10^{6.7}$; the critical upper pH limit is then $\backsim$ 5, which permits the use of Xylenol Orange as indicator.

The overall side-reaction coefficient of EDTA is

$$\alpha_{Y(X)} = \alpha_{Y(H)} + \alpha_{Y(N)} - 1 \qquad (1.30)$$

and so the apparent stability constant of MY is

$$K'_{MY}/\alpha_{Y(X)} \qquad (1.31)$$

and the value of $\alpha_{Y(X)}$ depends on the pH, the concentration of N and the value of K_{NY}.   If either $\alpha_{Y(H)}$ or $\alpha_{Y(N)}$ is significantly greater than the other (by a factor of say 100), then that α-value will be decisive for the value of K'_{MY}.   Then since the pH for the titration is chosen so that $\alpha_{Y(H)}$ does not cause K'_{MY} to be too low for a pre-selected degree of accuracy to be achieved, it follows that N will interfere only if $\alpha_{Y(N)} > \alpha_{Y(H)}$.   The value of $\alpha_{Y(N)}$ is given by

$$\alpha_{Y(N)} = 1 + [N]K_{NY}$$

and since it is incipient interference we are interested in, we can equate [N] with C_N, the total concentration of N in the solution. It follows that dilution of the solution will lower $\alpha_{Y(N)}$ and

increase K_{MY}'.   At the same time, it must be remembered that dilution
will also mean that K_{MY}' must be increased anyway if the accuracy of
titration is to be maintained.

$$K_{MY}' = [MY]/[M][Y']$$

At equivalence $[M] = [Y']$, so for 99.9% formation of MY we have $[MY] = 0.999C_M$ and $[M] = [Y'] = 0.001C_M$ so

$$K_{MY(99.9\%)}' \geq 0.999C_M/(0.001C_M)^2$$

$$\geq 10^6/C_M$$

Hence a 10-fold dilution will require K_{MY}' to be increased by a factor
of 10 if the same degree of complexation is required.

The end-point will be more fully considered in the next chapter.

REFERENCES

1.   P. Pfeiffer and W. Offermann, Berichte Deut. Chem. Ges., 1942,
      75B, 1.
2.   H. Brintzinger and G. Hesse, Z. Anorg. Chem., 1942, 249, 113.
3.   H. Brintzinger, H. Thiele and U. Müller, Z. Anorg. Chem., 1943,
      251, 285.
4.   H. Brintzinger and S. Munkelt, Z. Anorg. Chem., 1948, 256, 65.
5.   G. Schwarzenbach, Helv. Chim. Acta, 1949, 32, 839.
6.   G. Schwarzenbach and J. Sandera, Helv. Chim. Acta, 1953, 36,
      1089.
7.   G. Schwarzenbach and H. Ackermann, Helv. Chim. Acta, 1947, 30,
      1799.
8.   D. Chapman, J. Chem. Soc., 1955, 1786.
9.   G. Schwarzenbach and H. Flaschka, Die komplexometrische
      Titration, 5th Ed., Enke Verlag, Stuttgart, 1965;
      Complexometric Titrations, 2nd Ed., Methuen, London, 1969.
10.   A. Ringbom, Complexation in Analytical Chemistry, Interscience,
      New York, 1963.
11.   C.N. Reilley, R.W. Schmid and F.S. Sadek, J. Chem. Educ., 1959,
      36, 555.
12.   Ref. 9, English edition, p. 15.
13.   R. Pribil and V. Vesely, Talanta, 1962, 9, 23; 10, 1963, 1287.
14.   R. Belcher, M.A. Leonard and T.S. West, J. Chem. Soc., 1958,
      2390.
15.   J.B. Headridge, Analyst, 1958, 85, 381.

CHAPTER 2

Detection of the Titration End-point

2.1 COMPLEXOMETRIC INDICATORS

2.1.1 Introduction

The course of the titration and the change in free metal ion concentration in the neighbourhood of the equivalence-point can be calculated from the equation for the complexation equilibrium, and the end-point can be detected very simply from the colour change of the so-called complexometric indicators.

Although analytical chemistry has at its disposal a large number of very sensitive reagents for almost all metals, the choice of an indicator for complexation titrations, on account of certain special requirements, was originally rather difficult.   In the early days, only a few well known inorganic and organic compounds were used for complexometric titrations.   Schwarzenbach called these compounds "weak coloured" indicators, such as potassium thiocyanate for the titration of iron and cobalt, potassium iodide and thiourea for the determination of bismuth, salicylic acid and its derivatives for titration of iron and so on.

It was necessary to find new indicators among the organic dyes or to attempt the synthesis of entirely new compounds which would fulfil all requirements.   The first indicators proposed by Schwarzenbach – murexide and Eriochrome Black T - were not amongst the recognized analytical reagents.   The colour reaction of murexide with calcium was known to Beilstein [1] and was rediscovered nearly 100 years later [2].   Eriochrome Black T was a well known dye but had never been used for analytical purposes.

The most important task was to discover indicators suitable for use in acidic solutions.   The most useful are the so-called metallochromic indicators, which can satisfy all analytical needs.

2.1.2 Metallochromic Indicators

The term metallochromic indicator [3] was chosen for organic dyes

16

capable of forming with metal cations distinctly coloured chelates, with colours as different as possible from the original colour of the dye. When such indicators are used the titration end-point is shown by a pronounced colour change and not by simple appearance of colour in a colourless solution, or disappearance of colour to leave a colourless solution. Such compounds must possess a grouping capable of chelate formation, joined directly to the resonance system of a dye. Usually the indicators contain at least two oxygen atoms as well as two nitrogen atoms which can co-ordinate with the metal cation, forming five- or six-membered rings. Suitably disposed oxygen or nitrogen ligand atoms are possessed by the $-OH$ and $-COOH$ radicals, $-N=N-$ azo group, $-NH_2$ amino group, etc.

Particularly outstanding metallochromic properties are associated with indicators having an aromatic ring with an iminodiacetic acid group, $-N(CH_2COOH)_2$ ortho to a phenolic hydroxy group.

All metallochromic indicators can also behave as acid-base indicators, because changes in colour will accompany the dissociation of protons from the group attached to the resonance system. As a result, the colour produced by chelate formation is similar to the colour of the deprotonated (alkaline) form of the indicator. In other words there is competition between hydrogen ion and cation.

A full discussion of the mechanism of indicator reactions will be found in the book by Schwarzenbach and Flaschka [4].

2.1.3   The Properties of Metallochromic Indicators

A good metallochromic indicator must have a number of special properties.

(a) It must form water-soluble chelates with a colour (or fluorescence) which differs considerably from the colour (fluorescence) of the indicator itself. Ideally the change of colour occurs over a broad range of the visual spectrum - from yellow to blue for example. Methylthymol Blue fulfils this requirement in acidic medium.
(b) The reaction of the indicator with metals must be sufficiently sensitive for only a small amount of the indicator to be needed.
(c) The chelates formed must be less stable than the corresponding EDTA chelates and be formed at about the same pH as the EDTA chelates.
(d) The reaction between metal-indicator complex and EDTA must be instantaneous and reversible. Slow substitution reactions cause overtitration of the solution. Very slow reactions lead to blocking of the indicator.
(e) The indicator should be stable enough in aqueous or alcohol solution to be stored for reasonable periods.
(f) It should be pure.

Experience has shown that in many cases not all these requirements can be fulfilled. Several indicators are blocked by transition metal ions. PAN and its chelates are only slightly soluble in water, which causes difficulties in the end-point detection. However, the addition of alcohol and use of elevated temperature often eliminate such difficulties.

Many indicators are not stable in aqueous solution and it is necessary to use them as a solid mixture with a non-interfering salt such as potassium nitrate or sodium chloride. The indicators are often very impure.

The requirements mentioned are more or less qualitative in character. The most important property is the sensitivity of the indicator, which can be evaluated quantitatively (see next section). High selectivity (or specificity) of indicator reaction is of value only in special cases.

2.1.4   The Sensitivity of the Indicator and Accuracy of the Titration

The first attempts to find new indicators were very empirical. A solution of the new compound was added to solutions of various cations and the colour change during EDTA titration under optimal conditions (pH, temperature, low cation concentration etc.) was observed. The colour change was described purely subjectively as "very sharp", "sufficiently sharp", "distinct", "satisfactory". The consumption of EDTA was compared with the theoretical value. Not much attention was paid to the purity of the compound - especially of commercial dyes. Sometimes the structure of the new indicator and its complexes were not known satisfactorily, nor were the dissociation constants of the indicators, which were usually also acid-base indicators. All this led to doubtful results.

As already said, the most important property of the indicator is its sensitivity towards cations, expressed in terms of the logarithm of the inverse of the metal concentration, $-\log [M] = pM$.

Although this monograph is devoted to practical complexometry, it is necessary to mention the quantitative evaluation of the indicator sensitivity worked out by Ringbom [5]. His theoretical considerations can be summarized as follows.

(<u>a</u>)   The stability of the EDTA complex is given by its thermodynamic stability constant:

$$K_{MY} = [MY]/[M][Y] \tag{2.1}$$

(<u>b</u>)   The stability of the complex is influenced by a number of factors such as acidity of the solution, side-reactions of the metal with further complex-forming compounds (ammonia buffers, various anions etc.). Under these conditions the stability of the complex is given by the apparent stability constant:

$$K'_{MY} = [MY]/[M'][Y'] \tag{2.2}$$

where $[M']$ and $[Y']$ are the total (analytical) concentrations of M and Y not bound in MY.

At the equivalence point the concentration of free metal is equal to the concentration of free ligand:

$$[M']_{eq} = [Y']_{eq} \tag{2.3}$$

Because the complex is only slightly dissociated, its concentration can be taken as equal to the total concentration of the metal C_M. Then:

$$[M']_{eq} = \sqrt{C_M/K'_{MY}} \qquad (2.4)$$

and in logarithmic form:

$$pM'_{eq} = \tfrac{1}{2}(pC_M + \log K'_{MY}) \qquad (2.5)$$

The indicator chelates behave similarly.   Let us assume the formation of only the 1:1 chelate MIn.   A similar equation is valid for the apparent stability constant:

$$K'_{MIn} = [MIn]/[M'][In'] \qquad (2.6)$$

and in logarithmic form:

$$\log K'_{MIn} = pM' + \log([MIn]/[In]) \qquad (2.7)$$

The indicator colour-change is 50% complete when $[In'] = [MIn]$. Then we get the very simple relation:

$$\log K'_{MIn} = pM' \qquad (2.8)$$

The logarithm of the inverse of the free metal ion concentration at which the colour changes is equal to the logarithm of the apparent stability constant of the indicator complex, and we designate it pM'_{trans}.   To evaluate the titration accuracy we have to know both pM'_{trans} and pM'_{eq}.   The smaller the difference between these values the smaller is the titration error.   The difference

$$\Delta pM' = pM'_{trans} - pM'_{eq} \qquad (2.9)$$

is the basis for the calculation of titration errors.

Both pM' values are calculated from the stability constants of the individual metal complexes, which must be measured very carefully. In some cases they are rather uncertain.   Practically all indicators are polybasic acids forming a series of anions $H_{j-1}In$, $H_{j-2}In$, ... HIn and In (charges are omitted) and can form other complexes besides the 1:1 complex.   It is on account of this acid-base behaviour that the conditional constant K'_{MIn} must be used in the calculations, i.e. $K_{MIn}/\alpha_{In(H)}$.

From pM'_{trans} we can evaluate the minimum value for K'_{MY}.   The colour change will appear complete at $pM' = pM'_{trans} + 1$.   If this is to coincide with the equivalence point, then from equations (2.2) and (2.3)

$$K'_{MY} = [MY]/[M']^2 = C_M/[M']^2 \qquad (2.10)$$

or

$$\log K'_{MY} = \log C_M + 2pM'$$

$$= \log C_M + 2pM'_{trans} + 2$$

$$= 2 \log K'_{MIn} + \log C_M + 2 \qquad (2.11)$$

The uncertain purity of the indicator also leads to error, especially
if the impurity can also act as a metallochromic indicator (e.g.
Semi-Xylenol Orange). From this point of view much remains to be
done in the future.

2.1.5   A Brief Survey of Metallochromic Indicators

The possibility of "discovering" or preparing entirely new compounds
of good metallochromic properties became, and still is, very attractive for analytical chemists.   During the last two decades more than
120 compounds have been proposed as indicators, but only a few are of
practical use.   A good guide to the value of these indicators is the
fact that only a few are offered on the chemical market.   From a
practical point of view 5 or 6 suffice for all purposes.   Their
choice can be partly subjective.

Metallochromic indicators can be divided into three main groups
according to their chemical composition.

(i)   Azo-dyes
(ii)   Triphenylmethane dyes
(iii)   Indicators of various structures

Indicators which function on the basis of reactions other than
complexation will be described in section 2.1.6.

(i)   Azo-dyes

(a)   Hydroxyazo Compounds

The simplest structure in this group of indicators is formed by two
aromatic rings bound together by the azo-group $-N=N-$.   Such compounds must contain one or two auxochromic groups situated ortho to
one or both nitrogen atoms.   Typical auxochromic groups are $-OH$,
$-COOH$, $-NH_2$.   These have configuration (I) and the chelates
configuration (II):

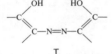   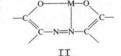

                I   II

1.   Eriochrome Black T

Alizarin Black T, Solochrome Black T, Erio T, C.I. 14645, sodium salt
of 1-(1-hydroxy-2-naphthylazo)-6-nitronaphthol-4-sulphonic acid (III).

The first indicator of this group was proposed by Schwarzenbach and
Biedermann [6] in 1948.

III: Erio T

Because of the similarity of other derivatives it will be described in some detail.

In aqueous solution, Erio T behaves as an acid-base indicator which at pH 6 changes from wine-red to blue and above pH 12 to orange-red. The change of colour is caused by deprotonation of the two phenolic groups:

$$H_2In^- \xrightarrow[pH = 6.5]{-H^+} HIn^{2-} \xrightarrow[pH = 11.5]{-H^+} In^{3-}$$

$$\text{red} \qquad\qquad \text{blue} \qquad\qquad \text{orange}$$

In a well-buffered solution at around pH 10 $(NH_4Cl + NH_3)$ Erio T forms a wine-red 1:1 complex with magnesium, calcium, zinc, cadmium, manganese:

$$HIn^{2-} + M^{2+} = MIn^- + H^+$$

$$\text{blue} \qquad\qquad \text{wine-red}$$

The reaction with magnesium is very sensitive - on addition of the indicator solution a distinct colour is developed in a $10^{-6} - 10^{-7}$M magnesium solution.   The colour with calcium is less intense $(10^{-4} - 10^{-5}$M calcium being required to give a reasonable colour$)$.

The reaction with magnesium has been thoroughly studied by many authors, with confusing results.   Harvey et al. [7] assumed the formation of a 2:1 complex at pH 10.1;   Young and Sweet [8] reported the formation of 1:1 and 2:1 complexes and even the existence of a 3:1 complex.   Diehl and Lindstrom [9] confirmed the formation of only a 1:1 complex.

Some stability constants of metal complexes of Erio T have been reported in the literature:

	Ba	Ca	Mg	Zn	Mn	Ca	Mg
$\log K_{MIn}$	3.0	5.4	7.0	12.9	9.6	3.72	5.75

The values for calcium and magnesium were reported by Diehl and Lindstrom [9].

The complexes of the metals with the indicator are weaker than the complexes with EDTA and therefore at pH $\sim$ 10 the intensely blue anion HIn^{2-} is formed at the end of the titration:

$$MIn^- + H_2Y^{2-} = MY^{2-} + HIn^{2-} + H^+$$

(the free hydrogen ions are absorbed by the buffer added to the solution titrated).

The colour change from wine-red to "steel blue" is very sharp and can be compared with the best colour change of acid-base indicators used in neutralization analysis.

Erio T cannot be used for direct EDTA titration of cobalt, nickel, aluminium or copper because it is blocked by these ions.   Iron(III) also interferes even if bound with triethanolamine (TEA). Manganese(III) also interferes and must be reduced with ascorbic acid or hydroxylamine in the presence of TEA.   The colour change in manganese titrations is particularly sharp.

Aqueous solutions of Erio T are not stable.   The indicator tends to polymerize.   A relatively stable solution can be prepared by addition of TEA and hydroxylamine hydrochloride, the stabilization being due to binding of traces of iron and copper, which activate the polymerization.   The usual form of indicator is a 1:100 solid mixture with potassium nitrate.

Erio T is supplied commercially by various firms under various names. It contains a large amount of sodium salts and its purity is usually only 50 - 70%.   For detailed study of the acid-base properties and determination of stability constants of various complexes, a pure compound is necessary.   Diehl and Lindstrom [9] described its purification in detail.   This was based on desalting with repeated washing with hydrochloric acid, extraction of the impurities with benzene, and crystallization as the dimethylamine salt from dimethyl-formamide.   The free acid was obtained by addition of hydrochloric acid.

Utilization: Erio T is no longer so widely used as at the beginning of complexometry, except for titration of magnesium because of the sharp end-point.   A pure solution of calcium can also be titrated after the addition of small amount of Mg-EDTA complex.

2.   Eriochrome Blue Black R

Calcon, Solochrome Dark Blue BS, Erio R, C.I. 15705, sodium salt of 1-(2´-hydroxynaphthylazo)-2-naphthol-4-sulphonic acid (IV).

As a complexometric indicator, Calcon was proposed at the same time by Belcher et al. [10] and Hildebrand and Reilley [11].

IV: Calcon

Calcon is not a suitable indicator for calcium itself but is
excellent in the presence of precipitated magnesium hydroxide.   It
behaves as an adsorption indicator, covering the whole surface of the
$Mg(OH)_2$ and therefore reducing the adsorption of calcium.   According
to the first authors it is the only indicator which gives accurate
results in the determination of $10^{-2}M$ calcium in the presence of an
equal concentration of magnesium.   For other determinations it is
practically worthless.

3. Patton-Reeder Indicator (Cal-Red)

2-Hydroxy-1-(2-hydroxy-4-sulpho-1-naphthylazo)-3-naphthoic acid (V).

This was proposed by Patton and Reeder [12] for EDTA titration of
calcium at pH 12 - 14 and for the determination of calcium + magnesium
in ammoniacal solution at pH 10.

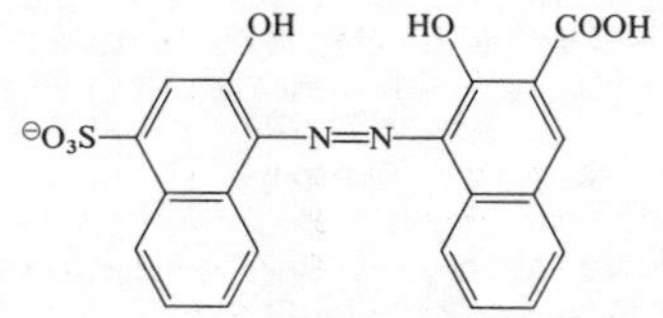

V: Cal-Red

The colour change from wine-red to blue in the titration of magnesium
is very sharp.   In the presence of magnesium in sodium hydroxide
medium the titration of calcium is also reliable.   Like Calcon this
indicator also functions more or less as an adsorption indicator.
It is not very stable in alkaline solution and its aqueous solution
also has poor stability.   It is therefore used as a solid mixture
with sodium chloride.   Further drawbacks (oxidation, blocking with
transition metals) are similar to those of Erio T.

4. Calmagite

1-(2-Hydroxy-5-methyl-1-phenylazo)-2-naphthol-4-sulphonic acid (VI).

This indicator was proposed by Lindstrom and Diehl [13] for the
titration of calcium and magnesium.

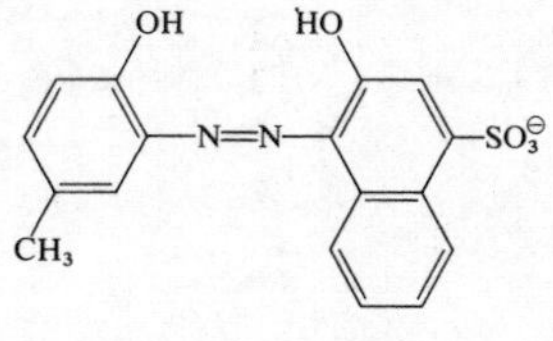

VI: Calmagite

It functions similarly to Erio T.   Its main advantage is that it is
more stable than the others just mentioned and can be stored in
aqueous solution for a long time.   The colour change in the
titration of calcium is not very distinct and can be improved by the
addition of traces of magnesium.

5. Further Eriochrome Dyes

The metallochromic properties of several relatively simple o,o'-
dihydroxyazo dyes were studied by Belcher et al. [14,15]. Among
them two compounds have shown very good end-points in the titration
of magnesium.   These are Eriochrome Blue Black 6B (which has the
Eriochrome Black T structure, without the nitro group) and Eriochrome
Black A, which is a derivative of Calcon, with a nitro-group in the
6-position.

The solubility of all these azo-dyes depends on the presence of the
sulphonic acid group.   The compounds without this group are insoluble in water but soluble in organic solvents (and so are their
chelates).   Such a compound is Erio OO (Erio T without the SO_3H
group), which has been recommended by Flaschka [16] for the detection
of some trace elements.

From the group of bis-azo and tris-azo dyes, two indicators have been
recommended for complexometric determination of calcium and other
elements: Acid Alizarin Black SN and Calcichrome.

6. Acid Alizarin Black SN (C.I. 21725)

Sodium salt of 1-hydroxy-2 (2-hydroxy-6-sulpho-1-naphthylazo)-6-(2-
hydroxy-1-naphthylazo)benzene-4-sulphonic acid.

This indicator has been proposed for EDTA titration of calcium,
cadmium, nickel, manganese, zinc and strontium in alkaline medium,
and of thorium in acetate medium [14].   According to Close and West
[17,18] it is the best indicator for titration of a very dilute "pure"
solution of calcium (<0.02M).   Higher concentrations of magnesium
considerably decrease the sharpness of the end-point.   Also, with
higher concentrations of calcium (> 0.02M)   the colour contrast at
the end-point is not very marked.   The indicator has two co-ordin-
ation centres and we can assume that more complicated complexes are
formed at higher concentrations of calcium.   In a further study the
authors assumed the formation of CaIn and $CaIn_2$ complexes [18].
Ross, Aikens and Reilley [19] have proved spectrophotometrically the
existence of the red binuclear complex Ca_2In_2 which is stable at pH
12 and pCa 4, and a violet-blue complex Ca_2In at higher concentrations
of calcium.   The existence of the further complexes CaHIn and $CaIn_2$
has not been reliably proved.

Acid Alizarin Black SN

7. Calcichrome

Calcion IREA.

Of the indicators described up till now, Calcichrome is the most

interesting, not only for its highly selective reaction with calcium
but also for its as yet unsolved structure.   It was prepared by
Lukin et al. and independently by Close and West [21], by diazotiz-
ation of H-acid (8-hydroxy-1-aminonaphthalene-3,6-disulphonic acid)
in hydrochloric acid solution.   The diazo compound becomes deep blue
in concentrated sodium hydroxide solution.   The British authors
assume that in this stage the self-coupling of three molecules of
diazotized H-acid takes place.   After acidification a red compound
is precipitated after some time.   These authors think that the
compound is a trimer with a cyclic structure: cyclo-tris-7-(1-azo-8-
hydroxynaphthalene-3,6-disulphonic acid) (VII).

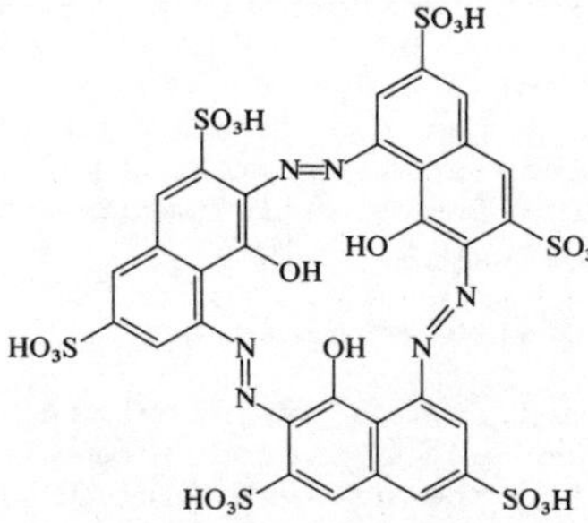

VII: Calcichrome

The reagent is crimson at pH 2 - 8.5, purple at pH 10 - 11.5, blue at
pH 13 and dark red in concentrated alkali.   In solution at pH 11 -
11.5 the indicator forms red chelates with zinc, cadmium, lanthanum,
calcium, barium and magnesium.   At pH 13 it gives a colour only with
calcium.   The reaction is very sensitive and allows detection of
0.25 μg of calcium at a dilution of 1:10$_6$.   This highly selective
reaction of calcium is explained by the authors as due to the fact
that calcium has a smaller ionic radius than barium and strontium and
is more suitable for the "cage" at the centre of mulecule.   Magnesium
also does not react, therefore we can detect calcium in the presence
of up to a 20:1 molar ratio of magnesium.   The titration of calcium
with DCTA in the presence of 10 - 12 times as much barium is also
possible.

Lukin et al. [20] called their product Calcion IREA, and assumed the
linear structure (VIII):

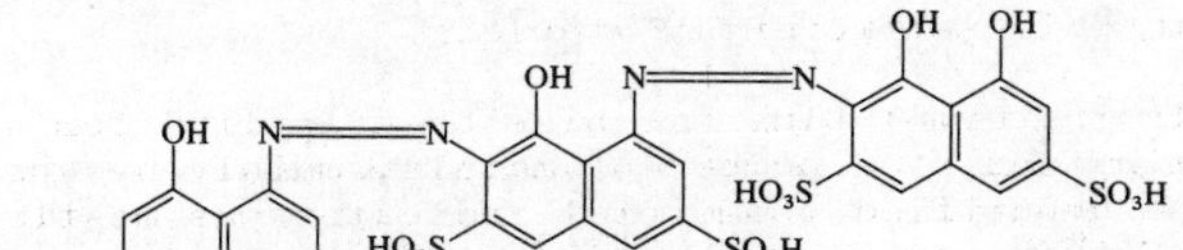

VIII: Calcion IREA

Mendes-Bezerra and Stephen [22] studied some commercial samples of
Calcichrome and of Calcion and proposed their purification by means
of chromatography on silica gel.   Elemental analysis did not give
reliable results.   The infrared spectra of both are practically
identical.   From the polarographic waves in 1M sulphuric acid medium
these authors assume that the indicator probably has the structure
(VIII) proposed by the Russian authors, but they think that this
problem is still open, and that the structure of the chelates can
hardly be proposed before the structure of the indicator is precisely
known.   This is mentioned in detail, because this indicator is
considered by the British authors as unique for its reaction with
calcium.

Remarks.   Both compounds have also been proposed for colorimetric
determination of calcium.   Herrero-Lancina and West [23] studied
thoroughly the conditions for the determination in the presence of
other metals, including barium and strontium.   Magnesium interferes.
Lukin et al. [24] carry out the reaction at pH 12.5 in 15% acetone
medium, where magnesium does not interfere.   They applied the
procedure for the determination of calcium in distilled water and in
ammonium chloride.   The error is $\pm$ 10%.

(b)   Pyridine derivatives with an azo-group

These indicators are formed from a pyridine ring with an azo-group
close to the nitrogen atom, and a second aromatic ring with at least
one hydroxyl group ortho to the azo-group.   The chelates are formed
with two 5-membered rings where the metal is co-ordinated to two
atoms of nitrogen and one atom of oxygen (IX):

IX

Indicators of this type are more selective than those in group (a)
and form more stable complexes with metals of the "ammine" group (Co,
Ni, Cu etc.).   They also form weak complexes with lead, manganese
and bismuth.   They are mostly slightly soluble in water and extract-
able into chloroform.   The two indicators described below are used
the most.

1.   1-(2-Pyridylazo)-2-naphthol (PAN)

PAN (X) was proposed by Cheng and Bray [25] as complexometric
indicator for the titration of copper, zinc and cadmium in slightly
acidic solution.

PAN forms red or violet chelates with most metals.   The chelate with
palladium is green.   It does not react with beryllium, alkaline-
earth metals and with other platinum metals.   A great number of
direct titrations with EDTA have been described.   Because the
chelates are only slightly soluble in water, very slow titration at
elevated temperature is recommended.   The colour change can be
considerably improved by addition of 25 - 50% of organic solvent such

as ethanol, isopropyl alcohol, dioxan, dimethylformamide etc. [26].

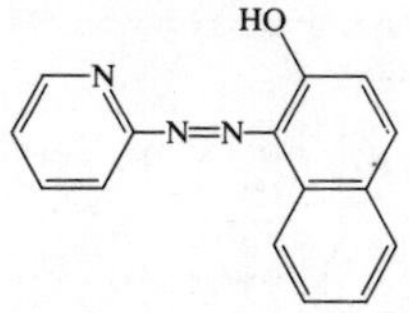

X: PAN

In practical application PAN is not much used.   Because of the
solubility of its 1:1 and 1:2 complexes it has been applied in
combination with EDTA to spectrophotometric determination of
palladium, cobalt and uranium [27].

2.   4-(2-Pyridylazo)-resorcinol (PAR)

PAR (XI) was recommended as an indicator by Wehber [28] and thoroughly studied by Sommer and Hniličková [29].

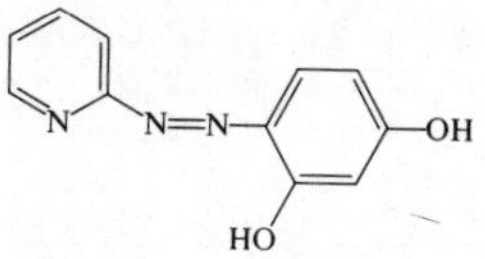

XI: PAR

It is superior to PAN owing to its solubility and the solubility of
its red chelates, which is caused by the ionization of the second
phenolic group.   It is considered as one of the best indicators for
the titration of metals at pH 5 - 7, buffered with hexamine.   The
colour change from red to yellow is very sharp.   Bismuth can be
titrated at pH 1 more selectively than with other indicators.

PAR has also been used as a spectrophotometric reagent for the
determination of cobalt, niobium, palladium, thallium, uranium and
vanadium [27].

Remarks.   Further compounds belonging to this group have in addition
further nitrogen and sulphur atoms (thiazolyl derivatives).   They
have similar properties and are seldom applied in practical analysis.
The indicators PAN and PAR are available commercially.

(ii)   Triphenylmethane dyes

The most important metallochromic indicators are derived from
triphenylmethane dyes and have superior properties to all the azo-
dyes.   Complexometry must be grateful to some of them for its
extension in practice.   The reader will find detailed explanations
of their indicator function in the literature [3,4].

In principle we derive these indicators from well-known acid-base
indicators of the phthalein- and sulphonphthalein series.   The acid-

base indicators are not themselves able to form chelates. If we
introduce into the position ortho to the phenol group a further group
which is capable of co-ordinating a metal atom with formation of a 5-
or 6-membered ring, then the metallochromic indicator is created.
Obvious chelate-forming groups are

$$-CH, \quad -COOH, \quad -CH_2NH-CH_2COOH \text{ and } \quad -CH_2N(CH_2COOH)_2$$

In the chelate formation, dissociation of the phenolic proton takes
place. The chelate thus formed is roughly of the same colour as the
deprotonated (alkaline) form of the parent acid-base indicator. This
piece of knowledge allowed prediction of the colour of the chelates,
before the indicators were synthesized [3].

These metallochromic indicators can be divided into three main sub-
classes:

(a) phenolic compounds
(b) indicators with two $-CH_2N(CH_2COOH)_2$ groups
(c) indicators with two $-CH_2NHCH_2COOH$ groups

(a) Phenolic compounds

These indicators contain two or more OH-groups in the molecule, some
of which can be replaced by the carboxyl group. Altogether five
indicators of this type have been used.

1. Pyrocatechol Violet, PV (Catechol Violet)

Pyrocatechol Violet or pyrocatecholsulphonphthalein (XII) has been
prepared in a pure form by Vodák and Leminger [30] and recommended
for complexometric determinations by Suk et al. [31] and by Suk and
Malát [32]. It was the first metallochromic indicator prepared by
the Czech school and one of the first used for EDTA titrations in
acidic medium.

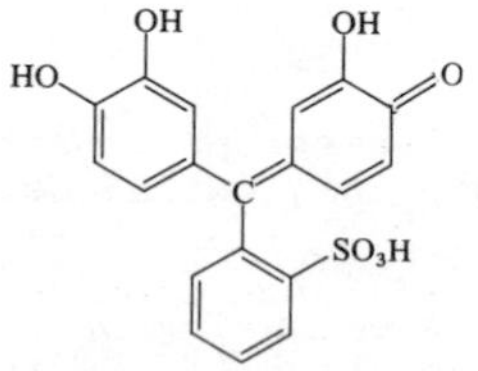

XII: Pyrocatechol Violet

Pyrocatechol Violet in pure form is a red-brown powder with a slightly metallic lustre. It is soluble in water, forming a yellow
solution of the singly charged anion H_3In^-. In alcohol solution PV
has a violet to red-violet colour. As an acid-base indicator it
changes its colour with increasing pH from yellow to violet and red-
violet, owing to stepwise dissociation of phenolic groups [32,33]:

$$H_3In^- \xrightarrow[pK_1 - 7.81]{-H^+} H_2In^{2-} \xrightarrow[pK_2 = 9.8]{-H^+} HIn^{3-} \xrightarrow[pK_3 = 11.7]{-H^+} In^{4-}$$

| | | | |
| yellow | violet | red-
violet | red-
violet |

Chelates with PV are intensely blue, with the exception of that of thorium, which is red.   In bismuth titrations the violet 1:1 bismuth complex can be observed only very close to the end-point.

Pyrocatechol Violet is a very good indicator for acidic medium.   Its colour changes in alkaline solution are not too distinct.   A great number of papers describe the complexometric determination of bismuth [31], thorium [34], copper [35] and indirect determination of indium [36], iron, aluminium, and titanium in pyridine solution by back-titration with a copper solution [37].   Many metals have been determined in alkaline medium [38].

Pyrocatechol Violet is still an outstanding indicator for the acid range of pH, but in practical applications is being steadily replaced by Xylenol Orange.

Remarks.   Pyrocatechol Violet has also been used as a spectrophoto-metric reagent for zirconium and antimony(III).   Great attention has been paid to both colour reactions [27].

2. Pyrogallol Red (PGR)

PR, pyrogallolsulphonphthalein (XIII) was also prepared by Vodák and Leminger [39] and proposed by Jeničková et al. [40,41] for direct complexometric determination of bismuth, lead, nickel and cobalt.

XIII: Pyrogallol Red

It can also be used, similarly to Pyrocatechol Violet, for a great number of direct and indirect determinations of nearly all metals.

Pyrogallol Red in acidic solution is orange-red and in alkaline solution violet to blue.   The colour change is usually from red to orange-red.   Solutions of PGR are stable for only a few days.

Remarks.   This indicator has been used as a specific reagent for silver [42] and later for its colorimetric determination [43].

3. Bromopyrogallol Red (BPGR)

This is a dibromo-derivative of Pyrogallol Red (XIV) and has been suggested by Jeničková et al. [44] for complexometric determination

of bismuth in acidic medium and of nickel, cobalt, cadmium and
magnesium in ammoniacal solutions.   A number of back-titrations with
bismuth nitrate have been also described [45].   Rare earths can be
determined in ammonium acetate medium with the colour change from
blue to red [44].

XIV: Bromopyrogallol Red

Remarks.   It is worthwhile mentioning the very interesting reaction
with silver.   In the presence of 1,10-phenanthroline (Phen) silver
forms with Pyrogallol Red at pH 7 an intensely blue ternary complex
$Ag(Phen)_2^+ BPGR^{2-}$.   In sensitivity this reaction is superior to other
known colorimetric reactions for silver.   The molar absorptivity at
635 nm is 5.0×10^4 $1.mole^{-1}cm^{-1}$.   With this reaction it is possible
to determine traces of silver in aqueous solutions [46] or after
extraction of the complex into nitrobenzene [47].

4.   Eriochrome Cyanine R and Chrome Azurol S

Both dyes are known as reagents for the detection and colorimetric
determination of aluminium.   Some attempts have been made to use
them as complexometric indicators, but with disputable results.
Therefore they are mentioned here only briefly.

Eriochrome Cyanine R is the trisodium salt of $3,3''$-dimethyl-2-
sulphofuchsonedicarboxylic acid (XV).   It was suggested as an
indicator for the titration of zirconium by Fritz and Fulda [48].
The direct titration of zirconium in hot solution was not good enough,
however, and the authors proposed the back-titration of EDTA with
zirconium solution.   The colour change is from colourless to pink.
The indirect determination of aluminium [49] by back-titration of
EDTA with a zinc or cadmium solution has never been applied in
practical analysis.   The acid-base and chelate-forming properties
have been studied by Suk and Mikeťuková [50].

XV: Eriochrome Cyanine R   XVI: Chrome Azurol S

Chrome Azurol S, the trisodium salt of 2,6-dichloro-3,3''-dimethyl-3-sulphofuchsonedicarboxylic acid (XVI) has been recommended in a number of papers for complexometric determination of aluminium [51], copper [52], iron[53], zirconium [54], and in strongly alkaline medium for the determination of alkaline earths [55]. Malát and Tenorová [56] applied this indicator also for the determination of thorium, nickel, cerium and lanthanum.

The indicator is orange-red in acidic solutions (pH 1 - 4) yellow in neutral solutions and yellow-green in alkaline medium. Theis [51] determined aluminium at pH 4 and at 80°C by direct titration with EDTA. The colour change here is from violet to red-orange. According to this author it is possible to determine aluminium in the presence of iron after reduction of the latter with hydrazine sulphate. In view of the small difference in the apparent stability constants of the iron(II) and aluminium EDTA complexes at pH 4 (log K'_{AlY} = 7.5 and log $K'_{Fe(II)Y}$ = 5.9), it is very doubtful whether this method is accurate enough.

Iron(III) itself can be determined at pH 2 and at 60°C with a colour change from blue to golden-orange. It seems likely that this indicator is somewhat underestimated for direct determination of iron.

(b)   Indicators with two methyliminodiacetate groups

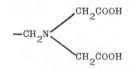

These indicators represent today the most valuable class of indicators. They are derived from phthalein and sulphonphthalein acid-base indicators. Their preparation is based on Mannich condensation of the acid-base indicator with formaldehyde and iminodiacetic acid. The synthesis of these indicators is covered by some patents [57,58].

Both groups differ considerably in their metallochromic properties. The phthalein indicators are lactones and the lactone ring opens in alkaline medium, giving a suitable resonance system for the formation of coloured chelates. In solutions at pH lower than 8 such indicators are able to form weak coloured chelates, with only the iminodiacetate group taking part. For example, Thymolphthalexone in large amounts masks traces of bismuth in acidic medium even from such a sensitive indicator as Xylenol Orange.

With sulphonphthalein indicators the acid-base changes lie in the acid range of pH and therefore the formation of coloured chelates takes place with the dissociation of phenolic protons.

Schwarzenbach was the first to suggest Mannich condensation for the synthesis of metallochromic indicators. After the preparation of the condensation product of p-nitrophenol with iminodiacetic acid, which had only slight metallochromic properties [59], the first real indicator of "complexone type" was prepared from o-cresolphthalein by his co-workers [60]. For this reason the indicators derived from phthalein acid-base indicators will be described first.

1. Phthaleincomplexone

3,3''-Bis{[N,N-di(carboxymethyl)amino]methyl}-o-cresolphthalein (XVII) is also called Phthalein Purple, Metalphthalein or Cresolphthalexone in the literature.

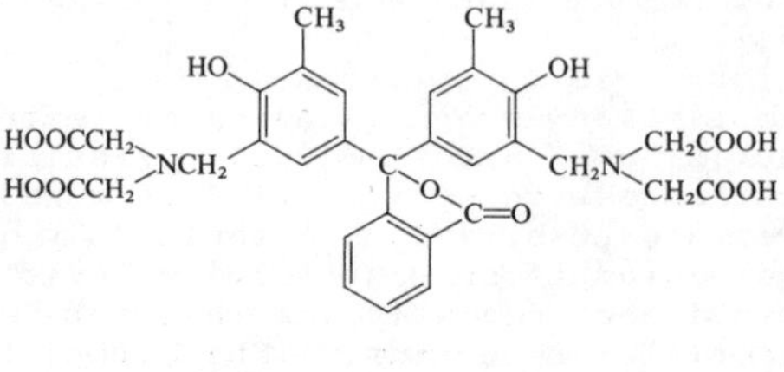

XVII: Cresolphthalexone

It was prepared by condensation of o-cresolphthalein with formalde-hyde and iminodiacetic acid [60]. As an acid-base indicator it is colourless in acidic medium, pink at pH 7 - 10 and intensely violet at pH above 11. In ammonia solution it forms intensely violet chelates with calcium, strontium, and barium. Most heavy metals do not form coloured complexes with it, with the exception of manganese and cadmium. In fact they block the indicator, forming colourless complexes. For example zinc as tetrahydroxozincate in strongly alkaline medium "decolorizes" the alkaline form of the indicator, without interfering with its reaction with calcium. This has been utilized for the indirect determination of cadmium and zinc in sodium hydroxide solution by back-titration with a calcium solution [61]. The hypsochromic effect produced by formation of the calcium chelate can be improved by suppressing the colour of free indicator by addition of 20 - 30% of ethanol. Under these conditions the colour change in the calcium titration is from violet to nearly colourless.

2. Thymolphthalexone

This indicator, 3,3''-bis{[N,N-di(carboxymethyl)amino]methyl}thymol-phthalein (XVIII) was first prepared by Körbl and Přibil [62]. In strongly alkaline medium it forms intensely blue complexes with calcium, barium and strontium. The colour change on titration is from blue to colourless or smoky grey. It is superior to the preceding indicator, as its colour changes occur at higher pH.

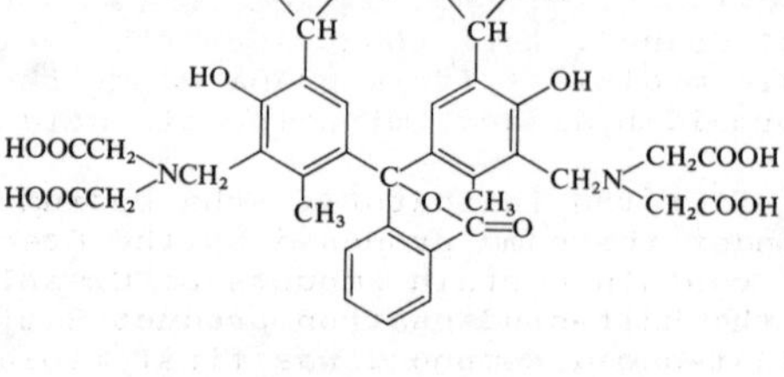

XVIII: Thymolphthalexone

The indicator is not too stable in aqueous solution and a solid
mixture with potassium nitrate is recommended for use. It has been
recommended for the determination, in ammoniacal solution, of
manganese(II) in the presence of hydroxylamine hydrochloride and
triethanolamine (TEA) [63].   Compared with the azo-dye indicators it
has one main advantage: it can be used in the presence of iron masked with TEA (the iron complex of which is colourless in sodium
hydroxide solutions).

Cadmium behaves similarly to the alkaline-earth metals, forming an
intensely blue complex in alkaline medium.   Zinc forms only a
slightly pink or colourless complex.   This is more stable than the
blue Mg-indicator complex, which is decolorized by addition of traces
of zinc, even if magnesium is present in very large concentration.
This has been utilized more recently for indirect determination of
calcium in the presence of magnesium.   The method is based on back-
titration of excess of EGTA with zinc solution in borax medium [64].

3.   Phenolphthalexone

Bis{N,N,-[di(carboxymethyl)amino]methyl} phenolphthalein (XIX) is an
analogue of Thymolphthalexone, and also forms complexes with many
cations.   Its complex with calcium is slightly orange and that with
zinc colourless.   In EDTA titration of zinc in alkaline medium, the
violet indicator form is freed at the equivalence point.   Mixing it
with Fluorexone (p. 45) improves its colour change in calcium
titrations.

XIX: Phenolphthalexone

Remarks. Calcein (Fluorexone) also belongs to the group of substitu-
ted phthaleins.   Because of its metallofluorescent properties it
will be described in the section on metallofluorescence indicators
(p. 45).

4.   Xylenol Orange (XO)

XO, 3,3''-bis[N,N-di(carboxymethyl)aminomethyl]-o-cresolsulphonphth-
alein (XX), Xylenol Orange, is without doubt the best complexometric
indicator for acidic medium.   It is prepared by the condensation of
Cresol Red with formaldehyde and iminodiacetic acid [65,66].

A number of firms offer the indicator as the sodium salt on the
chemical market, under the name proposed by the Czech originators.
The usual products contain certain amounts of Cresol Red, iminodiacetic acid and the half-condensation product Semi-Xylenol Orange,
SXO (XXI).   The last-named compound was first isolated from
commercial samples by Olson and Margerum [67].   These authors pointed out that Semi-Xylenol Orange is a more sensitive reagent than

Xylenol Orange for zirconium.   Murakami et al. [68] found that the
purification of XO by chromatography with various alcohol/acetic acid
mixtures as eluents led to decomposition of the XO to a red compound
which formed no chelates with metal ions, but they succeeded in
separation of Cresol Red and SXO from XO, and of XO and iminodiacetic
acid from SXO by column chromatography on cellulose powder.   Yamada
and Fugimoto [69] used thin-layer chromatography on cellulose powder
for the same purpose.   Nakada et al. [70] described the purification
of a large sample of Xylenol Orange (960 mg) by high-pressure liquid
chromatography (HPLC).

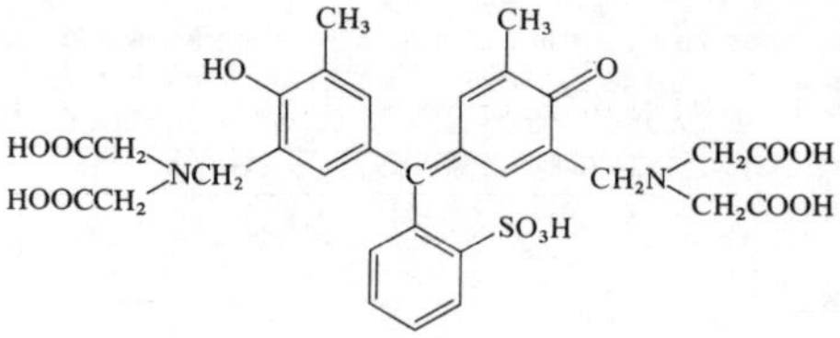

XX: Xylenol Orange

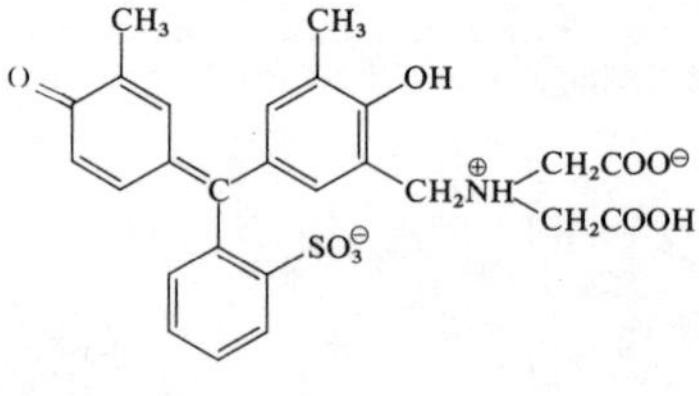

XXI: Semi-Xylenol Orange

Sato et al. [71] improved the synthesis of Xylenol Orange as well as
its chromatographic purification.   For the synthesis they used a
lower temperature (60 - 65°C) and rapid distillation of the solvent
(acetic acid) under low pressure.   They gradually eluted the DEAE
cellulose column with sodium chloride or nitrate solution.   First
eluted were iminodiacetic acid (ninhydrin test) and Cresol Red.
Very carefully purified Xylenol Orange was finally obtained as the
hydrochloride, with the composition $H_6XO.2H_2O.1\cdot8HCl$.   The method
also allowed the separation of pure Semi-Xylenol Orange in sufficient
amounts for the measurement of its formation constants and the
stability constants of its zirconium complex [72].

As an acid-base indicator Xylenol Orange shows a number of colour
changes, which have been thoroughly studied by many authors
[68,71 - 73].   The protonation scheme of Xylenol Orange is shown on
the next page.

The acid formation constants are very important for the calculation
of the stability constants of individual metal complexes.   The
values are very dependent on the purity of the compound.   Some
values given in the literature are summarized in Table 2.1.

$$H_9XO^{3+} \xrightarrow{pK_1} H_8XO^{2+} \xrightarrow{pK_2} H_7XO^+ \xrightarrow{pK_3} H_6XO \xrightarrow{pK_4} H_5XO^- \xrightarrow{pK_5} H_4XO^{2-} \xrightarrow{pK_6}$$

  red   red   red   pink   yellow   yellow

$$H_3XO^{3-} \xrightarrow{pK_7} H_2XO^{4-} \xrightarrow{pK_8} HXO^{5-} \xrightarrow{pK_9} XO^{6-}$$

  yellow   red   weakly   red
                                     red

TABLE 2.1   Dissociation constants of Xylenol Orange

		Murakami [66]	Řehák and Körbl [71]	Sato [69]	Sato [70]
H_9XO^{3+}					
H_8XO^{2+}	pK_1	-3.32^a	–	–	–
H_7XO^+	pK_2	-1.83^a	–	–	-1.24^a
	pK_3	-1.04^a	-1.74	–	1.43^b
H_6XO	pK_4	1.5^b	-1.09	–	1.45^b
H_5XO^-	pK_5	2.32^b	2.58	2.11	2.24^b
H_4XO^{2-}	pK_6	2.85^b	3.23	2.79	2.56^b
H_3XO^{3-}	pK_7	6.74^a 6.67^b	6.40	6.66	6.09^b
H_2XO^{4-}	pK_8	10.56^a 10.39^b	10.46	10.35	9.34^b
HXO^{5-}	pK_9	12.23^a	12.28	–	10.58^b
XO^{6-}					

^{a} spectrophotometrically ^{b} pH titration in 1M KNO_3

The red cationic forms of Xylenol Orange exist only in extremely acid solutions.   Only the yellow forms are analytically interesting, and these form violet-red chelates with at least 30 cations.   Some of these cations (Fe, Al, Co, Ni, Cu, Ga) form XO complexes that are stable towards ligand-exchange with EDTA and cannot be determined directly.   The aluminium complex can be used in a lecture demonstration of entropy effects [74].   Cobalt can be determined at elevated temperature, and copper after the addition of traces of 1,10-phenanthroline [75].   The sensitivity of the indicator is very high (a colour is produced by 1 μg or less of metal per ml) and is strongly influenced by pH.   Table 2.2 shows the lowest limits of pH at which the indicator gives a distinct colour with various cations.

Much attention has been paid to Xylenol Orange as a spectrophotometric reagent.   The conditions for the colour reactions of all the

TABLE 2.2   The lowest pH limit for the detection
of cations with XO

$[HNO_3]$ M	Cation	pH	Cation
1	Zr^{4+}, Hf^{4+}	3	Pb^{2+}, Al^{3+}, In^{3+}, Ga^{3+}
0.5	Bi^{3+}	3 - 4	Lanthanides, Y^{3+}
0.2	Fe^{3+}	4	Zn^{2+}, Co^{2+}, Tl^{3+}
0.1	Th^{4+}	5	Cu^{2+}
0.05	Sc^{3+}	5 - 5.5	Mn^{2+}, Ni^{2+}, Cd^{2+}, Hg^{2+}

The optimal pH for titration is somewhat higher
than in this table.   For example for bismuth it
is 1 - 3, for thorium 2.5 - 3.5 etc.

rare earths and 20 other cations have been studied thoroughly [27].
Information on the composition of the complexes, derived from
spectrophotometric measurements by various authors, is rather
inconsistent.   It is generally considered that 1:1 complexes are
formed in the presence of excess of Xylenol Orange.   The composition
of the complexes is strongly dependent on pH;   for example thorium
gives a 1:1 complex at pH 2 and a 1:2 (Th:Y) complex at pH 4 - 5 [76],
but Budĕšínský has also proved the existence of a 2:2 complex and
determined its apparent stability constants [77].   In complexometric
titrations, where the metal is always in excess, we can assume the
formation of 1:1 or 2:1 (Th:Y) complexes.

Aqueous solutions of Xylenol Orange are reasonably stable, even in
hot 1M nitric acid and in the presence of hydrogen peroxide at room
temperature, but on long storage often lose their metallochromic
properties, except for giving a specific reaction with copper at pH
5 - 5.5.   This is due to partial decarboxylation of the iminodiacetic
acid group:

$$-CH_2N\diagup^{CH_2COOH}_{\diagdown CH_2COOH} \quad CO_2 + \quad -CH_2N\diagup^{CH_3}_{\diagdown CH_2COOH}$$

A sarcosine derivative is formed, which is an excellent indicator for
copper [78].   It has similar properties to Glycinethymol Blue
(p. 40).   The slow decomposition of the indicator has been proved by
chromatography [79].   The decarboxylation can be accelerated by
oxidation of the indicator with lead dioxide in solutions acidified
with acetic or sulphuric acid.   The oxidized form of Xylenol Orange
prepared by boiling with hydrogen peroxide has been used for the
determination of copper in acid copper baths [80].

Xylenol Orange is an excellent indicator for all determinations in
acidic solutions (from 0.1M nitric acid up to pH 5.6).   Direct
determinations can be made of zirconium and hafnium, thorium,
scandium, indium, rare earths, yttrium, lead, zinc, cadmium, copper

(in the presence of 1,10-phenanthroline) and cobalt (at 60°C).
Cations which block the indicator, such as iron, aluminium, nickel,
cobalt, copper, gallium, are best determined indirectly, by back-
titration of excess of EDTA with zinc or lead solution at pH 5 - 5.5
(hexamine-buffered). The end-point colour change in titration with
lead nitrate is particularly sharp - from intensely red-violet to
lemon yellow.

Iron can be determined at pH 2 - 3 by back-titration with bismuth
nitrate at pH 2 - 3, and so can titanium after formation of the
peroxo - Ti-EDTA complex [81].

The difference in stability of EDTA-complexes as a function of pH
permits consecutive determinations, for example of bismuth (at pH 2)
and lead (at pH 5.5).

For all complexometric titrations a few drops of 0.2 - 0.5% solution
are used.

5. Methylthymol Blue (MTB)

MTB, 3,3´-bis[N,N-di(carboxymethyl)aminomethyl]thymolsulphonphthalein
(XXII) was prepared by Korbl and Přibil [82] by the usual Mannich
condensation from thymolsulphonphthalein. A relatively pure product
was prepared by Korbl [83].

XXII: Methylthymol Blue

The crystalline sodium salt is a black powder with good solubility in
water but insoluble in absolute alcohol. Its aqueous solutions are
not very stable. On boiling in 1M nitric acid MTB decomposes very
quickly. Similarly to its parent compound - thymolsulphonphthalein -
it behaves as an acid-base indicator with three colour changes. The
solutions are yellow up to pH 6.5, changing to pale blue in the pH
range 6.5 - 8.5 and then to grey at pH 10.7 - 11.5. The solutions are
dark blue above pH 12.7. This apparently anomalous colour pattern
can be explained by the presence of hydrogen bonds and alternating
formation of symmetrical and unsymmetrical deprotonated forms of the
indicator [84]. As in the case of Xylenol Orange (see Table 2.1 on
p. 35), MTB will be red at negative pH values owing to addition of
further protons (probably to form H_7I^+, H_8I^{2+} or H_9I^{3+}).

Remarks. The absorption spectra of free Methylthymol Blue as a
function of pH were measured by Srivastava and Banerji [85]. These
authors found two absorption maxima at pH 1 - 7 (275 and 426 nm), and
one at pH 8 (426 nm), pH 9 - 10 (580 nm), and pH 11 - 12 (590 nm).

The acid-base properties of MTB were studied by Körbl and Kakáč [84] by spectrophotometric and potentiometric methods, and the last three dissociation constants corresponding to the equilibria between the deprotonated forms H_3I^{3-}, H_2I^{4-}, HI^{5-} and I^{6-} were determined:

$$pK_4 = 7.2, \quad pK_5 = 11.5, \quad pK_6 = 13.4$$

The values of the first three dissociation constants, corresponding to the equilibria for H_6I, H_5I^-, H_4I^{2-}, H_3I^{3-} were published by Tereshin et al. [86]:

$$pK_1 = 3.0, \quad pK_2 = 3.3, \quad pK_3 = 3.8$$

These authors also reported a value of 7.4 for pK_4.

Methylthymol Blue forms intensely blue complexes with many cations. In acidic medium (pH 0 - 6.5) it can be used similarly to Xylenol Orange. The colour changes are always from an intense blue to lemon yellow. In alkaline medium (pH 10.7 - 11.5) it may mainly be used for the determination of calcium, strontium, barium and magnesium. The colour changes are from deep blue to colourless or grey. Many cations which tend to block the indicator can be determined indirectly.

For all titrations a small amount of a solid mixture of the indicator with potassium nitrate (1:100) is mostly used.

Remarks. Yoshino et al. [87] have thoroughly studied the purity of Methylthymol Blue and succeeded (by chromatography on cellulose and paper) in isolating the semi-condensation product Semi-Methylthymol Blue (STMB). It has similar properties to MTB but forms only 1:1 complexes. These authors assume that this compound would be a very good spectrophotometric reagent.

According to Kosenko et al. [88] commercial preparations of MTB can contain up to 50% of SMTB. Both indicators can be separated from such mixtures by gel filtration on a column packed with Molselect G-10 or G-25, by elution with water. Methylthymol Blue containing less than 6% of SMTB can be prepared by three successive filtrations. In this way SMTB containing less than 0.5% of Methylthymol Blue was prepared. This method was recommended for preparation of SMTB for use as a spectrophotometric reagent.

6. Methylxylenol Blue (MXB)

This very sensitive indicator (XXIII) can be prepared similarly to MTB and XO from the corresponding acid-base indicator, p-Xylenol Blue, by condensation with formaldehyde and iminodiacetic acid. It was first prepared by Cherkesov et al. [89] and named p-Xylenolphthalexon S.

With respect to structure, Methylxylenol Blue is more closely related to Methylthymol Blue than to Xylenol Orange; it behaves similarly to them as an acid-base indicator. All the dissociation constants for Methylxylenol Blue ($H_9In^{3+} \xrightarrow{pK_1} H_8In^{2+}$ etc.) were measured by Vytřas and Vytřasová [90]. The values are given in Table 2.3.

XXIII: Methylxylenol Blue

TABLE 2.3   Dissociation constants of Methylxylenol Blue

$pK_1 = -2.8^a$ $pK_2 = -1.90 \pm 0.08^a$ $pK_3 = 0.08 \pm 0.02^a$

$pK_4 = 2.0 \pm 0.8^b$ $pK_5 = 3.4 \pm 0.2^b$ $pK_6 = 4.3 \pm 0.1^b$

$pK_7 = 7.00 \pm 0.01^a$, -6.95 ± 0.02^b $pK_8 = 10.61 \pm 0.03^a$

$pK_9 = 12.17 \pm 0.06^a$

[a] spectrophotometrically   [b] potentiometrically

Methylxylenol Blue was proposed for complexometric determination of
bivalent metals by Vytřas et al. [91] who studied the influence of pH
and buffers (hexamine and acetate).   All the titrations were also
followed spectrophotometrically.   It was found that MXB is a very
sensitive indicator for zinc, lead, manganese and cadmium.   Two
examples are given below - the determination of copper and cobalt,
which cannot be carried out with other indicators such as Xylenol
Orange, Methylthymol Blue, Pyrocatechol Violet, etc.

Copper.   Copper blocks Xylenol Orange and Methylthymol Blue, which
makes their use in its EDTA titration impossible unless the blocking
is eliminated by addition of a small amount of 1,10-phenanthroline
(Phen) [92].   Vytřas et al. found that hexamine is not a suitable
buffer for copper determinations.   In solution buffered with acetate
(pH 5.6) the microtitrations proceed smoothly and even a small amount
of Phen suffices.   Methylxylenol Blue gives a sharp end-point in
pH 5.6 acetate medium.

Cobalt.   Like copper, cobalt cannot be determined directly, and for
the same reason.   According to Vytřas et al. cobalt in hexamine
medium at pH 5.4 - 5.7 can be titrated directly, on the microscale,
with 0.01M EDTA, with Methylxylenol Blue as indicator.

Methylxylenol Blue can also be used for the determination of rare
earths [93].   Consecutive titrations, for example, of scandium and
lanthanum, are also possible.

Vytřas et al. [94] have also made a computer study of colour-change quality in complexometric titrations.

Japanese chemists have paid great attention to Methylxylenol Blue as spectrophotometric reagent, and references to many of their publications will be found in the paper by Vytřas and Vytřasová [90].

Remarks. Like the preceding indicators, commercial MXB also contains the half-condensation product, Semi-Methylxylenol Blue (SMXB), which has been isolated in pure form and used for spectro-photometric determination of thorium [95] and iron and aluminium [96]. SMXB has similar properties to MXB but forms only 1:1 complexes.

(c) Indicators containing two carboxymethylaminomethyl groups

$$-CH_2-NH-CH_2COOH$$

To this class of indicators belong all those which can be prepared by condensation of sulphonphthaleins with formaldehyde and glycine. It was originally expected that their reactions with the copper ion would be very sensitive. Two of them, Glycinethymol Blue (GTB) and Glycinecresol Red (GCR), are worth mentioning.

1. Glycinethymol Blue (GTB)

3,3´-Di(N-carboxymethylaminomethyl)thymolsulphonphthalein (XXIV) was prepared by Mannich condensation from Thymol Blue [97].

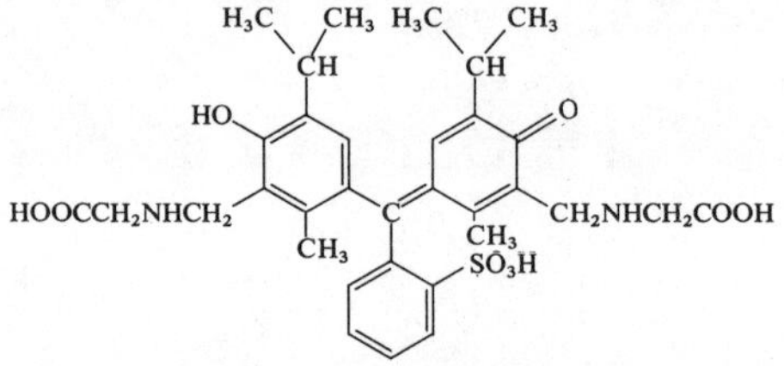

XXIV: Glycinethymol Blue

At pH 5 it forms an intensely blue complex with copper, and at higher pH it also reacts with nickel, cobalt and some other metals. Its practical use in complexometry is limited to the analysis of dilute solutions of copper (0.001M). The colour change is from blue to yellow. At higher copper concentration the end-point is less distinct because Cu-EDTA itself is slightly blue. Its slightly acid aqueous solution is sufficiently stable.

Remarks. Glycinethymol Blue has also been applied for spectrophoto-metric determination of a number of metals, mainly by Russian authors. The unpurified commercial indicator was mostly used. According to Vytřas and Langmyhr [98] such compounds contain 30 - 50% of the semi-condensation product - Semi-Glycinethymol Blue (SGTB). These authors described a new synthesis of GTB directly from Thymol Blue. The progress of the condensation was checked by paper chromatography. The GTB and its monosodium salt were purified by cation-exchange chromatography, and after evaporation yielded the

free acid H_4In. These authors described all the absorption spectra
and measured all the acid-base dissociation constants, starting from
triply protonated acid H_7In^{3+}:

$$pK_1 = -2.61, \ pK_2 = -1.42, \ pK_3 = -0.6, \ pK_4 = 1.2,$$

$$pK_5 = 6.98, \ pK_6 = 10.50, \ pK_7 = 12.62$$

In a further publication [99] they described the isolation of pure
SGTB and reported that its reactions with various metals showed that
it would not be useful for analytical application (in contrast to SXO
or SMTB).

2. Glycinecresol Red (GCR)

3,3'-Di(carboxymethylaminomethyl)-o-cresolsulphonphthalein (XXV)
[100] has similar properties to GTB.

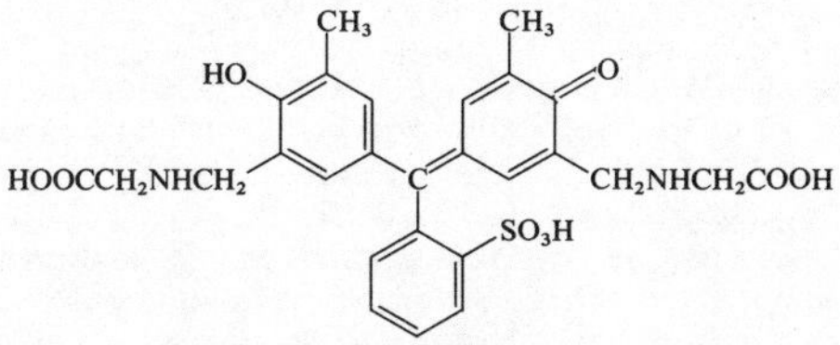

XXV: Glycinecresol Red

It generally forms red chelates but the copper chelate is particularly intense in colour.   Its reaction with copper is also more
sensitive than that of Glycinethymol Blue.   GCR has also been
proposed by Russian authors for spectrophotometric determination of
copper, iron and gallium [20], and also for determination of rare
earths, scandium, cadmium and zirconium.

Remarks.   Semi-Glycinecresol Red (XXVI) has been isolated from GCR
and proposed for complexometric determination of copper (pH 4 - 6) and
of lead, zinc and copper in alkaline medium [101];   both 1:1 and 1:2
complexes (metal:ligand) are formed.

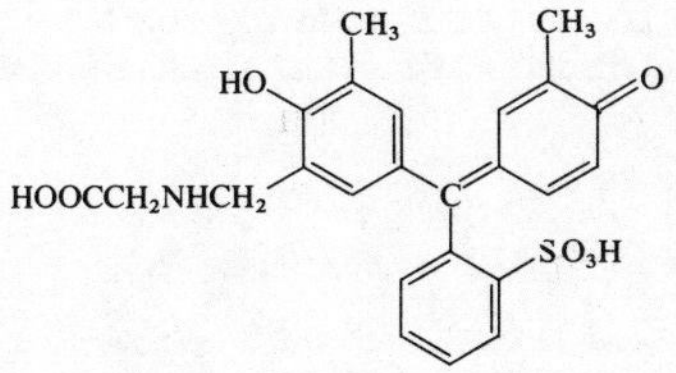

XXVI: Semi-Glycinecresol Red

(iii)   Indicators of various structures

This class of indicators is also numerous and includes not only
compounds already known as analytical reagents, such as dithizone,

haematoxylin, Alizarin S, but also new compounds specially prepared, among which the indicators of "complexone type" again predominate. For reasons of space only some will be mentioned here.

1. Murexide

Murexide, ammonium purpureate (XXVII), is a classical indicator proposed by Schwarzenbach [102] for complexometric determination of calcium, nickel, cobalt and copper.

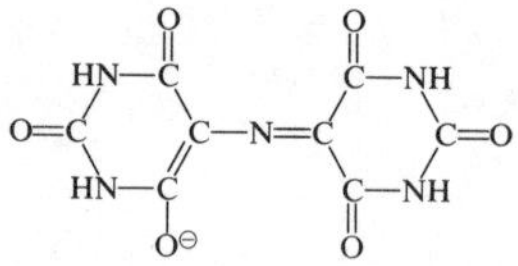

XXVII: Murexide

In strongly alkaline medium murexide is violet-blue, turning to red-violet below pH 9. Its complex with calcium is pink. The colour change in calcium titration is from pink to red-violet but is difficult to see visually, and is not to be compared with the colour change of other metallochromic indicators. Spectrophotometric titration is recommended. Magnesium forms only a very weak complex, which is destroyed in strongly alkaline medium, with precipitation of magnesium hydroxide. This allows the determination of calcium in the presence of magnesium, but there are several well-known dis-advantages. Many attempts have been made to improve the colour change in calcium titration by addition of "screening dyes" as an inner filter. At least ten such mixtures are described in the literature [4].

On the other hand, the colour change from yellow to violet in the titration of nickel in strongly ammoniacal solution is very sharp. Titration of copper in very slightly ammoniacal solution is also possible. Murexide can be used only in alkaline medium, and is applied as a solid mixture with KCl or KNO$_3$ (1:100). Murexide was very popular some years ago but has since lost much of its importance.

2. Glyoxal bis(2-hydroxyanil)

This indicator (XXVIII) was first described by Goldstein [103] and proposed for use in calcium titrations in sodium hydroxide medium.

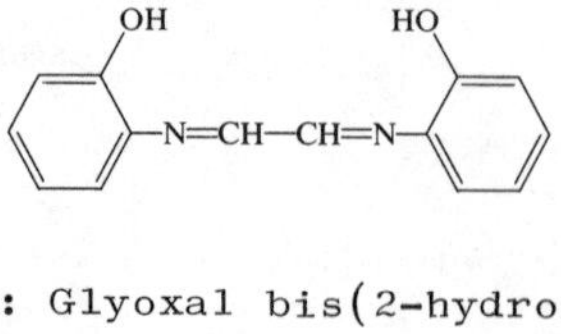

XXVIII: Glyoxal bis(2-hydroxyanil)

The indicator is slightly soluble in water, and therefore a 0.02% solution in alcohol is used. A freshly prepared solution is colour-less but slowly turns yellow owing to decomposition. The colour

change in the titration of calcium is only from pink to yellow, but
some authors consider it an excellent indicator for calcium [104].

3. Dithizone

Dithizone, diphenylthiocarbazone (XXIX), is one of the most sensitive
but also most unselective reagents ever used in analytical chemistry.

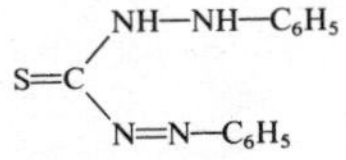

$$S=C \begin{matrix} NH-NH-C_6H_5 \\ \\ N=N-C_6H_5 \end{matrix}$$

XXIX: Dithizone

It has been used for colorimetric determination of a number of metals.
In combination with EDTA it has been proposed for the spectrophotometric determination of silver, mercury, copper, bismuth and tin [20].

Bovalini and Casini [105] have used dithizone and extractive
titration for complexometric determination of lead and cadmium.

Wänninen and Ringbom [106] applied dithizone for complexometric
determination of zinc at pH 4.5 in solutions containing 40 - 50%
ethanol.   Kotrlý [107] has used dithizone for complexometric microdetermination of lead.

4. Zincon

Zincon, 1-(2-hydroxy-5-sulphophenyl)-2-phenyl-5-(2-carboxyphenyl)-
formazan (XXX), has been proposed as a colorimetric reagent for zinc
and copper [108,109] with which it forms intensely blue chelates.

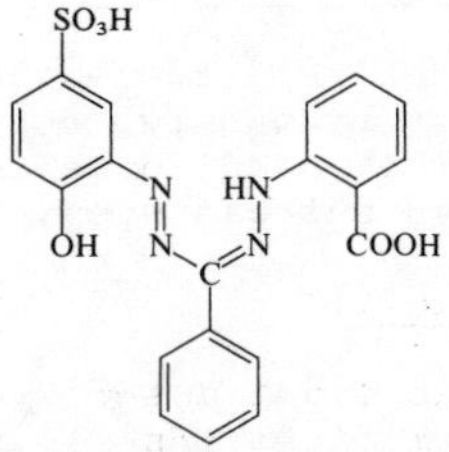

XXX: Zincon

It is suitable as complexometric indicator for the determination of
zinc in buffered ammoniacal solution (pH 9 - 10).   The colour change
from blue to yellow is very sharp.   It has also been applied for
indirect determination of iron, copper. indium, manganese, cadmium,
calcium, lead, nickel, cobalt etc. [110].

5. Alizarin Complexone

Belcher et al. [111] introduced a condensation product of dihydroxy-
anthraquinone with iminodiacetic acid and formaldehyde (XXXI), and
called it Alizarin Complexan.   The names Alizarin Complexone and

Alizarin Fluorine Blue are now more generally used.

XXXI: Alizarin Complexone

It is a fairly good indicator, with similar properties to Xylenol
Orange. Its aqueous solutions are yellow below pH 4.3, red at pH
6 - 10 and blue above pH 13 [112]. At pH 4.1 - 4.3 it forms red
chelates with cobalt, copper, indium, lead, zinc and mercury. The
lower limit of pH is about one unit lower than that for the formation
of the corresponding chelates with Xylenol Orange, but in titrations
it gives poorer response than XO. It cannot be used for the
determination of rare earths and thorium because their Alizarin
Complexone chelates are more stable than their EDTA chelates. The
colour changes from red to yellow with this indicator are sufficiently sharp if a high enough concentration of indicator is used.
Alizarin Complexone has also been used for the determination of
molybdenum(V) by Headridge [113].

Remarks. The most interesting reaction of this compound was discovered by Leonard [114] and exploited by Belcher, Leonard and West
[115]. The red complexes of Alizarin Complexone with cerium,
lanthanum and praseodymium form a blue ternary complex with fluoride,
which allows spectrophotometric determination of fluoride. It was
the first direct colour reaction for fluoride and has been widely
applied [20].

6. Naphthol Violet and Glycinenaphthol Violet

Both indicators (XXII-A, XXXII-B) are again of complexone type. They
were prepared by Budě̆šínský by Mannich condensation of 4-hydroxy-1-
(4-nitrophenylazo)naphthalene with formaldehyde and iminodiacetic
acid [116] or glycine [117] respectively.

XXXII-A: Naphthol Violet   XXXII-B: Glycinenaphthol Violet

R = CH$_2$N(CH$_2$COOH)$_2$.   R = CH$_2$NHCH$_2$COOH

Both indicators can be used in ammoniacal medium for the determination of cobalt, nickel, cadmium, zinc, manganese and copper, with
colour change from violet-red to blue. Bismuth can be determined at
pH 1 with Naphthol Violet. Glycinenaphthol Violet does not react
with Bi, Hg, Sr, Ba, Sc, Al, Zr, Th, Sn, Fe, Pd or Ag. It is

recommended for determination of nickel [117].

Neither indicator, in spite of the good properties, is widely used.

2.1.6   Metallofluorescent Indicators

With these indicators the end-point of an EDTA-titration is indicated by the appearance or disappearance of fluorescence in the solution. Sometimes diffused light is sufficient for excitation of the fluorescence but the fluorescence appears more pronounced if light of the optimal wavelength is used, e.g. light from a mercury lamp.

Like metallochromic indicators, the fluorescent indicators must have good chelate-forming properties.   These indicators have some value when the normal indication of the end-point cannot be used, e.g. in strongly coloured solutions, but the titrations are not so convenient as those with visual indicators.   A dark room and a suitable illuminating source have to be used, and after a time optical fatigue arises in the operator.

With the exception of morin [118] such indicators have all been specially synthesized, and it is obvious that they belong to the "complexone type" compounds.

1.   Calcein, Fluorexone

In the history of this indicator we find a number of conflicting opinions and statements about its structure and purity, and even its indicator function.   Because of its remarkable properties and the fact that its synthesis led to the preparation of further compounds, it will be discussed at some length.

The indicator was first prepared by Diehl and Ellingboe [119] by Mannich condensation from fluorescein, and named Calcein (XXXIII).

XXXIII: Calcein (Fluorexone)

3,3'-bis[N,N-di(carboxymethyl)aminomethyl]fluorescein

These authors recommended Calcein as indicator for calcium in strongly alkaline medium.   They described only its metallochromic properties.   The colour change at the end-point of the calcium titration was, according to the authors, from yellow-green to brown and they did not mention the fluorescence of the solutions.

As expected from previous considerations [3] the indicator must have some fluorescence from its parent compound – fluorescein.   This has been proved by preparation of pure products [120].   The indicator

itself has a green fluorescence in solutions of pH 3 - 11.   Above
this pH the fluorescence weakens, and disappears at pH above 12.   A
small residual fluorescence is caused by traces of unreacted
fluorescein [121].   In solutions of pH above 12 the green fluorescence reappears on the addition of calcium, strontium or barium.
Magnesium does not react under these conditions.   The end-point of
the calcium titration is indicated by the disappearance of the
brilliant green fluorescence, the solution becoming pink.

A residual fluorescence is also caused by sodium and lithium [122].
Therefore the titration of calcium is best done in 0.1M potassium
hydroxide medium.

In slightly acidic solutions the fluorescence is quenched by addition
of copper or manganese ions, and reappears at the end-point of their
EDTA titration [123].   This apparently anomalous behaviour of the
indicator can be explained in terms of different structures of the
indicator complexes.   This problem has also been a subject of
discussion [124 - 126].

Some doubts have been uttered about the structure of Calcein.
Wallach et al., in their detailed study [127], assumed the structure
was unsymmetrical (XXXIV).

XXXIV

3,6-Dihydroxy-2,4-bis-N,N,-di(carboxymethyl)aminomethylfluoran

In spite of their convincing work, little attention has so far been
paid to their opinion.

Although end-point indication with Calcein is reliable enough, some
attempts have been made to improve the visual contrast in the calcium
titration, especially if the indicator used contains traces of
fluorescein.   Tucker [128] recommends a mixture of 0.2 g of Calcein
with 0.12 g of thymolphthalein and 20 g of potassium chloride.   When
10 mg of this mixture are used for 50 ml of solution the end-point is
given by a change from the green fluorescence to a purple colour.
Addition of a few drops of 0.01% acridine solution is recommended by
Kirkbright and Stephen [129].

A mixture of Calcein and murexide in the ratio 5:2 was proposed by
Bauman [130].   According to our experience a very suitable combination for improvement of the visual end-point is a mixture of Calcein
and Phenolphthalexone (see p. 33).   Svoboda [131] recommends a 4:1
mixture as most useful.

The purity of commercial samples is sometimes doubtful.   Keattch
[132] analysed various batches of the indicator from two suppliers by
chromatography.   He recommends purification of a solution in aqueous
ethanol (1 + 2) of the indicator by column or batch chromatography on
γ-alumina.   According to him, such purified samples allow the
titrimetric determination of calcium at 0.00025M concentration.

The indicator is strongly debased by traces of fluorescein, which
cause residual fluorescence in the titration of calcium.   Bezhevolnov
and Kreinbold [133] have studied the fluorescence spectra of Calcein-
fluorescein mixtures and worked out a method for determination of the
fluorescein in the concentration range $5 \times 10^{-5} - 5 \times 10^{-4}\%$.

The stability of aqueous solutions of the indicator has been studied
thoroughly by Hoyle and Diehl [134] by following the development of
residual fluorescence.   According to those authors, solutions at pH
7 are stable for not more than 7 days.   Alkaline solutions can be
used reliably for up to 60 days.   In this interesting work, the
authors describe the preparation of co-called Statocalcein, of
composition $K_2Ca_5Calcein_2$ $(K_2Ca_5C_{60}H_{40}O_{26}N_4)$.   A solution of this
compound does not deteriorate even after 210 days and does not show
residual fluorescence.   We can have little doubt about its utility
in practice, although good quality Calcein in 1:100 solid mixture
with potassium nitrate, or mixed with Phenolphthalexone, is perfectly
satisfactory.

2. Calcein Blue

This indicator was first prepared by Eggers [135] by Mannich condensation from 4-methylumbelliferone and named Umbellicomplexone.
Shortly afterwards the indicator was prepared by Wilkins [136] under
the name Calcein Blue (XXXV).

XXXV: Calcein Blue

The indicator is a colourless compound, slightly soluble in water,
alcohol and acetone.   It dissolves readily in dilute alkali.   In
solution at pH 4 - 11 it exhibits a brilliant blue fluorescence, which
disappears at pH 12 and reappears after addition of calcium, strontium or barium.   In acidic solution the fluorescence is extinguished
by addition of copper, and this has been used by Wilkins for the
indirect determination of nickel and chromium [136].

According to Wilkins, Calcein Blue is a more efficient indicator than
Calcein because its excitation wavelength of 370 nm is very close to
an emission wavelength from a mercury lamp (366 nm).   Calcein itself
needs light of longer wavelength for pronounced fluorescence.

Similarly to Calcein, the indicator has a residual fluorescence in
fairly concentrated sodium hydroxide solution;   this disappears on
warming to 80°C and appears again after cooling.   Potassium hydroxide is the best medium for calcium determination.

Remarks. During recent years a series of fluorescent indicators containing two iminodiacetic acid residues has been proposed. These "tetra-acetates" were derived, for example, from dianisidine [137], dihydroxybenzidine [138] and diaminostilbene [139]. Also worth mention are two indicators prepared by Holzbecher, namely salicylidene acetylhydrazone [140] and salicylidene-o-aminophenol [141].

2.1.7   Redox Indicators

Cations having two oxidation states always form redox systems, for example the couples Fe^{3+}/Fe^{2+} and Cu^{2+}/Cu^+. Potentiometric titration of tervalent iron with EDTA (platinum electrode) proceeds similarly to a redox titration, according to the Nernst equation:

$$E = E° + 0.059 \log ([Fe^{3+}]/[Fe^{2+}])$$

During the titration with EDTA only the Fe^{3+} ion is bound and the concentration of Fe^{2+} remains constant. At the equivalence point the potential of the electrode suddenly drops, even in titration of a solution of "pure" iron(III), because a sufficient amount of ferrous ion is always present [142] (but in the presence of 1,10-phenanthroline such a titration fails [143] because the ferrous ion is also complexed, by the phenanthroline).

Similarly we can determine copper potentiometrically in the presence of a small amount of potassium thiocyanate [144].

For visual detection of the end-point of such titrations we need a suitable redox indicator, for example Variamine Blue [4-methoxy-4′-aminodiphenylamine (XXXVI)], which is oxidized by iron(III) to a deep violet product (XXXVII):

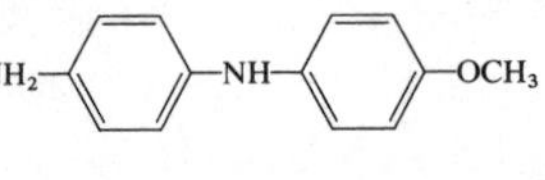

XXXVI: Colourless

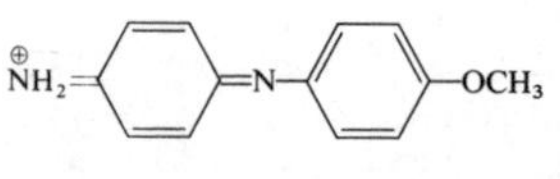

XXXVII: Dark violet

This reaction is reversible. At the end-point of a titration of iron(III) with EDTA the traces of iron(II) present reduce the violet product to the colourless leuco-form.

Variamine Blue was proposed by Erdey and Bodor [145] as an indicator for the iron titration. It can similarly be used for copper determination [146,147], and for the determination of aluminium by back-titration with copper solution [148], and for further applications [149-151].

For indication of EDTA titration end-points the reaction between vanadium(V) and diphenylcarbazide has also been proposed [152].

These redox indicators do not seem to be widely used in analytical
practice.

2.1.8   Chemiluminescent Indicators

It is beyond the scope of this monograph to describe in detail the
function of luminol and lucigenin [153] as indicators for EDTA
titration.   Details can be found in West's monograph [104].   For
routine work this type of end-point indication is rather impractical.

2.1.9   Blocking of Indicators

In the description above of the properties of metallochromic indicators, it has been mentioned many times that some cations block the
indicator.   This blocking - the formation of indicator complexes
that are stable towards ligand exchange with EDTA - makes the direct
titrations impossible.   This kind of reaction appears with all
indicators derived from azo-dyes or triphenylmethane-dyes, such as
Xylenol Orange or Methylthymol Blue.   It has been found in the
determination of copper with EDTA and Xylenol Orange that the blocking is eliminated in the presence of minute traces of 1,10-phenan-
throline [75].   It was originally thought that the complex $Cu(Phen)_3^-$
is formed, which reacts quickly with EDTA.   The Phen liberated
reacts with a further amount of copper ion.

Nakagawa and Wada [154], however, from comparison of the apparent
stability constants of the Cu-XO complex $(\log K' = 8.1)$ and the Cu-
Phen complex $(\log K' = 9)$ conclude that the simple reaction does not
exist, and that the ternary Cu-XO-Phen complex is formed, which
reacts with EDTA:

$$Cu-XO-Phen + EDTA = Cu-EDTA + Phen + XO$$

Berndt and Šara [145] have found that 2,2'-bipyridyl has similar
deblocking action.   Other complexing agents also give the effect,
but only if present at relatively high concentration.   They consider
that Phen is best for the EDTA titration of copper, when Xylenol
Orange or Methylthymol Blue is used as indicator.   They also
observed that hexamine, which is the buffer usually used, has a
retarding effect on this substitution reaction.   These reactions
have been thoroughly studied by Nakagawa and Wada [154].   By
spectrophotometric measurements they also found that phenanthroline
and bipyridyl are the best, and active even at $10^{-6}M$ concentration.
Hexamine has a considerable retarding effect at 0.1M concentration.
This effect was studied by Wada et al. [156].   The substitution
reaction of Cu-XO + EDTA was followed spectrometrically (at pH 6) in
2-(N-morpholino)ethanesulphonic acid buffer and in hexamine buffer,
and the hexamine was found to retard the reaction even at $10^{-3}M$
concentration, because of the formation of the ternary complex Cu_2-
XO-hexamine.   The stability constants of the complexes formed and
the conditional rate constants of the substitution reactions were
determined.

The problem of blocking is not limited to Xylenol Orange.   Nakagawa
et al. [157,158] have studied the effect in the reaction between

copper-PAN-Phen and EDTA, and between Ni-Phen-TAC and EDTA [TAC = 2-
(2-thiazolylazo)-4-methylphenol].

These works have enriched our knowledge of indicator reactions.

REFERENCES

1.   F. Beilstein, Liebig's Ann. 1858, 107, 186.
2.   G. Schwarzenbach and H. Flaschka, Complexometric Titrations, 2nd
     Ed., Methven, London, 1969, p. vii.
3.   J. Körbl and R. Přibil, Coll. Czech. Chem. Commun., 1957, 22,
     1122.
4.   G. Schwarzenbach and H. Flaschka, Die komplexometrische
     Titration, 5th Ed., Enke Verlag, Stuttgart 1965; Complexometric
     Titrations, 2nd Ed., Methuen, London, 1969.
5.   A. Ringbom, Complexation in Analytical Chemistry, Interscience,
     New York, 1963.
6.   G. Schwarzenbach and W. Biedermann, Helv. Chim. Acta, 1948, 31,
     678.
7.   A.E. Harvey, J.M. Komarny and G. Wyatt, Anal. Chem., 1953, 25,
     498.
8.   A. Young and T.R. Sweet, Anal. Chem., 1966, 27, 418.
9.   H. Diehl and F. Lindstrom, Anal. Chem., 1959, 31, 414.
10. R. Belcher, R.A. Close and T.S. West, Chemist-Analyst, 1957, 46,
     86.
11. G.P. Hildebrand and C.N. Reilley, Anal. Chem., 1957, 29, 258.
12. J. Patton and W. Reeder, Anal. Chem., 1957, 29, 238.
13. F. Lindstrom and H. Diehl, Anal. Chem., 1960, 32, 1123.
14. R. Belcher, R.A. Close and T.S. West, Chemist-Analyst, 1957, 46,
     86.
15. R. Belcher, R.A. Close and T.S. West, Chemist-Analyst, 1958, 47,
     2.
16. H. Flaschka, Mikrochim. Acta, 1956, 784.
17. R.A. Close and T.S. West, Anal. Chim. Acta, 1960, 23, 261.
18. R.A. Close and T.S. West, Anal. Chim. Acta, 1960, 23, 370.
19. G. Ross, D.A. Aikens and C.N. Reilley, Anal. Chem., 1962, 34,
     1766.
20. A.M. Lukin, G.B. Zavarikhina and N.S. Syslev, Zavod. Lab., 1954,
     23, 1524; USSR Patent 110996, 3 March 1957.
21. R.A. Close and T.S. West, Talanta, 1960, 5, 224.
22. A.E. Mendes-Bezerra and W.I. Stephen, Analyst, 1960, 94, 1117.
23. M. Herrero-Lancina and T.S. West, Anal. Chem., 1963, 35, 2131.
24. A.M. Lukin, K.A. Smirnova and G.B. Zavarikhina, Zh. Analit.
     Khim., 1963, 18, 444; Anal. Abstr., 1964, 11, 1637.
25. K.L. Cheng and R.H. Bray, Anal. Chem., 1955, 27, 782.
26. K.L. Cheng, Anal. Chem., 1958, 30, 243.
27. R. Přibil, Analytical Applications of EDTA and Related Compounds,
     Pergamon Press, Oxford, 1972.
28. P. Wehber, Z. Anal. Chem., 1957, 158, 10.
29. L. Sommer and M. Hniličková, Naturwissenschaften, 1958, 45, 544.
30. Z. Vodák and G. Leminger, Coll. Czech. Chem. Commun., 1954, 19,
     925.
31. M. Malát, V. Suk and O. Ryba, Coll. Czech. Chem. Commun., 1954,
     19, 258.
32. V. Suk and M. Malát, Chemist-Analyst, 1956, 45, 30.
33. M. Malát, Coll. Czech. Chem. Commun., 1961, 26, 1877.

34. V. Suk, M. Malát and O. Ryba, Coll. Czech. Chem. Commun., 1955, 19, 679.
35. V. Suk, M. Malát and A. Jeníčková, Coll. Czech. Chem. Commun., 1955, 20, 158.
36. J. Doležal, Z. Šír and K. Janáček, Coll. Czech. Chem. Commun., 1956, 21, 1300.
37. Z. Šír and R. Pribil, Coll. Czech. Chem. Commun., 1956, 21, 886.
38. M. Malát, V. Suk and A. Jeníčková, Coll. Czech. Chem. Commun., 1954, 19, 1056.
39. Z. Vodák and O. Leminger, Chem. Listy, 1955, 50, 943.
40. V. Suk, M. Malát and A. Jeníčková, Coll. Czech. Chem. Commun., 1956, 21, 418.
41. A. Jeníčková, M. Malát and V. Suk, Coll. Czech. Chem. Commun., 1956, 21, 1599.
42. Z. Vodák and O. Leminger, Chem. Listy, 1956, 50, 2028.
43. R.M. Dagnall and T.S. West, Talanta, 1961, 8, 711.
44. A. Jeníčková, V. Suk and M. Malát, Coll. Czech. Chem. Commun., 1956, 21, 1257.
45. M. Malát, V. Suk and M. Tenorová, Chem. Listy, 1958, 52, 2405; Anal. Abstr., 1959, 6, 3381.
46. R.M. Dagnall and T.S. West, Talanta, 1964, 11, 1533.
47. R.M. Dagnall and T.S. West, Talanta, 1964, 11, 1627.
48. J.S. Fritz and M.O. Fulda, Anal. Chem., 1954, 26, 1206.
49. I. Sajó, Magy. Kem. Foly., 1953, 59, 319; 1954, 60, 268; Anal. Abstr., 1955, 2, 1474, 1475.
50. V. Suk and V. Miketuková, Coll. Czech. Chem. Commun., 1959, 24, 311.
51. M. Theis, Z. Anal. Chem., 1955, 144, 106.
52. M. Theis, Z. Anal. Chem., 1955, 144, 275.
53. M. Theis, Z. Anal. Chem., 1955, 144, 351.
54. M. Theis, Z. Anal. Chem., 1955, 144, 427.
55. M. Theis, Radex-Rundschau, 1955, 333.
56. M. Malát and M. Tenorová, Coll. Czech. Chem. Commun., 1959, 24, 632.
57. A. Emr, J. Körbl and R. Přibil, Czech. Patent 89388, 15 April 1959.
58. A. Emr and R. Přibil, Czech. Patent 140865, 15 April 1971.
59. G. Schwarzenbach, G. Anderegg and R. Sallmann, Helv. Chim. Acta, 1952, 35, 1794.
60. G. Anderegg, H. Flaschka, R. Sallmann and G. Schwarzenbach, Helv. Chim. Acta, 1954, 37, 114.
61. R. Přibil and V. Veselý, Chemist-Analyst, 1966, 55, 4.
62. J. Korbl and R. Přibil, Coll. Czech. Chem. Commun., 1958, 23, 1213; Chem. Ind. (London), 1957, 233.
63. R. Přibil and M. Kopanica, Chemist-Analyst, 1959, 48, 35.
64. R. Přibil and J. Adam, Talanta, 1977, 24, 177.
65. J. Körbl and R. Přibil, Chemist-Analyst, 1956, 45, 102.
66. J. Körbl, R. Přibil and A. Emr, Coll. Czech. Chem. Commun., 1957, 22, 961.
67. D.C. Olson and D.W. Margerum, Anal. Chem., 1962, 34, 1299.
68. M. Murakami, T. Yoshino and S. Harasawa, Talanta, 1967, 14, 1293.
69. M. Yamada and M. Fujimoto, Bull. Chem. Soc. Japan, 1971, 44, 294.
70. S. Nakada, M. Yamada, T. Ito and M. Fukimoto, Bull. Chem. Soc. Japan, 1977, 50, 1887.
71. H. Sato, Y. Yokoyama and K. Momoki, Anal. Chim. Acta, 1977, 94, 217.
72. H. Sato, Y. Yokoyama and K. Momoki, Anal. Chim. Acta, 1978, 99, 167.
73. B. Řehák and J. Körbl, Coll. Czech. Chem. Commun., 1960, 25, 797.

74.   F.I. Miller and H.M. Fog, J. Chem. Educ., 1973, 50, 147.
75.   R. Přibil, Talanta, 1959, 3, 91.
76.   B.I. Nabivanets and L.N. Kubritskaya, Ukr. Khim. Zh., 1963, 29, 1198.
77.   B. Buděšínský, Z. Anal. Chem., 1962, 188, 266.
78.   R. Přibil, Talanta, 1959, 3, 200.
79.   V. Svoboda, Personal communication.
80.   O. Borchert, Massanalytische Schnellmethoden für Beizen und galvanische Elektrolyte, Deutscher Verlag für Grundstoff-industrie, Leipzig, 1966.
81.   B. Bieber and V. Večeřa, Coll. Czech. Chem. Commun., 1961, 26, 2081.
82.   J. Korbl and R. Přibil, Coll. Czech. Chem. Commun., 1958, 23, 873.
83.   J. Korbl, Chem. Listy, 1957, 51, 1304; Anal. Abstr., 1958, 5, 771.
84.   J. Körbl and B. Kakáč, Coll. Czech. Chem. Commun., 1958, 23, 889.
85.   K.C. Srivastava and K. Banerji, Chem. Age, India, 1969, 20, 606.
86.   G.S. Tereshin, A.R. Rubinshtein and I.V. Tananaev, Zh. Analit. Khim., 1965, 20, 1082.
87.   T. Yoshino, H. Imada, K. Iwasa and T. Kuwano, Talanta, 1969, 16, 151.
88.   N.F. Koshenko, T.V. Mal'kova and K.B. Yatsimirskii, Zh. Analit. Khim., 1975, 30, 2245; Anal. Abstr., 1976, 31, 3A11.
89.   A.I. Cherkesov, N.K. Astakhova and A.A. Cherkesov, Ftaleksony, 1970, 19; Chem. Abstr., 1972, 76, 87121f.
90.   K. Vytřas and J. Vytřasová, Chem. Zvesti, 1974, 28, 779.
91.   K. Vytřas, V. Mach and S. Kotrlý, Chem. Zvesti, 1975, 29, 61; Anal. Abstr., 1975, 29, 4B18.
92.   R. Přibil, Talanta, 1959, 3, 91.
93.   K. Vytřas, M. Malcová and S. Kotrlý, Chem. Zvesti, 1975, 29, 599; Anal. Abstr., 1976, 31, 2B86.
94.   K. Vytřas, V. Mach and M. Malcová, Chem. Zvesti, 1976, 30, 640.
95.   J. Ueda, Nippon Kagaku Kaishi, 1977, 350.
96.   J. Ueda, Bull. Chem. Soc. Japan, 1978, 51, 773.
97.   J. Korbl, E. Kraus and R. Přibil, Coll. Czech. Chem. Commun., 1958, 23, 1219.
98.   K. Vytřas and F.J. Langmyhr, Anal. Chim. Acta, 1977, 92, 155.
99.   K. Vytřas and F.J. Langmyhr, Anal. Chim. Acta, 1977, 94, 429.
100. J. Korbl, V. Svoboda, D. Terzijská and R. Přibil, Chem. Ind. (London), 1957, 1624.
101. T. Yoshino, S. Murakami, K. Arita and K. Ishizu, Talanta, 1979, 26, 479.
102. G. Schwarzenbach and H. Hysling, Helv. Chim. Acta, 1949, 32, 1314.
103. G. Goldstein, Anal. Chim. Acta, 1959, 21, 339.
104. T.S. West, Complexometry with EDTA and Related Reagents, 3rd Ed., British Drug Houses Ltd., Poole, 1969.
105. E. Bovalini and A. Casini, Ann. Chim. (Roma), 1953, 43, 287; Anal. Abstr., 1954, 1, 53.
106. E. Wänninen and A. Ringbom, Anal. Chim. Acta, 1955, 12, 308.
107. S. Kotrlý, Coll. Czech. Chem. Commun., 1957, 22, 1765.
108. J.H. Yoe and R.M. Rush, Anal. Chim. Acta, 1952, 6, 526.
109. R.M. Rush and J.H. Yoe, Anal. Chem., 1954, 26, 1345.
110. J. Kinnunen and B. Merikanto, 1955, 44, 50.
111. R. Belcher, M.A. Leonard and T.S. West, J. Chem. Soc., 1958, 2390.
112. M.A. Leonard and T.S. West, J. Chem. Soc., 1960, 4477.

113. J.B. Headridge, Analyst, 1960, 85, 379.
114. M.A. Leonard, Ph.D. Thesis, Birmingham University, 1959.
115. R. Belcher, M.A. Leonard and T.S. West, Talanta, 1959, 2, 92.
116. B. Buděšínský, Chem. Listy, 1957, 51, 726; Anal. Abstr., 1957,
      4, 3528.
117. B. Buděšínský, Chem. Listy, 1958, 52, 247; Anal. Abstr., 1958,
      5, 3595.
118. V. Patrovský, Chem. Listy, 1953, 47, 1338.
119. H. Diehl and J.L. Ellingboe, Anal. Chem., 1956, 28, 882.
120. J. Korbl and F. Vydra, Coll. Czech. Chem. Commun., 1958, 23,
      622.
121. J. Korbl, F. Vydra and R. Přibil, Talanta, 1958, 1, 138.
122. J. Korbl, F. Vydra and R. Přibil, Talanta, 1958, 1, 281.
123. F. Vydra, R. Přibil and J. Korbl, Coll. Czech. Chem. Commun.,
      1959, 24, 2623.
124. D.H. Wilkins, Talanta, 1959, 2, 277.
125. J. Korbl and V. Svoboda, Talanta, 1960, 3, 370.
126. D.H. Wilkins, Talanta, 1960, 4, 80.
127. D.F.H. Wallach, D.M. Surgenor, J. Soderberg and E. Delano,
      Anal. Chem., 1959, 31, 456.
128. B.M. Tucker, Analyst, 1957, 82, 284.
129. G.F. Kirkbright and W.I. Stephen, Anal. Chim. Acta, 1962, 27,
      294.
130. A. Bauman, Arch. Hig. Rada Toksihol., 1967, 18, 155.
131. V. Svoboda, V. Chromý, J. Korbl and L. Dorazil, Talanta, 1961,
      8, 249.
132. C.J. Keattch, Talanta, 1963, 10, 1303.
133. E.A. Bezhevolnov and S. Yu. Kreingol'd, Zh. Analit. Khim.,
      1962, 17, 291.
134. W.C. Hoyle and H. Diehl, Talanta, 1972, 19, 206.
135. J.H. Eggers, Talanta, 1960, 4, 38.
136. D.H. Wilkins, Talanta, 1960, 4, 182.
137. R. Belcher, D.I. Rees and W.I. Stephen, Talanta, 1960, 4, 78.
138. G.F. Kirkbright and W.I. Stephen, Anal. Chim. Acta, 1963, 28,
      327.
139. G.F. Kirkbright, D.I. Rees and W.I. Stephen, Anal. Chim. Acta,
      1962, 25, 558.
140. Z. Holzbecher, Chem. Listy, 1958, 52, 430; Anal. Abstr., 1959,
      6, 77.
141. Z. Holzbecher, Chem. Listy, 1958, 52, 1822; Anal. Abstr.,
      1959, 6, 2071.
142. R. Přibil, Z. Koudela and B. Matyska, Coll. Czech. Chem.
      Commun., 1951, 16, 80.
143. R. Přibil, unpublished observation.
144. R. Přibil, Chimia, 1951, 2, 185.
145. L. Erdey and A. Bodor, Z. Anal. Chem., 1953, 137, 410.
146. P. Wehber, Mikrochim. Acta, 1955, 927.
147. P. Wehber, Z. Anal. Chem., 1956, 149, 244.
148. W. Johannsen, E. Bobowski and P. Wehber, Metall, 1956, 10, 211;
      Anal. Abstr., 1957, 4, 3258.
149. H. Flaschka, Mikrochim. Acta, 1954, 361.
150. L. Erdey and L. Pólos, Z. Anal. Chem., 1960, 174, 333.
151. L. Erdey and L. Pólos, Anal. Chim. Acta, 1957, 17, 458.
152. I. Sajó, Magy. Kem. Foly., 1956, 62, 176; Anal. Abstr., 1956,
      3, 3257.
153. L. Erdey and I. Buzás, Anal. Chim. Acta, 1960, 22, 524.
154. G. Nakagawa and H. Wada, Talanta, 1973, 20, 829.
155. W. Berndt and J. Šára, Talanta, 1960, 5, 281.
156. H. Wada, T. Ishizuki and G. Nakagawa, Talanta, 1976, 23, 669.

157. G. Nakagawa and H. Wada, Talanta, 1975, 22, 563.
158. G. Nakagawa, H. Wada and O. Nakazawa, Talanta, 1976, 23, 155.

2.2   INSTRUMENTAL METHODS IN COMPLEXOMETRY

Analytical chemistry uses a considerable number of physico-chemical
methods to follow the course of reactions in solution.   These
methods have found use in complexometry with varying success.   All
of them have the same aim - to determine the titration end-point with
high precision.   Only some of the methods directly monitor changes
in the concentration of the cation in question in the vicinity of the
end-point.   However, the changes in concentration of individual
species in the solution may be accompanied by changes in other
properties of the solution, e.g. conductivity, surface tension.
Most of these methods require special - sometimes very costly -
instrumentation and are all more time-consuming than simple visual
titrations.

The principles of these methods, e.g. potentiometry, amperometry,
conductimetry, are well known and their detailed description is outside the scope of this monograph, which is concerned mainly with the
practical application of visual complexometric titrations.   Therefore only a brief illustrative survey of these methods will be given.
It is left for the reader to judge which of them may be important in
practical complexometry.

Spectrophotometric titrations using common metallochromic indicators
are the closest to visual titrations.   These titrations seem to be
the most promising, with the widest field of practical application.
Their importance lies mainly in their use for the determination of
trace elements.   Much attention has been paid to the theory of
spectrophotometric titrations, with the main objective the precise
localization of an end-point.   Mathematical solution of the problem
has been approached by various authors from different points of view
and is based on different assumptions.

Older publications all presumed there was 1:1 stoichiometry in EDTA-
complex formation and the same ratio in a complex formed by an
indicator with a metal [1 - 4].   Still and Ringbom [5] solved the
problem of construction of titration curves for different indicator
sensitivities (low, medium, high).   Skrifvars and Ringbom [6]
derived equations for consecutive titrations of several metals, and
demonstrated the titration of manganese and magnesium with 0.005M
EDTA.   Kragten [7] solved precisely the problem of location of the
end-point when the indicator forms a binuclear complex M_2I.   According to Kragten the photometric titration of bismuth and thorium with
Xylenol Orange as indicator is subject to certain systematic errors
because of this behaviour.   A similar effect is observed with other
indicators derived from triphenylmethane dyes.   For these titrations
Kragten recommends use of PAR or indicators based on dihydroxyazo-
dyes that form only 1:1 complexes.   He has also drawn attention to
the importance of correct adjustment of the solution conditions, not
only in terms of pH, reagent concentration etc., but more importantly
in terms of localized formation of kinetically stabilized hydroxo-
complexes or other species, which may cause incomplete titration [8].
The problem of indicator complexes of M_2I, MI_2, M_2I_2 composition is
dealt with in various publications [9,10].   Readers with an inclin-
ation for mathematics may be interested in the work published by Sato

and Monoki [11,12].

Adequate instrumentation is essential for practical use of spectrophotometric titrations.   Some commercial spectrophotometers have
been adapted for this purpose [13,14], and home-made apparatus has
been constructed by various authors [15-19].

Electrometric methods have been extended by the recent development of
methods other than the classical amperometry with a dropping mercury
electrode.   The field of potentiometric titrations has been considerably widened by the use of various metal electrodes or platinum
electrodes coated with metal oxides.   A detailed study of metal
oxide electrodes was published by Kainz and Sontag [20].   It is not
surprising that ion-selective electrodes have been extensively used;
see for example, Carr [21].   Biamperometric, chronopotentiometric
and coulometric titrations are also applied in complexometry, some of
them on the microscale.   Alimarin and Petrikova [22] for instance,
devised for biamperometric titration a cell of volume only 3-5 μl.
They used a pair of platinum electrodes to determine down to 0.02 μg
of calcium by direct titration with EDTA.

Recently some other methods have been proposed that are very interesting from the theoretical point of view but seem to be of rather
problematic utility, and so will be only briefly described here.

Tensammetric titrations.   Adsorption of a surface-active compound on
the surface of a dropping mercury electrode in a.c. polarography is
manifested by production of a tensammetric wave.   An example of such
compounds is Alizarin Complexan (p. 44) which gives a tensammetric
wave at about 0.1 V vs. S.C.E.   The wave disappears when the
indicator is involved in complex formation, but reappears when the
indicator is released by EDTA at the end-point.   Nakagawa and Nomura
[23] used this effect for the complexometric determination of copper,
cadmium, zinc, nickel and lead at pH 6, manganese at pH 6.5 and
indium, thorium and bismuth at pH 3.   Amounts of 0.1-0.4 mg of
metal in 100 ml can be determined by using 2 x 10^{-6}M indicator
solution.

Stalagmometric titrations.   This type of titration again uses the
dropping mercury electrode.   Surface-active compounds change the
surface tension of the mercury and consequently the mercury drop-time.
The titration end-point is indicated by an abrupt jump or peak on a
plot of the drop-time vs. concentration of surface-active compound.
Kiba et al. [24] titrate EDTA solution with a solution of cupric ions
in the presence of chloride or bromide.   A surface-active chloride
or bromide complex is formed with the first surplus of cupric ions.
This results in a sharp decrease of drop-time at 0.0 V vs. S.C.E.
Determination of aluminium and copper and iron by titration with DCTA
[25] is based on the same principle.

Dilatometric titrations.   Every reaction is accompanied by a volume
change.   This occurs also during chelate formation.   Wiese et al.
[26] used this effect in so-called dilatometric EDTA-titration and
gave a detailed description of the apparatus used.   A sharp break in
the plot of total volume vs. volume of added titrant indicates the
end-point.   Wiese and Wunsch [27,28] also studied the protonation
and deprotonation properties of complexones and applied dilatometric
methods for the determination of the stability constants of EDTA-
complexes with alkaline earth metal cations.

Spectropolarimetric titrations. The end-point of a complexometric
titration can also be detected by means of a change in the optical
rotation of the solution in the vicinity of the end-point, when the
titration is performed with optically active complexones, e.g. $D(-)$-
trans-1,2-diaminocyclohexanetetra-acetic acid $[29,30]$ or with $D(-)$-1,
2-propylenediaminetetra-acetic acid $[31 - 36]$.

Radiometric titrations. These titrations can be recommended to
especially patient analysts. One example taken from Schwarzenbach
and Flaschka's monograph $[37]$ may suffice for illustration. Zinc
can be titrated with EDTA in the presence of labelled silver iodate.
When all the zinc is complexed the dissolution of silver iodate
begins (formation of the Ag-EDTA complex). Therefore the mixture
must be filtered and its radioactivity measured after each addition
of EDTA. The filtrate and precipitate must be recombined after the
measurement, and the operation is repeated till radioactivity is
detected in the filtrate. Other examples can be found in the same
monograph $[37]$.

REFERENCES

1.   J.M.H. Fortuin, P. Karsten and H.L. Kies, Anal. Chim. Acta,
     1954, 8, 356.
2.   A. Ringbom and E. Wänninen, Anal. Chim. Acta, 1954, 11, 153.
3.   H. Flaschka and S. Khalafallah, Z. Anal. Chem., 1957, 156, 401.
4.   C.N. Reilley and R.W. Schmid, Anal. Chem., 1959, 31, 887.
5.   E. Still and A. Ringbom, Anal. Chim. Acta, 1965, 33, 50.
6.   B. Skrifvars and A. Ringbom, Anal. Chim. Acta, 1966, 36, 105.
7.   J. Kragten, Analyst, 1971, 96, 106.
8.   J. Kragten, Talanta, 1977, 24, 483; Atlas of Metal-ligand
     Equilibria in Aqueous Solution, Horwood, Chichester, 1977.
9.   S. Kotrlý, Anal. Chim. Acta, 1963, 29, 552.
10. J. Kragten, Talanta, 1973, 20, 937.
11. H. Sato and K. Momoki, Anal. Chem., 1970, 42, 1477.
12. H. Sato and K. Momoki, Anal. Chem., 1971, 43, 938.
13. H. Thomann, Z. Anal. Chem., 1961, 184, 241.
14. W.E. Clarke, Brit. Cast Iron Res. Assoc. J., 1960, 351.
15. R.A. Chalmers and C.A. Walley, Analyst, 1957, 82, 329.
16. H. Flaschka and P. Sawyer, Talanta, 1961, 8, 521.
17. J.P. Phillips and R.C. Crowley, Talanta, 1962, 9. 178.
18. S. Miyake, Talanta, 1966, 13, 1253.
19. S. Miyake and I. Sakamoto, Japan Analyst, 1967, 16, 328.
20. G. Kainz and G. Sontag, Z. Anal. Chem., 1964, 269, 267.
21. P.W. Carr, Anal. Chem., 1972, 44, 452.
22. I.P. Alimarin and M.N. Petrikova, Zh. Analit. Khim., 1969, 24,
     1136.
23. S. Nakagawa and T. Nomura, Anal. Lett., 1972, 5, 723.
24. T. Kiba, E. Okiyama and T. Kambara, Japan Analyst, 1974, 23, 81.
25. T. Osaki and T. Kambara, Japan Analyst, 1974, 23, 664.
26. G. Wiese, E. Pietsch and S. Nix, Z. Anal. Chem., 1974, 270, 104.
27. G. Wiese and J. Wunsch, Z. Anal. Chem., 1975, 275, 110.
28. G. Wiese and J. Wunsch, Z. Anal. Chem., 1975, 275, 275.
29. J.R. Baker and K.H. Pearson, Anal. Chim. Acta, 1970, 50, 255.
30. K.H. Pearson, J.R. Baker and P.E. Reinbold, Anal. Chem., 1972,
     44, 2090.

31. R.J. Palma, P.E. Reinbold and K.H. Pearson, Chem. Commun., 1969, 254; Anal. Lett., 1969, 2, 553; Anal. Chim. Acta, 1970, 51, 329; Anal. Chem., 1970, 42, 47.
32. R.J. Palma and K.H. Pearson, Anal. Chim. Acta, 1970, 49, 497.
33. D.L. Caldwell, P.E. Reinbold and K.H. Pearson, Anal. Chim. Acta, 1970, 49, 505; Anal. Chem., 1970, 42, 416; Anal. Lett., 1970, 3, 93.
34. K.W. Street and K.H. Pearson, Anal. Chim. Acta, 1973, 63, 107; 1974, 71, 405.
35. P.E. Reinbold and K.H. Pearson, Anal. Chem., 1971, 43, 293.
36. S.J. Simon and K.H. Pearson, Anal. Chem., 1973, 45, 620.
37. G. Schwarzenbach and H. Flaschka, Complexometric Titrations, 2nd Ed., Methuen, London, 1969.

CHAPTER 3

Masking (Screening) Reagents

EDTA is very unselective as a titrimetric reagent and is nearly universal, forming more or less stable chelates with almost all metal cations. It becomes very valuable if we can increase its select- ivity in such a way that we make only one cation determinable, even in the presence of others. To this end, the influence of pH on the formation of EDTA-chelates is of prime importance. With decreasing pH the degree of dissociation of a particular chelate increases to the point where the conditional stability constant $K'(= K_{MeY}/\alpha_{Y(H)})$ approaches zero and the chelate virtually ceases to exist. The choice of the optimal pH for the titration of certain mixtures of cations is very helpful, but not sufficient for all possible mixtures of metals to be determined, as we can see from Table 6.1 on p. 92.

At the beginning of complexometry the interfering metals were separated from the solution by the classical precipitation methods. Before the determination of calcium and magnesium the heavy metals were precipitated with hydrogen sulphide, aluminium and iron with ammonia etc., but these methods make the analysis slower.

A very ingenious and quick method for the elimination of interference is to bind the interfering elements in undissociated water-soluble complexes directly in the solution. Such an operation is called masking (screening, sequestering) and the compound used is called a masking (screening) reagent.

It is well known that any element (M) can be titrated with EDTA provided the conditional stability constant of the complex does not drop below $10^6/C_M$ where C_M is the total concentration of M. Any other element present will not interfere if the conditional stability constant of its EDTA complex is less than about 10^2, and its concentration is about the same as that of the analyte. The aim of masking is therefore to make the conditional stability constants of the EDTA complexes of interfering elements as low as possible. Many masking agents have been proposed, both inorganic and organic in nature.

In the narrow sense of the term we understand masking to mean binding of metals in the form of chelates that are more stable than the

58

corresponding EDTA chelates, and that these complexes should be as
colourless as possible and soluble in water. The masking reagent
may react with one cation only or with several, but not with the
cation to be titrated (or at any rate not sufficiently to make the
conditional constant for its EDTA complex too low). The selectivity
of masking agents varies considerably, if considered in terms of the
number of cations which can be masked. Potassium cyanide is less
selective than potassium iodide or sodium sulphate, for example.

Masking sometimes results in the formation of a precipitate. For
example, aluminium can be masked as AlF_6^{3-} but on boiling the
aluminium is often precipitated as Na_3AlF_6. Similarly, iron(III)
can be masked as insoluble K_3FeF_6 [1]. When copper in large
concentrations is masked with thiourea the solution becomes cloudy,
but this does not affect the EDTA titration of the element of
interest.

Some authors include precipitation as a masking reaction if the EDTA
titration can be done with the precipitate still present. Traces of
heavy metals can be precipitated with sodium sulphide or diethyldithiocarbamate, barium with sodium sulphate, etc.

Redox reactions can also be used for masking. Thallium(III) can be
reduced with ascorbic acid to thallium(I), which forms only a very
weak complex with EDTA. The conditional stability constants of
Fe(III)-EDTA and Fe(II)-EDTA differ by a factor of about 10^{10}.
Therefore in the determination of thorium or bismuth the interference
of iron(III) can be removed by reduction to Fe(II). Interference of
chromium(III) can be eliminated by oxidation to chromate.

In so-called "kinetic masking" we utilize the difference in rate of
EDTA reaction with some cations. For example, chromium(III) reacts
so slowly with EDTA that its presence does not affect the direct
titration of cobalt, zinc or copper in acetate medium, with Naphthyl
Azoxine S as indicator [2]. DCTA reacts with aluminium at room
temperature, but not with chromium(III), and this makes possible the
stepwise titration of aluminium and chromium [3].

3.1   CLASSIFICATION OF MASKING REAGENTS

Schwarzenbach [4] divided masking reagents into groups, in order of
increasing selectivity: O-donors and fluoride, N-donors, S-donors,
cyanide and iodide. Within these groups there are compounds of
widely differing properties. For example, the group of O-donors
contains sodium hydroxide (for masking aluminium), ammonium carbonate
(for masking uranium), pyrophosphate, acetylacetone, tiron, organic
hydroxy-acids and so on.

West [5], however, recognizes only two groups: inorganic and organic
masking reagents.

I hold it more convenient - especially with regard to the development
of complexometry - to divide masking agents according to their
activity in alkaline and acidic medium. For a long time we were
able to use EDTA titration only in alkaline medium, because of a lack
of indicators. It was the development of new metallochromic indicators for use in the acid range of pH that offered further possibilities for use of EDTA, and of masking reagents suitable for acidic

solutions.

Classification of masking agents into these two groups is sufficiently sharp, because most of them function in only the one medium.
Potassium cyanide and triethanolamine (TEA) have excellent masking
properties in alkaline medium, but are useless in acidic solution,
and indeed potassium cyanide becomes very dangerous (liberation of
HCN).   Mercaptoacetic acid in acidic solution precipitates copper as
a yellow compound, which dissolves in ammonia to form a colourless
complex.   Thiourea masks copper in slightly acidic medium, forming a
colourless solution, but in alkaline medium black copper sulphide is
slowly precipitated.   The influence of pH on masking is sometimes
considerable.   For example, sodium sulphate in large amounts masks
thorium at pH 2, but not pH 5.

3.2   MASKING REAGENTS FOR ALKALINE MEDIA

The usual medium for masking and titrations is Schwarzenbach's
ammoniacal buffer of pH 10 $(NH_3 + NH_4Cl)$.   Calcium can also be
determined in sodium hydroxide medium (murexide as indicator) or in
potassium hydroxide medium (Calcein as indicator).   Ammonia itself
is only weakly complex-forming.   In sufficient concentration it
keeps the cations of the "ammine group" in solution and prevents
formation of their insoluble hydroxides.   Similarly magnesium,
aluminium and chromium can be kept in solution if a very large amount
of ammonia is used.   Complexation side-reaction with ammonia causes
inconvenience only in the determination of copper with murexide as
indicator, because the copper-indicator complex is rather weak.

3.2.1   Potassium Cyanide

Cyanide was the first masking agent used in complexometry [6].   It
forms very stable complexes with mercury, zinc, cadmium, cobalt and
nickel.   Manganese(II) forms much weaker complex with cyanide than
with EDTA and can be determined directly with EDTA even in the
presence of potassium cyanide (see below).   Silver forms the complex
$Ag(CN)_2^-$.   Copper(II) is reduced to copper(I) which forms a very
stable colourless cyanide complex (more easily in the presence of
hydroxylamine or ascorbic acid).   All these complexes are formed
smoothly by addition of potassium cyanide to an ammoniacal solution
of the cation.   The cyanide complexes of tervalent cobalt and
manganese are particularly stable.   Those of iron(II) and iron(III)
are also very stable but are difficult to prepare.   The best way is
to boil an alkaline triethanolamine solution of iron(III) with an
excess of potassium cyanide; $Fe(CN)_6^{3-}$ is formed, which can be
reduced to $Fe(CN)_6^{4-}$.

Potassium cyanide makes possible the direct titration of calcium or
magnesium in the presence of all these cations [except manganese(II)].
Some other metals, for example lead, can also be determined indirectly in presence of these cations and cyanide, by back-titration of
excess of EDTA with magnesium.   All such determinations belong to
the earlier period of complexometry.

Potassium cyanide is very poisonous and must be handled with great
care.   It must be added only to alkaline or slightly acidic
solutions (pH 5 - 7), and the solutions must be well buffered to avoid
the consequences of hydrolysis:

$$CN^- + H_2O = HCN + OH^- \tag{3.1}$$

Otherwise a large excess will considerably increase the alkalinity of the solution.

Great attention has been paid to the determinations of the individual cations in this "cyanide group".

Manganese can be determined directly with EDTA or DCTA even in the presence of cyanide if it is kept in the bivalent state by the presence of triethanolamine and a small amount of hydroxylamine [7]. The cyanide complex of manganese(III), $Mn(CN)_6^{3-}$, is very stable and can be used for masking manganese [8].   (See pp. 63, 180, 181.)

Nickel and cobalt.   The Ni-EDTA and Co(II)-EDTA complexes both react with potassium cyanide, which liberates all the EDTA, but the red-violet Co(III)-EDTA complex does not react with KCN.   Both nickel and cobalt are determined indirectly.   First the sum of Ni + Co is determined by back-titration of excess of EDTA with magnesium sulphate.   Then, after oxidation of the cobalt with hydrogen peroxide, cyanide is added and the EDTA liberated from the nickel complex is determined in the same way [9].   Because of the intense colour of the Co(III)-EDTA complex, the method is suitable only for low concentrations of cobalt - less than 20 mg/100 ml.

It is more convenient to form the nickel and cobalt(II) cyanide complexes, oxidize the cobalt complex with hydrogen peroxide to the light yellow $Co(CN)_6^{3-}$, then add silver ions, which displace nickel but not cobalt(III):

$$Ni(CN)_4^{2-} + 2Ag^+ + 4NH_3 = 2Ag(CN)_2^- + Ni(NH_3)_4^{2+} \tag{3.2}$$

The nickel liberated is directly titrated, with murexide as indicator [10].   An excess of silver does not affect the titration (the Ag-EDTA complex is too weak).

Zinc and cadmium.   Flaschka [11] studied in great detail the reaction of formaldehyde with cyanide and its metal complexes:

$$CN^- + HCHO + H_2O = HCH(OH)CN + OH^- \tag{3.3}$$
$$\text{cyanhydrin}$$

$$M(CN)_x^{n-} + xHCHO + xH_2O = M^{(x-n)+} + xHCH(OH)CN + xOH^-$$

Only the cyanide complexes of zinc and cadmium give this reaction. Other complex cyanides are more stable and do not react with form-aldehyde.   This makes it possible to determine zinc or cadmium in the presence of copper, nickel and cobalt.   Chloral hydrate (tri-chloroacetaldehyde, Cl_3CHO) reacts similarly [12], but a little slowly; however it is very pure, does not contain traces of metals and does not polymerize, whereas the purity of commercial formalde-hyde solutions is sometimes questionable.

Demasking of zinc and cadmium from their cyanide complexes is not so simple in practice as it may seem at first sight.   If copper is present, it is also demasked in traces to a very slight extent, but sufficiently to block Erio T used as indicator.

The acidity of the solution may change considerably after the
addition of formaldehyde. This problem has been studied thoroughly
by Berák [13]. Formaldehyde reacts with the ammonium chloride in
the buffer as well as with free cyanide [reaction (3.2)] and the zinc
or cadmium cyanide complex:

$$4NH_4Cl + 6HCHO = N_4(CH_2)_6 + 6H_2O + 4HCl \qquad (3.4)$$

hexamine

With reaction (3.3) the pH shifts to the alkaline side, by up to 2
units according to Berák. The solution may then be so alkaline that
Erio T loses its metallochromic function and behaves only as an acid-
base indicator. The formation of hydrochloric acid by reaction
(3.4) might neutralize the alkalinity, but it proceeds very slowly
under the conditions for titration of zinc and cadmium (pH 10) and so
has no real effect.

Berák's conclusions can be summarized as follows.

(i)   The solutions prepared for demasking must be well buffered.
(ii) Demasking should be done at a temperature lower than 15°C.
(iii) Potassium cyanide should not be used in great excess, because
otherwise a very large amount of formaldehyde would be needed for
demasking, since the free cyanide will react preferentially.
(iv) The concentration of formaldehyde must be kept as low as
possible. It is best to add the formaldehyde in small portions
until the colour of the zinc (or cadmium) indicator complex appears.

The difficulties described by Berák have also been observed by other
authors.

Študlar and Janoušek [14] described a simpler demasking procedure.
Zinc can be demasked by acidification of cyanide solutions to pH 6,
and titrated directly with EDTA, with Xylenol Orange or Methylthymol
Blue as indicator. According to these authors, the danger from
liberation of toxic HCN is very small. Under the conditions used,
only 2% of the total HCN is liberated in the first minute after
acidification of the solution. After 5 minutes – sufficient time to
do the titration two or three times over – only 4% of the HCN has
escaped from the solution. For safety, the authors recommend
performing the titrations in a well ventilated place (fume-cupboard).

Summary. Although potassium cyanide is a very powerful masking
agent, it has comparatively little value and applicability in
complexometry, partly on account of its toxicity, but mainly because
it can be replaced by other compounds which are equally powerful and
can be used in acidic media, where EDTA titrations are more selective.
For example, zinc can be determined in the presence of copper very
easily by masking the copper with thiourea at pH 5 - 6, and the masking of iron with cyanide in alkaline medium [15] is now replaced by
masking with triethanolamine.

3.2.2   Triethanolamine (TEA)

Triethanolamine, $N(CH_2CH_2OH)_3$, has been proposed by Přibil for masking of aluminium, iron and manganese in the determination of calcium
with murexide as indicator [16] and for the determination of
magnesium in the presence of aluminium (with Erio T) [17].   In

ammoniacal solutions TEA forms an iron(III) complex which should be colourless, but the solution is usually yellow because of formation of colloidal $Fe(OH)_3$. This colour practically disappears on standing. In hydroxide medium a colourless solution is obtained. For larger amounts of iron it is best to add EDTA equivalent to about a third of the iron, and then TEA and sodium hydroxide [18]. Generally it is best to add TEA to a solution sufficiently acid for no precipitate to form and then make the solution alkaline with ammonia or sodium hydroxide.

For titrations in ammoniacal solution in the presence of the Fe(III)-TEA complex we should not use Erio T as indicator, because it is oxidized by the Fe-TEA. Thymolphthalexone and Methylthymol Blue are suitable for such titrations, however.

Simultaneous masking with TEA and potassium cyanide (masking of iron and heavy metals) gives intensely yellow ammoniacal solutions, but the yellow colour fades on warming and disappears after sufficient dilution. However, warming the Fe-TEA complex with cyanide in strongly ammoniacal solution converts it into $Fe(CN)_6^{3-}$, and if manganese is also present it is then oxidized [19]:

$$Mn(CN)_6^{4-} + Fe(CN)_6^{3-} = Mn(CN)_6^{3-} + Fe(CN)_6^{4-} \qquad (3.5)$$

The cyanide complexes of iron(II) and manganese(III) are both inert towards EDTA. Reaction (3.5) is quantitative if a nearly equivalent amount of iron is present [traces of manganese(II) are oxidized by air]. This reaction allows combined masking of iron and manganese, but in the determination of calcium difficulty arises because of precipitation of calcium hexacyanomanganate(III), which reacts very slowly with EDTA.

In the presence of a reductant (e.g. hydroxylamine) manganese does not form a complex with TEA and can be reliably determined in the presence of iron; this has been used in the analysis of ferromanganese [20].

In the presence of TEA, manganese is slightly oxidized by air to give the Mn(III)-TEA complex, which is green and in amounts greater than a few milligrams makes the determination of calcium impossible by masking the indicator colour change. The manganese(III) complex can be reduced with hydroxylamine. This has been used for stepwise determination of calcium and traces of manganese. The green Mn(III)-TEA complex is measured spectrophotometrically. The oxidation to the Mn(III)-TEA complex can be accelerated by addition of dioxan [21].

The Mn(III)-TEA complex reacts at pH 10 with potassium cyanide, forming the stable slightly yellow $Mn(CN)_6^{2-}$ complex. This reaction allows determination of calcium in the presence of up to 80 mg of manganese [8].

Aluminium forms a colourless stable complex with TEA, permitting EDTA titration of magnesium and the sum of Mg + Ca when Thymolphthalexone or Methylthymol Blue is used as indicator. The first titration has also been done in ammoniacal buffer, with Erio T as indicator [17]. Ritchie [22] has reported that in this titration the blue colour of the indicator at the end-point quickly turns to wine-red unless the solution has been cooled to below 5°C. He attributes this to

displacement of traces of magnesium by aluminium.   I have never
observed this colour change except when the titrated solution was
warmed, and then the colour change was irreversible, most probably
because the Al-TEA complex reacted directly with the Erio T.   The
same change can be observed with Thymolphthalexone if the solution is
warmed.   Chromium(III), on boiling with TEA; forms an intensely
coloured complex that is inert to EDTA.   Because of the strong
colour, only very small amounts of the Cr-TEA complex can be
tolerated in the determination of magnesium, and even then considerable dilution is needed [23].   Cobalt can be masked with TEA after
oxidation with hydrogen peroxide, but the complex creates the same
difficulties as the chromium complex [24].

TEA has also been proposed for the masking of bismuth in the determination of calcium and magnesium [25], and of tin in the analysis of
its alloys [26].

Demasking of TEA-complexes.   Only one method is known from the
literature.   The Fe-TEA complex can be demasked by formaldehyde for
the successive determination of calcium, magnesium and iron [27], but
this micromethod is rather complicated, requiring three volumetric
solutions (EDTA, EGTA and Cu-EDTA) and is probably of limited use.

Summary.   Triethanolamine is still the best masking reagent for iron,
aluminium and traces of manganese in alkaline medium.   It can be
combined with other masking reagents such as potassium cyanide or
ammonium fluoride.   Thioglycollic acid (see next paragraph) cannot
be used in ammoniacal solution in the presence of Fe-TEA, because the
Fe-TEA complex reacts with it to form an intensely red solution.   In
sodium hydroxide medium Fe-TEA is much more stable and does not give
this reaction, thus allowing simultaneous masking of iron, bismuth,
lead, tin, zinc, cadmium etc.

3.2.3   Thioglycollic Acid (TGA)

Thioglycollic acid (mercaptoacetic acid, $HSCH_2COOH$) was proposed by
Přibil and Veselý [28] as masking agent for all metals which react
with hydrogen sulphide.   These metals mostly form colourless
complexes soluble in ammoniacal solution.   Copper in acid solution
is precipitated quantitatively as a canary yellow compound, soluble
in ammonia to give a colourless solution.   The yellow salt can be
filtered off and dissolved in acid (with addition of hydrogen
peroxide) and determined highly selectively by EDTA titration.

For masking, either a 10% solution of the acid or solid sodium thio-
glycollate·can be used.   Because of its versatility, its reactions
are summarized in Table 3.1.

Thioglycollic acid has been used for masking zinc, cadmium, bismuth,
indium, copper, lead and silver in the determination of calcium and/
or magnesium.   Manganese can be determined selectively in the
presence of iron, aluminium, copper, mercury, zinc and cadmium.   In
combination with triethanolamine, potassium cyanide and hydrogen
peroxide, TGA has been applied in the analysis of iron-nickel-cobalt-
copper alloys [29].   TGA is not suitable for direct titration of
nickel and cobalt because their TGA complexes are coloured.   Both
metals can be determined indirectly, because their EDTA complexes do
not react with TGA.

Demasking of TGA.   It is easily oxidized by hydrogen peroxide in
acid medium.

Summary.   Thioglycollic acid is a powerful masking agent for heavy
metals, including lead (difference from KCN) in alkaline medium.   It
replaces the very toxic KCN.   It has some disadvantages: unpleasant
smell, corrosive action on the skin and eyes.   Its solutions must be
kept in brown glass bottles.   Polyethylene bottles are quickly
destroyed.   Old dilute solutions contain traces of hydrogen sulphide,
which interferes in the masking (formation of insoluble sulphides).

TABLE 3.1   Reactions of thioglycollic acid (TGA)

Cation	Remarks
Ag^+	Light yellow precipitate insoluble in excess of TGA, easily soluble in ammonia. Ammoniacal solution of TGA dissolves AgCN and AgI.
Hg^{2+}	White precipitate, soluble in mineral acids and ammonia. Hg-EDTA reacts with TGA quantitatively.
Bi^{3+}	Yellowish precipitate soluble in excess of TGA and ammonia. Bi-EDTA complex reacts with TGA in acidic and ammoniacal solutions.
Cd^{2+}	Yellowish precipitate soluble in TGA and ammonia. Cd-EDTA reacts with TGA quantitatively.
Pb^{2+}	White precipitate soluble in excess of TGA and ammonia. Pb-EDTA reacts with TGA quantitatively. $PbSO_4$ and $PbCrO_4$ are soluble in ammoniacal solution of TGA.
Cu^{2+}	With a small amount of TGA a brown colour appears which on addition of more TGA turns to a canary yellow precipitate, which dissolves in ammonia to give a colourless solution. The yellow precipitate can be filtered off and copper determined complexometrically after the dissolution of precipitate with acid and H_2O_2. Cu-EDTA reacts with TGA quantitatively in ammonia.
Tl^+	Is masked with TGA in acidic and ammoniacal solutions. In more concentrated solutions, a crystalline precipitate is formed, soluble in ammonia.
As^{3+}, Sb^{3+}, Sn^{4+}	No colour reaction. In ammonia stable complexes are formed.
Fe^{3+}	First forms unstable intense blue colour [reduction to Fe(II)]. Resulting colourless solution turns intensely red on addition of ammonia even in the presence of triethanolamine (TEA). In sodium hydroxide medium this red Fe-TGA complex is converted into the colourless Fe-TEA complex.

TABLE 3.1 (continued)

Cation	Remarks
Ni^{2+}	Brown colour only in alkaline solution; colour disappears on addition of EDTA. The Ni-EDTA complex <u>does not</u> react with TGA.
Co^{2+}	Yellow-brown colour in alkaline medium. It <u>does not</u> change with EDTA. Co-EDTA complex <u>does not</u> react with TGA.
Mn^{2+}	A green or yellow-brown colour (according to Mn concentration) in alkaline medium, and is due to aerial oxidation of $Mn(II)$ to $Mn(III)$. After the addition of KCN an intense red colour appears [either the $Mn(CN)_6^{3-}$ complex or possibly a mixed ligand complex], which disappears on standing, because of reduction by TGA. On stirring, the red colour reappears because of aerial reoxidation. The sequence continues until the TGA is exhausted. The Mn-EDTA complex does not react with TGA.
Zn^{2+}	Quantitative masking in alkaline medium.
In^{3+}	Colourless complex in ammonia. The In-EDTA complex reacts with TGA.
Ga^{3+}	Does not react with TGA.
U^{4+}	Intense red colour in alkaline medium, also in the presence of EDTA.
$Mo(VI)$, $W(VI)$	Yellow colour (more intense with Mo), disappears in ammonia solution.
Cr, Th, Al, Ti, Be, Ca, Ba, Mg	No colour reactions.

3.2.4   3-Mercaptopropionic Acid (MPA)

3-Mercaptopropionic acid, $HSCH_2CH_2COOH$, is also used for masking of heavy metals. Its advantage is that it forms colourless or faintly coloured complexes with copper, nickel, cobalt and iron [30]. It has been used for masking of lead at pH 3 - 6 in the determination of zinc, with dithizone as indicator [31]. It has been used for masking copper and cadmium at pH 5 in the determination of zinc with Xylenol Orange as indicator. Cadmium in the same solution can be determined indirectly by addition of DCTA and back-titration with zinc solution (the Cu-MPA complex does not react with DCTA [32]).

3.2.5   2,3-Dimercaptopropan-1-ol (BAL)

BAL (British anti-lewisite, Dimercaprol - I) is a very active anti-dote to arsenic and mercury poisoning. Its activity is based on the formation of very stable complexes, which are eliminated from the body.

$$CH_2OH$$
$$CHSH \qquad I: BAL$$
$$CH_2SH$$

BAL is a colourless liquid with a characteristic smell and is easily oxidized by air.   Its alcoholic solution is more stable.   In many cases it can replace potassium cyanide as masking agent.   It has been proposed for masking all heavy metals, including lead and bismuth.   Its reactions are as follows [33].

(a) Colourless cations (Hg, Ag, Cd, Zn, Sb, Sn) in slightly acidic solutions form white precipitates with BAL, soluble in ammonia. Lead and bismuth form faintly yellow precipitates soluble in ammonia.

(b) Copper, cobalt and nickel form brown (Cu) and olive-green precipitates, which dissolve in ammonia to give the same colour. BAL is not suitable for masking these metals in high concentrations. From their EDTA complexes only copper and cobalt are displaced.   The nickel complex is inactive towards EDTA.   This can be used for very selective determination of nickel in the presence of all metals listed in group (a).

(c) Manganese in ammoniacal solution forms with BAL a dark green Mn(III)-BAL complex.   The Mn-EDTA complex reacts with BAL very slowly.

(d) In ammoniacal solution iron(III) forms with BAL an intense red colour even in the presence of EDTA.   The Fe-TEA complex in sodium hydroxide solution does not react with BAL.   Therefore TEA can be used for masking of iron, simultaneously with masking of other metals with BAL, in the determination of calcium.

(e) BAL prevents precipitation of aluminium with ammonia, but the complex is too weak for masking aluminium against reaction with EDTA and the usual indicators.

BAL can be combined with other more selective masking agents, either simultaneously or successively.   In this way more metals can be determined successively with EDTA in the same solution.   We introduce two examples [33].

(i) Determination of calcium in the presence of iron, aluminium, lead and traces of magnesium.   Add to the acidic solution enough triethanolamine to prevent formation of the red Fe-BAL complex.   Then add 20% alcoholic BAL solution to precipitate lead.   Make alkaline with 1M sodium hydroxide and titrate the calcium with EDTA, using murexide (or Thymolphthalexone) as indicator.

(ii) Determination of magnesium, nickel and zinc.   Add excess of EDTA to the solution and back-titrate the surplus with magnesium solution at pH 10 (Schwarzenbach buffer).   The amount of EDTA corresponds to the sum of Ni + Zn + Mg.   To the same solution add BAL and titrate the EDTA liberated from the Zn-EDTA complex.   The

amount of EDTA corresponds to Zn.   To this solution add potassium
cyanide to destroy the Ni-EDTA complex and titrate the liberated EDTA
as above, with magnesium solution, using Erio T as indicator.   The
concentration of magnesium is obtained by difference.

Summary.   BAL is a very convenient masking agent, but it is not too
stable and is probably more expensive than thioglycollic acid.

3.2.6   Unithiol

2,3-Dimercaptopropane-1-sulphonic acid (sodium salt) (II) has been
proposed as an efficient masking agent for arsenic, antimony, tin,
bismuth, lead, cadmium, zinc and mercury [34 - 37].

$$
\begin{array}{l}
CH_2SH \\
CHSH \qquad \text{II: Unithiol} \\
CH_2SO_3Na
\end{array}
$$

The complexes of bismuth and antimony are intensely yellow in
alkaline medium and their masking is limited to a maximum of 20 mg/
100 ml [36].   Unithiol does not displace nickel and manganese from
their EDTA complexes, and these cations may be determined in the
presence of other metals masked with Unithiol [37].   It is used for
masking zinc, cadmium, lead, tin and mercury [34] in the determination of strontium and barium.   Similarly calcium and magnesium can
be determined in the presence of zinc, lead and mercury [35].

Summary.   Unithiol is more stable than BAL and also more soluble in
water.   It is probably unsuitable for masking iron and cobalt, since
it should give reactions similar to those of BAL.   Its complexes are
easily destroyed (demasked) with hydrogen peroxide.

3.2.7   Dimercaptosuccinic Acid

Dimercaptosuccinic acid (III) was proposed by Mekada et al. [38] for
masking of lead, nickel and cobalt in the determination of alkaline-
earth metals.

$$
\begin{array}{l}
HSCHCOOH \\
HSCHCOOH
\end{array}
\qquad \text{III: Dimercaptosuccinic acid}
$$

It cannot be used in the presence of iron, and its copper complex
(because of its colour) causes negative errors in the titration of
magnesium.

Remarks.   A number of other compounds - thiols and thio-ethers -
were studied as masking reagents by a group of Japanese authors [39].
For example, thiodipropionic acid (IV), TDPA, can be used for masking
of manganese in the titration of copper at pH 5 - 6 (Pyrocatechol
Violet as indicator).   Another compound, carboxymethylmercapto-
succinic acid (V), CMMS, may be used for masking of manganese and

mercury in the determination of cadmium, zinc and lead at pH 5 - 6
(Xylenol Orange as indicator) as well as for masking of thorium in
the determination of bismuth at pH 2 - 3.

$$
\begin{array}{ccc}
\begin{array}{l} CH_2CH_2COOH \\ | \\ S \\ | \\ CH_2CH_2COOH \end{array} & IV: \; TDPA & \begin{array}{l} CH_2COOH \\ | \\ S \\ | \\ CHCOOH \\ | \\ CH_2COOH \end{array} \quad V: \; CMMS
\end{array}
$$

3.2.8   Ammonium Fluoride

Ammonium fluoride was proposed for masking aluminium, calcium and
manganese in ammoniacal medium[40] and applied for simple titrations
of zinc and manganese in ammoniacal buffer, with Erio T as indicator.
These methods were later replaced by more convenient procedures.

Fluoride forms very stable complexes or precipitates with a number of
metals (Ti, Zr, Al, Fe, Th, rare earths). It is also suitable for
masking these metals in acidic solutions (see p. 75).

3.2.9   Hydrogen Peroxide

Hydrogen peroxide in alkaline medium is a highly selective masking
agent for titanium and uranium. The masking of titanium(IV) was
proposed by Wehber [41]. It forms a colourless complex and permits
EDTA determinations, e.g. of zinc and magnesium, with Erio T as
indicator. The titration must be performed quickly because of the
susceptibility of the indicator to oxidation.

Uranium(VI) forms an intensely yellow complex with hydrogen peroxide.
Only low concentrations of uranium (<0.05%) can be masked with it in
the determination of zinc, cadmium, magnesium and calcium at pH 10
with Erio T as indicator [42]. According to the authors the
indicator is not attacked by hydrogen peroxide at room temperature.
Back-titration is necessary, because the formation of a precipitate
makes direct titration impossible.

Hydrogen peroxide also has an interesting function in acidic medium
(see p. 74).

3.2.10   Sodium or Potassium Hydroxide

The masking action of the hydroxide ion is limited to amphoteric
elements. For complexometry formation of the hydroxo-complexes of
aluminium, zinc and lead has been utilized, for example, in the
determination of cadmium with EGTA in the presence of large amounts
of zinc and lead [43]. The Cd-EGTA complex is stable in sodium
hydroxide whereas the zinc and lead complexes are converted into
$Zn(OH)_4^{2-}$ and $Pb(OH)_3^-$. Only the EGTA bound to cadmium remains
complexed. The excess of EGTA is titrated with calcium solution,
with Cresolphthalexone as indicator. The determination can be
extended to solutions also containing iron and aluminium if we mask

them with triethanolamine.

Hydroxide is also used for "masking" magnesium in the determination
of calcium [by precipitation of $Mg(OH)_2$]. In the determination of
calcium only potassium hydroxide may be used, if Fluorexone is used
as indicator; sodium ions produce a slight fluorescence.

3.3   MASKING REAGENTS FOR ACID SOLUTIONS

This group of masking agents is divided roughly equally between
inorganic and organic ompounds. These reagents are often very
selective, reacting with only a few cations. Complexometric
titrations can be performed in acid solutions over a wide range of
pH $(0 - 7)$, depending on the properties of the individual cations.
The masking action of some·compounds is limited to a certain pH
range, which must be maintained (cf. masking of thorium with sulphate,
section 3.3.1, below). Inorganic compounds will be described first,
and as far as possible in order of decreasing selectivity.

3.3.1   Sodium or Ammonium Sulphate

Sodium sulphate in large concentrations $(1 - 2$ g in 100 ml) at pH 1 - 2
reliably masks thorium, even releasing it from its EDTA complex [44 -
46]. This reaction is reversible - at pH 4 - 5 thorium can be determined even in the presence of sulphate. There are two factors
involved in the masking. At pH <1 the conditional stability
constant of the Th-EDTA complex is extremely low, so the complex does
not form anyway, but the sulphate is also > 90% protonated, which
reduces its complexing ability also. At pH 1 - 2, however, there is
sufficient free sulphate (remember that a high concentration is used)
to complex the thorium and reinforce the effect of protonation of the
EDTA, since $K'_{ThY} = K_{ThY}/\alpha_{Th(SO_4)}\alpha_{Y(H)}$. At higher pH, the protonation of the EDTA is sufficiently decreased for the Th-EDTA complex
to be formed preferentially, since $\alpha_{Th(SO_4)}$ remains constant at
pH > 4.

Masking of thorium has been used in the determination of zirconium,
and of iron, uranium and other metals. The sum of Th + Zr is
determined indirectly by titration of excess of EDTA with bismuth
nitrate solution at pH 2 - 3, with Xylenol Orange as indicator. After
adjustment of the pH to 1.5 - 2 and addition of sodium sulphate, the
EDTA liberated from its thorium complex is also titrated with the
bismuth solution. Ammonium sulphate has been used for. masking of
thorium by Milner and Edwards [46] in the analysis of binary Zr-Th
alloys. It was also recommended by Fritz and Johnson [47] for masking of thorium and a smaller amount of titanium in the determination
of zirconium.

Sulphate has also been used for precipitation "masking" of barium and
strontium (see section 3.4).

3.3.2   Potassium Iodide

Potassium iodide in large excess masks mercury(II) as the HgI_4^{2-}
complex, which is more stable than the Hg-EDTA complex [48].

Potassium iodide also forms complexes with cadmium, bismuth and lead. Much attention has been paid to the masking of large concentrations of cadmium in the determination of zinc [49 - 51]. An enormous amount of potassium iodide (30 - 40 g per 100 ml) has to be used for this purpose, and the resulting high ionic strength has an adverse influence on the colour change of the indicator, e.g. Xylenol Orange. However, Flaschka and Butcher [50] obtained reliable results for Cd/Zn ratios up to 3.3 x 10³ in the photometric titration of zinc with DTPA, using Xylenol Orange as indicator.

Potassium iodide precipitates copper as Cu_2I_2 with the liberation of iodine. In the presence of hydroxylamine or ascorbic acid the Cu_2I_2 is directly formed, and does not react with EDTA. Such masking of copper is not used in practice, however. The determination of copper and mercury has been solved by Ueno [52]. The sum of Hg + Cu is determined indirectly by titration of excess of EDTA in ammoniacal solution with copper sulphate, with murexide as indicator. On addition of potassium iodide the mercury is removed from its EDTA complex and masked, and the EDTA liberated is titrated with copper sulphate. Other cations present are bound with EDTA and do not react with the iodide, so we can consider the method as highly selective for mercury.

Mercury/cadmium or mercury zinc mixtures can be analysed by titrating the sum and then masking the mercury with iodide and titrating the liberated EDTA. The theory of this method has been described in detail by Ramsay [53].

3.3.3   Sodium Thiosulphate

Sodium thiosulphate forms complexes with silver, platinum(IV), mercury(II) and thallium(III). Copper is reduced and then masked as a complex of copper(I). This has been used for the determination of zinc and cadmium in copper alloys [54] with PAN as indicator, and of nickel at pH 8.5 - 9.9 with murexide as indicator [55]. Thiosulphate in high concentration partly masks zinc in its determination in the presence of copper. Therefore care must be taken not to add too much. The addition can easily be monitored. Addition of thiosulphate to copper solution gives a brownish colour but this disappears when a small excess of thiosulphate is present.

Summary. Thiosulphate is no longer used, and has been replaced by thiourea in practical applications.

3.3.4   Thiourea

Thiourea behaves similarly to thiosulphate, but is more convenient for masking copper, especially large concentrations of copper in the determination of zinc and nickel [56]. Thiourea has also been used to mask copper in the determination of iron and aluminium in copper alloys [57]. Both elements are determined indirectly, by back-titration of an excess of DCTA with 0.5M lead or zinc solution, with Xylenol Orange as indicator at pH 5 - 5.5. In one aliquot, after masking of copper, the sum Al + Fe is determined. In a second aliquot, after masking of copper and aluminium with ammonium fluoride, the iron is determined.

By suitable masking with thiourea and ammonium fluoride we can determine successively copper, tin and lead [58].

Summary. Because thiourea masks copper over the pH range 2 - 6 it
can be used in practically all analyses of samples containing copper
(alloys, ores, concentrates etc.).

3.3.5   Thiosemicarbazide

Thiosemicarbazide (VI) is a highly selective masking agent for
mercury, forming a colourless and highly soluble 1:2 complex [59].

$$S=C \overset{\displaystyle NH-NH_2}{\underset{\displaystyle NH_2}{\big<}}$$ VI: Thiosemicarbazide

It also reacts with silver. With copper it gives an intense blue
colour. It has been used for masking of mercury during the EDTA
titration of zinc, cadmium and lead at pH 5 - 5.5 and of bismuth at
pH 1 - 2, with Xylenol Orange as indicator. The Hg-EDTA complex
reacts with thiosemicarbazide quantitatively, releasing a corresponding amount of EDTA. This permits determination of mercury in the
presence of several metals bound with EDTA. Such back-titrations
are done with 0.05 - 0.02M zinc or lead solution, with Xylenol Orange
as indicator.

For masking, only the stoichiometric amount of reagent is necessary.
A 0.1M solution prepared by dissolving 9.1 g of the compound in
200 ml of warm water and diluting to 1 litre is very suitable.

Summary. Thiosemicarbazide is more useful than potassium iodide for
masking mercury, since iodide also reacts with copper, lead and
silver. It is not suitable for masking copper, because of the
intense colour of the complex. It can be combined with 1,10-phenan-
throline in the determination of mercury, lead and zinc or cadmium.

3.3.6   1,10-Phenanthroline (Phen)

This well known spectrophotometric reagent (VII) for iron(II) forms
very stable cationic complexes $[M(Phen)_2^{2+}]$ with other bivalent metals
(except Pb, Ca, Mg).

VI: Phenanthroline

It has been proposed for masking of copper, zinc, cadmium, cobalt,
nickel and manganese in slightly acidic medium. After masking of
these elements, lead or aluminium can be determined complexometrically at pH 5 - 5.5, either directly (Pb) or by back-titration with

lead nitrate (Al) in hexamine-buffered solutions, with Xylenol Orange
as indicator [60]. We can also determine lead and zinc successively.
First we determine the sum Pb + Zn, then add Phen to destroy the Zn-
EDTA complex, and finally titrate the liberated EDTA with lead
nitrate. Indium can be determined directly at pH 3 after addition
of Phen to mask the elements mentioned [61]. Nakahara et al. [62]
have similarly determined thorium (pH 3) and bismuth (pH 2).
Phenanthroline has been applied as masking agent in the analysis of
low-melting alloys containing lead, bismuth, cadmium and tin [63].

The optimal masking activity of phenanthroline is limited to acidic
solutions (pH 3 - 7). In alkaline medium some metal-Phen complexes
are less stable. For example the zinc and cadmium complexes
dissociate and react with indicators. The manganese complex behaves
similarly. This has been used by Yurish [64] for complexometric
determination of manganese at pH 8 - 9 with Methylthymol Blue as
indicator, after masking of copper, nickel and cobalt with Phen.
For 5 - 10 mg of each element, the error in the manganese determination is less than 1%.

Summary. Phenanthroline is a very efficient masking agent for
cations of the "ammine group" and totally replaces the unpleasant
potassium cyanide, which functions only in alkaline medium.

An interesting book dealing with the analytical use of phenanthroline
was published some years ago by Schilt [65].

3.3.7   Derivatives of Salicylic Acid

Salicylic acid and sulphosalicylic acid were first used in complexometry as indicators for iron, giving an intensely violet colour with
it. Fritz et al. [66] used sulphosalicylic acid to mask aluminium
and uranium in the determination of bivalent metals, thorium and rare
earths, by potentiometric titration with EDTA at pH 4.5. Chernikov
[67] also masked aluminium with sulphosalicylic acid, in the determination of thorium at pH 1.6.

Salicylic acid and p-aminosalicylic acid are suitable for complexometric analysis of a mixture of iron, aluminium and titanium [68].
The titrations are done in acid medium with Xylenol Orange as indicator. If the titanium is determined indirectly by formation of the
peroxo-Ti-EDTA complex and back-titration of excess of EDTA with
bismuth nitrate at pH 1 - 2, high concentrations of aluminium interfere. The aluminium can be masked if we use the fact that in slightly alkaline medium aluminium forms with salicylic acid a complex
stable enough not to dissociate after acidification to pH 1 - 2, and
this permits the titration of titanium.

Summary. From the available literature it seems that these
compounds are not widely used for masking of aluminium.

3.3.8   Acetylacetone

Acetylacetone (2,4-pentanedione) was proposed for masking of aluminium
by Fritz et al. [66] in the potentiometric determination of various
metals. Jablonski and Johnson [69] have studied its use in certain
titrations with Xylenol Orange as indicator. Iron(III) forms an

intensely red complex with acetylacetone, uranium forms a yellow complex. Both interfere in visual titrations and must be extracted with nitrobenzene. Palladium, beryllium and molybdenum are precipitated with acetylacetone. The colourless precipitates of palladium and beryllium do not interfere in the determination, for example, of zinc or lead. Molybdenum is "masked" (as a precipitate) in the determination of bismuth [69].

Yamaguchi and Ueno [70] re-examined these methods and for the determination of zinc and lead recommended use of acetylacetone to mask aluminium (present in up to ratios of 30 and 100 to zinc and lead respectively). The rare earths can also be determined in the presence of aluminium and acetylacetone.

Summary. Acetylacetone can be considered as a "true" masking reagent only for aluminium. Palladium, beryllium and molybdenum are precipitated, and although iron(III) is also truly masked its acetylacetonate complex must be removed by extraction.

3.3.9   Hydrogen Peroxide

The masking action of hydrogen peroxide on titanium and uranium in alkaline medium has already been described (see p. 69). In acidic medium it is a very welcome "stabilizer" for the titanium complexes of EDTA and DCTA. After addition of hydrogen peroxide the very stable complexes H_2O_2-Ti-EDTA and H_2O_2-Ti-DCTA are formed. Details are given in section 6.1.12.

The formation of the mixed-ligand complexes increases the conditional stability constant, since $K_{Ti(Y)}' = K_{Ti(Y)}\alpha_{TiY(H_2O_2)}/\alpha_{Y(H)}$. According to Jurczyk [71] the peroxo-Ti-EDTA complex is sufficiently stable towards ammonium fluoride, and this allows indirect determination of up to 5 mg of aluminium in the presence of titanium.

Formation of peroxo-complexes with niobium and tantalum (also in the presence of EDTA, DCTA) has also been observed [72,73].

The masking and stabilization effect of hydrogen peroxide on vanadium(V) has been used in the analysis of binary vanadium-titanium alloys [74]. The vanadium-peroxo complex is formed in alkaline medium, but is stable in acidic solutions and does not react with an excess of DCTA. Titanium in this medium is firmly bound as the H_2O_2-Ti-DCTA complex and can be determined by back-titration of excess of DCTA with lead nitrate at pH 5 - 5.5, with Xylenol Orange as indicator (EDTA cannot be used because it partly reacts with the vanadium - peroxo complex). In a second aliquot vanadium is determined indirectly after reduction with ascorbic acid and masking of titanium with ammonium fluoride.

Summary. Hydrogen peroxide has a rather special position among masking reagents. It specifically masks titanium in alkaline medium, as a colourless complex [41]. On the other hand, it also functions as a stabilizer of the Ti-EDTA and Ti-DCTA complexes in acidic medium. It also reacts directly with EDTA, oxidizing it on heating - at above 40°C according to Jurczyk [71], a fact used in the precipitation of bismuth phosphate from homogeneous solution [75].

3.3.10   Ammonium Fluoride

Fluoride (see also p. 69) in acid solution forms complexes with all
quadrivalent and tervalent cations.   Some fluorides are less soluble
in water (e.g. Na_3AlF_6).   It has great value for masking aluminium
and titanium in the determination of iron.   The large excess of
ammonium fluoride usually recommended and used for masking of
aluminium is not, according to our experience, necessary [76].   In
the cold, aluminium can be masked with an amount of fluoride which
corresponds to an Al:F ratio between 1:5 and 1:6.   It is only for
displacement from its EDTA or DCTA complexes that an excess is
necessary (about 30 - 50%) and that boiling of the solution is
required.   Direct masking of aluminium (or titanium) permits the
determination of iron by back-titration of excess of EDTA (or DCTA)
with lead nitrate or zinc chloride at pH 5 - 5.5, in a solution well
buffered with hexamine.   The colour change of Xylenol Orange (used
as indicator) is very sharp.   Proper masking of iron seems rather
problematical.   Solutions of iron(III) decolorized with fluoride
become yellow after addition of EDTA (formation of Fe-EDTA).   Also
iron, as FeF_6^{3-}, reacts with Xylenol Orange and blocks it.   Nevertheless ammonium fluoride in high concentration has been used for masking iron.   Theis [77] recommends ammonium fluoride for masking iron
in direct EDTA titration of copper, with Chrome Azurol S as indicator.
Wehber [78] masks up to 100 mg of iron with 1 g of ammonium fluoride
in the determination of copper with Variamine Blue as indicator.
Lassner et al. [79] similarly mask iron in the determination of
copper and nickel.   Such apparent discrepancies can be explained
since all these procedures have used direct titration with EDTA and
less sensitive indicators for iron.   Any occasional overtitration
must lead to formation of the Fe-EDTA complex even in the presence of
fluoride.   This is clear from our experiments with the masking of
copper with thiourea, where we had to "protect" iron initially with
ammonium fluoride [80].   After masking of copper and addition of
excess of EDTA, iron was determined indirectly by back-titration of
EDTA with lead nitrate, with Xylenol Orange as indicator.   We
consider this method the best for the determination of iron in the
presence of large amounts of copper.

A further interesting reaction of fluoride is the precipitation of
iron in the presence of a potassium salt.   The precipitate K_3FeF_6
does not react with sensitive reagents for iron [1].   This has been
used for complexometric determination of lead, nickel, zinc and
cadmium in the presence of iron.   We can also successively mask
aluminium and iron in the determination of a third element Me.   In
one aliquot we mask aluminium with fluoride and after addition of
excess of EDTA we determine the sum of Fe + Me by back-titration.   In
a second aliquot we precipitate iron (and mask Al) and determine Me.

Ammonium fluoride has also been proposed for masking of quadrivalent
tin in the determination of bivalent tin [81].

Demasking of fluoride complexes.   Boric acid can be used, forming
the complex anion BF_4^-.   Beryllium also forms a very stable complex,
BeF_4^{2-}.   This can be applied for determination of calcium in calcium
fluoride.   The sample is boiled with 1 g of beryllium nitrate.
After dissolution, the calcium is determined by the usual method.
Aluminium chloride can be used for the same purpose.

3.4   MASKING BY PRECIPITATION

Formation of precipitates in titration solutions is not very welcome
in complexometry, especially if they are coloured, because they more
or less obscure the indicator colour change.   In addition, adsorption on a voluminous precipitate with a large active surface must
also be taken into account.   The deleterious effect of precipitation
of large amounts of $Mg(OH)_2$ in the determination of calcium is well
known.   Precipitates are often formed spontaneously during the
titration.   For example, careless neutralization of a bismuth
solution in the presence of chloride leads to the formation of the
insoluble oxychloride BiOCl, which dissolves slowly in EDTA and makes
the end-point uncertain (see section 6.1.5 on p. 106).   Large amounts
of hexamine added in pH adjustment may cause precipitation of zinc,
aluminium etc.

We therefore rarely use such "masking" reactions where the precipitate remains in the solution.   For example, barium sulphate precipitated from a hot solution and then cooled can be tolerated in a direct
EDTA titration, but barium sulphate precipitated from cold solution
is more soluble in EDTA.

However, small amounts of precipitates need not be separated, for
example, in the determination of water hardness.

Generally we use masking by precipitation only if it offers some
advantage over other methods.   Only one compound, from the early
period of complexometry, is worth mention - sodium diethyldithiocarbamate, well known as "Cupral".

In spite of its low selectivity, sodium diethyldithiocarbamate (VII)
is very often used for isolation of small amounts of metals by
precipitation or extraction.

$$S{=}C \overset{\textstyle N(C_2H_5)_2}{\underset{\textstyle SNa}{\Big\langle}}$$

VII: Cupral - DDCNa

In slightly ammoniacal solution, in the presence of EDTA, it reacts
only with mercury, silver, lead, copper, bismuth and cadmium.   In
this medium EDTA masks iron, nickel, cobalt, manganese and zinc.
The precipitates are white, except the yellow $Bi(DDC)_3$ and brown
$Cu(DDC)_2$.   These last two elements can be determined very specifically by colorimetry.

Diethyldithiocarbamate has been proposed for selective precipitation
of cadmium in the presence of zinc [82].   The sum of both is determined by back-titration of excess of EDTA with magnesium solution,
with Erio T as indicator.   After addition of Cupral the EDTA liberated from its cadmium complex is titrated in the same way.   We can
determine still more components in a single solution if we combine
diethyldithiocarbamate and potassium cyanide masking.   For example,
the combination Ni-Cd-Mg can be analysed as follows.   The sum of all
three elements is determined as for the cadmium/zinc determination
above.   After addition of Cupral, the EDTA liberated is titrated,

giving the amount of cadmium.   Then addition of potassium cyanide masks the nickel, and the $Cd(DDC)_2$ dissolves, and further titration with magnesium solution gives the amount of nickel.   The content of magnesium is obtained by calculation from the results of all three titrations.   Alternatively we can determine magnesium in another aliquot by direct titration with EDTA after masking of zinc and nickel with potassium cyanide.

The principle described has been used by other authors in the determination of zinc in ores [83] and in the determination of zinc and cadmium in cadmium-based solders [84].   Similarly Kinnunen and Merikanto [85] determined lead and manganese in lead brass.   After masking of copper with potassium cyanide they determined the sum of lead and manganese by EDTA titration.   After precipitation of lead with Cupral they determined the liberated EDTA by titration with manganese solution.

3.5   CONCLUSIONS

In the field of masking reagents we come to the same situation as with complexometric indicators.   A large number of reagents (perhaps too many) have been proposed for masking of metals, and several are simply minor variants of reagents already in use (e.g. various thiocompounds).

The masking agents for use in acidic solutions are of greatest value, since complexometric titrations in acidic media are more selective. According to the literature, triethanolamine and ammonium fluoride are the most useful - the first for the masking of iron, aluminium and traces of manganese, the second for the analysis of titanium - iron - aluminium mixtures.   For masking of heavy metals glycollic acid is suitable, or the more expensive BAL.   Phenanthroline is still neglected on account of price, but it is superior to potassium cyanide, especially in acid solutions.

Other compounds such as hydrogen peroxide, thiourea and thiosemicarbazide are also useful for special cases.

In this chapter only a selection of the most important masking agents has been given.   The reader will meet some other compounds at the appropriate places in the text.   Further information can also be found in Perrin's monograph [86].

REFERENCES

1.   R. Přibil and V. Veselý, Talanta, 1965, 12, 385.
2.   J.S. Fritz, J.E. Abbink and M.A. Payne, Anal. Chem., 1961, 33, 1381.
3.   R. Přibil and V. Veselý, Talanta, 1963, 10, 1287.
4.   G. Schwarzenbach and H. Flaschka, Die komplexometrische Titration, 5th Ed., Enke Verlag, Stuttgart 1965;   Complexometric Titrations, 2nd Ed., Methuen, London, 1969.
5.   T.S. West, Complexometry with EDTA and Related Reagents, British Drug Houses Chemicals Ltd., Poole, 1969.
6.   H. Flaschka and F. Huditz, Z. Anal. Chem., 1952, 137, 162.
7.   R. Přibil, Coll. Czech. Chem. Commun., 1954, 19, 465.

8.   P. Povondra and R. Přibil, Coll. Czech. Chem. Commun., 1964, 26, 2164.
9.   R. Přibil, Coll. Czech. Chem. Commun., 1954, 19, 1171.
10.   R. Přibil and V. Veselý, Talanta, 1966, 13, 515.
11.   H. Flaschka, Z. Anal. Chem., 1953, 138, 332.
12.   R. Přibil, Coll. Czech. Chem. Commun., 1953, 18, 783.
13.   L. Berák, Hutn. Listy, 1957, 12, 817.
14.   I. Janoušek and K. Študlar, Coll. Czech. Chem. Commun., 1959, 24, 3799.
15.   H. Flaschka and R. Püschel, Chemist-Analyst, 1955, 44, 71.
16.   R. Přibil, Coll. Czech. Chem. Commun., 1954, 19, 58.
17.   R. Přibil, Coll. Czech. Chem. Commun., 1954, 19, 64.
18.   R. Přibil, J. Körbl, B. Kysil and J. Vobora, Chem. Listy, 1958, 52, 243.
19.   R. Přibil and V. Veselý, Talanta, 1961, 8, 270.
20.   R. Přibil and M. Kopanica, Chemist-Analyst, 1959, 48, 35.
21.   J. Adam, J. Jírovec and R. Přibil, Hutn. Listy, 1969, 24, 739.
22.   J.A. Ritchie, Analyst, 1955, 80, 402.
23.   R. Přibil and V. Veselý, Talanta, 1961, 8, 565.
24.   R. Přibil and V. Veselý, Chemist-Analyst, 1965, 54, 114.
25.   L. Barcza, Acta. Chim. Acad. Sci. Hung., 1962, 29, 156.
26.   J. Jankovský, Coll. Czech. Chem. Commun., 1957, 22, 1052.
27.   A.M. Escarrilla, Talanta, 1966, 13, 363.
28.   R. Přibil and V. Veselý, Talanta, 1961, 8, 880.
29.   R. Přibil and V. Veselý, Chemist-Analyst, 1963, 52, 5.
30.   K. Yamaguchi and K. Ueno, Talanta, 1963, 10, 1195.
31.   S. Hara, Japan Analyst, 1961, 10, 633.
32.   R. Přibil and V. Veselý, Talanta, 1965, 12, 475.
33.   R. Přibil and Z. Roubal, Coll. Czech. Chem. Commun., 1954, 19, 1162.
34.   L.A. Volf, Zavod. Lab., 1960, 26, 1353; Anal. Abstr., 1961, 8, 2779.
35.   L.A. Volf, U.S.S.R. Patent 128191 (28th April 1960); Anal. Abstr., 1962, 8, 1857.
36.   Yu. V. Morachevskii and L.A. Volf, Uch. Zap. Leningrad Gos. Univ., 1960 No. 297, 144; Anal. Abstr., 1962, 9, 560.
37.   Yu. V. Morachevskii and L.A. Volf. Zhr. Analit. Khim., 1960, 15, 656; Anal. Abstr., 1962, 9, 562.
38.   T. Mekada, K. Yamaguchi and K. Ueno, Talanta, 1964, 11, 1461.
39.   H. Yamaguchi, K. Iwasaki, K. Yamaguchi and K. Ueno, Japan Analyst, 1967, 16, 703; Anal. Abstr., 1969, 16, 16.
40.   R. Přibil, Coll. Czech. Chem. Commun., 1954, 19, 64.
41.   P. Wehber, Z. Anal. Chem., 1957, 154, 182.
42.   R. Lassner and R. Scharf, Z. Anal. Chem., 1958, 159, 212.
43.   R. Přibil and V. Veselý, Chemist-Analyst, 1966, 55, 4.
44.   R. Přibil and K. Burger, Magy. Kem. Foly., 1959, 65, 204; Anal. Abstr., 1960, 7, 447.
45.   R. Přibil and K. Burger, Talanta, 1960, 4, 8.
46.   G.W.C. Milner and J.W. Edwards, Anal. Chim. Acta, 1959, 20, 31.
47.   J.S. Fritz and M. Johnson, Anal. Chem., 1955, 27, 1653.
48.   H. Biedermann and G. Schwarzenbach, Chimia, 1948, 2, 56.
49.   H. Flaschka and J. Butcher, Microchem. J., 1963, 7, 407.
50.   H. Flaschka and J. Butcher, Talanta, 1964, 11, 1067.
51.   H. Flaschka and J. Butcher, Chemist-Analyst, 1963, 54, 36.
52.   K. Ueno, Anal. Chem., 1957, 29, 1668.
53.   C.G. Ramsay, J. Chem. Ed., 1977, 54, 714.
54.   K.L. Cheng, Anal. Chem., 1958, 30, 243.
55.   I.I. Kalinichenko, Zavod. Lab., 1958, 24, 266; Anal. Abstr., 1959, 6, 185.

56. J.S. Fritz, W. Lane and A.S. Bystroff, Anal. Chem., 1957, 28, 821.
57. R. Přibil and V. Veselý, Chemist-Analyst, 1965, 54, 46.
58. W.J. Ottendorfer, Chemist-Analyst, 1958, 47, 96.
59. J. Korbl and R. Přibil, Coll. Czech. Chem. Commun., 1957, 22, 1771.
60. R. Přibil and F. Vydra, Coll. Czech. Chem. Commun., 1959, 24, 3103.
61. M. Kopanica and R. Přibil, Coll. Czech. Chem. Commun., 1960, 25, 2230.
62. K. Nakahara, K. Hyashi, T. Danzuka, S. Ngamura and T. Imamura, Japan Analyst, 1963, 12, 854; Anal Abstr., 1965, 12, 525.
63. R. Přibil and M. Kopanica, Chemist-Analyst, 1959, 48, 87.
64. I.M. Yurist, Zh. Analit. Khim., 1967, 22, 442; Anal. Abstr., 1968, 15, 6641.
65. A.A. Schilt, Analytical Application of 1,10-Phenanthroline and Related Compounds, Pergamon Press, Oxford, 1969.
66. J.S. Fritz, M.J. Richard and S.K. Karraker, Anal. Chem., 1958, 30, 1347.
67. Y.A. Chernikov, R.S. Tramm and K.S. Pevzner, Zavod. Lab., 1960, 26, 921.
68. R. Přibil and V. Veselý, Talanta, 1963, 10, 283.
69. W.Z. Jablonski and E.A. Johnson, Analyst, 1960, 85, 297.
70. K. Yamaguchi and K. Ueno, Japan Analyst, 1963, 12, 55, Anal. Abstr., 1964, 11, 4748.
71. J. Jurczyk, Chem. Anal. (Warsaw), 1965 10, 441.
72. E. Lassner and R. Scharf, Talanta, 1960, 7, 12.
73. E. Lassner and R. Puschel, Mikrochim. Acta, 1963, 950.
74. R. Přibil and V. Veselý, Hutn. Listy, 1973, 28, 661; Anal. Abstr., 1974, 26, 2624.
75. P.F.S. Cartwright, Analyst, 1960, 85, 216.
76. R. Přibil and V. Veselý, Talanta, 1962, 9, 23.
77. M. Theis, Z. Anal. Chem., 1955, 144, 175.
78. P. Wehber, Z. Anal. Chem., 1956, 149, 244.
79. E. Lassner, R. Scharf and P.L. Reiser, Z. Anal. Chem., 1959, 165, 88.
80. R. Přibil and V. Veselý, Talanta, 1958, 8, 743.
81. I. Dubský, Coll. Czech. Chem. Commun., 1959, 24, 4045.
82. R. Přibil, Coll. Czech. Chem. Commun., 1954, 19, 58.
83. M. Hisada and K. Kashikawa, Japan Analyst, 1959, 8, 235; Anal. Abstr., 1964, 11, 4748.
84. E. Yu. Zel'tser, Zavod. Lab., 1959, 25, 798; Anal. Abstr., 1960, 7, 1682.
85. J. Kinnunen and B. Merikanto, Chemist-Analyst, 1954, 43, 93.
86. D.D. Perrin, Masking and Demasking of Chemical Reactions, Wiley-Interscience, New York, 1970.

CHAPTER 4

Separation Methods

The methods described in the last chapter for the masking of metals
are very attractive for use in fast routine analysis.   Nevertheless
they cannot solve all the problems encountered by an analyst.   In
the complete analysis of alloys, silicates etc. it is often necessary
to analyse diverse materials containing various numbers of components
present in a wide range of concentrations.   Frequently, we are not
interested in the main component (e.g. in iron in steel) but only in
certain components present in amounts ranging from a few per cent
downward.   For solving such problems analytical chemistry has a
number of methods available, that can be successfully combined with
complexometry.

In the course of time many procedures have been worked out that make
it possible to determine complexometrically metals present even at
very low levels, but they necessitate the use of large samples.   In
such cases, after dissolution of the sample the interfering components
may be present in concentrations high enough to exclude the use of
complexometry unless a suitable separation method is applied.   These
methods will be briefly disucssed in this chapter.

It should be kept in mind that the reliability of complexometric
titrations is to some extent dependent on the atomic weight of the
element to be determined.   Determination of 2 mg of magnesium or
aluminium (0.2% in 1 g of sample) can be carried out more reliably,
for example, than that of 2 mg of lead.   This follows from the
volumes of EDTA solution required - 1.64 and 1.49 ml of 0.05M
solution for magnesium and aluminium respectively, and only 0.19 ml
for lead.

4.1   SEPARATION DURING DISSOLUTION

Even during the dissolution of samples some components can be
quantitatively separated, silicic acid in the analysis of silicates,
for example.   The insoluble residue obtained in the dissolution of
alloys by evaporation with acids may contain, for example, tantalum,
niobium, tungsten, tin and antimony, depending on the acid used.
Similarly barium, strontium and lead can be separated as their

sulphates during digestion of the sample with sulphuric acid.   On
the other hand, interfering fluoride can be removed by evaporation
with sulphuric acid, mercury volatilized by heating of the original
sample etc.

Sample dissolution is often performed in such a way that the resulting solution is suitable for complexometric titrations.   Determination of small amounts of magnesium in aluminium can be given as an
example.   The sample is dissolved in sodium hydroxide solution to
remove most of the aluminium (plus zinc).   The magnesium is undissolved and relatively free from the matrix metal, and after dissolution can be determined complexometrically.   Some authors dissolve
the aluminium in a mixed solution of sodium hydroxide, triethanolamine and potassium cyanide.   Both masking reagents will bind a
number of trace elements in firm complexes.   Dissolution in sodium
hydroxide/sodium sulphide solution is recommended for the determination of zinc in aluminium.   The precipitated zinc sulphide (contam-
inated with heavy metal sulphides) is then determined complexometrically.

4.2   SEPARATION BY PRECIPITATION

Iron, aluminium and titanium are separated most frequently by precipitation with ammonia in the analysis of silicates, ores and concentrates.   The precipitation must usually be repeated at least once to
avoid adsorption of other cations on the precipitates, which makes
the analysis rather tedious.

Small amounts of heavy metals can be separated as sulphides are
removed by solvent extraction, for example as diethyldithiocarbamates
or oxinates.   Sometimes even apparently simple mixtures, for example
iron-titanium-aluminium oxides (or hydroxides) may need selective
preparation for complexometric determination or colorimetry.   Iron
and titanium are separated from aluminium by precipitation with
sodium hydroxide.   Titanium is separated from iron and aluminium
with sodium hydroxide and triethanolamine.   These and many other
procedures are of great benefit in complexometry

4.3   SEPARATION BY ELECTROLYSIS

Classical electroanalysis is nowadays practically limited to the
separation and gravimetric determination of copper (on a platinum
gauze electrode) or zinc (on a copper-plated platinum gauze electrode).
Lead (at low levels) can be separated as PbO_2 on a platinum wire
anode.   Electrolysis at a mercury cathode for separation of smaller
amounts of all heavy metals is more often used.

4.4   SEPARATION ON ION-EXCHANGERS

Some thousands of publications and a number of comprehensive books
have been devoted to the chromatography of inorganic compounds on
ion-exchangers.   Thanks to its wide scope, this method has become
an independent analytical technique that make possible many selective
separations.   It is especially advantageous when the eluted cations
can be determined by simple complexometric titration.   The separation
process frequently takes several hours, however, so the procedure is

AC - D

not very advantageous for a single analysis.   For routine series
analysis a battery of columns can be used, with overlap of batches to
economise on time.   Another drawback of this method is that it
frequently produces large volumes of eluate, which must be evaporated.
As complex-forming compounds (fluoride, organic acids etc.) are very
often used in chromatographic separations, they may interfere in the
final determination, whether it is photometric, polarographic or
complexometric.   This method of separation is, of course, justified
in cases where other methods either fail completely or are even more
tedious or subject to interference.

4.5   SEPARATION BY LIQUID-LIQUID EXTRACTION

Attempts at metal extraction date back to the end of the last century.
As an example we can mention the analytically useful extraction of
iron with ether from hydrochloric acid (Rothe, 1892).

During the past thirty years there has been extensive investigation
of extraction of ion-association complexes and insoluble metal
chelates formed with organic reagents (oxine, dithizone, cupferron,
etc.).   Such extractions have mostly been used for colorimetric
determination of trace metals.

For complexometric work, priority should be given to such extraction
methods suitable for semimicro work, e.g. extraction of several
centigrams of metal.   In this respect high molecular-weight amines
are very promising, especially the tertiary and quarternary amines
like trioctylamine (TOA), and trioctylmethylammonium chloride (TOMA),
produced commercially as Alamine 336-S or Aliquat 336-S.   There are
many other such amines on the market today (e.g. Amberlite LA-1,
Amberlite LA-2, Arquad 2C etc.).

The basic property of these amines is their ability to react with
acids to form ion-pairs that are easily extracted into organic
solvents such as chloroform, benzene and xylene.   Extraction by
tertiary amines can be expressed by the equation:

$$R_3N_{org} + H^+ + A^- = (R_3NH^+A^-)_{org}$$

where A^- is the anion of the acid.   The solution of the salt in an
organic solvent can then be equilibrated with other anions in aqueous
medium, for instance with $ZnCl_4^{2-}$, CdI_4^{2-}, $Mn(NCS)_4^{2-}$.   With Aliquat in
$R_4N^+Cl^-$ form the extraction of cadmium from iodide medium proceeds
according to the equation:

$$2(R_4N^+Cl^-)_{org} + CdI_4^{2-} = [(R_4N^+)_2CdI_4^{2-}]_{org} + 2Cl^-$$

With respect to ease of exchange of anions such amines resemble
chromatographic ion-exchangers and hence are often called "liquid
anion-exchangers".   Most studies have dealt with the extraction of
trace amounts of elements, especially radioactive isotopes.

Some extractions used in connection with spectrophotometry can be a
useful supplement in total or special analysis.   This can be illustrated by the extraction of gold with a chloroform solution of trioctylamine [1], or of chromium(VI) with subsequent measurement of its
own yellow colour in the extract [2] or of the colour after reaction

with diphenylcarbazide [3], extraction of molybdenum (as thioglycollate) [4] and of uranium from sulphuric acid medium [5]. A modified
method has been used for the extraction of large amounts of uranium
in the polarographic determination of traces of lead [6]. Cobalt
can also be selectively determined after extraction with nitroso-R-
salt [7]. Chromium(III) can be extracted as the Cr-(DCTA) complex
by TOMA in chloroform solution, and determined spectrophotometrically
[8,9].

Among the extractions suitable for complexometry we may mention
extraction of zinc and cadmium and their mutual separation from
manganese, nickel, cobalt, indium, gallium, aluminium and copper [10]
and the separation of manganese by extraction as the $Mn(NCS)_4^{2-}$
complex from calcium and magnesium [11].

Titanium can be extracted from weakly acid solution as its ascorbate
complex, with a chloroform solution of TOMA, and determined complexometrically, after stripping [12].

A great deal of attention has been paid to phenylacetic acid and
other carboxylic acids as high-capacity extractants that give more
selective extraction than such reagents as oxine [13,14]. In this
way copper can be extracted with a chloroform solution of phenylacetic acid from a solution buffered with hexamine, after masking of
iron and aluminium with fluoride and determined by direct spectrophotometry of the green extract [15] or complexometrically after
stripping [16].

The applicability of these methods has been proved by a series of
analyses of various materials such as silicates, alloys, etc.
Uranium(VI) can be extracted very selectively with phenylacetic acid
from solutions buffered with hexamine in the presence of nitrilotriacetic acid. This extraction can be used for the separation of
uranium from thorium, zirconium, rare earths, and various bivalent,
tervalent and quadrivalent cations [17]. After stripping, the
uranium is reduced to the quadrivalent state with ascorbic acid and
determined indirectly by EDTA titration.

Suitable extraction systems may be found by consulting handbooks of
solvent extraction analysis [18 - 21].

REFERENCES

1.   J. Adam and R. Přibil, Talanta, 1071, 18, 405.
2.   J. Adam and R. Přibil, Talanta, 1971, 18, 91.
3.   J. Adam and R. Přibil, Talanta, 1974, 21, 616.
4.   R. Přibil and J. Adam, Talanta, 1971, 18, 349.
5.   J. Adam and R. Přibil, Coll. Czech. Chem. Commun., 1972, 37, 129.
6.   J. Adam and M. Štulíková, Talanta, 1974, 21, 1203.
7.   J. Adam and R. Přibil, Talanta, 1971, 18, 733.
8.   J. Adam and R. Přibil, Talanta, 1974, 21, 1206.
9.   G.M. Kinhikar and S.S. Dara, Talanta, 1974, 21, 1208.
10.   R. Přibil and V. Veselý, Coll. Czech. Chem. Commun., 1972, 37,
      13.
11.   R. Přibil and J. Adam, Talanta, 1973, 20, 49.
12.   J. Adam and R. Přibil, Talanta, 1975, 22, 905.
13.   J. Adam, R. Přibil and V. Veselý, Talanta, 1972, 19, 825.

14.   F.I. Miller, Talanta, 1974, 21, 685.
15.   J. Adam and R. Přibil, Talanta, 1972, 19, 1105.
16.   J. Adam and J. Jírovec, Coll. Czech. Chem. Commun., 1973, 38, 507.
17.   J. Adam and R. Přibil, Talanta, 1973, 20, 1344.
18.   A.K. De, S.M. Khopkar and R.A. Chalmers, Solvent Extraction of Metals, Van Nostrand Reinhold, London, 1970.
19.   G.F. Morrison and H. Freiser, Solvent Extraction in Analytical Chemistry, Wiley, New York, 1957.
20.   J. Starý, The Solvent Extraction of Metal Chelates, Pergamon Press, Oxford 1964.
21.   T. Sekine and Y. Hasegawa, Solvent Extraction Chemistry, Dekker, New York, 1977.

CHAPTER 5

Apparatus and Solutions

Before systematic description of the complexometric determination of individual cations and anions, a short description will be given of preparation and standardization of the necessary standard solutions and of the preparation of other reagent solutions (indicators, buffers, etc.).

5.1 GLASS AND PLASTIC WARE

Complexometric titrations are so simple that no special laboratory equipment is required. Generally, burettes, microburettes, pipettes and standard flasks are the only graduated equipment necessary. Any glass titration vessels, with a volume of 200 – 250 ml, or sometimes 500 ml, will do. Vessels used for storing reagent and EDTA solutions should be of good quality glass, though polyethylene vessels are being used more frequently, as they do not have the main disadvantage of glass, namely the possible contamination of solutions by dissolution of calcium from the container, e.g. with ammonia solutions, which is of special importance in microtitrations. Flaschka [1] recommends using bottles made of best quality Jena glass, previously exposed to the effect of water vapour, for micro-titrations. He also recommends that titration flasks be steamed out and then boiled in a dilute ammonia solution of EDTA. This sort of treatment is useful only for microtitrations.

Swirling by hand is perfectly satisfactory for mixing the titration solution for titrations on the ordinary scale, but magnetic stirring is especially useful for microtitrations or when free movement of the titration vessel is impossible, e.g. during spectrophotometric titrations.

Under optimum conditions normal visual titrations are sufficiently precise, so expensive equipment is usually not necessary. More complicated equipment is required for spectrophotometric, polaro-graphic and other physicochemical methods for end-point indication.

5.2   REAGENTS

5.2.1   Ethylenediaminetetra-acetic Acid - EDTA

Standard solutions are generally prepared from the disodium salt,
$Na_2H_2Y.2H_2O$ (M.W.372.16), which is manufactured in sufficiently pure
form by a number of companies and sold under various names.   According
to Blaedel and Knight [2] and Schwarzenbach [3], the commercial
products contain about 0.3 - 0.5% hygroscopic water.   The first two
authors recommend removing this by drying at 80°C in a hygrostat
giving 50% relative humidity at 25°C.   Equilibrium is complete after
several days [2].   The anhydrous salt can be prepared by drying at
130 - 150°C, but its high hygroscopicity prevents its use for the
preparation of standard solutions.

EDTA solutions of the most generally used concentration (0.05M) can
be prepared by dissolving 18.61 g of EDTA per litre of solution.
More dilute solutions (0.02 - 0.005M) can be prepared by dissolving
the appropriately scaled down amounts.

Schwarzenbach [3] suggests that it is especially useful to employ the
free acid (M.W. 292.13), which can readily be prepared from the
commercial disodium salt by precipitation with hydrochloric or
sulphuric acid.   This involves a certain loss of the acid as a
result of its solubility in water.   (The solubility in pure water is
0.050 g/l, increasing to 0.6 - 2.5 g/l in 0.01 - 1.0M hydrochloric
acid [4].)

After drying at 110°C the free acid is very pure and is not
hygroscopic.   A 0.05M solution can be prepared by dissolving
29.215 g of the acid and 9 g of sodium hydroxide in 400 ml of
redistilled water and, after cooling, diluting to 2 litres.

In the early days of complexometry, a great deal of attention was
paid to the purity of reagents from various sources, since experience
showed that some were more suitable than others for analytical use.
Blaedel and Knight [2] recommend simple purification of the disodium
salt by precipitation from a 10% solution by addition of ethanol.
The salt is collected, washed with ethanol and diethyl ether and
dried at 80°C for 4 days.   Drying at 100°C is not recommended
because of the resultant slight dehydration.

Barnard et al. have evaluated the quality of very pure standard
substances [5,6] and considered the preparation and purity control of
highly pure EDTA.   Even traces of metals are very important when
using EDTA in preparation of very pure chemicals and in enzymological
work, biochemical studies, etc.   In addition to the preparation of
EDTA, these authors describe methods - generally spectrophotometric -
for determining traces of metals in the product.   The amounts of
impurities they found are listed in Table 5.1 for reference.

These authors also give the elemental analysis, complexometric assay
and data on mass loss after drying for two days at 105°C, determination of sulphated ash (0.003%) and the content of nitriloacetic acid
(0.02%).   The results require no comment, except that the amounts of
germanium and gold found are rather surprising.

The authors emphasize that carrying out these tests is rather
expensive.   The consumption of EDTA in all the tests is 80 g for

each batch prepared (typically 300 g). The purity attained is
extremely high, and is certainly not necessary for routine complexometry.   In practice, analytical-reagent grade preparations are
sufficiently pure.

TABLE 5.1   Barnard's analysis of very pure EDTA [6]

Maximum ppm	Elements Found	Maximum ppm	Elements Found
0.01	Ag, Be, Si	0.20	B, Cd, Ge, Au
0.02	Cu, In, Mg, Mo, Ti, V	1.00	Sb
0.03	Al	2.00	Zn
0.04	Ba	25.00	Na
0.05	As	30.00	Cl (halogens)
0.10	Bi, Ca, Cr, Co, Ga, Fe, Ni, Sr, Sn, Zr		

5.2.2   The Stability of Standard EDTA Solutions

No observable changes are found for standard solutions of EDTA stored
in vessels made of good-quality borosilicate glass [7].   Blaedel and
Knight [2] found titre changes of less than 0.05%.   Soft glass,
however, has undesirable effects, storage for 4 months being found to
give a change of up to 1% [8].   The decrease in titre is produced
primarily by dissolution of calcium and magnesium.   This effect is
especially marked with very dilute EDTA solutions, where even tiny
changes in the titre have large effects in microtitrations.   Flaschka
and Sadek [9] recommend that polyethylene bottles be used for all
solutions.   The titre of 0.001M EDTA stored in a polyethylene bottle
did not change during 44 days, while the same solution stored in
glass bottles decreased in titre by 3 - 8.5%, depending on the quality
of the glass.

5.2.3   Standardization of EDTA Solutions

A number of substances have been recommended for this purpose.
Calcium carbonate has been especially recommended, as it is available
in very pure form, can be readily dried and is not hygroscopic.
Magnesium sulphate heptahydrate is not suitable as it readily changes
in composition on exposure to the air.   A well-defined product can
be obtained by storing it in a desiccator over a mixture of 5 parts
of the heptahydrate and 1 part of water (w/w) [10].   Very pure
metals, such as copper, zinc, nickel, etc., have also been suggested,
but they have a small equivalent weight and their surfaces are not
free from oxide.   Zinc oxide and mercuric oxide have been recommended, as they have well-defined compositions.

Lead chloride or nitrate is probably the most suitable substance [11].
We have found lead nitrate to be most useful because of its high
solubility and large molecular weight [250.00 mg of $Pb(NO_3)_2 \equiv 15.699$

ml of 0.05M EDTA]. Lead nitrate is readily recrystallized from hot
water acidified with several drops of 1M nitric acid. It can
readily be dried and is completely non-hygroscopic.

The titre of a 0.05M EDTA solution is found by dissolving a precisely
weighed amount of 200 - 220 mg of lead nitrate in 100 - 150 ml of water
acidified with several drops of 1M nitric acid. After addition of
Xylenol Orange, the pH is adjusted by addition of solid hexamine
until a purple colour appears and the solution is titrated with the
EDTA solution to the appearance of a lemon yellow colour.

A standard 0.05M stock solution of lead nitrate is very useful and
can be prepared by dissolving 16.562 g of $Pb(NO_3)_2$ in distilled water
and diluting to 1 litre. It is an excellent standard and a good
standard reagent for back-titration of excess of EDTA (in the determination of Ni, Co, Cu, Fe, Al, etc.). It also satisfies the
condition that individual complexometric determinations should be
carried out under the same conditions as the standardization (see
below).

Schwarzenbach [3] recommends use of a 0.01M solution of bismuth
nitrate for titrations in relatively acid solutions. This solution
is prepared by dissolving 2.090 g of the pure metal in 7 ml of
concentrated nitric acid and diluting to 1 litre. A 0.05M solution
can also be prepared from the solid nitrate, and is also useful in
some back-titrations of free EDTA (indirect titration of Th, Fe or
Ti).

Many other substances have been recommended as primary standards.
Those described here are, however, completely sufficient for basic
complexometric requirements.

As in all analytical work, it is assumed that sufficiently pure
chemicals are used, including acids, bases, buffer solutions,
redistilled water, etc. Trace impurities become especially important in microdeterminations and also affect the standardization. It
can be assumed that a 0.001M EDTA solution will contain zinc and
calcium as impurities. In standardization with bismuth nitrate (pH
1 - 2), calcium and zinc will not interfere, but when zinc is titrated
with Xylenol Orange as indicator, the factor will be somewhat lower·
because of the zinc present in the reagent. The factor found in
titrations of calcium will be lower still. This example is extreme.
In normal titrations, the traces of metals present do not affect the
consumption of EDTA, but may block the indicators, e.g. traces of
copper in ammoniacal buffer solutions during the titration of
magnesium with Erio T as indicator.

It was mentioned above that the same conditions should be used for
both determinations and standardizations. This is because EDTA (or
other complexone used) may contain impurities that are themselves
chelating agents, such as nitrilotriacetic acid. The impurities may
give metal-ion complexes that are too weak to be formed in acid
medium, but can be formed in alkaline solution. The complexing
titrant therefore will have a higher titre if it is standardized by a
titration at pH 10 than it will for standardization with the same
metal at pH 5 or lower. Impurities of this nature can cause errors
of 0.1% or more [12,13]. Complexometric standards have been
reviewed [14].

5.2.4   Preparation of Other Standard Solutions

Solutions of other complexing agents are also used, especially when
the use of EDTA is not favourable or is less than satisfactory.
These reagents include, DCTA, DTPA, EGTA and recently also TTHA
(which is less readily available).   The structural formulae of these
substances are given in Section 1.1.

All these acids can be obtained commercially in sufficiently pure
form.   Preparation of their 0.05M solutions is simple.   The weighed
substance is dissolved in 100 ml of 1M sodium hydroxide and 200 ml of
water, if necessary with slight heating, and after cooling is diluted
to 1 litre.   The following amounts are required for preparation of
1 litre of 0.05M solution:

DCTA	DTPA	EGTA	TTHA
17.317 g	19.659 g	19.018 g	24.723 g

The importance of these substances in complexometry will be discussed
later.

When a complexometric determination involves a back-titration, an
appropriate stock solution is prepared, preferably of the same molar
concentration as the EDTA, generally 0.05M.   Bismuth nitrate and
lead nitrate have already been mentioned.   In alkaline media, a
0.05M solution of calcium or magnesium can be used.   For the first,
5.005 g of pure calcium carbonate is dissolved in a minimal amount of
hydrochloric or nitric acid and diluted to 1 litre with redistilled
water.   A 0.05M solution of magnesium is best prepared from a suitably pure salt, and standardized by direct titration with the EDTA
solution.

A 0.05M zinc solution is also useful and is generally prepared by
dissolving 3.269 g of pure zinc in hydrochloric or nitric acid and
diluting to 1 litre.   It is useful for some back-titrations in
weakly acid medium (buffered with hexamine, Xylenol Orange as
indicator).   It is preferable to lead nitrate when the solution
contains larger amounts of sulphate, a mixture of chloride and
fluoride, or acetate.   If it is to be used with Erio T, however, and
has been prepared by dissolving zinc in nitric acid, it is essential
to remove any nitrous acid with urea, or the indicator may be
decomposed [12].

5.2.5   Preparation of Indicators

With very few exceptions, indicators are strongly coloured substances
and are thus used in very dilute form.   Only a few are sufficiently
stable in aqueous solution.   Some are only poorly soluble in water
(e.g. PAN).   Those that are soluble and stable are used as 0.1 - 0.5%
solutions.   These are primarily Pyrocatechol Violet, Xylenol Orange
and Chrome Azurol S (0.4% solution).   Ethanol solutions are preferable for dithizone (25 mg in 100 ml) and PAN or PAR (50 mg in 100 ml).

Although the other indicators can be kept for a short time in aqueous
solution, it is preferable to use then as a solid 1% w/w mixture with
potassium chloride or nitrate.

AC - D*

5.2.6   Buffers

For titrations in alkaline medium with Erio T as indicator it is
necessary that the solution be well buffered to pH 10.
Schwarzenbach's buffer is mainly used for this purpose (70 g of
ammonium chloride + 570 ml of concentrated ammonia solution per
litre).

In the acid pH region, Xylenol Orange is mainly used as indicator, at
pH 5 - 5.5.   This value is readily attained with hexamine (urotropine)
added as the solid or a 20% solution.   The appearance of the colour
of the indicator complex of the metal to be titrated shows when the
appropriate pH has been reached, but additional buffer must be added
to provide adequate capacity during the titration.   The pH can also
be adjusted with 1M sodium hydroxide and 1M nitric acid.   In practical analyses, pH adjustment presents no difficulty;   in more complicated cases, details are given in the procedure.

REFERENCES

1.   H. Flaschka, Mikrochemie, 1952, 39, 38.
2.   W.J. Blaedel and H.T. Knight, Anal. Chem., 1954, 26, 741.
3.   G. Schwarzenbach and H. Flaschka, Die komplexometrische
      Titration, 5th Ed., Enke Verlag, Stuttgart, 1965, p. 125.
4.   P.N. Palei and N.I. Udaltsova, Zh. Neorgan. Khim., 1960, 5, 2315.
5.   A.J. Barnard Jr., E.F. Joy and F.W. Michelotti, Clin. Chem.,
      1971, 17, 841.
6.   A.J. Barnard Jr., E.F. Joy, K. Little and J.D. Brooks, Talanta,
      1970, 17, 785.
7.   J.D. Betz and C.A. Noll, J. Am. Water Works Assoc., 1950, 42,
      749.
8.   C.A. Goetz, T.C. Loomis and H. Diehl, Anal. Chem., 1950, 22, 750.
9.   H. Flaschka and F. Sadek, Z. Anal. Chem., 1957, 156, 23.
10.   G. Brunisholz, M. Genton and W. Plattner, Helv. Chim. Acta,
      1952, 36, 782.
11.   J. Vřešťál, J. Havíř, J. Brandštetr and S. Kotrlý, Coll. Czech.
      Chem. Commun., 1959, 24, 360.
12.   R.N.P. Farrow and A.G. Hill, Analyst, 1965, 90, 210.
13.   R.G. Monk, Analyst, 1966, 91, 597.
14.   R.A. Chalmers, in Comprehensive Analytical Chemistry, Vol. III,
      G. Svehla (ed.), p. 204, Elsevier, Amsterdam, 1975.
15.   R.A. Chalmers, unpublished observation.

CHAPTER 6

Classification of EDTA Complexes

Careful consideration of the numerical stability constant values and their trends reveals that the complexes of tervalent and quadrivalent cations are the most stable; the complexes of most bivalent cations are much weaker and those of the alkaline-earth metals are weakest of all. The more stable the complex, the lower the pH value at which complete dissociation occurs. In other words: the stability constant (or, more exactly, the conditional stability constant) determines the lowest pH value at which the given cation can be determined by EDTA titration. In this connection, it should be pointed out that the chelate complexes of metallochromic indicators are subject to the same rule: their stability constants also determine the pH values at which their dissociation is complete. As they are always less stable than the corresponding EDTA complexes, they dissociate at higher pH values than do the EDTA complexes. The difference is usually 1 - 2 pH units. An example is Xylenol Orange, used at pH 5 - 5.5, for the most determinations of bivalent metals, although other detection methods can be used for the determination of most of these metals at lower pH values.

Classification into three groups (Table 6.1) is most useful for evaluation of the analytical utility of EDTA complexes.

The first group consists of all complexes with stability constants (K_{MY}) greater than 10^{20}. These include those of all quadrivalent and tervalent metals (except aluminium), as well as bivalent mercury and tin. All these elements can be determined complexometrically at pH 1 - 3.

The second group consists of cations with K_{MY} values of $10^{12} - 10^{19}$. These are all the rare earths and all the bivalent metals except the alkaline-earth metals. These elements can be determined by titration with EDTA in the pH region 5 - 6.

The third group contains silver, beryllium and the alkaline-earth metals. Their K_{MY} values lie in the interval $10^7 - 10^{11}$ and they can be determined with very low selectivity and only in alkaline media, usually at pH 8 - 10.

TABLE 6.1   Complexometric groups

| Group I | | Group II | | | | Group III | |
Cation	log K_{MY}	Cation	log K_{MY}	Cation	log K_{MY}	Cation	log K_{MY}
Co^{3+}	36	Cu^{2+}	18.8	Lu^{3+}	19.8	–	
Zr^{4+}	29.5	VO^{2+}	18.8	Yb^{3+}	19.5	Ca^{2+}	10.7
Bi^{3+}	27.9	Ni^{2+}	18.6	Tm^{3+}	19.3	Be^{2+}	9.8
V^{3+}	25.9	VO_2^+	18.1	Er^{3+}	18.8	Mg^{2+}	8.7
Fe^{3+}	25.1	Pd^{2+}	18.5†	Ho^{3+}	18.7	Sr^{2+}	8.6
U^{4+}	25.5	Y^{3+}	18.1	Dy^{3+}	18.3	Ba^{2+}	7.8
In^{3+}	24.9	Pb^{2+}	18.0	Tb^{3+}	17.9	Ra^{2+}	7.1
Th^{4+}	23.2*	TiO^{2+}	17.3	Gd^{3+}	17	Ag^+	7.3
Sc^{3+}	23.1	Cd^{2+}	16.5	Eu^{3+}	17.3		
Cr^{3+}	23	Zn^{2+}	16.5	Sm^{3+}	17.1		
Sn^{2+}	22.1	Co^{2+}	16.3	Nd^{3+}	16.6		
Hg^{2+}	21.8	Al^{3+}	16.1	Pr^{3+}	16.4		
Tl^{3+}	21.5	Fe^{2+}	14.3	Ce^{3+}	16.0		
$Ti(O_2)^{2+}$	20.4	Mn^{2+}	13.8	La^{3+}	15.5		
Ga^{3+}	20.3						

Group I: at pH 2 – 4 log K_{MY} is smaller by 14 – 19.
Group II: at pH 4 – 6 log K_{MY} is smaller by 5 – 9.
Group III: at pH 8 – 10 log K_{MY} is smaller by 0.5 – 2.3.
*See page 98.   †See page 222.

The following conclusions can be drawn from Table 6.1.

(i) The determination of cations of the first group is not disturbed by the presence of cations of the third group.   Cations from the second group also do not interfere at lower concentrations.
(ii) The determination of ions of the second group is not disturbed by ions of the third group.   All other cations are titrated simultaneously.
(iii) All the cations in the first two groups interfere in the determination of cations from the third group.
(iv) The determination of individual cations within a single group is disturbed by all the other members of that group unless masking reactions etc. are used.

6.1   COMPLEXOMETRY OF GROUP I CATIONS

6.1.1   Zirconium
$$\left(\log K_{ZrY} = 29.5\right)$$

Experience has shown that zirconium forms one of the most stable of

all EDTA complexes. The literature contains descriptions of EDTA
titrations of zirconium in strong acid medium and even a direct
titration of zirconium in boiling 2M hydrochloric acid (with Erio T
as indicator) has been described [1].

Consequently, the first data obtained on the stability constant of
the Zr-EDTA complex were rather surprising.   The value log K_{ZrY} =
19.4, given by Morgan and Justus [2], and the value 21.9 found by
Iwase [3] are disproportionately small, considering that they would
indicate that the Zr-EDTA complex would have to be practically
completely dissociated in solutions with pH below 1.   The value log
K_{ZrY} = 29.5 given by Kyrš and Caletka [4] and found from study of the
sorption of zirconium on silica gel from strongly acid solutions in
the presence of EDTA, is much more acceptable.   The low values found
earlier for the stability constant were probably due to hydrolysis or
polymerization of the zirconium solutions.

The behaviour of aqueous zirconium solutions presents an important
problem in complexometry.   It is not completely clear under what
conditions a solution of simple Zr^{4+} cations or of zirconyl ions
ZrO^{2+} can be expected the hydrolysis is always accompanied by polymerization.   A number of works have been devoted to this problem.
For example, Matijević et al. [5] assumed the formation of the dimer,
$Zr_2(OH)_2^{3+}$, or of the trimer, $Zr_3(OH)_9^{3+}$.   Kragten and Parczewski [6]
consider that tetrameric units play a dominant role and discuss the
effect on titration of small amounts of zirconium.

It is generally agreed that complications in the complexometric
determination of zirconium are produced solely by polymerization and
that only completely depolymerized solutions yield correct results in
the direct titration of zirconium.   The polymerization process is
relatively rapid and depends not only on the acidity but also on the
nature of the acid and on the zirconium content.   Polymerization
begins during dissolution of zirconium salts.   After preparation of
a 0.1M solution of zirconium sulphate in 0.1M sulphuric acid,
Pilkington and Wilson [7] found only 47.8% titratable zirconium.
This value decreased to 24.8% after 6 days and to 20.8% after 13 days.
Other zirconium salts behave similarly.

Direct determination

The introduction of Xylenol Orange, which has been found to be the
most suitable indicator for zirconium, made direct titrations very
attractive, as this determination is highly selective because of the
high acidity at which the solution can be titrated.   Thus considerations of polymerization and depolymerization became analytically
interesting.   We have found [8] that zirconium solutions can be
titrated directly with an EDTA solution in 0.3 - 0.6M nitric acid at
temperatures close to the boiling point, but heating for more than
one minute leads to low results.   After the change in the indicator
colour from red to yellow, the original colour returns and the
titration has to be carried out very carefully.   The maximal zirconium content is 35 mg.   Similar zirconium solutions in hydrochloric
or sulphuric acid are completely unsuitable for complexometric determination under these conditions.   An interesting phenomenon was
observed during this study.   After precipitation with sodium hydroxide from a triethanolamine medium, the zirconium hydroxide was
dissolved in nitric acid (0.3 - 0.6M) and titrated.   The results were
very low, but repetition of the titration of these solutions of

zirconium nitrate yielded ever increasing results.   Apparently,
depolymerization occurred.   Under these conditions, depolymerization
was complete after 15 - 20 hr.   Depolymerization of a zirconium
solution in 0.2M perchloric acid took 8 - 10 days.   Solutions in 0.2N
sulphuric acid remained almost completely polymerized.   The content
of "titratable" zirconium was only about 60%.   It is unfortunate
that these studies were not extended to higher acid concentrations.
Our work was repeated to a certain extent by Sinha and Gupta [9] who
came to the conclusion that a zirconium solution, whatever its degree
of polymerization, can be completely depolymerized by boiling for 5
min in 3N acid medium, preferably nitric acid.

To prevent repolymerization during the titration, the solution is
best titrated in a medium of 1M nitric acid at a temperature of 90°C,
or in 1M hydrochloric acid or 0.5N sulphuric acid.

Pilkington and Wilson [7] came to interesting conclusions in their
study of strongly polymerized solutions of zirconium solphate.
Depolymerization can be reliably achieved by boiling for 15 min in 5N
sulphuric acid medium.   Evaporation of the solution to white fumes
is not necessary.   Depolymerization is not quantitative in solutions
of more dilute sulphuric acid.   Completely depolymerized solutions
are clear and no precipitate is formed on addition of EDTA.   The
formation of a precipitate was considered by various authors to be an
unwelcome phenomenon, rendering the direct determination of zirconium
impossible.   These authors found that depolymerization by other
acids (e.g. hot 11M hydrochloric acid or fuming with perchloric acid)
is not successful.   The complexometric determination of zirconium
after depolymerization can be carried out in 1 - 2N sulphuric acid,
which reliably prevents repolymerization.   In principle, these
authors used a procedure involving titration of a solution of the
given acidity, with Xylenol Orange as indicator, up to a change in
the colour.   This transition is, however, premature.   After the
colour change a calculated amount of 7M ammonia should be added to
decrease the acidity and the solution should be slowly titrated to
the end-point.   The authors give a total of four modifications of
the determination of zirconium, for amounts from 0.1 to 600 mg, with
use of appropriately diluted EDTA solutions.

Klygin and Kolyada [10] also describe depolymerization of zirconium
by boiling in 1N sulphuric acid with addition of 1.5 g of sodium
sulphate;   they titrated 2 - 20 mg of zirconium directly with 0.02M
EDTA, (Xylenol Orange as indicator).

A comparison of their method with that of the earlier one is interesting.   The difference in the concentration of the sulphuric acid used
can be explained by the fact that the Russian authors used a solution
obtained by dissolving a Zr - Al or Zr - U alloy, and this is probably
only slightly polymerized, while Pilkington and Wilson [7] used a
solution artificially polymerized by heating for 30 min at pH 1.8.

An interesting work on the photometric microtitration of zirconium
was published by Kotrlý et al. [11].   They titrated 7 - 30 µg of
zirconium in cold 0.5M nitric acid, using Xylenol Orange as indicator.
The zirconium solution was prepared by precipitating the hydroxide
from triethanolamine medium and dissolving it.   They did not consider possible polymerization of this solution [8].   This is probably because they used very dilute solutions ($10^{-5} - 10^{-6}$M) in which
polymerization is negligible.   At very low zirconium concentrations,

the titration can be carried out at room temperature.   Their work
should be considered in conjunction with that of Kragten and
Parczewski [6].

Indirect determination

Older methods recommend back-titration of excess of EDTA with ferric
chloride, bismuth nitrate or thorium nitrate.   For the last two
titrations, Pyrocatechol Violet and especially Xylenol Orange are
useful.

Interferences

Direct titrations of zirconium in sufficiently acid solutions are
very selective.   Primarily bismuth and large concentrations of iron
interfere;   the latter must be reduced with ascorbic acid.   Phosphate,
fluoride, oxalate and high concentrations of tartrate and citrate
interfere.   In back-titrations, e.g. with bismuth nitrate solution
(pH 2 - 3), the presence of bismuth, thorium, scandium, iron, etc.
produces interference.

Separation methods

In practice, materials are frequently encountered which have unfavourable composition;   then the zirconium must be isolated or the interfering components removed.   The first method is preferable and
frequently more rapid.

Separation of zirconium by precipitation

A number of methods have been proposed for separation of zirconium.
It can be precipitated as barium hexafluorozirconate, $BaZrF_6$ [12],
which after collection must be freed from fluoride by fuming with
sulphuric acid.   Precipitation of zirconium with mandelic acid
[13,14] or other acids has been recommended.   Precipitation with
sodium hydroxide has also been suggested in some cases [15,16].
Endo and Takagi [17] precipitated zirconium together with magnesium
as the hydroxides, with simultaneous masking of other elements by
addition of EDTA (and hydrogen peroxide for masking titanium).   In
the presence of triethanolamine, zirconium can be separated with
sodium hydroxide from a number of elements, including iron.   Titanium,
thorium and hafnium are, however, also precipitated [18].

Separation of zirconium by extraction

Zirconium can be selectively extracted from nitric acid with a
chloroform solution of diantipyrinylmethane.   1,2-Dichloroethane can
be used instead of chloroform [19].

Separation of zirconium on ion-exchangers

The sorption of zirconium on various ion-exchangers was studied
systematically by Korkisch and Farag.   Zirconium forms a negatively
charged complex in 0.1N sulphuric acid, which is very strongly sorbed
on Amberlite IRA-400.   Elution with 4M hydrochloric acid allows
separation of zirconium from a number of elements and its complexometric determination [20].   In the presence of phosphate, zirconium
can also be sorbed on Amberlite IRA-400 or on Dowex 1 or Dowex 2 from
0.1N sulphuric acid containing sodium fluoride [21].   Zirconium is

sorbed on strongly basic ion-exchangers as its complex with ascorbic
acid [22].

Korkisch also employed Dowex 1 X8 for sorption of zirconium from
fluoride medium [23].

Masking of interferents

One of the most frequently used methods is masking of iron with
ascorbic acid.   Lukyanov and Knyazeva [24] used thiourea for masking
copper in the determination of zirconium in copper alloys.

Masking of zirconium

If this is necessary, ammonium fluoride can be used.   Attempts to
mask zirconium with hydrogen peroxide have not been very successful.

Determination of zirconium

In view of the polymerization problem, it is not always easy to
choose a complexometric method for determination of zirconium.
Choice of an indicator is much simpler, as Xylenol Orange is by far
the best choice for both direct and indirect determinations.
Titrations in sulphuric acid medium are probably the most reliable.
In his monograph, West gives the method published by Klygin and
Kolyada [10] as an example of the determination of zirconium.

Procedure.   Add 50 ml of 1N sulphuric acid and 1.5 g of sodium
sulphate to a solution containing 2 - 20 mg of Zr (in a volume less
than 50 ml).   Boil the solution for 10 min, dilute it with 80 ml of
water and again boil.   After cooling, add 1 ml of 0.05% Xylenol
Orange solution and titrate slowly with 0.02M EDTA until the solution
is yellow.

The indirect determination of zirconium can best be carried out by
back-titration with bismuth nitrate solution at pH 1 - 3 (Xylenol
Orange indicator).   Stepwise determination of zirconium and thorium
or zirconium and titanium [18] is relatively simple (see also p. 101).

REFERENCES

1.   V.G. Gorushina and E.V. Romanova, Zavod. Lab., 1957, 23, 781;
      Anal. Abstr., 1958, 5, 1484.
2.   L.O. Morgan and N.L. Justus, J. Am. Chem. Soc., 1956, 78, 38.
3.   A. Iwase, Nippon Kagaku Zasshi, 1959, 80, 1142;   Chem. Abstr.,
      1961, 55, 4250.
4.   M. Kyrš and R. Caletka, Talanta, 1963, 10, 1115.
5.   E. Matijević, K.G. Mathal and M. Kerker, J. Phys. Chem., 1962,
      66, 1799.
6.   J. Kragten and A. Parczewski, Talanta, 1981, 28, 149.
7.   E.S. Pilkington and W. Wilson, Anal. Chim. Acta, 1965, 33, 577.
8.   R. Přibil and V. Veselý, Z. Anal. Chem., 1964, 200, 332.
9.   B.C. Sinha and S. Das Gupta, Analyst, 1967, 92, 558.
10.   A.E. Klygin and N.S. Kolyada, Zavod. Lab., 1961, 27, 23; Anal.
      Abstr., 1961, 8, 3224.

11.   S. Kotrlý, V. Štaffa and M. Zimáková, Sb. Ved. Pr., Vys. Sk.
      Chemickotechnol., Pardubice, 1970, 23, 37;   Chem. Abstr., 1972,
      77, 42776c.
12.   G.W.C. Milner and G.A. Barnett, Anal. Chim. Acta, 1956, 14, 414.
13.   G.W.C. Milner and P.J. Pennah, Analyst, 1954, 79, 475.
14.   Sao-sin Su, Anal. Chem., 1965, 37, 1067.
15.   E. Lassner and R. Scharf, Chemist-Analyst, 1960, 49, 22.
16.   V.F. Lukyanov and E.M. Knyazeva, Zh. Analit. Khim., 1960, 15, 69;
      Anal. Abstr., 1960, 7, 4204.
17.   Y. Endo and H. Takagi, Japan Analyst, 1960, 9, 503;   Anal. Abstr.,
      1962, 9, 3147.
18.   R. Přibil and V. Veselý, Talanta, 1964, 11, 1197.
19.   V.P. Zhivopiskev, B.I. Petrov, Zh. Analit. Khim., 1968, 23, 1634;
      Anal. Abstr., 1970, 18, 3070.
20.   J. Korkisch and A. Farag, Z. Anal. Chem., 1959, 166, 81.
21.   J. Korkisch and A. Farag, Z. Anal. Chem., 1959, 166, 170.
22.   J. Korkisch and A. Farag, Z. Anal. Chem., 1959, 166, 181.
23.   J. Korkisch, Z. Anal. Chem., 1960, 176, 403.
24.   V.F. Lukyanov and E.M. Knyazeva, Zavod. Lab., 1962, 28, 804;
      Anal. Abstr., 1063, 10, 582.

6.1.2   Hafnium
        $\overline{(\log K_{HfY}} = 19.5)$

Hafnium behaves similarly to zirconium and can be determined under
similar conditions.   Nonetheless, analysis of a Zr + Hf mixture is
possible if the marked difference in the atomic weights of these
elements is exploited (Hf = 178.50, Zr = 91.22).   Ottendorfer [1]
recommends precipitating both elements with mandelic acid or p-chloro-
mandelic acid, igniting the product and weighing the mixture of ZrO_2 +
HfO_2.   After fusion of this mixture with sodium pyrosulphate,
both elements are determined by addition of excess of EDTA and back-
titration with thorium nitrate solution (indicator Xylenol Orange,
pH 2.5 - 3.5).   The contents of both elements can be found from the
weight of the oxides and the consumption of EDTA.   Kononenko and
Poluektov [2] precipitated hafnium and zirconium with phenylarsonic
acid and ignited to the oxides to 1000°C.   The sum of Zr + Hf was
then determined complexometrically.   Hoshino [3] separated small
amounts of hafnium (0.2 mg) from zirconium (500 mg) with Amberlite
IR-120.   He also recommended separation of hafnium from zirconium by
a very selective extraction procedure employing cyclohexanone, from
solutions containing hydrochloric acid, sulphate and ammonium thiocyanate [4].

REFERENCES

1.   L.J. Ottendorfer, Chemist-Analyst, 1959, 48, 97.
2.   L.I. Kononenko and N.S. Poluektov, Zavod. Lab., 1962, 28, 794;
      Anal. Abstr., 1963, 10, 584.
3.   Y. Hoshino, J. Chem. Soc. Japan Pure Chem. Sect., 1959, 80, 393;
      Anal. Abstr., 1960, 7, 443.
4.   Y. Hoshino, Japan Analyst, 1962, 11, 1032;   Anal. Abstr., 1964,
      11, 1675.

6.1.3   Thorium
         $(\log K_{ThY} = 23.2)$*

Thorium forms one of the most stable complexes ThY, which has been
demonstrated by electrophoresis to be uncharged [1]. In spite of
the high stability of its EDTA complex, thorium can be precipitated
quantitatively with ammonia and hydrogen peroxide, permitting various
separations from other elements [2].

Direct determination

Thorium is always determined in acid medium.   About 20 more or less
suitable indicators have been recommended.   Broad practical application has been made of Pyrocatechol Violet [3] and either Xylenol
Orange [4] or Methylthymol Blue [5].

At present, Xylenol Orange is generally used at pH 2.5 - 3.5.

Indirect determination

This can be carried out in various ways but has rather limited
application.   Back-titration with bismuth nitrate, with Xylenol
Orange as indicator permits some stepwise determinations, e.g. of
thorium and zirconium or thorium and titanium, but the pH must be
carefully controlled or the end-point is sluggish.

Interferences

All the members of the first complexometric group, i.e. all tervalent
and quadrivalent cations, interfere.   Of the bivalent ions, manganese and the alkaline-earth metals, uranyl ions and, under certain
conditions, tungstate and molybdate do not interfere.   The interference of the other bivalent ions has been the subject of several conflicting reports.   Suk et al. [3] determined small amounts of
thorium (3 - 4 mg), using Pyrocatechol Violet, in the presence of 1-g
amounts of the nitrates of lead, copper, zinc, cobalt, etc., whereas
other authors [6,7] found that these elements interfere.   Ringbom
[8] analysed the spectrometric curves for use of Alizarin S in the
titration of thorium and found errors in the determination of thorium
in the presence of as little as 10 mg of lead in 200 ml of solution.
The titration of thorium in the presence of lead with Xylenol Orange
as indicator also yielded high results.   Costa and Tavares [9] have
returned to this problem in a study of the titration of thorium, with
Methylthymol Blue as indicator.   They also found that these elements
at the 0.05 mmole level interfere during titration under the optimum
acidity conditions (pH 2.4 - 2.8).   The excellent results obtained by
Suk et al. [3] suggest that comparison of the spectrophotometric
characteristics of the titrations using Pyrocatechol Violet and
Xylenol Orange would be useful.

Separation methods

The analysis of materials from various sources should include preliminary isolation of thorium, or possibly separation of the interfering
components.

---

*Kragten (personal communication) prefers the value log K = 25.3
given by Bottari and Anderegg, Helv. Chim. Acta, 1967, 50, 2355.

Separation of thorium by precipitation

A great deal of attention has been paid to the isolation of thorium
as the iodate, $Th(IO_3)_4$. Kitahara and Hara [10] precipitated thorium
with sodium iodate in a sulphuric acid medium which also contained
oxalic acid. After isolation and decomposition of the iodate, they
determined thorium complexometrically, using Pyrocatechol Violet as
indicator. However, small amounts of thorium remain in the filtrate
and the precipitated thorium iodate always contains small amounts of
the rare earths. The authors used the method for determining thorium in monazite sands [11]. Busev et al. [12] precipitated thorium
with iodate in nitric acid medium. After dissolving the precipitate
they determined thorium indirectly by back-titration of excess EDTA
with 0.01M cupric nitrate, with PAN as indicator (pH 3 - 4). Titanium
and zirconium are coprecipitated but the determination is not disturbed by bismuth, iron, mercury, vanadium, nickel, tin and indium.
Chernikov et al. [13] recommend precipitation of thorium with meta-
periodate in the presence of oxalic acid. After mineralization of
the precipitate, they also determine thorium complexometrically.
The error in the determination of thorium in monazite sands (5 - 7%
Th) was less than 0.04% (absolute).

Among the other possible methods, a procedure in which thorium is
precipitated as the oxalate should be mentioned, but this method is
not suitable for low thorium contents [14]. Willard et al. [14]
recommended analysing uranium - thorium - titanium alloys by precipitating thorium (together with lanthanum) as the fluoride. Small amounts
of lanthanum act as a carrier. The further treatment of the precipitate before complexometric determination of the thorium is tedious
and has been replaced by more convenient methods (see below). Sometimes (in the presence of tungsten) it is suitable to precipitate
thorium as the hydroxide. Precipitation with sodium hydroxide from
triethanolamine medium for the separation of thorium from iron and
aluminium [15] is also suitable.

Separation of thorium by extraction

An older extraction method [16], based on the extraction of thorium
with mesityl oxide from 1M nitric acid containing 1.9 g of aluminium
nitrate trihydrate per ml, was used by Fritz and Ford [17] in the
complexometric determination of thorium. Their data indicate that
zirconium, vanadium, uranium and small amounts of aluminium are also
extracted with the thorium. Banks and Edwards [18] replaced the
aluminium nitrate with lithium nitrate in the extraction of thorium
with mesityl oxide. Zirconium, iron, uranium and tin are extracted
simultaneously. Phosphates interfere. Neither of these methods is
used much at the present time.

Chromatographic separation

Chromatography has been used in special cases, either for selective
separation of thorium or for removing interferents. Chernikov [19]
sorbed thorium on KU cation-exchanger before the actual determination
of thorium. Milner et al. [20] recommended analysing thorium - lead
alloys (0.025 - 9% Th) by separating lead in 2M hydrobromic acid
medium on Deacidite FF anion-exchanger and then determining thorium
in the eluate either complexometrically or photometrically. Bismuth
can be separated from thorium in a similar manner [21,22].

Masking of interferents

A wide range of masking agents is available for use when concentration ratios are unfavourable. For example, iron, thallium and copper can be reduced with ascorbic acid. Costa and Tavares [9] found thioglycollic acid useful for masking bismuth, antimony and nickel and used 2,3-dimercaptopropanol (BAL) for masking larger amounts of bismuth, copper, lead, mercury and antimony. These authors also used bis(2-hydroxyethyl)dithiocarbamate for masking bismuth, mercury, lead, antimony and nickel.

Direct titration

Several drops of 0.5% Xylenol Orange solution are added to an acid solution of thorium chloride or nitrate and the pH is adjusted to 2.5 - 3.5. The solution is slowly titrated with 0.02 - 0.05M EDTA to a sharp change in colour from red-purple to clear lemon yellow.

Note. If the indicator colour fades during the titration, the pH should be adjusted with several drops of saturated hexamine solution.

Indirect titration

A small excess of 0.05M EDTA is added to the thorium solution, the pH is adjusted to 2 - 3 and the solution is back-titrated with 0.05M bismuth nitrate, with Xylenol Orange as indicator, until the lemon yellow colour changes back to red-purple.

The nature of natural thorium and its practical use have led to a number of studies of the determination of thorium in the presence of a variety of elements - uranium, zirconium, titanium, the rare-earth elements and some heavy metals. This application is a good illustration of the development of complexometry from less suitable, tedious methods to simple, precise methods. Some examples will be given below.

Thorium and uranium

Uranyl ions form a very weak complex with EDTA. Uranium(VI) thus does not interfere in the determination of thorium, provided that its yellow colour at higher concentrations does not affect the colour change of the indicator.

Direct determination of small amounts of thorium (0.1%) in uranium was developed by Ryabinin and Afanas'ev [23], who used tributyl phosphate to extract the uranium. The method is suitable for uranium contents up to 1 g.

Procedure. A uranium sample containing at least 1 mg of thorium is dissolved in nitric acid and diluted to 50 - 100 ml with pH adjustment to 2 - 3. Several drops of Xylenol Orange indicator are added along with 8 g of potassium nitrate (as a salting-out agent) and 1 ml of tributyl phosphate per 0.25 g of uranium. The solution is mixed for 5 min with a magnetic stirrer and then titrated with sufficiently dilute EDTA solution until the red colour changes to yellow-green. The authors state that the relative error is less than $\pm5\%$.

Stepwise determination of thorium and uranium was developed by de Heer et al. [24]. The method is suitable for smaller uranium contents and has been applied to the sintered oxide mixture of ThO_2 + UO_2 (86% + 14%).

Procedure. A suitable amount of solution is diluted to give a 0.001M concentration of both elements and the pH adjusted (pH-meter) to 3.5.   After addition of Xylenol Orange, the thorium is titrated with 0.05M EDTA.   Then a measured volume of EDTA is added, about 100% excess relative to the uranium present, along with 10 drops of 35% formaldehyde solution and 100 mg of sodium dithionite to reduce the uranium.   As soon as the brown colour of the solution changes to green, the pH is adjusted to 3.5 - 4.0 and the excess of EDTA is titrated with thorium nitrate solution from yellow to red.

Note.   Sulphates do not interfere in the determination of thorium, provided that their concentration is not more than three times the molar concentration of the metal.   Otherwise, according to the authors, arsenazo is a far more suitable indicator.   The optimum pH for both titrations is 1.8.   Sulphates can be present in 3 - 8 times the molar concentration of the metal.

Thorium and zirconium

The sum of the two elements can be determined by back-titrating the excess EDTA at pH 2.5 with bismuth nitrate solution.   Then after adjustment of the pH to 1, ~1 g of sodium sulphate is added to the solution (to mask thorium) and the EDTA freed is back-titrated with bismuth nitrate [25 - 27].   Thorium and zirconium can also be determined by direct stepwise titrations - zirconium in 0.6M nitric acid and thorium at pH 2.5 - 3.5 [28].   The pH needs careful adjustment to avoid a sluggish end-point.

Thorium and the rare-earth elements (RE)

This combination has received a great deal of attention.   Onosova [29] determines the sum of Th + RE in alkaline medium by back-titration with a nickel solution, with murexide as indicator.   In a second portion, Th is determined in acid medium (Pyrocatechol Violet as indicator).

The rare-earth elements form much weaker complexes than thorium.   The stability of these complexes increases with the atomic weight of the element (from $\log K_{LaY}$ = 15.5 to $\log K_{LuY}$ = 19.8).   It can be expected that stepwise determination of thorium and the lighter rare earths will be readily feasible, provided that the difference in the stability constants is not less than 10^4.   It has been found that Xylenol Orange is not suitable for the stepwise determination of thorium (pH 2.5) and the rare-earth elements (pH 5 - 5.5) as the thorium complex reacts with the Xylenol Orange (XO) at this pH.   It is assumed that a ternary complex with the composition Th - EDTA - XO is formed.   Other indicators, such as Methylthymol Blue or arsenazo, react similarly.   Recently, a simple method has been found for preventing formation of this coloured ternary complex [30].   The Th-EDTA complex does not react at pH 5 - 6 with Xylenol Orange in the presence of acetylacetone (AcAc) and acetone.   This can be explained as due to formation of a further colourless ternary complex, Th - EDTA - AcAc.   Because of the simplicity of the method, the procedure will be given.

Procedure.   A solution containing less than 35 mg of thorium and 40 - 50 mg of a mixture of rare-earth metals is diluted to 250 - 300 ml and

the pH carefully adjusted to 2 $\pm$ 0.1 with sodium hydroxide or nitric
acid.   After addition of several drops of Xylenol Orange indicator,
the solution is slowly titrated with 0.02 - 0.05M EDTA from an intense
red colour to lemon yellow.   Then the solution is heated to 35 - 40°C
and 20 ml of a 1:1 v/v mixture of acetylacetone and acetone are added.
A saturated hexamine solution is then added until an intense red
colour forms, then several more drops are added.   The solution is
slowly titrated with EDTA to the formation of a light orange-yellow
colour.

Note.   The amount of thorium is limited to 35 - 40 mg.   At greater
thorium contents a larger amount of acetylacetone would have to be
added to suppress formation of the Th - EDTA - XO complex.   However,
higher acetylacetone concentrations interfere in the subsequent
titration of the rare-earth elements (formation of AcAc complexes).
Heating the solution to 40°C hastens the dissolution of the acetylacetone - acetone mixture and substantially improves the sharpness of
the indicator transition.

As expected, the determination of thorium and the light lanthanides
yields very good results.   With the higher members, beginning with
dysprosium, the results for thorium are higher as a result of partial
co-titration.   The colour transition for the titration of thorium is
also less sharp.   It is however, very probable that in the analysis
of monazite sands.(Th + RE) these errors will be negligible because
of the very low concentrations of higher members of the rare-earth
group.

Diethylenetriaminepenta-acetic acid (DTPA) has also been recommended
as a standard reagent for stepwise titration of thorium and the rare-
earth elements [31].   The Th-DTPA complex does not form a coloured
ternary complex with Xylenol Orange at pH 5 - 6, so thorium and the
lanthanides can readily be determined stepwise.

Procedure.   A solution containing 10 - 200 mg of thorium and 25 - 30
mg of lanthanides is diluted to 150 - 200 ml and the pH adjusted to
2.5 - 3.0 with ammonia or nitric acid.   Thorium is determined by
titration with 0.05M DTPA, with Xylenol Orange as indicator.   Then
excess of DTPA is added to bond all the lanthanides and the pH is
adjusted to 5 - 5.5 with solid hexamine (universal indicator paper)
and the excess of DTPA is titrated with 0.05M lead nitrate from
yellow to red-purple.

Note.   In the second determination an EDTA solution can be used
instead of DTPA and the excess is titrated in the same way with lead
nitrate.

Triethylenetetraminehexa-acetic acid (TTHA) behaves similarly.   We
have studied only thorium and the light rare earths (La, Ce, Pr, Nd),
however, primarily because of a lack of sufficiently pure rare-earth
compounds.   Mukherji [32] used TTHA for the stepwise determination of
thorium and the heavier rare-earth elements (Ho, Er, Tm, Yb and Lu)
with surprisingly good results - practically identical with the
expected values.   Mukherji explains these results by the large difference between the stability of the Th-TTHA complex and that of the
other TTHA complexes, the K_{MY} values of which were not known at that
time.   This assumption was confirmed by the later work of Harju and
Ringbom [33,34] who list unusually high values for the stability

constants of thorium, lanthanum, neodymium and erbium: $\log K_{MY}$ = 31.9 (Th). 22.3 (La), 22.8 (Nd), and 23.2 (Er).   The difference between the stability constants of the TTHA complexes of thorium and erbium is $10^{9.1}$ for example, and is much greater than that for the EDTA complexes (in this case $10^{4.5}$).

Gupta and Powell [35] also used DTPA for the determination of thorium and all the rare-earth elements, including lutetium, and consider our procedure very satisfactory.   They chose the following procedure for the direct titration of thorium and the rare-earth elements: thorium is determined at pH 2.5 - 3.0 with 0.05M DTPA;   after pH adjustment to 5.0 - 5.5, the sum of the rare-earth elements is determined with 0.5M N-hydroxyethylethylenediaminetriacetic acid (HEDTA).

Among older works, that of Chernikov should be mentioned [36];   he titrated thorium (Xylenol Orange as indicator, at pH 1.6) and then the rare-earth elements in hot solution at pH 4.5.   In both cases the pH appears to be somewhat lower than is usual.   Peshkova et al. [37] titrated thorium and the rare-earth elements spectrophotometric-ally, with arsenazo I as indicator (which must be added in an amount equal to that of the thorium).   The authors give only the example of determination of thorium and neodymium in ratios from 10:1 to 1:100.

REFERENCES

1.   K. Macek and R. Přibil, Coll. Czech. Chem. Commun., 1955, 20, 715.
2.   P. Schneider, Coll. Czech. Chem. Commun., 1956, 21, 1054.
3.   V. Suk, M. Malát and O. Ryba, Coll. Czech. Chem. Commun., 1954, 19, 679.
4.   J. Korbl and R. Přibil, Chemist-Analyst, 1956, 45, 102.
5.   J. Korbl and R. Přibil, Coll. Czech. Chem. Commun., 1958, 23, 873.
6.   J.S. Fritz, R.T. Oliver and D.J. Pietrzyk, Anal. Chem., 1958, 30, 1111.
7.   D.R. Rogers and W.B. Brown, Anal. Chem., 1963, 35, 1261.
8.   A. Ringbom, Comlexation in Analytical Chemistry, pp. 132 - 134. Interscience, New York, 1963.
9.   A.C. Spinola Costa and T.M. Tavares, Anal. Lett., 1970, 3, 549.
10.   S. Kitahara and S. Hara, Rept. Sci. Res. Inst. Tokyo, 1957, 33, 340;   Anal. Abstr., 1959, 6, 4348.
11.   S. Hara and S. Kitahara, Rept. Sci. Res. Inst. Tokyo, 1957, 33, 343;   Anal. Abstr., 1959, 6, 4348.
12.   A.I. Busev, V.M. Ivanov and V.G. Tiptsova, Zavod. Lab., 1962, 28, 799;   Anal. Abstr., 1963, 10, 585.
13.   Yu. A. Chernikov, V.F. Lukyanov and A.B. Kozlova, Zh. Analit. Khim., 1959, 14, 567;   Anal. Abstr., 1960, 7, 2680.
14.   H.H. Willard, A.W. Mosen and R.D. Gardner, Anal. Chem., 1958, 30, 1614.
15.   R. Přibil and V. Veselý, Chemist-Analyst, 1964, 53, 77.
16.   H. Levine and F.S. Grimaldi, U.S. At. Energ. Comm., AECD-3168, 1950.
17.   J.S. Fritz and J.J. Ford, Anal. Chem., 1953, 25, 1640.
18.   C.V. Banks and R.E. Edwards, Anal. Chem., 1953, 27, 947.
19.   Yu. A. Chernikov, V.F. Lukyanov and A.B. Kozlova, Zh. Analit. Khim., 1960, 15, 452;   Anal. Abstr., 1961, 8, 1468.

20. G.W.C. Milner, J.W. Edwards and A. Paddon, U.K. At. Energ. Res. Est., AERE C/R 2612, 1958.
21. G.W.C. Milner and J.H. Nunn, Anal. Chim. Acta, 1957, 17, 494.
22. G.W.C. Milner and J.H. Edwards, Anal. Chim. Acta, 1957, 17, 259.
23. A.I. Ryabinin and Yu. A. Afanas'ev, Zh. Analit. Khim., 1966, 21, 374; Anal. Abstr., 1967, 14, 7410.
24. B.H.J. de Heer, Th. van der Plas and M.E.A. Hermans, Anal. Chim. Acta, 1965, 32, 292.
25. R. Přibil and K. Burger, Mag. Kem. Foly., 1959, 65, 204.
26. R. Přibil and K. Burger, Talanta, 1960, 4, 8.
27. G.W. Milner and J.W. Edwards, Anal. Chim. Acta, 1959, 20, 31.
28. R. Přibil and V. Veselý, Talanta, 1964, 11, 1197.
29. S.P. Onosova, Zavod. Lab., 1962, 28, 271; Anal. Abstr., 1962, 9, 5103.
30. R. Přibil and V. Veselý, Talanta, 1972, 19, 1448.
31. R. Přibil and V. Veselý, Talanta, 1963, 10, 899.
32. A.K. Mukherji, Talanta, 1966, 13, 1183.
33. L. Harju and A. Ringbom, Anal. Chim. Acta, 1970, 49, 205.
34. L. Harju and A. Ringbom, Anal. Chim. Acta, 1970, 49, 221.
35. A.K. Gupta and J.E. Powell, Talanta, 1964, 11, 1339.
36. Yu.A. Chernikov, R.S. Tramm and K.S. Pevzner, Zavod. Lab., 1960, 26, 921; Anal. Abstr., 1961, 8, 1433.
37. V.M. Peshkova, M.I. Gromova and N.M. Alexandrova, Zh. Analit. Khim., 1962, 17, 218; Anal. Abstr., 1962, 9, 4632.

6.1.4   Scandium
$$\overline{(\log K_{ScY}} = 23.1)$$

Scandium forms EDTA complexes similar to those formed by thorium and all that was said of thorium is valid here.   The high stability of the Sc-EDTA complex has led to its determination in alkaline media being of little importance.

Direct determination

The determination of scandium, with Xylenol Orange as indicator was described by Körbl and Přibil [1].   Kinnunen and Wennerstrand [2] also determined scandium under the same acidity conditions (pH 3 - 4, XO indicator).   Methylthymol Blue has also been found to be very useful for the determination of scandium [3].

Indirect determination

Cheng and Williams [4] determined 3 - 10 mg of scandium indirectly by back-titration of excess of EDTA with a copper solution at pH 2.5, using PAN as indicator.   The colour transition of PAN in direct titration of scandium is insufficiently sharp.

Interferences

The situation here is again roughly the same as that for thorium. All tervalent and quadrivalent cations are co-titrated.   Under the given acidity conditions, higher concentrations of bivalent metals also interfere.   However, their interference can be considerably decreased by titrating scandium at a low pH value wherever possible. Fritz and Pietrzyk [5] recommend spectrophotometric titration of scandium at pH 2.7 (745 nm) in the presence of traces of copper.

The Sc-EDTA complex, which does not absorb light at this wavelength, is formed first. At the equivalence point the strongly absorbing Cu-EDTA complex is formed. The authors state that most bivalent metals do not interfere in this determination.

Recommended procedures

With Xylenol Orange [1]. A few drops of the indicator are added to an acid solution of scandium(III) nitrate and the pH is adjusted to 3 - 4 with solid hexamine. The solution is titrated with 0.05 - 0.01M EDTA solution, from red-purple to lemon yellow.

With Methylthymol Blue [3]. One ml of 1M nitric acid is added to a solution containing up to 60 mg of scandium and pyridine is added dropwise to adjust the pH to 5 - 6 (universal indicator paper). Solid indicator is added (1:100 mixture with KNO_3) and the solution is titrated with 0.05M EDTA from blue through grey to clear yellow.

Determination of thorium and scandium [6]

Both elements form EDTA complexes of the same stability and are always titrated together. As no suitable agent is known for masking one of these elements in the presence of the other, their individual determination would seem impossible. It has, however, been found that scandium can readily be freed from its Sc-TTHA complex in the presence of phosphate by use of a zinc solution.

Procedure. The pH of a solution containing up to 100 mg of Th and 50 mg of Sc is adjusted to 2.5 - 4 and the solution is titrated at 50°C with 0.05M DCTA, with Xylenol Orange as indicator. This gives the sum of Th + Sc.

An excess of TTHA solution (0.05M) is added to a second portion of the original solution and the pH is adjusted to 5 - 5.5 with hexamine. After addition of 10 - 20 ml of 1M disodium hydrogen phosphate (adjusted to pH 4 - 5) the titration is carried out at room temperature with 0.05M zinc nitrate, with Xylenol Orange as indicator, from yellow to an intense red colour. The TTHA consumption is equivalent to the thorium content; the scandium content is found by difference. It should be noted that back-titration of free TTHA with zinc proceeds according to the equation

$$TTHA + 2Zn^{2+} = Zn_2TTHA$$

Thus the amount of free TTHA corresponds to half the consumption of 0.05M zinc solution.

Note. The zinc titration must be done slowly, especially close to the equivalence point, as the last traces of scandium are freed very slowly from its TTHA. The red colour is stable for 5 min after the end of the titration and then slowly fades (freeing of traces of thorium and precipitation by phosphate). During the back-titration with zinc, white scandium phosphate is precipitated, but does not affect the colour transition of the indicator.

The use of a combination of EDTA, DCTA and TTHA yields further possibilities for analysis of mixtures of thorium, scandium and rare-earth elements or even some bivalent metals. Some further examples are given in Section 6.2.11.

REFERENCES

1.   J. Körbl and R. Přibil, Chemist-Analyst, 1956, 45, 102.
2.   J. Kinnunen and B. Wennerstrand, Chemist-Analyst, 1957, 46, 92.
3.   J. Körbl and R. Přibil, Coll. Czech. Chem. Commun., 1958, 23, 873.
4.   K.L. Cheng and T.R. Williams, Chemist-Analyst, 1955, 44, 96.
5.   J.S. Fritz and D.J. Pietrzyk, Anal. Chem., 1959, 31, 1357.
6.   R. Přibil and V. Veselý, Talanta, 1967, 14, 266.

6.1.5   Bismuth
        $\log K_{BiY}$ = 22.8 - 27.9)

Bismuth also forms a very stable complex with EDTA, which spectrophotometric measurements have shown to have very high stability [1]
at pH 2 - 9.   Kotrlý and Vřešťál [2] studied the spectrophotometric
titration of bismuth and gave a value of $\log K_{BiY}$ = 22.8 for the
conditional stability constant of the Bi-EDTA complex at pH 1.5.
Miklós and Szegedi [3] found a value of $10^{27.94}$ from polarographic
measurements.   The stability constant of the Bi-EDTA complex is
greater than that given for the Bi-DCTA complex (log K = 23.9 - 24.5)
by Selmer-Olsen [4].   This is rather surprising, as the stability
constants of DCTA complexes are generally larger than those of the
corresponding EDTA complexes by about two orders of magnitude.

It is assumed that at higher pH values the $Bi(OH)Y^{2-}$ complex is formed, which is much less stable, as reflected in the reactions of Bi-
EDTA complexes in ammoniacal media - bismuth hydroxide is quantitatively precipitated by addition of calcium ions.   This was once used
for separation of bismuth from lead [5].

Direct determination

Because of the high stability of the Bi-EDTA complex and the ready
hydrolysis of bismuth solutions, only titrations in acid solutions at
pH 1 - 3 are of practical importance.   Up to 20 indicators have been
recommended for the titration, some of which are not readily available.   The most suitable are Pyrocatechol Violet, Xylenol Orange and
Methylthymol Blue.   The titration proceeds smoothly at room temperature.   Gattow and Schott [6] tested the precision of the titration
of bismuth, using a total of 11 indicators, and concluded that use of
Xylenol Orange as indicator yields slightly higher results that use
of Pyrocatechol Violet (PV) or Methylthymol Blue.   Xylenol Orange is,
however, a more sensitive indicator than PV and can be used at lower
pH values.   The difference in the sensitivity of the two indicators
can be demonstrated by a simple experiment: 1 - 3 drops of 0.05M
bismuth nitrate and 1 - 3 drops of 0.2% Xylenol Orange solution are
added to 100 ml of 0.01M nitric acid.   An intense red colour is
formed, which does not disappear on addition of a small amount of
tiron.   If this experiment is repeated with PV, then the blue colour
of the Bi-PV complex disappears immediately on addition of tiron.
Formation of the complex of bismuth with tiron results in a decrease
in the concentration of free bismuth ions to a level where there is

no reaction with PV; whereas this concentration is still sufficient
to yield a colour reaction with Xylenol Orange.

Adjustment of the solution pH is very important in the direct determination of bismuth.   Gattow and Schott studied this factor [7].
Bismuth solutions have a marked tendency to hydrolyse, yielding poly-
cations (polynuclear hydroxo-complexes), which are not dissolved
immediately on addition of acid.   Bismuth can be lost through local
increases in the pH during the titration.   The authors give tables
demonstrating the unsuitability of neutralizing bismuth solutions
with ammonia or sodium hydroxide.   They recommend pH adjustment with
1M sodium hydrogen carbonate or 1M sodium acetate.

Indirect determination

Bismuth can be determined indirectly in acid medium, e.g. by back-
titration of EDTA with iron(III) (with salicylic or sulphosalicylic
acid as indicator), or thorium nitrate solution at pH 2.5 (Xylenol
Orange), or zinc nitrate solution at pH 5 - 5.5 (Xylenol Orange).   In
the last case, however, elements in complexometric group II (p. 92)
are co-titrated and must be masked.

Bismuth nitrate solution can also be used for back-titration of EDTA.
This subject has been studied in detail by Flaschka and Sadek [8],
who developed an indirect determination of bismuth, thorium, iron,
indium and gallium at pH 2 - 3 by back-titration with bismuth solution,
with Pyrocatechol Violet as indicator.   Other cations were partially
or completely displaced from their complexes by bismuth and thus
could not be determined in this way.   Even zirconium cannot be determined, as it forms a stronger complex with PV than with EDTA.   In
the back-titration with bismuth, difficulties were encountered from
the presence of sulphate and chloride.   We have found that sulphate
and chloride do not interfere in sufficiently acid solutions, as used
in the successive determination of thorium and zirconium (Section
6.1.3).

Interferences

All cations of the complexometric group I (p. 92) interfere in the
determination of bismuth.   Of the bivalent cations, the alkaline-
earth metal cations, manganese, cadmium, zinc, lead, etc. do not
interfere, nor does aluminium.   Malát et al. [9] titrated fractions
of a milligram of bismuth (0.002M EDTA solution) in the presence of
5 g of lead nitrate, using Pyrocatechol Violet as indicator.   They
stated that large amounts of nitrates of various other metals do not
interfere.   Kékedy and Balogh [10] obtained similar results in the
titration of bismuth at pH 2.3, using gallein (4,5-dihydroxyfluor-
escein) in the presence of cadmium and lead in ratios of up to 10^4 to
bismuth and of iron (after reduction with ascorbic acid) up to a
ratio of 400 to bismuth.   Antimony(III), tin(II) and tin(IV) interfered by blocking the indicator.   Halides also interfered.   Gattow
and Gotthardt [11] obtained somewhat different results.   They chose
the titration of bismuth at pH 1.6, with PAR as indicator (suitable
for 0.4 - 600 µg of Bi per ml) as a control method.   They found that
the alkaline-earth metals, beryllium, thallium(I), germanium, zinc,
cadmium, scandium, yttrium, the rare-earth elements, and manganese up
to a ratio of 1000 to bismuth did not interfere, nor did nitrate and
borate.   All the other elements, including lead, arsenic, antimony,
tin, gold, niobium, tantalum, etc. interfere to a greater or lesser

degree.   In the presence of non-interfering elements, the error of
the determination was within the limits of the statistical error
$\pm$0.04%.

The determination of bismuth in the presence of lead or of cadmium
and indium is still of analytical importance.   The bismuth-lead
combination has been studied in a number of works, but the results do
not agree well.   For example, Busev [12] determined bismuth at pH
1 - 1.4, using PAN as indicator.   In the titration of 2.09 mg of
bismuth in the presence of 360 mg of lead the error was +5.7%, but in
presence of ~ 1 g of lead was only +3.8%.   On the other hand, 1.25 g
of calcium produced an error of +9%.   The considerable differences
in the results of other authors [9 - 11] can be explained by experimental errors, different indicator sensitivity and different ionic
strengths of the solutions titrated, which influences the colour
transitions of the indicators.

Various electrometric methods appear to be much more reliable for the
bismuth-lead combination.   One rather forgotten method [13] deserves
mention: the amperometric titration of bismuth, which can be carried
out even in saturated or supersaturated lead nitrate solutions [the
solubility of $Pb(NO_3)_2$ at 20°C is 56.6 g in 100 g of water].   The
determination is carried out at pH 2 at a potential between -180 and
-200 mV vs. a saturated calomel electrode.   Lead nitrate is used as
an electrolyte.   In this way, 0.004 - 0.04% Bi can be determined in
the presence of 5 - 50 g of lead [13].   Kraft and Dosch [14] published a very interesting work on the potentiometric determination of
traces of bismuth in lead.

So far, only extreme concentration ratios of bismuth and other metals
have been considered.   The marked difference in the stabilities of
the complexes of bismuth and lead and other metals (cadmium, zinc,
etc.) permits stepwise determination in alloys, where the ratios of
the components is much more favourable.   Xylenol Orange is again
found suitable for the stepwise determination of bismuth and lead or
cadmium or zinc.   Bismuth is determined by direct titration at pH
2 - 2.5, then the pH is raised to 5 - 5.5, preferably with hexamine,
and the other element is determined [15].   Further variations of
this titration will be discussed in the section on lead alloys
(pp. 280 - 283).

Masking of interferents

Interferences (especially from elements of group I) can sometimes be
dealt with by masking.   The reduction of iron(III) to iron(II) with
ascorbic acid is simple [16].   Thallium(III) can similarly be
reduced, to thallium(I).   Mercury can be reduced to the metal with
ascorbic acid or formic acid [16], or can be masked with potassium
thiocyanate [17].   Tin can be masked with sodium fluoride [18], and
thorium with sodium sulphate [19].

Separation of bismuth

Kinnunen and Wennerstrand [20] recommend that small amounts of bismuth,
especially in the presence of high concentrations of copper and lead,
should be determined by precipitation with sodium diethyldithiocarbamate (Cupral) and extraction of the precipitate with chloroform
from an ammoniacal solution containing tartaric acid and potassium
cyanide.   The extract is decomposed with nitric acid and bismuth is

determined by titration with EDTA, with Pyrocatechol Violet as
indicator.   Other authors [11,14] also used Cupral for separation of
bismuth from lead.

Ripan et al. [21] separated small amounts of bismuth from lead (5 -
12 g) by precipitation with cupferron;   after decomposition of the
precipitate with nitric acid, bismuth was titrated at pH 2 with
Pyrocatechol Violet as indicator.   In this way, bismuth can be
determined in presence of 10^4 times as much lead, with an error of
only $\pm$ 0.2%.

Bismuth can be separated from a number of elements as the tribromide
by evaporation of a hydrobromic acid solution containing 15% bromine.
The evaporation is done in an air-bath at 350°C.   This method has
been used in the analysis of Bi-U alloys [22] and of alloys containing bismuth, iranium, praseodymium or neodymium [23].

Busev and Shvedova [24] separated bismuth from indium and other
metals by precipitation of tetraiodobismuthate with o-tolylthiourea.
After filtration, the precipitate is dissolved in dimethylformamide
and the bismuth titrated with 0.01M EDTA from a yellow colour to
colourless.   The method can be used for 8 - 67 mg of Bi in the
presence of 100 mg of indium.   The determination is free from interference by other metals: Al (400 mg), Ni (200 mg), Ti (100 mg), Ga
(100 mg), and Fe (40 mg).   In the analysis of binary Bi-In alloys,
the filtrate left after separation of bismuth can be used for the
determination of indium: excess of EDTA is titrated with a zinc
solution, Xylenol Orange being used as indicator.

Chromatographic separation

In special cases bismuth is separated on ion-exchangers.   Milner and
Edwards [23] separated bismuth from praseodymium or neodymium with
Deacidite FF resin.   Taketatsu [25] used Amberlite IR-112 for separation of bismuth from lead, zinc and cadmium.   In EDTA medium at pH
1.3 - 1.5, the EDTA complexes of lead, cadmium and zinc do not form,
so the metal ions are retained on the resin, while the Bi-EDTA
complex is stable even at pH 1 and passes through the column.   Zinc
and cadmium can be eluted stepwise with 3M hydrochloric acid and lead
with 2% ammonium acetate solution.   A number of other works also
disucss the use of ion-exchangers for separation of bismuth from lead
or other elements [14,26].

Masking of bismuth

In acid medium bismuth can be masked by 2,3-dimercaptopropanol (BAL),
Unithiol, or thioglycollic acid.   Barcza [27] recommends masking
bismuth (up to 30 mg) with triethanolamine at pH 11 for the determination of e.g. calcium, which is then determined directly by titration
with EDTA, with Methylthymol Blue or murexide as indicator.   In the
determination of iron, Flaschka and Garrett [28] masked bismuth with
large amounts of ammonium chloride (20 - 30 g in 150 ml) and titrated
iron spectrophotometrically at pH 2 - 3, using sulphosalicylic acid as
indicator.   Iron can be determined in this way in the presence of
2000 times as much bismuth.

Recommended procedures

With Xylenol Orange.   The pH of a bismuth nitrate solution (up to

30 mg of Bi) is adjusted to 1 - 2.5, with indicator paper for control.
Several drops of indicator are added and the solution is titrated at
room temperature with 0.01 - 0.05M EDTA from a red colour to lemon
yellow.

Note. BiOCl will probably form if a high chloride concentration is
present, and is dissolved only slowly. The addition of several
milligrams of tartaric acid, which rapidly dissolves any precipitate,
is recommended.

With Pyrocatechol Violet. Several drops of indicator are added to
an acid solution containing up to 50 mg of Bi in 100 ml and the
colour formed is observed. The solution acidity is correct if the
colour is deep blue. A purple colour indicates that the solution is
too acid and sodium hydrogen carbonate should be added (for details
see [7]) until the colour turns blue (pH 2.5). Then the solution is
titrated with 0.002 - 0.05M EDTA from blue to clear yellow.

Stepwise determinations

Bismuth and lead. Bismuth is determined directly by titration at
pH 2.5, with Xylenol Orange as indicator. After completion of the
titration, the acidity is decreased by adding small portions of solid
hexamine until a red-purple colour appears and the solution is
titrated immediately with EDTA solution to a sharp colour change from
red-purple to lemon yellow. The method is useful for analyses of
fusible alloys, etc.

Bismuth and cadmium (or zinc). This determination is carried out in
the same way as for bismuth and lead. The solution should be well
buffered after titration of bismuth.

Bismuth and the rare-earth elements [29]. Bismuth is determined by
EDTA titration at pH 1.1, with Xylenol Orange as indicator. After
pH adjustment to 5.1 - 5.3 with hexamine, the titration is continued
till the colour changes from red to clear yellow.

REFERENCES

1.   P.W. West and H. Coll, Anal. Chem., 1955, 27, 1221.
2.   S. Kotrlý and J. Vřešťál, Coll. Czech. Chem. Commun., 1960, 25,
     1148.
3.   I. Miklós and R. Szegedi, Acta Chim. Acad. Sci. Hung., 1961, 26,
     365
4.   A.R. Selmer-Olsen, Acta Chem. Scand., 1961, 15, 2052.
5.   R. Přibil and L. Fiala, Coll. Czech. Chem. Commun., 1953, 18,
     309.
6.   G. Gattow and D. Schott, Z. Anal. Chem., 1962, 188, 10.
7.   G. Gattow and D. Schott, Z. Anal. Chem., 1962, 188, 81.
8.   H. Flaschka and F. Sadek, Z. Anal. Chem., 1956, 149, 345.
9.   M. Malát, V. Suk and O. Kyba, Coll. Czech. Chem. Commun., 1954,
     19, 258.
10. L. Kékedy and G. Balogh, Stud. Univ. Babes-Bolyai, Ser. Chem.,
     1962, 1, 109; Anal. Abstr., 1964, 11, 5437.
11. G. Gattow and B. Gotthardt, Z. Anal. Chem., 1964, 206, 331.
12. A.I. Busev, Zh. Analit. Khim., 1957, 12, 386; Anal. Abstr.,
     1958, 5, 462.

13. R. Přibil and B. Matyska, Coll. Czech. Chem. Commun., 1951, 16, 139.
14. O. Kraft and H. Dosch, Erzmetall, 1972, 25, 26; Anal Abstr., 1973, 24, 2141.
15. R. Přibil, Komplexometrie I, SNTL, Prague, 1957.
16. J. Cífka, M. Malát and V. Suk, Coll. Czech. Chem. Commun., 1956, 21, 412.
17. L. Barcza and E. Köros, Chemist-Analyst, 1959, 48, 94.
18. E.A. Shteiman, Z.G. Dobrynina and E.A. Mordovskaya, Zavod. Lab., 1964, 30, 1200; Anal. Abstr., 1966, 13, 531.
19. A. Bacon and G.W.C. Milner, U.K. A.E.R.E. Rept., C/R 1992, 1956.
20. J. Kinnunen and B. Wennerstrand, Chemist-Analyst, 1954, 45, 109.
21. R. Ripan, N. Pascu and I. Todorut, Stud. Cercet. Chim. Cluj, 1962, 1, 49; Anal. Abstr., 1964, 11, 2552.
22. G.W.C. Milner and G.A. Barnett, Anal. Chim. Acta, 1957, 17, 220.
23. G.W.C. Milner and J.W. Edwards, Anal. Chim. Acta, 1958, 18, 513.
24. A.I. Busev and N.V. Shvedova, Zavod, Lab., 1968, 34, 140; Anal. Abstr., 1969, 16, 2431.
25. T. Taketatsu, J. Chem. Soc. Japan, Pure Chem. Sect., 1957, 78, 151; Anal. Abstr., 1957, 4, 3530.
26. E. Wohlmann, Z. Angew. Geol., 1961, 7, 242.
27. L. Barcza, Acta Pharm. Hung., 1962, 32, 156; Anal. Abstr., 1964, 11, 1219.
28. H. Flaschka and J. Garrett, Talanta, 1964, 11, 1651.
29. V.B. Chernogorenko and L.L. Vereikina, Ukr. Khim. Zh., 1970, 36, 1296; Chem. Abstr., 1971, 74, 150821.

6.1.6   Iron
$(\log K_{FeY} = 25.1)$

Iron(III) forms a yellow complex FeY^- with EDTA in acid media; this complex is sufficiently stable for titrations to be done at pH 1.5. As the pH is increased, the hydroxo-complexes $Fe(OH)Y^{2-}$ and $Fe(OH)_2Y^{3-}$ are formed; the formation of these complexes was studied in detail by Gustafson and Martell [1], who also demonstrated the presence of the binuclear complex $[FeY(OH)_2FeY]^{4-}$. The hydroxo-complexes are quite stable, so ferric hydroxide is not precipitated on addition of ammonia.   However, calcium and magnesium ions slowly release iron from the complex in ammoniacal medium.   Ferric hydroxide is quantitatively precipitated from the EDTA complex by sodium hydroxide.   A solution of the Fe-EDTA complex is temporarily coloured purple by hydrogen sulphide in alkaline medium [2].   The reaction of the Fe-EDTA complex with hydrogen peroxide in ammoniacal medium is also interesting.   The purple peroxo-complex is formed and can be decomposed by boiling.   The reaction proceeds in the presence of fluoride and has been used for the colorimetric determination of traces of iron in metallic zirconium [3].   Ringbom et al. [4] studied this reaction in detail, and assumed the formation of the $Fe(O_2)Y^{3-}$ complex, in which the hydroxo group is replaced by the peroxo group, O_2^{2-}.   In the presence of precipitated ferric hydroxide and hydrogen peroxide, ammoniacal EDTA solutions are also purple.   Kochanny and Timnik [5] demonstrated the formation of ferrate ions $[Fe(VI), FeO_4^{2-}]$.

The complexometric determination of iron cannot, in itself, be considered as a special contribution to analytical chemistry, since bivalent and tervalent iron can be determined by a number of suffic-

iently sensitive titrimetric methods.   The importance of titration
of iron with EDTA appears in practical analyses, where combinations
such as Fe-Al, Fe-Mn or Fe-Al-Mn-Ca-Mg, can readily be determined
complexometrically, and the classical separation with ammonia or
sodium hydroxide frequently becomes unnecessary.

Direct determination

The formation of the Fe-EDTA complex can readily be studied by
potentiometric titration using a platinum electrode [6].   At pH 1 - 3
the potential stabilizes slowly; potential stabilization at pH 4 - 5
in acetate buffer is almost instantaneous.   Titrations using indicators are also slow at low pH but fast at the higher pH.   Iron is
generally determined at pH 1 - 3, preferably with gently heating to up
to 50°C, though this is associated with the danger of hydrolysis of
iron.

A number of indicators have been recommended.   In the early stages
of complexometry, potassium thiocyanate [7], tiron [8], and particularly sulphosalicylic acid [9] were recommended.   Similarly, salicylic acid or p-aminosalicylic acid can be used.   Of the other indicators, Chrome Azurol S has quite a sharp colour transition [10].

The colour transition in the direct titration of iron is much less
sharp than that for other titrations using metallochromic indicators,
especially at high iron concentrations, where the yellow colour of
the Fe-EDTA complex is also an important factor.

DCTA forms a very stable complex with iron(III)   (log K_{FeY} = 29.3).
It has been recommended for the successive determination of iron
(with salicylic acid as indicator) and manganese (Erio T as indicator)
[11].   A similar determination with EDTA is not possible, as the Fe-
EDTA complex reacts with the Erio T.   Kőrös [12] also recommends
DCTA for the stepwise determination of iron (sulphosalicylic acid,
pH 2) and manganese (Methylthymol Blue, pH 6 - 6.5).

"Complexone type" indicators form strong complexes with iron, and
cannot be used in the direct titration of iron.   The reaction of
Xylenol Orange is interesting [13]:   it forms a blue complex with
iron, which is inert towards EDTA.   However, tervalent iron can be
titrated at pH 1.5 in the presence of large amounts of ferrous
sulphate (up to 2 g) with a sharp colour transition from blue to
yellow.   The following explanation has been proposed for this
reaction: in the vicinity of the equivalence point small amounts of
the Fe(II)-EDTA complex are formed, which immediately react with the
iron(III)-XO complex according to the equation

$$Fe(III)XO + Fe(II)Y = Fe(III)Y + Fe(II)XO$$

The Fe(II)XO complex immediately decomposes and releases the indicator, the yellow colour of which indicates the end of the titration.

The ease of oxidation of ferrous sulphate makes it necessary to carry
out the titration in a carbon dioxide atmosphere.   The iron(II)
content in the solution is sufficient for several titrations of
tervalent iron;   the sample solutions are simply pipetted into the
ferrous sulphate solution.   Preliminary experiments have demonstrated that various substances that quantitatively oxidize iron(II) can

be determined in this way (MnO_4^-, VO_3^-). However, attempts to determine small amounts of dichromate in this manner are unsuccessful, as the Cr^{3+} formed blocks the Xylenol Orange [14].

Indirect determination

Only determinations in acid medium have any practical importance. Back-titration of EDTA with thorium nitrate at pH 2.5 - 3.5, bismuth nitrate at pH 1 - 2, and lead or zinc nitrate at pH 5 - 5.5 has been found useful.   Xylenol Orange is the best indicator in all cases. Back-titration is useful when several elements can be determined in a single solution, e.g. determination of the sum Fe + Al, followed by masking of the aluminium and titration of the freed EDTA to obtain the amount of aluminium.

Interferences

The determination of iron suffers from interferences from all the cations of group I and from a number of group II cations, such as copper, nickel, etc.   The Fe-Al combination has been studied extensively.   The difference in stability of the Fe-EDTA and Al-EDTA complexes and the slow formation of the Al-EDTA complex permit successive titration, with various indicators. Some of these methods are discussed in detail in the section on aluminium (6.2.7).

Masking of Interferents

As fluoride does not disturb the determination of tervalent iron, it can be used to mask aluminium, thorium, zirconium etc.   The masking of bismuth has been mentioned in Section 6.1.5. Copper in high concentrations also interferes in the determination of iron, but can be reliably masked with thiourea, but the reaction of thiourea with iron must be prevented by addition of fluoride [15].   Frequently, no suitable masking agent is available for interferents and indirect determination must be used.   For example, the Fe + Ni combination can be analysed by finding the sum of the two by back-titration of excess of EDTA with lead, using Xylenol Orange as indicator at pH 5 - 5.5, and then determining nickel (murexide as indicator) after masking iron with triethanolamine [16].

Masking of iron

The masking of iron has presented a substantial problem in complexometry, especially for determinations in alkaline media.   Most useful and most frequently used in practice is masking with triethanolamine in either ammoniacal medium or sodium hydroxide or potassium hydroxide medium [11,16-18].   Flaschka and Püschel [19] recommend that, in special cases, iron can first be reduced with ascorbic acid and then bonded in a complex with potassium cyanide, which also masks other cations. In acid medium, iron(III) is generally reduced to iron(II) for the determination of all metals which form sufficiently strong complexes at pH 1 - 3.   Iron can also be masked with thioglycollic acid or pyrophosphate, etc.

Separation of iron

Sometimes, especially when the sample composition is unfavourable, with large iron contents, preliminary separation of iron is necessary. Frequently, simultaneous precipitation of iron and aluminium with

hexamine or ammonia is used. Several methods are based on the
chromatographic separation of iron as the chloro-EDTA mixed-ligand
complex or as the EDTA complex. Depending on the nature of the
sample, extraction methods can be used, such as extraction with amyl
acetate, acetylacetone, tributyl phosphate, long-chain amines, etc.

Recommended procedures

Direct determination. The pH of an acid solution containing up to
70 mg of iron is adjusted to 2 - 3 with 10% sodium acetate solution
and 10 ml of 2% sulphosalicylic acid solution are added. The
solution is heated to 40 - 50°C and titrated with 0.05M EDTA to a
change from red-purple to clear yellow.

Indirect determination. A sufficient amount of 0.05M EDTA is added
to a solution of iron(III) and the pH is adjusted to 2 - 3 with nitric
acid or solid hexamine (pH paper). Several drops of 0.2% Xylenol
Orange solution are added and the solution is titrated with thorium
nitrate or bismuth nitrate solution to a colour change from yellow to
red.

In an alternative procedure the pH is adjusted to 5 - 5.5 and free
EDTA is back-titrated with 0.05M lead or zinc nitrate, with Xylenol
Orange as indicator. The colour transition with lead is from clear
yellow to purple-red, and to red with zinc.

Successive determinations

Iron and aluminium. The difference between the stability constants
of the Fe-EDTA and Al-EDTA complexes (about 8 - 9 orders of magnitude)
led at the beginning of complexometry to attempts to determine these
two metals by stepwise titration. However, in spite of the large
number of papers published on it, this problem has not been satisfactorily solved.

The first publications described primarily the determination of iron
in the presence of aluminium by direct titration using salicylic acid
[20], sulphosalicylic acid [21] or potassium thiocyanate [22] as
indicator. Wehber used various oxidation-reduction indicators [23].
Theis [24] used Chrome Azurol S as indicator for the successive
determination of iron and aluminium; it gives a sufficiently sharp
transition in the titration of iron, but not for the titration of
aluminium.

The results published by different authors are frequently contradictory. Cheng et al. [25] suggest that aluminium in suitable concentrations does not interfere in the determination of iron. On the
other hand, Sweetser and Bricker [26] found an interfering effect of
aluminium even in the titration of iron in quite acid solutions (pH
1.7 - 2.3). Davis and Jacobsen [27] came to the same conclusion and
recommended spectrophotometric titration of iron at pH 1.0, using
sulphosalicylic acid as indicator. Their results were satisfactory
for the determination of iron in the presence of up to 100 mg of
aluminium.

Other works on this subject will not be discussed here, as simple
mixtures of iron and aluminium can be analysed by better methods,
which will be discussed below.

Indirect determination of iron and aluminium.   It is known from older works that sulphosalicylic acid is useful for masking small
concentrations of aluminium.   Salicylic and p-aminosalicylic acid
behave similarly.   However, attempts to mask large amounts of
aluminium have not been successful.   Detailed experiments have
demonstrated that a sufficiently strong complex of salicylate with
aluminium is formed in alkaline medium.   Once it is formed, the
complex does not dissociate, even when the pH is lowered to 1 [28].

Procedure.   An acid solution containing iron and aluminium is
diluted to 100 ml with water and 10 ml of 10% sodium salicylate
solution and 20 ml of 20% triethanolamine solution are added.
Several drops of phenolphthalein indicator are added and the solution
is alkalized with 2M sodium hydroxide until an intense purple colour
is formed.   After about 10 min the pH is adjusted to 2 with dilute
nitric acid (1 + 3).   A sufficient excess of 0.05M EDTA is added and
the solution is titrated with 0.05M bismuth nitrate, with Xylenol
Orange as indicator.

Note.   It is necessary to add triethanolamine to prevent precipitation of ferric hydroxide during alkalization of the solution.

An alternative is to mask the aluminium with fluoride.   An excess of
0.05M EDTA is added to an acid solution of iron and aluminium and the
pH is adjusted to 4.   The solution is boiled for 1 min, then cooled
and adjusted to pH 5 - 5.5 with solid hexamine.   The free EDTA is
back-titrated with 0.05M lead nitrate, with Xylenol Orange as indicator.   Ten ml of 10% ammonium fluoride solution are then added and
the solution is boiled for 3 - 4 min.   A slight alkaline reaction
results from hydrolysis of the hexamine (pink colour of XO).   The pH
is adjusted with 1M nitric acid (to the yellow colour of XO) and the
solution is again titrated with lead to an intense red colour.

The results for iron are very good even in the presence of 100 mg of
aluminium.   Titanium can be separated by precipitation from triethanolamine medium, before the titration.

Iron and copper.   In spite of the large difference in the stabilities of the two EDTA complexes, high concentrations of copper interfere in the determination of iron with visual end-point detection.
However, both elements can be determined by successive spectrophotometric titration.   Underwood [29] titrated iron at pH 2, without an
indicator, using a wavelength of 745 nm, where the Fe-Y complex does
not absorb light.   After titration of the iron, the strongly absorbing Cu-EDTA complex is formed.   The absorbance increases until all
the copper is titrated.   Two sharp breaks on the absorbance curve
indicate the ends of the titrations of iron and copper.   This method
was recommended for determination of small amounts of iron and copper
in aluminium alloys.

Both elements can be determined very precisely with visual end-point
detection if thiourea is used to mask copper in one portion of the
solution and the sum Fe + Cu is determined in another portion [30].

Procedure.   An excess of 1M ammonium fluoride is added to a weakly
acid solution containing 2 - 150 mg of Fe and up to 160 mg of Cu,
until the blue-green colour changes to clear blue (formation of
colourless FeF_6^{3-}).   Then 20% thiourea solution is slowly added until
the solution becomes colourless, followed by a further 2 ml.   A

sufficient amount of 0.02 - 0.05M EDTA is added to the colourless
solution and the pH is adjusted to 5 - 5.5 with hexamine.   The
solution is now clear yellow (formation of Fe-EDTA).   Ecess of EDTA
is back-titrated with 0.02 - 0.05M lead nitrate (Xylenol Orange as
indicator).   The colour transition from clear yellow to purple is
very sharp.

For determination of Fe + Cu, excess of EDTA is added to a weakly
acid solution, the pH is adjusted to 5 - 5.5 with hexamine, Xylenol
Orange is added and the titration is carried out as above.

Note.   Copper cannot be masked well without addition of ammonium
fluoride, because of the formation of hydrolytic iron compounds.
The fluoride also prevents reaction of iron with thiourea.

Iron, aluminium and copper.   Here the following principle is employed.   In one portion of solution iron is determined after masking of
aluminium with ammonium fluoride and copper with thiourea.   In a
second portion, only aluminium is masked with ammonium fluoride, and
the sum Fe + Cu is determined as given above.   Ina third portion of
the solution the sum of all three metals is determined by back-
titration of excess of EDTA with lead solution (Xylenol Orange as
indicator).   The amounts of all three metals can be found from the
different consumptions of titrant.   The third solution can be used
for a check determination of aluminium by addition of fluoride and
titration (with lead) of the EDTA liberated from the Al-EDTA complex
(see procedure on p. 115).

REFERENCES

1.   R.L. Gustafson and E.A. Martell, J. Phys. Chem., 1963, 67, 567.
2.   R. Přibil, Komplexony v chemické analyse, Publishing House of
     the Czechoslovak Academy of Sciences, Prague, 1957.
3.   P. Schneider and J. Janko, Chem. Listy, 1956, 50, 899;   Anal.
     Abstr., 1957, 4, 1544.
4.   A. Ringbom, S. Siitonen and B. Saxén, Anal. Chim. Acta, 1957,
     16, 541.
5.   G.L. Kochanny Jr., and A. Timnick, J. Am. Chem. Soc., 1961, 83,
     2777.
6.   R. Přibil, Z. Koudela and B. Matyska, Coll. Czech. Chem. Commun.,
     1951, 16, 80.
7.   D. Lyderson and O. Gjems, Z. Anal. Chem., 1953, 138, 249.
8.   G. Schwarzenbach, Komplexon-Methoden, Siegfried A.G., Zofingen,
     1948.
9.   H. Flaschka, Mikrochim. Acta, 1952, 39, 38.
10. M. Theis and A. Musil, Z. Anal. Chem., 1955, 144, 351.
11. R. Přibil, Coll. Czech. Chem. Commun., 1955, 20, 162.
12. E. Körös and I. Paczók, Mag. Kem. Foly., 1958, 64, 230.
13. F. Vydra, R. Přibil and J. Korbl, Talanta, 1959, 2, 311.
14. R. Přibil, unpublished results.
15. R. Přibil and V. Veselý, Talanta, 1961, 8, 743.
16. R. Přibil, Coll. Czech. Chem. Commun., 1954, 19, 58.
17. R. Přibil, Coll. Czech. Chem. Commun., 1955, 20, 162.
18. R. Přibil, J. Korbl, B. Kysil and J. Vobora, Coll. Czech. Chem.
     Commun., 1959, 24, 2266.
19. H. Flaschka and R. Puschel, Z. Anal. Chem., 1954, 143, 330.
20. A. Ringbom and E. Wänninen, Nord. Kem. Forh., 1953, 96.

21. I. Sajó, Mag. Kem. Foly., 1954, 60, 268.
22. K. ter Har and J. Bazen, Anal. Chim. Acta, 1954, 10, 23.
23. P. Wehber, Z. Anal. Chem., 1957, 158, 321.
24. M. Theis, Radex-Rundschau, 1955, 333.
25. K.L. Cheng, R.H. Bray and T. Kurtz, Anal. Chem., 1953, 25, 347.
26. P.B. Sweetser and C.E. Bricker, Anal. Chem., 1953, 25, 253.
27. D.G. David and W.R. Jacobsen, Anal. Chem., 1960, 32, 215.
28. R. Přibil and V. Veselý, Talanta, 1963, 10, 383.
29. A.L. Underwood, Anal. Chem., 1953, 25, 1910.
30. R. Přibil and V. Veselý, Talanta, 1961, 8, 743.

6.1.7   Uranium
$$\overline{(\log K_{UY}} = 25) \; [1,2]$$

Generally speaking, two of the oxidation states of uranium are analytically important. In aqueous solutions uranium(VI) is present as the uranyl ion, UO_2^{2+}, reduction of which yields the uranium(IV) cation, U^{4+}. Complexometric determination of uranium is rather limited. Determinations based on the formation of the UO_2-EDTA or $(UO_2)_2$-EDTA complexes, which have rather low stability constants (log K_{UO_2Y} = 10.4 ± 0.2, log $K_{(UO_2)_2Y}$ = 15.2 ± 0.3, as found by Kozlova and Krota [3]) are of little practical importance.

Because of the rarity of uranium and its importance for nuclear energy, micromethods are of considerable analytical importance. On the macroscale, the two oxidation states of uranium can be determined by a number of oxidation-reduction titrations with various reagents. The high atomic weight of uranium (1 ml of~0.05M EDTA $\cong$ 11.90 mg of U) is disadvantageous for complexometry. Thus complexometric determinations are important only in the analysis of special binary or ternary uranium alloys or of uranium-rich raw materials.

Direct determination

The high stability of the U(IV)-EDTA complex has led to the development of a number of determinations of uranium in acid solution after suitable reduction of uranium(VI). Korkisch [4] titrated uranium in 0.01 - 0.2M hydrochloric acid medium, using Solochrome Black 6 BN (C.I. Mordant Black 3) as indicator, the colour transition being from blue to red. Uranium(VI) is reduced either with granulated zinc in the titrated solution or by use of a suitable reductor column (Zn, Pb, Ag). Bismuth, copper, zirconium, hafnium, thorium and tin(II) interfere in the determination of uranium by blocking the indicator, making the solution permanently blue.

Thorin at pH 2 - 5 [5] and arsenazo I at pH 1 ± 0.1 [6] have also been recommended as indicators for similar titrations.

For the sake of completeness, two methods for the direct determination of uranium(VI) will be given. Lassner and Scharf [7] titrated uranium(VI) at pH 4 - 5 in solutions containing at least 50% isopropyl alcohol, using PAN as indicator. The titration must be done at 80 - 90°C. The colour transition from red to yellow is not very sharp.

Brück and Lauer [8] titrated from 12 µg to 9 mg of uranium(VI) spectrophotometrically with 0.003M EDTA at pH 4.4 at a wavelength of 555 nm, using PAN indicator. The titration can be followed at 340 nm and pH 4 in the absence of indicator.

Indirect determination

The method developed by Kinnunen and Wennerstrand [9] is most suitable.
Uranium is reduced to U(IV) at pH 3 - 4 in the presence of EDTA by
boiling for 5 - 10 min with ascorbic acid.   After cooling, the pH is
adjusted to 2 - 3 and the excess of EDTA is titrated with 0.02 - 0.05M
thorium nitrate (Xylenol Orange as indicator).

Budešínský et al. [10] reduced uranium(VI) to uranium(IV) with sodium
bisulphite and titrated the free EDTA with thorium solution, using
Methylthymol Blue as indicator.

Separation methods

A large number of methods have been described for separation of
uranium from other elements, especially by use of ion-exchanger and
extraction.   Uranium can be extracted quite selectively from sulphuric acid medium with a chloroform solution of Aliquat 336-S (trioctylmethylammonium chloride) [11].   It appears, however, that these
methods have not been applied in the complexometry of uranium.

Recently, a very selective method has been developed for the extraction of uranium from NTA medium with a chloroform solution of phenylacetic acid [12].   After stripping with slightly acidic potassium
nitrate solution, the uranium is determined complexometrically.   The
method is also useful for small uranium contents in large samples.

Vinogradov [13] suggested an interesting separation method.   The
carbonato complex of uranium is precipitated with hexa-ammine-
cobalt(III) chloride·solution.   Cobalt is determined complexometrically after decomposition of the isolated precipitate.   This method
was, in principle, used earlier by Monk and Exelby [14] for the
isolation and complexometric determination of beryllium (Section
6.3.4).

Masking of uranium

Provided it does not react with the indicator or form a precipitate
with other metals, uranium(VI) does not interfere in the determination of other elements.   A number of substances can be used for
masking uranium.   In alkaline medium, uranium can be masked with
carbonate or hydrogen peroxide, in acid medium with fluoride, phosphate, citrate, sulphosalicylic acid, etc.   Some examples are given
in Chapter 3.

As mentioned above, the complexometry of uranium is important in
special cases.   A number of examples are given in Section 6.1.3 on
thorium, and some in the chapters on practical applications.

Recommended procedures

Back-titration using Xylenol Orange indicator.   Acetate buffer (pH
4.3) and 1 g of ascorbic acid are added to a solution containing 50 -
60 mg of uranium per 100 ml.   After addition of 25 ml of 0.02M EDTA,
the solution is boiled gently for 10 min, cooled, adjusted to pH 2 - 3
with hydrochloric acid and after addition of 8 - 10 drops of indicator
is titrated with 0.02M thorium nitrate from green to red.

Determination after isolation with phenylacetic acid [12].   Phenylacetic acid (PAA) is a much more selective extractant than the more
frequently used oxine.   From weakly acid and sufficiently dilute
solutions, a chloroform solution of PAA extracts only iron, copper,
lead, gallium, indium and uranium [15,16].   Only uranium is extracted in the presence of nitrilotriacetic acid (NTA) in solutions buffered with hexamine.

Procedure.   A weakly acid solution containing 10 - 20 mg of uranium
is pipetted into a 200 ml separating funnel.   The solution is
diluted to 70 ml with distilled water and 10 ml of 1M solution in
chloroform are added;   the solution is mixed by swirling.   Depending
on the expected interferents, 10 - 25 ml of 0.5M NTA and 3 ml of
saturated hexamine solution are added, and the mixture is shaken for
1 min.   The yellow extract which separates is transferred into
another separating funnel.   The extraction is repeated with a
further 10 ml of PAA solution and the extract generally light yellow,
is added to the first extract.   The aqueous phase is extracted twice
more, with 5 ml of pure chloroform, all the extracts being combined.
The combined chloroform extract is stripped with two 20-ml portions
of water containing 3 ml of 1M hydrochloric acid, these aqueous
extracts being transferred into a 500-ml titration vessel.   After
addition of excess of 0.05M EDTA and 1 g of ascorbic acid, the pH is
adjusted to 3 - 4 with 1M sodium hydroxide and the solution is boiled
for 10 min.   After cooling the solution is diluted to 200 - 250 ml
and adjusted to pH 2 - 3 by addition of 10 - 20 ml of 1% acetic acid
solution.   The surplus EDTA is titrated with 0.05M thorium nitrate,
with Xylenol Orange as indicator.   Zirconium, thorium, the rare-
earth elements, bismuth, iron, aluminium, and copper and other
bivalent metals do not interfere.

REFERENCES

1.   A.F. Klygin, I.D. Smirnova and N.A. Nikolskaya, Zh. Neorgan.
      Khim., 1959, 4, 2766.
2.   N.N. Krot and N.P. Ermolayev, Zh. Neorgan. Khim., 1962, 7, 2054.
3.   A.G. Kozlov and N.N. Krot, Zh. Neorgan. Khim., 1960, 5, 1959.
4.   J. Korkisch, Anal. Chim. Acta, 1961, 24, 306.
5.   P.N. Paley and L.Y. Hsu, Zh. Analit. Khim., 1961, 16, 61; Anal.
      Abstr., 1961, 8, 3257.
6.   A.E. Klygin, N.A. Nikolskaya, N.S. Kolyada and D.M. Zavrazhnova,
      Zh. Analit. Khim., 1961, 16, 110; Anal. Abstr., 1961, 8, 3256.
7.   E. Lassner and R. Scharf, Z. Anal. Chem., 1958, 164, 398.
8.   A. Bruck and F. Lauer, Anal. Chim. Acta, 1967, 37, 325.
9.   J. Kinnunen and B. Wennerstrand, Chemist-Analyst, 1957, 46, 92.
10.   B. Budĕšínský, A. Bezdĕková and D. Vrzalová, Coll. Czech. Chem.
      Commun., 1962, 27, 1528; Anal. Abstr., 1963, 10, 1011.
11.   R. Přibil and J. Adam, Coll. Czech. Chem. Commun., 1972, 37, 129.
12.   J. Adam and R. Přibil, Talanta, 1973, 20, 1344.
13.   A.V. Vinogradov, R.M. Apirina and I.V. Pavlova, Zh. Analit. Khim.,
      1967, 22, 1323; Anal. Abstr., 1969, 16, 2891.
14.   R.G. Monk and K.A. Exelby, Talanta, 1964, 11, 1163;   1965, 12,
      91.
15.   J. Adam, R. Přibil and V. Veselý, Talanta, 1972, 15, 825.
16.   J. Adam and R. Přibil, Talanta, 1972, 19, 1105.

6.1.8   Indium
        $(\log K_{InY} = 24.9)$

The high stability of the InY^- complex makes direct titrations in
acid medium at pH 2 - 4 preferable. There are only a few suitable
indicators. Cheng [1] titrated indium directly, using PAN indicator
at pH 2.3 - 2.8. Flaschka and Abdine [2] and Gusev and Nikolayeva [3]
used the same indicator. Cheng also used PAN in neutral or slightly
alkaline medium (pH 7 - 8). Addition of tartrate prevented hydrolysis of indium, and potassium cyanide was used to mask up to 10 mg
each of copper, nickel, cobalt, silver, mercury, etc. The sharpness
of the colour transition from red to yellow can be increased by adding a mixture of Cu-EDTA-PAN during titration of solutions at pH 2.5 -
3 [1].

Other useful indicators are Xylenol Orange [4] and Methylthymol Blue
[5]. Gusev and Nikolayeva [6] studied various pyridylazoaminophenyl
derivatives and found 5-(2-pyridylazo)-4-ethoxy-2-monomethylamino-1-
methylbenzene (PAMB) very useful, permitting very selective titration
of indium at pH 1.95 - 2.0. In this way indium can be determined in
the presence of 100-fold amounts of lead or cadmium and 200-fold
amounts of zinc. A fivefold amount of iron and cobalt and a 30-fold
amount of nickel also do not interfere.

Busev and Talipova [7] studied 30 substances derived from 8-hydroxy-
quinoline for determining indium at pH 2 and found that 7-(1-naphthyl-
azo)-8-hydroxyquinoline-5-sulphonic acid and 7-(5,6-sulpho-2-naphthyl-
azo)-8-hydroxyquinoline-5-sulphonic acid are the most useful. The
colour transition in these titrations (pH 2.2 - 2.8) is from yellow to
red, and the optimal temperature for the titration is 70 - 80°C. The
alkaline-earth metals, cadmium and zinc do not interfere. Aluminium
must be masked with·fluoride.

Flaschka used Erio T for direct determination of indium in alkaline
medium, titrating a boiling solution containing tartrate at pH 10.
The results for macrodetermination [8] and microdetermination [9]
were satisfactory in the presence of various heavy metals which can
be masked with potassium cyanide.

Indirect determination

Titrations in acid medium are again important because of their high
selectivity. Excess of EDTA can be titrated with thorium nitrate or
thallium nitrate, with Xylenol Orange as indicator [4]. Flaschka
and Sadek [10] recommend back-titration with bismuth nitrate at pH 2
(Pyrocatechol Violet as indicator), where the alkaline-earth metals,
zinc (up to 350-fold amount), lead (50-fold), nickel, cobalt and
copper (10-fold) do not interfere. The authors also state that
"small concentrations" of aluminium do not interfere if the excess of
EDTA is titrated with bismuth solution at 60°C. Indium can be
determined by using Xylenol Orange indicator in the same manner [11].

Indium can be determined indirectly in slightly acid medium by back-
titration with zinc solution at pH 5 - 5.5, using Xylenol Orange or
Methylthymol Blue as indicator [11]. Indium can be determined (in
the presence of some other metals) in alkaline medium containing
hydroxylammonium chloride and potassium cyanide by back-titration
with manganese solution [11].

Interferences

The determination of indium is disturbed by the presence of all quadrivalent and tervalent metals except small concentrations of aluminium. Tervalent iron and thallium can be reduced with ascorbic acid,
aluminium can be masked with sulphosalicylic acid (followed by back-
titration with bismuth nitrate) [5], bismuth can be masked with thioglycollic acid [11] or sodium thiosulphate. Fluoride has been
recommended for masking zirconium [12], tin [13], aluminium and some
other metals.

In back-titrations, e.g. with zinc (pH 5 - 5.5), all group I and II
elements are co-titrated and must be masked. In back-titrations
with lead, other elements can be masked with 1,10-phenanthroline [11].

Masking of indium

Indium is masked in alkaline medium with thioglycollic acid, permitting determination of nickel by back-titration of excess of DCTA with
calcium (Methylthymol Blue as indicator) [14].

Separation methods

A wide choice of methods is available to simplify the composition of
samples. For example, indium can be separated from aluminium, zinc
and cadmium by double precipitation with a large excess of 1,6-di-
aminohexane. The precipitate is collected and dissolved, and the
indium in it determined complexometrically; aluminium, zinc or
cadmium in the filtrate can be determined by direct titration with
EDTA [15].

A great deal of attention has been paid to the indium-gallium
combination. Paskelev [16] employed the easy hydrolysis of gallium
to separate it from indium. Gallium is precipitated quantitatively
from 5M sodium chloride solution at pH 4 - 5.5, while indium is
precipitated as the basic chloride at a higher pH. Both elements
can be determined complexometrically after the separation.

Kocheva [17] studied the chromatographic separation of gallium and
indium, as well as of aluminium, in detail, using the chloride form
of Dowex 1-X8 and a sample solution in 0.5M potassium iodide medium
(pH 1). Gallium is eluted first with 0.5M potassium iodide and then
indium with 1M ammonium chloride [17]. Gallium can be separated
from indium in a similar manner by using the sulphate form of Dowex
1-X8 and eluting indium with 1M ammonium sulphate [18].

The determination of gallium and indium with EDTA and TTHA solutions
will be discussed in detail in Section 6.1.9.

Recommended procedures

Direct determinations. After adjustment to pH 3 - 3.5, a solution
containing up to 50 mg of In is heated to 50 - 60°C and titrated with
0.05M EDTA, with Xylenol Orange as indicator, from red to clear
yellow [19].

An acid solution containing 5 - 30 mg of In in 50 ml is neutralized
with sodium hydroxide until formation of a white precipitate, which
is dissolved in 2 ml of glacial acetic acid. Several drops of 1%

PAN solution in ethanol are added and the solution is titrated with
EDTA from red to yellow [1].

Indirect determination

A sufficient excess of 0.05M EDTA is added to an acid solution of
indium, the pH is adjusted to 2 - 2.5 and the solution is back-
titrated with bismuth nitrate, with Pyrocatechol Violet [10] or
Xylenol Orange [11] as indicator.

REFERENCES

1.   K.L. Cheng, Anal. Chem., 1955, 27, 1582.
2.   H. Flaschka and H. Abdine, Chemist-Analyst, 1956, 45, 58.
3.   S.I. Gusev and E.M. Nikolayeva, Zh. Analit. Khim., 1966, 21, 166;
      Anal. Abstr., 1967, 14, 6024.
4.   J. Kinnunen and B. Wennerstrand, Chemist-Analyst, 1957, 46, 92.
5.   Y. Horiuchi and O. Ichijyo, Rept. Technol. Iwate Univ., 1967,
      20, 43.
6.   S.I. Gusev and E.M. Nikolayeva, Zh. Analit. Khim., 1966, 21, 281;
      Anal. Abstr., 1967, 14, 6714;   see also Anal. Abstr., 1970, 18,
      3820.
7.   A.I. Busev and L.L. Talipova, Zh. Analit. Khim., 1962, 17, 447;
      Anal. Abstr., 1963, 10, 961;   see also Anal. Abstr., 1963, 10,
      4094.
8.   H. Flaschka and A.M. Amin, Z. Anal. Chem., 1953, 140, 6.
9.   H. Flaschka and A.M. Amin, Mikrochim. Acta, 1953, 410.
10.   H. Flaschka and F. Sadek, Z. Anal. Chem., 1956, 149, 345.
11.   R. Přibil and V. Veselý, Chemist-Analyst, 1965, 54, 12.
12.   R.S. Volodarskaya, N.A. Kanayev and G.N. Derevyanko, Zavod. Lab.,
      1966, 32, 413;   Anal. Abstr., 1967, 14, 4569.
13.   D.G. Biechler, Chemist-Analyst, 1963, 52, 48.
14.   R. Přibil and V. Veselý, Chemist-Analyst, 1966, 55, 38.
15.   I.M. Korenman and S.F. Chelysheva, Tr. Khim. Technol. Gorky,
      1966, 3, No.14, 103;. Anal. Abstr., 1967, 14, 4590.
16.   N. Paskalev, D. Mihaylov and D. Trendafelov, C.R. Acad. Bulg.
      Sci., 1968, 21, 237; Anal. Abstr., 1969, 17, 80.
17.   L.L. Kocheva, C.R. Acad. Bulg. Sci., 1969, 22, 447;   Anal. Abstr.,
      1970, 19, 197.
18.   L.L. Kocheva, God. Sof. Univ. Khim. Fak., 1970, 62, 135; Anal.
      Abstr., 1971, 20, 2351.
19.   M. Kopanica and R. Přibil, Coll. Czech. Chem. Commun., 1960, 25,
      2230.

6.1.9   Gallium
         $\overline{(\log K_{GaY}} = 20.3)$

In acid medium gallium forms the sufficiently strong GaY⁻ complex.
With increasing pH the Ga-EDTA complex hydrolyses to form the hydroxo-
complex.   By pH 10.5 the Ga-EDTA complex is completely converted
into the tetrahydroxo-complex $Ga(OH)_4^-$, similarly to aluminium.   Cheng
and Goydish [1] used this reaction to mask gallium and for its
indirect determination.

Direct determination

At present, only determinations in acid medium are of practical importance. Many excellent metallochromic indicators cannot be used because gallium blocks them; the reaction between the indicator complexes of gallium and EDTA is very slow, even at higher temperatures. Two indicators can be used for the direct determination of gallium - morin and gallocyanine.

Patrovský [2] first used morin as a fluorescence indicator in ultraviolet light for the titration of gallium in acetate-buffered solutions. The determination can be carried out at pH 3.5 in the presence of acetate and fluoroborate [3]. Gregory and Jeffery [4] found that the fluorescence of gallium with morin increases substantially when the pH is decreased to 1.7 or the solution is acidified with acetic acid to pH 2. Addition of an equal volume of methanol or acetone substantially increases the sharpness of the fluorescence quenching, and the work can then be carried out in daylight. At pH 2 fluoride quenches the fluorescence and must be eliminated.

Milner [5,6] suggested gallocyanine for titration of gallium. Gallocyanine is a well-known analytical reagent, which was suggested by Dubský et al. [7 - 9] for detection of certain cations. It forms an intensely blue compound with gallium at pH 2.8, but this reacts slowly with EDTA so the titrations must be done slowly. The minimum amount of indicator necessary to form the blue colour must be used; the colour transition is from blue to red.

Crawley [10] critically evaluated the two methods and found that morin does not give a precisely defined end-point and the indicator has substantial residual fluorescence. The purity of the morin used was not specified. It is not suitable for titration with 0.001M EDTA. Gallocyanine yields a good colour transition for titration with 0.02M EDTA and is useful for determining amounts of gallium above 2 mg. Crawley [10] recommends 8-hydroxyquinoline as an indicator for gallium. The fluorescence is observed under ultraviolet radiation, and its quenching is very sharp (0.1 ml of 0.001M EDTA solution for 20 ml of titrated solution). Methanol quenches the fluorescence and chloroform should also not be present.

Indirect determination

Milner [11] recommends that the microdetermination of gallium in the presence of up to 2 g of uranium be carried out by back-titration of excess of EDTA at pH 2.2 with 0.02M ferric chloride, with potassium benzohydroxamate as indicator. Back-titration with thorium or thallium [12], bismuth [13,14] or zinc [15] can be used.

Interferences

All group I cations and also most group II cations interfere in the determination of gallium. Back-titration in acid medium with a bismuth nitrate solution improves the selectivity considerably.

Japanese authors [13] titrated gallium, using Methylthymol Blue indicator, and found that 5 mg of silver, alkaline-earth metals, lead, cobalt, manganese or zinc will not interfere. Other elements (in the same amount) are partially or completely co-titrated.

Masking of interferents

Tervalent iron or thallium can be reduced with ascorbic acid.
Tungsten and molybdenum are masked by phosphate in titration of
gallium (Erio T indicator) at pH 7.5 [16]. Small amounts of aluminium can be masked with sulphosalicylic acid or fluoride, or with
fluoroborate [2].

Separation of gallium

In some cases, gallium can be separated by hydrolytic precipitation
as the hydroxide at pH 3. Busev and Tiptsova [17] precipitated
gallium from 5.5 - 6M hydrochloric acid with diantipyrinylpropylmeth-
ane. The precipitate $(C_{20}H_{30}O_2N_2.HGaCl_4)$ can be collected and
weighed or the gallium in it determined complexometrically.

The extraction of gallium with diethyl ether is highly recommended.
Milner [6] extracted gallium from ammonium chloride and sulphuric
acid medium. Cocozza [18] separated gallium from indium by extraction of gallium from 6M hydrochloric acid and determined both
elements complexometrically. Gregory and Jeffery [4] used a similar
procedure. Mizuno [19] recommends extraction of traces of gallium
(30 μg) with di-isopropyl ether from 7M hydrochloric acid. In this
way, gallium can be separated from 10 - 20 mg of iron and aluminium
and from small amounts of other elements. Cherkashina [20] extracted gallium together with molybdenum and tin and small amounts of
other metals with butyl acetate from 6M hydrochloric acid.

Masking of gallium

It is sometimes possible to mask gallium with sodium citrate. The
demasking of gallium from the EDTA complex with sodium hydroxide [1]
has already been mentioned.

Recommended procedures

Gallocyanine as indicator. Glacial acetic acid (10 ml) is added to
a solution containing up to 50 mg of Ga in 100 ml, and the pH is
adjusted with ammonia to 2.8 - 3.0 (pH-meter). Then 5 drops of a
saturated solution of gallocyanine in glacial acetic acid are added
and the solution is thoroughly mixed. The blue solution is titrated
very slowly with 0.02M EDTA with thorough shaking, to a colour
transition from blue to red. Small amounts can be titrated with
0.002M EDTA.

Pyrocatechol Violet as indicator [3]. Acetate buffer (pH 3.8) is
added to a weakly acidic solution containing up to 40 mg of Ga in
100 ml and the solution is slowly titrated with 0.05M EDTA solution
from blue to yellow.

Indirect titration, Xylenol Orange as indicator. A small excess of
0.05M EDTA is added to the acid solution and the pH is adjusted to
5 - 5.5 with hexamine. Several drops of indicator are added and the
solution is titrated with 0.05M zinc from yellow to an intense red
[15]. Alternatively, an excess of 0.05M EDTA is added to an acidic
gallium solution and the pH is adjusted to 3. The solution is back-
titrated with bismuth nitrate; the colour change is clear yellow to
red.

Some separation determinations

Determination of gallium and indium. Although indium alone can be
determined directly by titration at pH 3 - 5 (Xylenol Orange indicator), In + Ga cannot be determined in this manner because gallium
blocks the indicator. The sum can be determined by back-titration
with a zinc solution at pH 5 - 5.5. Determination of the individual
concentrations is made possible by addition of excess of TTHA, which
forms a 2:1 complex with gallium and a 1:1 complex with indium [15],
and back-titration.

$$2 \ Ga + TTHA = Ga_2TTHA$$

$$In + TTHA = InTTHA$$

Procedure. A sufficient amount of 0.05M EDTA is added to an acidic
solution of gallium and indium and the pH is adjusted to 5 - 5.5 with
solid hexamine; the solution is then heated almost to boiling point.
The solution is titrated with a zinc solution (Xylenol Orange indicator) to an intense red colour; the EDTA consumption corresponds to
Ga + In.

An amount of TTHA solution corresponding to the EDTA consumption in
the first titration is added to another portion of the solution.
After pH adjustment, the excess of TTHA is titrated as above.   The
TTHA consumption corresponds to ½Ga + In.

The colour transitions during the titrations are sufficiently sharp
provided that the indium to gallium ratio is between 10:1 and 1:10.

Note.   Vydra and Vorlíček [21] studied the reaction mechanism for
the formation of the TTHA complexes of indium and gallium, and found
that, at a ratio of In:Ga of up to 12:1, only the Ga-TTHA complex is
formed during titration with TTHA solution (pH 2).   Formation of the
InGaTTHA complex seems more probable.   With excess of gallium the
Ga_2TTHA complex is first formed and then the mixed InGaTTHA complex.

Wiersma and Lott [22] also studied this determination of indium and
gallium.   The back-titration of EDTA and TTHA was carried out with
gallium instead of zinc nitrate, morin being used as indicator.   The
determination of microgram amounts of gallium and inium was accompanied by an error of 2.2 - 2.8%.

Determination of gallium in the presence of other elements [1].   The
determination is based on finding the sum of all the elements by back-
titration with lead solution, using PAR as indicator.   Gallium is
then demasked by making the solution strongly alkaline and the EDTA
liberated is determined indirectly.

Procedure.   A sufficient excess of 0.025M EDTA and 1 ml of 1M tartaric acid are added to a solution containing gallium and other
metals (Pb, Cu, Ni, Zn, Cd, In, La, Zr) and the pH is adjusted to
6.5 ± 0.2 with hexamine and hydrochloric acid.   Several drops
of 0.1% PAR indicator solution are added and the free EDTA is
titrated with a lead solution to formation of a red colour.

After completion of the titration, a known slight excess of standard
lead is added, the pH is adjusted to 10.5 ± 0.1 with ammonia or

sodium hydroxide, and the excess of lead is titrated with EDTA.   The
difference between the two titrations corresponds to the gallium
present.

REFERENCES

1.   K.L. Cheng and B.L. Goydish, Talanta 1966, 13, 1161.
2.   V. Patrovský, Chem. Listy, 1953, 25, 1338; Anal. Abstr., 1955,
     2, 54.
3.   J. Doležal, V. Patrovský, Z. Šulcek and J. Švasta, Chem. Listy,
     1955, 49, 1517; Anal. Abstr., 1967, 3, 3297.
4.   G.R.E.C. Gregory and P.G. Jeffery, Talanta, 1962, 9, 800.
5.   G.W.C. Milner, A. Wood and J.L. Woodhead, Analyst, 1954, 79, 272.
6.   G.W.C. Milner, Analyst, 1955, 80, 77.
7.   J.V. Dubský, Chem. Listy, 1940, 34, 1.
8.   J.V. Dubský and L. Chodák, Coll. Czech. Chem. Commun., 1939, 11,
     523.
9.   J.V. Dubský and J. Trtílek, Chem. Obzor., 1934, 9, 68.
10. R.H.A. Crawley, Anal. Chim. Acta, 1959, 19, 540.
11. G.W.C. Milner, Analyst, 1956, 81, 367.
12. J. Kinnunen and B.Wennerstrand, Chemist-Analyst, 1957, 46, 92.
13. Y. Horiuchi and O. Ichijyo, Rep. Technol. Iwate Univ., 1967, 20,
     43; Chem. Abstr., 69, 32695 w.
14. H. Flaschka and F. Sadek, Z. Anal. Chem., 1956, 149, 345.
15. R. Přibil and V. Veselý, Talanta, 1964, 11, 1319.
16. O.W. Rollins and B.J. Haynes, Chemist-Analyst, 1967, 56, 98.
17. A.I. Busev and V.G. Tiptsova, Zh. Analit. Khim., 1960, 15, 698.
18. E.P. Cocozza, Chemist-Analyst, 1960, 49, 46.
19. K. Mizuno, Japan Analyst, 1965, 14, 410; Anal. Abstr., 1967,
     14, 59.
20. T.V. Cherkashina, Zavod. Lab., 1956, 22, 276; Anal.Abstr., 1956,
     3, 3298.
21. F. Vydra and J. Vorliček, Proc. Anal. Chem. Conference, Budapest
     1966.
22. J.H. Wiersma and P.F. Lott, Anal. Lett., 1968, 1, 603.

6.1.10   Thallium
$(\log K_{Tl(III)Y} = 22.5$ [1])

In contrast to thallium(I) $(\log K_{Tl(I)Y} = 5.8)$, thallium(III) forms a
very stable complex with EDTA.   Thallium can be determined by
titration of thallium(III) with EDTA in acid medium.

Direct determination

The complexometric behaviour of thallium is similar to that of indium
and the same indicators can be used; the most useful is Xylenol
Orange.   Kinnunen and Wennerstrand [2]· titrated thallium directly at
pH 4 - 5, with a very sharp transition, which led to their recommend-
ing using a standard 0.05M thallium solution for a number of titrations.   Back-titrations with thallium(III) have the further advantage that they are not disturbed by the presence of sulphate, fluoride
and phosphate.

Strelow and Toerien [3] titrated 5 - 100 mg of Tl at pH 7 - 10 in the presence of tartrate, using Methylthymol Blue as indicator.   They oxidized thallium(I) with bromine;   the bromide formed does not interfere in this medium.   These authors found an error of ± 2 µg for the determination of 1 mg of thallium by titration with 0.001M EDTA.

Of the other possible indicators, PAN and PAR [4] have been recommended for the titration of thallium.   Busev and co-workers studied a number of substances which can be used as indicators for thallium, such as 8-hydroxy-7-(pyridylazo)quinoline [5], suitable for pH 1.8 - 3.5 medium with a sharp transition from purple to yellow, and a number of derivatives of 8-hydroxyquinoline-5-sulphonic acid [6] (see Section 6.1.1).

The high stability of the Tl-EDTA complex has led to attempts to titrate thallium at very low pH.   Busev and Tiptsova [4] state that the minimum pH at which thallium can be reliably titrated, with Xylenol Orange as indicator, is 2, with PAN pH 1.8 and PAR pH 1.7. Talipov et al. [7] recommend 4-(2-quinolylazo)resorcinol as indicator, titrating thallium in solutions with pH 1, acidified with sulphuric or nitric acid, with a colour transition from purple to yellow.   They titrated thallium and cadmium successively (pH 8, acetate buffer).

Babayev et al. [8] recommended 4-(6-methyl-2-pyridylazo)resorcinol as a very sensitive indicator for thallium, titrating at pH 1.25 - 1.45 with 0.01M EDTA;   the colour transition is from bright red to yellow. A number of elements do not interfere in the determination, including zirconium, hafnium, gallium and titanium, at low concentration. Copper, iron and indium interfere and the colour of chromium, nickel and cobalt also interferes.   Chloride and reducing agents (ascorbic acid) interfere, but fluoride, nitrate and phosphate do not.

The only determination of thallium in alkaline medium is that described by Flaschka [9].   First a sufficient amount of Mg-EDTA complex must be added to a solution of thallium(III) and then the solution is alkalized with buffer to pH 10.   The magnesium freed from its complex is then titrated, with Erio T as indicator.   The transition is sufficiently sharp for 30 - 2000 µg of thallium to be determined.

Indirect determination

Thallium can be determined in acid medium indirectly by back-titration with thallium(III) nitrate (Xylenol Orange as indicator) or with a thorium solution at pH 3.5 (Alizarin S indicator [10] or Xylenol Orange).

Interferences

Most bivalent metals do not interfere in titrations at pH 2, but all cations forming EDTA complexes with stability constants greater than 10^{18} interfere.   Zirconium and thorium can be masked by addition of fluoride.   Frequently, it is best to proceed in reverse, i.e. first the sum of the elements present is determined and then thallium is masked and the liberated EDTA titrated.

Masking of thallium

This is generally based on reduction of thallium, e.g. with ascorbic acid or sodium sulphite.   Thallium can also be masked in acid medium

with sodium thiosulphate and potassium bromide. For example, a
mixture of thallium and bismuth can be determined by titrating the
sum of the elements at pH 1.8 - 2, using Busev's indicator [5] and,
after reduction of thallium with sulphite, the EDTA freed is titrated,
the same indicator being used. The thallium-iron combination is
determined by first masking thallium with potassium bromide and
titrating iron at pH 2 (sulphosalicylic acid as indicator) and then
increasing the pH to 4 - 5 with sodium acetate to release the thallium
from the TlBr$_4^-$ complex and determining it in a further titration, with
PAR as indicator [11].

Separation of thallium

A very simple method is based on reduction of thallium and precipitating it as TlI with potassium iodide. EDTA is used to bind all
elements which react with iodide. After filtration, the thallium
iodide is decomposed by evaporating to dryness twice with aqua regia
and the tervalent thallium produced is determined complexometrically.
Nitric acid alone is not suitable for oxidation of thallium [12].

Many extraction methods can be used for microamounts of thallium,
which are than determined by a suitable photometric or radiometric
method.   Levin and Rodina [13,14] studied extraction of macroamounts
of thallium, using various esters of phosphoric acid, such as the
2-ethylhexyl and the di-2-ethylhexyl esters.   Stripping from the
organic phase into hydrochloric acid or sodium chloride allows
separation of thallium from gallium, indium, antimony, bismuth and
iron.

Chromatographic methods can be used for the separation of thallium
from gallium, indium or lead, based on adsorption on Amberlite XAD-2
impregnated with isobutyl methyl ketone or di-isopropyl ether.   The
sample solution is in hydrobromic acid [15].   Gallium passes through
the column and indium and thallium are eluted stepwise by more dilute
solutions of hydrobromic acid.

Recommended procedure

The pH of a solution containing up to 100 mg of Tl is adjusted to
about 4 (acetate buffer), the solution is heated to 60 - 80°C, and
titrated with 0.02 - 0.05M EDTA solution from red to yellow (Xylenol
Orange as indicator).

REFERENCES

1.   A.I. Busev, V.G. Tiptsova and T.A. Sokolova, Zh. Neorgan. Khim.,
      1960, 5, 2749.
2.   J. Kinnunen and B. Wennerstrand, Chemist-Analyst, 1957, 46, 92.
3.   F.W.E. Strelow and F.S. Toerien, Anal. Chim. Acta, 1966, 36, 189.
4.   A.I. Busev and V.G. Tiptsova, Zh. Analit. Khim., 1960, 15, 573.
5.   A.I. Busev, L.L. Talipova and V.M. Ivanov, Zh. Vses. Khim.
      Obshch., 1961, 6, 589; Anal. Abstr., 1962, 9. 5099.
6.   A.I. Busev and L.L. Talipova, Uzb. Khim. Zh., 1962, 3, 24;
      Anal. Abstr., 1963, 10, 4094.
7.   Sh. T. Talipov, K. Rakhmatullaev, N. Babaev and R. Barkanskaya,
      Tr. Tashk. Gos. Univ., 1968, 323, 50; Anal. Abstr., 1970, 18,
      80.

8. N. Babaev, N. Kulmuratov and Sh. T. Talipov, Vest. Karakalp. Fil.
    Akad. Nauk. Uzb. SSR, 1969, 3, No. 37, 15; Anal. Abstr., 1971,
    20, 1573.
9. H. Flaschka, Mikrochim. Acta, 1952, 40, 42.
10. W.T. Foley and R.F. Pottie, Anal. Chem., 1956, 28, 1101.
11. A.I. Busev and V.G. Tiptsova, Zh. Analit. Khim., 1961, 16, 275;
    Anal. Abstr., 1962, 9, 61.
12. R. Přibil, V. Veselý and M. Kratochvíl, Talanta, 1961, 8, 52.
13. I.S. Levin and T.F. Rodina, Zh. Analit. Khim., 1968, 23, 673;
    Anal. Abstr., 1970, 18, 82.
14. I.S. Levin and T.F. Rodina, Zh. Analit. Khim., 1968, 23, 1315;
    Anal. Abstr., 1970, 18, 2295.
15. J.S. Fritz, R.T. Frazee and G.L. Latwesen, Talanta, 1970, 17,
    857.

6.1.11   Chromium
$$(\log K_{CrY} = 23)$$

The strongly hydrated chromium(III) reacts with EDTA very slowly at
room temperature; the complex formation is much faster in boiling
solution.   Hamm [1] and Cellini and Valiante [2] studied the
kinetics of the complex formation.   Altogether, four different
complexes of tervalent chromium are formed, depending on the pH of
the solution: in strongly acid medium (pH 0.33) the protonated
complex CrHY(H$_2$O), the purple complex CrY(H$_2$O)$^-$, and in alkaline medium the blue Cr(OH)Y^{2-} and green Cr(OH)$_2$Y^{3-}[3] hydroxo-complexes. The
stability constants of these hydroxo-complexes were also determined.
Irving and Al-Jarrah [4] found that the purple complex CrY(H$_2$O)$^-$ is
formed by boiling the solution at pH 4.88.

A great deal of attention has been paid to conditions for formation
of the Cr-EDTA complexes.   The formation is dependent not only on
the pH and the length of boiling, but also on the presence of certain
anions - sulphate, chloride, etc.   The boiling time given by various
authors varies considerably - from 1 to 15 min.   Most authors have
found that 10 min is sufficient for quantitative formation at pH 3-6.

Formation of the purple complex at room temperature has also been
observed, e.g. in the reduction of chromium(VI) with iodide in the
presence of EDTA.   After reduction of the iodine with thiosulphate
the solution is dark purple, complicating the iodometric determination of chromium(VI) in the presence of iron and copper, both of
which can readily be masked with EDTA against reaction with iodide at
pH 1.5 [5].   DCTA does not undergo this reaction [6] and a green
chromium(III) complex is formed after the reduction of chromate.
Both reactions have been used for quantitative differentiation
between EDTA and DCTA [7].   Přibil [5] explained the formation of
the purple Cr-EDTA complex during reduction of chromate as due to
production of unhydrated chromium(III), which immediately reacts with
EDTA.   Aikens and Reilley [8] reduced chromate in 0.12M triethanolamine medium at pH 6.1 - 6.9 with sodium sulphite and explained the
formation of the purple Cr-EDTA complex by the formation of a very
labile intermediate EDTA complex of chromium(IV) or (V).   Beck and
Bardi [9] suggested that the complex formation is connected with
formation of a chromate complex with EDTA - (CrO$_2$)$_2$Y.

The formation of the Cr-EDTA complexes can be catalysed by

chromium(II) [10]. Irving and Tomlinson [11] used this reaction for preparation of a Cr-EDTA complex for photometric determination. They added traces of powdered zinc to a solution of tervalent chromium and EDTA at pH 4.5; the purple complex was formed within seconds.

The reaction of tervalent chromium with EDTA can also be catalysed at pH 5.3 - 6.0 by addition of sodium bicarbonate. The complex is formed within 10 min [12]. The molar absorptivity is only about 95% of that obtained by boiling a Cr-EDTA solution. The same authors later published the optimum conditions for the complexometric determination of chromium [13]. As with the photometric determination, the catalytic reaction took 10 min at pH 5.3 - 6.0 in the presence of a tenfold excess of EDTA (maximum of 6 mg of Cr in 100 ml). The excess of EDTA is then determined in ammoniacal medium at pH 9.5 with a standard nickel solution and murexide as indicator.

Indirect determination

The direct determination of chromium with EDTA is not practical because of its slowness. All indirect methods are based on boiling a chromium solution with excess of EDTA or DCTA. The intense colour of the Cr-EDTA complex complicates the titrations, so the determination is limited to a few milligrams of chromium, even with dilution. On the other hand, the strength of the complex permits determination in relatively acidic solutions, making it quite selective. Various standard solutions and indicators have been used for the back-titration. In acid medium, the excess of EDTA can be back-titrated with thorium at pH 4, with Xylenol Orange as indicator [14]. Bismuth nitrate has been used with "masking" of the purple colour of the Cr-EDTA complex with Bromophenol Blue. At pH 1 - 3, up to 7.5 mg of Cr can be determined·in 250 ml of solution, with Xylenol Orange as indicator [15]. Liteanu et al. [16] recommended back-titration with ferric chloride solution, using tiron as indicator. Calcein (Fluorexone) seems much more useful; Wilkins [17] titrated the excess of EDTA in acetate medium at pH 4.5 with a copper solution, using Calcein as indicator in ultraviolet light. The end of the titration was indicated by disappearance of the bright green fluorescence. In this way, Wilkins determined up to 40 mg of chromium in 150 ml of solution. Verma and Agarwal [18] determined excess of DCTA in a similar manner. The use of DCTA in the determination of chromium has the advantage that it is not oxidized by boiling with chromium(VI) and thus chromium(III) can be determined in the presence of chromate. Lead nitrate and Xylenol Orange have also been recommended for back-titration of the EDTA. However, the method is useful only for small amounts of chromium (not more than 2.5 mg in 100 ml of solution).

Solutions of manganese (Erio T indicator) [19] or of nickel (murexide) [20,21] have been recommended for the back-titration in alkaline media. The light green $Cr(OH)_2Y^{3-}$ complex is formed in alkaline media; this complex interferes far less with the colour transition of the indicator. Calcium (Thymolphthalexone or Calcein as indicator) [22] has also been used for the back-titration.

Because the sharpness of the colour transition is always affected by the colour of the Cr-EDTA complex, some authors have suggested using extraction indicator systems. For example, Cameron and Gibson [23] titrated the excess of EDTA with cobalt(II) in solutions containing potassium thiocyanate and triphenylmethylarsonium chloride. The blue

ion-association complex $[(C_6H_5)_3As(CH_3)^+]_2[Co(CNS)_4^{2-}]$ is formed at
the equivalence point and is extracted into chloroform.   The end of
the titration is marked by formation of a blue colour in the organic
phase.

Similarly, Irving and Al-Jarrah [24] titrated EDTA with cobalt
sulphate solution in the presence of thiocyanate, and the $Co(SCN)_4^{2-}$
formed was extracted with tetrahexylammonium chloride in dichloroethane.   The end of the titration is again marked by formation of a
blue colour in the organic phase.   These authors determined up to
50 mg of chromium by back-titration with 0.1M cobalt sulphate at pH
9.3.   The extraction could also be done with a chloroform solution
of Aliquat S-336 (trioctylmethylammonium chloride).   The titration
was carried out in specially designed conical flasks or ordinary
separatory funnels.

Interferences

The interferences are similar to those already described for determinations in weakly acid medium.   The very slight reaction of
chromium with EDTA allows determination of a number of elements at
normal temperature without interference from tervalent chromium.
Chromium is then determined after addition of further EDTA and boiling for an appropriate time.   DCTA reacts with aluminium in cold
solutions, so aluminium can be determined in the same solution as
chromium, which is then determined after boiling.   Chromium(VI) does
not interfere in the determination [25].   Iron or nickel can be
determined in the presence of chromium by back-titration of EDTA with
lead nitrate (Xylenol Orange as indicator) [22].

Masking chromium

Chromium(VI) does not interfere in a number of determinations.   In
alkaline medium calcium or magnesium can be determined without
difficulty, with Thymolphthalexone as indicator, in the presence of
large amounts of chromate.   Triethanolamine masks chromium in
alkaline medium.   On boiling the solution, the ruby red Cr-TEA
complex is formed.   The determination of calcium with Thymolphthalexone as indicator is sufficiently precise in the presence of 40 mg
of Cr in 500 ml of solution.   Masking of chromium with ascorbic acid
is much better.   On boiling a weakly acidic chromium solution with
1 - 2 g of ascorbic acid, a rather pale blue-green complex is formed,
enabling determination of nickel, manganese or calcium, for example
in ammoniacal medium [26].

Determination of chromium(VI)

A number of titrations can be used for the determination of
chromium(VI) with sufficient selectivity.   The reduction is
frequently followed potentiometrically.

In complexometric determinations, however, chromium(VI) must be
reduced by a suitable reducing agent.   Two methods can, in principle,
be used.   Either chromium(VI) can be reduced under optimal conditions
to tervalent chromium and then EDTA added to form the Cr(III)-EDTA
complex or the chromium(VI) can be reduced in the presence of EDTA so
that the Cr(III)-EDTA complex is formed in situ.   It has been found
that this seemingly simple reaction involves a number of compli-

cations. Aikens and Reilley [8], as already mentioned, reduced the chromium(VI) (up to 5 mg of Cr) in triethanolamine medium at pH 6.1 - 6.9 with sodium bisulphite by boiling for 2 min. The solution was cooled, diluted to 250 ml, and adjusted to pH 2.5 - 3.5, then excess of EDTA was titrated with thorium nitrate (Xylenol Orange). The authors found that the reduction is quantitative if 100% excess of bisulphite is used, but yield of Cr-EDTA decreases with increasing amounts of bisulphite; with 300% excess the results are 8% low. The authors explain this phenomenon as due to shortening of the lifetime of the intermediate complex. The Cr(III) formed reacts too slowly with EDTA under the given experimental conditions. The authors do not consider other reducing agents such as formaldehyde, ascorbic acid and hydrogen peroxide as suitable. Horiuchi and Iimura [27] studied the reduction of chromium(VI) in detail. They recommend ascorbic acid or hydroxylammonium chloride, but regard sodium bisulphite as quite unsuitable. They found the following optimal pH values for reduction of chromium(VI) with and without EDTA present.

Reductant	Without EDTA	With EDTA
$NaHSO_3$	1 - 1.80	6.25 - 6.35
$NH_2OH.HCl$	2.0 - 7.0	4.00 - 6.50
Ascorbic acid	0.60 - 1.20	5.20 - 6.50
Ethanol	0.0 - 0.60	0.4 - 0.45

The authors determined chromium (up to 5 mg) by back-titration with bismuth nitrate at pH 1.8 - 1.9, using Xylenol Orange as indicator. The relative error for reduction of up to 5 mg of Cr in the absence of EDTA was ±0.06 mg, and 0.08 mg in its presence.

The insolubility of barium chromate has also been used for the complexometric determination of chromium(VI). Isagai and Tateshita [28] precipitated chromate with a standard barium chloride solution and determined the excess of barium in the filtrate complexometric-ally. The method is unnecessarily complicated, as it is sufficient to collect the $BaCrO_4$, dissolve it in excess of EDTA and back-titrate the EDTA in ammoniacal solution with magnesium, using Thymolphthalex-one indicator.

Recommended procedures

Calcein (Fluorexone) as indicator [17]. A small excess of 0.05M EDTA is added to a weakly acidic solution of chromium(III) (up to 40 mg of Cr in 150 ml) and the pH is adjusted to 4.5 with acetate buffer; the solution is diluted to about 150 ml and boiled for 15 min. After cooling, several drops of indicator (0.1% in 0.001M NaOH) are added and the solution is titrated in ultraviolet light with 0.05M copper sulphate, to disappearance of the green fluor-escence.

Note. Because of the relatively high pH used, only the alkaline-earth metals do not interfere in the determination.

Xylenol Orange as indicator.   Excess of EDTA is added to a weakly
acidic chromium(III) solution and the mixture is boiled for 15 min.
After cooling, the pH is adjusted to 4 - 4.5 and the solution is back-
titrated with thorium nitrate, with Xylenol Orange as indicator [14].

Alternatively, excess of EDTA is added to a solution containing up to
5 mg of Cr at pH 1.8 - 1.9, the mixture is boiled for 10 min then
cooled, indicator is added, and the solution is titrated with 0.05M
bismuth nitrate to a purple-red colour [27].

REFERENCES

1.   R.E. Hamm, J. Am. Chem. Soc., 1953, 75, 5670.
2.   R.F. Cellini and E.A. Valiante, An. Real. Soc. Esp. Fis. Quim.,
      1955, 51B, 47; Anal. Abstr., 1955, 2, 2088.
3.   C. Furlani, G. Morpurgo and G. Sartori, Z. Anorg. Chem., 1960,
      303, 1.
4.   H.M.N.H. Irving and R.H. Al-Jarrah, Anal. Chim. Acta, 1973, 63,
      79.
5.   R. Přibil and J. Sýkora, Chem. Listy, 1951, 45, 105.
6.   R. Přibil, unpublished results.
7.   R. Přibil and V. Veselý, Chemist-Analyst, 1967, 56, 51.
8.   D.A. Aikens and C.N. Reilley, Anal. Chem., 1962, 34, 1707.
9.   M.T. Beck and I. Bardi, Acta Chim. Acad. Sci. Hung., 1961, 29,
      283.
10.   G. Schwarzenbach, personal communications; H. Irving, unpublish-
      ed findings.
11.   H.M.N.H. Irving and W.R. Tomlinson, Chemist-Analyst, 1966, 55,
      14.
12.   V.K. Rao, D.S. Sundar and M.N. Sastri, Chemist-Analyst, 1965,
      54, 86.
13.   V.K. Rao, D.S. Sundar and M.N. Sastri, Z. Anal. Chem., 1966,
      218, 93.
14.   J. Kinnunen and B. Wennerstrand, Chemist-Analyst, 1957, 46, 92.
15.   Ya. Makarov-Zemlyanski, N.N. Pavlov and G.A. Arbuzov, Nauch. Tr.
      Isled. Mosk. Tekhnol. Inst., 1970, 36, 99; Anal. Abstr., 1971,
      21, 1069.
16.   C. Liteanu, I. Crișan and C. Calu, Stud. Univ. Babes Bolyai,
      1959, 1, 105.
17.   D.H. Wilkins, Anal. Chim. Acta, 1959, 20, 324.
18.   M.R. Verma and K.C. Agarwal, J. Sci. Indust. Res., 1960, 19B,
      319.
19.   J. Kinnunen and B. Wennerstrand, Chemist-Analyst, 1955, 44, 33.
20.   R. Weiner and E. Ney, Z. Anal. Chem., 1957, 157, 105.
21.   N.N. Pavlov, A.R. Kuznetsov and G.A. Arbuzov, Izv. Vyssh. Ucheb.
      Zaved. Tekhnol. Leg. Prom., 1960, No. 1, 54; Anal. Abstr.,
      1961, 8, 1934.
22.   R. Přibil and V. Veselý, Chemist-Analyst, 1961, 50, 100.
23.   A.J. Cameron and N.A. Gibson, Anal. Chim. Acta, 1961, 25, 429.
24.   H.M.N.H. Irving and R.H. Al-Jarrah, Chem. Anal. Warsaw, 1962,
      17, 799.
25.   R. Přibil and V. Veselý, Talanta, 1963, 10, 1287.
26.   R. Přibil and V. Veselý, Talanta, 1961, 8, 565.
27.   Y. Horiuchi and I. Iimura, Technol. Rept. Iwate Univ., 1966, 2,
      49; Anal. Abstr., 1968, 15, 2604.
28.   K. Isagai and N. Tateshita, Japan Analyst, 1955, 4, 222; Anal.
      Abstr., 1956, 3, 98.

29. A. De Sousa, Chemist-Analyst, 1961, $\underline{50}$, 9.

6.1.12   Titanium
$$\overline{(\text{Log K}_{\text{TiOY}} = 17.3; \quad \log K_{\text{TiO(H}_2\text{O}_2)\text{Y}} = 20.4)}$$

The ready hydrolysis of titanium(IV) solutions is evident in the
formation of the Ti-EDTA complex.   Polarographic study has indicated
[1,2] that the TiY complex exists only in the narrow pH interval 1 -
2.5, and is reduced reversibly at a half-wave potential of -0.22 V
(vs. NCE), corresponding to the reaction TiY + e = TiY⁻ (these
measurements yield a stability constant for the Ti(III)-EDTA complex
of log K = 21.3).   It is assumed that the titanyl complex
TiOY²⁻ (log K = 17.3) exists at pH values above this range.   The
stability of the complex decreases very rapidly with increasing pH.
Titanium is precipitated quantitatively by ammonia, after a prolonged
time [3].

Direct determination

Direct determination is rarely, if ever, used for visual titrations
of titanium.   However, several methods have been developed, which
are lent a certain importance by their selectivity.   Musha and Ogawa
[4] titrated 0.3 - 3 mg of titanium in 0.05M sulphuric acid medium in
the presence of several drops of hydrogen peroxide.   They followed
the titrations spectrophotometrically at a wavelength of 450 nm.   The
method is based on the formation of the yellow peroxo mixed-ligand
complex [TiO(H₂O₂)Y]²⁻.   Aluminium, alkaline-earth metals and
zirconium (up to 2 mg/ml) do not interfere.

An interesting method was developed by Khadeyev et al. [5], who
titrated titanium amperometrically in 0.2 - 0.4N sulphuric acid medium
containing a 3 - 7-fold excess of EDTA, with a hydrogen peroxide
solution, using a platinum electrode at +0.9 V.   Titanium was also
titrated biamperometrically under the same conditions at 0.7 - 0.8 V.
The peroxide solution must be standardized with a titanium solution.
The authors state that chloride, nitrate, acetate, tartrate, citrate,
a 50-fold amount of oxalate or a 10-fold amount of phosphate do not
interfere.   Of the cations, Al, Cr, Zn, Co, Ni, Mn, Th, Cd, U and
Ag, a 50-fold amount of bismuth or cerium and a 10-fold amount of
zirconium, tungsten or iron do not interfere.   Niobium, tantalum,
vanadium, molybdenum and antimony(III) interfere.

Takao and Musha [6] titrated titanium amperometrically at pH 2 in
hydrochloric acid medium at -0.22 V.   Interferences are marked,
however, nickel, cobalt, copper, iron and vanadium yielding the same
titration curves.

Indirect determination

In the early stages of complexometry a number of methods were
developed based on formation of the poorly stable colourless TiOY²⁻
complex.   Excess of EDTA or DCTA was titrated with various standard
metal solutions, various indicators being used.   This method was not
very successful from the point of view of practical complexometry.

The stabilizing effect of hydrogen peroxide on the Ti-EDTA complex
made a marked contribution to the complexometry of titanium.   The

history of this discovery is interesting.   In a study of the
conditions for the indirect determination of zirconium by back-
titration with ferric chloride, Sweetser and Bricker [7] found that
titanium interfered to a certain degree.   They considered masking
titanium with hydrogen peroxide, but found that the opposite effect
occurred: the Ti-EDTA complex was so stabilized that it could be
determined by spectrophotometric back-titration with ferric chloride,
with salicylic acid as indicator.   Musha and Ogawa demonstrated
spectrophotometrically [4] the existence of this peroxo-complex;   its
stability constant is very high (log K = 20.4).   The high stability
of the complex permits indirect determination of titanium even in
relatively acid solutions.   Sweetser and Bricker [7] determined
titanium by spectrophotometric back-titration at pH 1.9 with ferric
chloride, with salicylic acid as indicator, at a wavelength of 540 nm.
Lieber [8] determined titanium and iron;   in one portion of solution
he titrated the iron (salicylic acid as indicator) without taking
account of possible cloudiness from precipitated Ti(OH)$_4$.   The sum
Fe + Ti was then determined in a second portion by back-titration of
excess of EDTA with iron solution after addition of hydrogen peroxide.
The colour transitions at the equivalence point are not very marked
because of the yellow colour of the peroxo-Ti-EDTA complex.   Wilkins
used copper sulphate solution and PAN [9] or Calcein [10] as
indicator for back-titration of EDTA at pH 4 - 5.   The methods are
not very selective because of the relatively high pH used.

Lassner and Scharf [11] also titrated excess of EDTA, using PAN as
indicator, at pH 5 - 5.2 and 70 - 80°C.   They do not even mention
possible oxidation of EDTA by hydrogen peroxide.   In contrast,
Jurczyk [12] found a marked effect beginning at 40°C.   He recommended
using a maximum of 2 ml of 1% peroxide solution for 10 mg of titanium.

Bieber and Večeřa developed a much better method [13], determining
titanium indirectly by back-titration of excess of EDTA at pH 2 with
bismuth nitrate, using Xylenol Orange as indicator and temperatures
below 20°C.   The yellow colour of the peroxo-Ti-EDTA complex does
not interfere in the colour transition of the indicator, up to 25 mg
of Ti per 100 ml.   The effect of high concentrations of chloride and
sulphate can be eliminated by dilution.   Hirouchi and Osamu [14]
determined titanium similarly, using Methylthymol Blue as indicator.
They used a zinc or lead solution for the back-titration at pH 5 - 5.5.
DCTA also forms a mixed peroxo-complex with titanium, which Lassner
and Scharf [15] used for its determination.   After formation of the
mixed complex, they back-titrated DCTA with copper solution, using
PAN as indicator.   In a later work they recommended using Methylcalcein or Methylcalcein Blue indicator in ultraviolet light [16].
DCTA has the advantage that it does not form a mixed peroxo-complex
with niobium as EDTA does, and thus can be used for determination of
titanium in the presence of niobium.

Interferences

Except for the alkaline-earth metals, most bivalent metals do not
interfere in the determination of titanium at pH 2.   The determination is, however, disturbed by the presence of iron, bismuth,
thorium, zirconium and high concentrations of aluminium.   Titrations
at pH 4 - 5 are far less selective, as all elements of group II
interfere.

Masking of interferents

Titanium determinations are usually needed in analysis of various
alloys, ores, minerals, slags, etc. The methods which have been
developed are concerned with these materials, and especially with the
determination of titanium in the presence of aluminium and iron.
Sulphosalicylic acid is generally recommended for masking small
amounts of aluminium. Large concentrations of aluminium (up to
100 mg per 100 ml of solution) can be masked with salicylic acid or
p-aminosalicylic acid, which form Al-complexes quantitatively in
alkaline medium in the presence of triethanolamine. The Al-complex
formed is sufficiently stable even at pH 1 - 2, at which titanium can
be determined by back-titration with bismuth solution [17]. Jurczyk
[12] found that the Ti-EDTA complex is sufficiently stable with
respect to fluoride, so this can be used to mask small amounts of
aluminium in the presence of up to 5 mg of titanium.

Masking of titanium

Fluoride is an excellent masking agent for titanium and has been
successfully used in the determination of some bivalent elements.
It is, however, not very selective, as it also masks a number of
quadrivalent elements (Zr, Hf), as well as aluminium and the rare-
earth metals. It can be used in indirect titrations only for the
Fe - Ti combination. Sajó [18] demasked the Ti-EDTA complex by
addition of ammonium hydrogen phosphate (pH 6 - 6.5). The EDTA freed
was back-titrated with zinc acetate, with the benzidine-Fe(CN)$_6^{3-}$ /
Fe(CN)$_6^{4-}$ redox system as indicator. The author claimed that
titanium could be determined in the presence of iron and aluminium
with an error of 0.2 mg. In contrast, de Languyon [19] found that
the precipitation of titanium is not quantitative if the phosphate is
not added in large amount (50 ml of concentrated solution); however,
the buffering capacity of this solution prevents optimal pH adjustment for back-titration with zinc. Nestoridis [20] has used a
different indicator for this method. After demasking titanium with
a sufficient amount of phosphate (10 ml of 20% NaH$_2$PO$_4$ solution), he
titrated the freed EDTA with a mixed 0.01M Cu(II) - Zn(II) sulphate
solution, using Xylenol Orange and PAN as indicator. He also
determined titanium and aluminium by successive demasking with
phosphate and fluoride, and considered the method suitable for
determining these elements in ores and alloys.

Chen and Li [21] freed titanium (up to 20 mg) from the Ti-EDTA
complex by use of lactic acid. Aluminium and titanium were
determined by back-titration with lead solution, with Xylenol Orange
as indicator; an excess of 20% lactic acid solution (pH 5.25 - 5.30)
and of lead nitrate were added and after 15 min the surplus lead was
titrated with EDTA.

In alkaline media, titanium forms a colourless complex, Ti(O$_2$)$_4^{4-}$,
with hydrogen peroxide, which has been found useful for masking
titanium in complexometric determinations in alkaline media [22].

Separation methods

It is sometimes necessary to analyse materials containing trace
amounts of titanium, usually photometrically. When titanium is
determined complexometrically together with aluminium, a correction
is applied for the aluminium. When the titanium is present in mg

amounts, complexometric determination is attempted.   Frequently,
however, separation from other components is necessary.   Titanium
can, for example, be extracted as the cupferronate (together with
iron and zirconium) or more selectively from a medium containing EDTA.
Precipitation with sodium hydroxide in the presence of triethanolamine is a very simple method [23].   Very pure $Ti(OH)_4$ is obtained
in this way, and iron and aluminium remain in solution. Simultaneous
addition of EDTA prevents co-precipitation of nickel, cobalt and
copper.   Zirconium, hafnium, thorium, bismuth and the rare-earth
metals are co-precipitated.   After dissolution of the $Ti(OH)_4$,
traces of adsorbed aluminium are masked with salicylic acid.
Combination of this precipitation with masking of aluminium [17]
allows reliable determination of titanium in the presence of up to
1 g of aluminium [24].

There are many extraction and chromatographic methods for the
separation of titanium.   Extraction with a benzene solution of
2-ethylhexyl phosphate (or higher esters) has been described in
detail in connection with the complexometric determination of
titanium.   Kletenik and Bykhovskaya [25] extracted up to 25 mg of
titanium from 2 - 3N sulphuric acid medium, with a 0.3N solution of
this reagent in benzene.   After back-extraction into 5N sulphuric
acid containing hydrogen peroxide and butyl phosphate, they determined titanium complexometrically by back-titration of EDTA with zinc
acetate solution at pH 5.2, using Xylenol Orange as indicator.   They
state that the extraction is not subject to any large interference.

Recommended procedures

Back-titration with bismuth nitrate [12].   Two ml of 30% hydrogen
peroxide solution and sufficient 0.02 - 0.05M EDTA to bind all the
titanium are added to an acid solution containing up to 25 mg of Ti
per 100 ml.   The pH is adjusted to 1 - 2 with ammonia and nitric or
acetic acid and, after addition of several drops of 0.1% Xylenol
Orange solution, the solution is cooled to below 20°C and titrated
with 0.02 - 0.05M $Bi(NO_3)_3$ from yellow-orange to red.

Back-titration with lead nitrate.   Excess of EDTA (or DCTA) is added
to an acid titanium solution, along with 1 ml of hydrogen peroxide,
and the pH is adjusted with solid hexamine to 5 - 5.5.   After
addition of Xylenol Orange indicator, the solution is titrated with
0.02 - 0.05M $Pb(NO_3)_2$ to a red-purple colour.

Analysis of mixtures

Iron and titanium.   Excess of EDTA is added to the acidic solution,
the pH is adjusted to 2, and the solution is back-titrated with
bismuth nitrate (Xylenol Orange as indicator).   The net consumption
of EDTA corresponds to the iron content.   Further EDTA is then added
along with 2 ml of hydrogen peroxide to bind the titanium, and after
cooling the solution is again titrated with bismuth nitrate.   The
net EDTA consumption corresponds to the titanium content.

Titanium in the presence of aluminium [17].   Sodium salicylate
solution (10%, 10 - 20 ml), 20 - 40 ml of 20% triethanolamine solution
(depending on the Al content) and 2 drops of phenolphthalein are
added to an acid solution containing up to 110 mg of Al and 40 mg of
Ti.   Then 2M sodium hydroxide is added dropwise until a purple
colour appears, and then the pH is adjusted to 1 - 2 with nitric acid

(1 + 1).   Then 2M sodium hydroxide is added with constant stirring
until the solution becomes colourless (formation of the Fe-TEA
complex).   The mixture is heated for 1 min and the precipitate of
$Ti(OH)_4$ is left to settle out.   If the aluminium concentration is
high, the mixture should be digested on a hot-plate for 15 min.   The
$Ti(OH)_4$ is filtered off hot, and the filter is washed five times with
1% triethanolamine solution and twice with hot water.   The precipitate is then dissolved in 25 - 30 ml of hot nitric acid (1 + 3).   The
solution is diluted to 400 ml and cooled to below 20°C, 0.05M EDTA
and 10 ml of 10% hydrogen peroxide solution are added, and after
adjustment to pH 1 - 2 the solution is titrated with a bismuth
solution as described above for iron - titanium mixtures.

Titanium in the presence of 0.15 - 1.5 g of aluminium.   At such high
aluminium concentrations, adsorption on the precipitated titanium
hydroxide is inevitable.   Thus the method just given for the
separation of titanium is combined with masking of aluminium with
salicylic acid [17], as described above.

REFERENCES

1.   M. Blumer and I.M. Kolthoff, Experimentia, 1952, 8, 138.
2.   R.L. Peczok and E.F. Maverick, J. Am. Chem. Soc., 1954, 76, 358.
3.   R. Přibil and P. Schneider, Coll. Czech. Chem. Commun., 1950,
       15, 886.
4.   S. Musha and K. Ogawa, J. Chem. Soc. Japan, Pure Chem. Sect.,
       1957, 78, 1686.
5.   V.A. Khadevev, M.V. Krivosheina and D. Mukhamedzhanova, Tr.
       Tashkent. Gos. Univ., 1968, 323, 186;   Anal Abstr., 1969, 17,
       3390.
6.   H. Takao and S. Musha, Japan Analyst, 1961, 10, 160;   Anal.
       Abstr., 1963, 10, 2214.
7.   P.B. Sweetser and C.E. Bricker, Anal. Chem., 1954, 26, 195.
8.   W. Lieber, Zement-Kalk-Gibs, 1956, 9, 216;   Z. Anal. Chem.,
       1960, 177, 429.
9.   D.H. Wilkins, Anal. Chim. Acta, 1959, 20, 113.
10.   D.H. Wilkins, Talanta, 1959, 2, 12.
11.   E. Lassner and R. Scharf, Chemist-Analyst, 1961, 50, 69.
12.   J. Jurczyk, Chem. Anal. Warsaw, 1965, 10, 441;   Anal. Abstr.,
       1966, 13, 4722.
13.   B. Bieber and Z. Večeřa, Coll. Czech. Chem. Commun., 1961, 26,
       2081.
14.   Y. Horiuchi and I. Osamu, Rept. Technol. Ivate Univ., 1967, 20,
       49.
15.   E. Lassner and R. Scharf, Talanta, 1960, 7, 12;   Chemist-Analyst,
       1961, 50, 69.
16.   E. Lassner and R. Scharf, Chemist-Analyst, 1962, 51, 49.
17.   R. Přibil and V. Veselý, Talanta, 1963, 10, 383.
18.   I. Sajó, Acta Chim. Acad. Sci. Hung., 1955, 6, 251;   Anal.
       Abstr., 1956, 3, 732, III.
19.   G.L. de Languyon, Ber. Deutsch. Keram. Ges., 1958, 35, 155.
20.   A. Nestoridis, Anal. Chim. Acta, 1970, 49, 335.
21.   Chen Yung-chao and Huan-Yan Li, Acta Chim. Sin., 1965, 31, 391.
22.   P. Wehber, Z. Anal. Chem., 1957, 154, 182.
23.   R. Přibil and V. Veselý, Talanta, 1963, 10, 233.
24.   R. Přibil and V. Veselý, Chemist-Analyst, 1963, 52, 43.

25.   Yu. B. Kletenik and I.A. Bykhovskaya, Zh. Analit. Khim., 1966,
      21, 1499.

6.1.13   Tin
         $(\log K_{Sn(II)Y} = 22.1$ [1])

Tin(II) and tin(IV) form sufficiently stable complexes with EDTA for
both forms to be determined complexometrically in acid medium.
Determination of the stability constants is considerably complicated
both by the ready hydrolysis of tin salts and by the formation of
hydroxo-complexes.   The stability constant given above was found by
polarographic study of the ThY + Sn(II) = Sn(II)Y + Th equilibrium
[1].

It seems highly probable that the tin(II) complex is more stable at
higher pH values than the tin(IV) complex, since tin is readily and
quantitatively freed and precipitated from the latter by ammonia [2]
whereas tin(II) can be determined in alkaline media.   The greater
stability of the tin(II) complex is also reflected in the slower
oxidation of Sn(II) in EDTA medium by atmospheric oxygen [3], as a
result of a shift of the redox potential of the Sn(II)/Sn(IV)-EDTA
system to more positive values.

However, Kragten [4] has shown from a study of the hydrolysis of
tin(IV) that the thermodynamic stability constant for the Sn(IV)-EDTA
complex should be log $K_{Sn(IV)Y} = 34.5$.

Direct determination

Only one method for determination of tin(II) has been described.
Dubský [5] titrated tin(II) at pH 5.5 - 6 (pyridine/acetic acid) in
tartrate medium containing fluoride, using Methylthymol Blue
indicator.   Tin(IV) is masked by the tartrate and fluoride, and
antimony(III) is masked by the tartrate.

Indirect determination

Takamoto [6] recommends determining tin(II) by back-titration at pH
4 - 5 in acetone medium with cobalt, using potassium thiocyanate as
indicator.   Kubišta [7] determined tin(II) by back-titration of EDTA
with magnesium, using Erio T as indicator.   All other methods
involve determination of tin(IV).   Sajó back-titrated EDTA with zinc,
using the redox indicators 3,3´-dimethylnaphthidine [8] and benzidine
[9] in the presence of ferrocyanide and ferricyanide.   Kinnunen and
Wennerstrand [10] used back-titration with thorium nitrate solution
at pH 2 - 3, and Xylenol Orange as indicator.   Kragten [4] also
recommended back-titration with thorium nitrate, at pH 1.8 - 2.1, but
preferred Semi-Xylenol Orange as indicator;   the method was applied
to analysis of organo-tin compounds.   Dixon et al. [11] developed a
procedure involving binding excess of EDTA by addition of excess of
bismuth nitrate, the surplus of which was then titrated with 0.005M
EDTA (Xylenol Orange indicator).   They claimed that as little as
4 mg of tin could be determined with "excellent precision and
accuracy".   Antimony up to a fifth of the amount of tin did not
interfere, but bismuth and iron were co-titrated.   The method is
suitable for the analysis of solder with trace contents of nickel,
copper or silver.   Jankovský [12] recommended Pyrocatechol Violet as

indicator and zinc acetate for back-titration.   Krleža and Vuletič
[13] employed the replacement reaction between Zn-EDTA complex and
tin(IV).   The freed zinc was titrated with 0.02M EDTA solution in
hexamine medium (pH 5 - 6), with Xylenol Orange as indicator.   Crişan
and Boloş [14] used lead nitrate for back-titration in hexamine
medium, with Catechol Violet as indicator.   They found errors of
0.04 - 0.1 mg in the determination of 50 mg of tin.   Ottendorfer [15]
also titrated excess of EDTA with lead nitrate, but used Xylenol
Orange as indicator.   He also determined lead and copper, using
masking of tin with ammonium fluoride and of copper with thiourea.
Chromý and Vřešt'ál [16] analysed various organic tin compounds by
decomposing them and then determining tin(IV) by back-titration of
excess of EDTA, in acetic acid medium, with copper nitrate solution,
and various derivatives of 1-(2-thiazolylazo)-2-naphthol (TAN) as
indicator.

A determination of both oxidation states of tin in a single solution
was developed by Konishi [17] who applied the Dubský method for
tin(II) [5] and the Kinnunen method [10] for tin(IV).

Compared with the titration at pH 2 - 3, the determination in weakly
acid medium is poorly selective.   Back-titrations with thorium
nitrate suffer from interference from sulphate.   The colour transition in titrations with zinc acetate, using Pyrocatechol Violet as
indicator, is also rendered less favourable by large concentrations
of sulphate [16].   A general difficulty is the pH adjustment, which
must be done very carefully to prevent local alkalization and
precipitation of tin during the neutralization.   Chromý and Vřešt'ál
first add acetic acid (10 ml per 100 ml of solution) to the tin-EDTA
solution, then 2,5-dinitrophenol as indicator, carefully neutralize
with ammonia until appearance of a faint yellow colour, and then add
10 ml of 10% ammonium acetate solution.

Titrations in acid medium have the advantage of simultaneous determination of both oxidation states.

Interferences

All anions which form sufficiently strong complexes with tin, such as
oxalate, tartrate, citrate, fluoride, phosphate and a number of thio-
compounds, such as thioglycollic acid, interfere in the determination
of tin(IV).   Dubský states that tartrate and fluoride do not interfere in the determination of tin(II).   Interference by cations has
not been studied in great detail, as only a small number of quite
simple alloys of tin are generally analysed.   In the indirect determination of tin, copper can be masked with thiourea, and antimony
with tartrate as mentioned above.   Krleža and Vuletič [13] masked
chromium(III) with glycerol.   Traces of iron can be reduced with
ascorbic acid.

All group II cations interfere in the determination of tin in weakly
acid medium, where they are co-titrated.   Tin can be determined by
titrating the sum of all the elements in one portion of the solution
and then masking tin selectively in another portion.   It is also
possible to arrange successive titrations for suitable mixtures of
metals with tin.

Masking of tin

Jankovský [12] recommends using triethanolamine for masking tin;
Chen et al. [18] demasked tin from the Sn(IV)-EDTA complex by using
lactic acid, with quite selective results.   Lactic acid does not
form complexes with bismuth, lead, copper, zinc, cadmium, cobalt or
nickel.   In special cases, tin is masked by a number of other
substances, such as Unithiol, BAL, thioglycollic acid.

Separation methods

Tin can readily be separated during analysis of alloys by dissolving
them with nitric acid and evaporating to low bulk.   The metastannic
acid formed is determined gravimetrically.   Adsorbed traces of
metals are removed by washing the precipitate on the filter with an
EDTA solution [3].

Cheng [19] recommends precipitation of tin in the presence of EDTA
for separation from vanadium.   In the determination of tin in iron
and steels, Wakamatsu [20] also used a procedure involving precipitation of tin in the presence of EDTA and beryllium sulphate by
addition of ammonia.   The mixture of tin and beryllium hydroxides is
dissolved in nitric acid and tin is determined complexometrically.

The extraction of the tin-iodide complex with a chloroform solution
of diantipyrinylmethane [21] is among the more selective extraction
methods;   tin is extracted from a solution containing 20% potassium
iodide and 1 - 2N sulphuric acid.   Cadmium, lead, germanium, titanium
and antimony(III) do not interfere in the extraction.   The tin is
stripped from the organic phase with strongly alkaline EDTA solution
and the excess of EDTA is titrated with bismuth nitrate, after pH
adjustment to 1.5 - 2, the end-point being indicated by appearance of
the yellow BiI_4^- complex.

Recommended procedures

Xylenol Orange as indicator [10].   An approximately two-fold excess
of 0.01M EDTA is added to a hydrochloric acid solution containing up
to 30 mg of tin(IV) and the pH is adjusted to 2 - 2.5 by careful
addition of 20% ammonium acetate solution with thorough stirring.
Several drops of 0.1% indicator solution are added and the excess of
EDTA is titrated with 0.01M thorium nitrate.

Pyrocatechol Violet as indicator [12].   An excess of 0.05M EDTA,
2 ml of glacial acetic acid and several drops of Metanil Yellow or
Thymol Blue are added to an acid tin(IV) chloride solution containing
up to 50 mg of tin.   Ammonia is added dropwise until the indicator
just changes colour.   The pH is then adjusted to the optimum by
addition of 10 ml of 3M sodium acetate and the solution is diluted to
100 ml and heated to 70 - 80°C.   After cooling and addition of
indicator the solution is titrated with 0.05M zinc acetate till blue.

Note.   Furuya and Tajiri [22] recommended that the Sn-EDTA complex
be formed by heating the solution to 30 - 50°C for 3 min.   If copper
is to be masked with thiourea, then the solution must be cooled to a
temperature below 15°C to prevent decomposition of the thiourea.

Successive determinations

Tin and lead [15].   Excess of EDTA is added to a solution containing both metals and the pH to 5 - 6 is adjusted with hexamine.   The solution is diluted to 150 ml, several drops of Xylenol Orange solution are added and the solution is titrated with 0.05M lead nitrate from yellow to red-purple.   The EDTA consumption corresponds to the total lead and tin contents.

Then 2 g of sodium fluoride are added to the titrated solution to bind the tin, and the freed EDTA is slowly titrated with lead nitrate. Towards the end of the titration, the red colour of the Pb-XO complex appears prematurely, and thus it is necessary to wait for the yellow colour to reappear before continuing the titration.   The titration is complete when the red colour lasts for more than 1 min.

Tin, lead and copper [15].   The sum of the three elements is found by the procedure given in the first paragraph of the method above. Then lead and tin are determined in a second portion of solution by masking the copper with thiourea and applying the whole of the procedure above.

Tin in the presence of Pb, Cu, Cd, Zn, Ni, Co [12].   The sum of all the components, including tin, is determined by the recommended Xylenol Orange procedure [10] given above.   Then tin is masked in a second portion of the solution and the other elements are determined in the following manner.

An amount of 0.05M EDTA corresponding to that consumed by all the elements in the first titration is added, followed by 5 ml of 0.5M triethanolamine solution.   The solution is neutralized with ammonia, to Methyl Red as indicator.   If the solution is strongly acidic, it can be partially neutralized with sodium hydroxide.   Then 2 - 3 ml of Schwarzenbach buffer and a small amount of Erio T are added, and the excess of EDTA is titrated with 0.05M zinc acetate from blue to wine-red.   The difference in the EDTA consumptions corresponds to the tin content.

Note.   The method cannot be used in the presence of alkaline-earth metals, which are co-titrated in alkaline medium.

Tin by masking with lactic acid [19].   The sum of tin(II) and other bivalent metals is determined indirectly by back-titration of excess of EDTA at pH 5 - 5.5 with lead nitrate (Xylenol Orange as indicator). Then 10 ml of 20% lactic acid solution and 10 ml of 0.025M lead nitrate (to free tin from its EDTA complex) are added and the excess of lead is determined directly by titration with 0.025M EDTA.

Note.   The first consumption of 0.025M lead nitrate need not be noted, as this titration simply adjusts the conditions for the further determination.   All the metals are then present as their EDTA complexes and only the Sn-EDTA is demasked with lactic acid.

REFERENCES

1.   T.D. Smith, J. Chem. Soc., 1961, 2554.
2.   B. Bieber and Z. Večeřa, Slevarenstvi, 1957, 4, 48.
3.   R. Přibil, Chimia, 1950, 4, 160.

4.   J. Kragten, Talanta, 1975, 22, 505.
5.   I. Dubský, Coll. Czech. Chem. Commun., 1959, 24, 4045.
6.   S. Takamoto, J. Chem. Soc. Japan, Pure Chem. Sect., 1955, 76, 1342; Anal. Abstr., 1956, 3, 2496-III.
7.   Z. Kubišta, Chem. Prumysl, 1954, 4, 175.
8.   I. Sajó, Acta Chim. Acad. Sci. Hung., 1956, 6, 251.
9.   I. Sajó, Magy. Kem. Foly., 1956, 62, 56.
10. J. Kinnunen and B. Wennerstrand, Chemist-Analyst, 1957, 46, 92.
11. M. Dixon, P.J. Heinle, D.E. Humlicek and R.L. Miller, Chemist-Analyst, 1962, 51, 42.
12. J. Jankovský, Coll. Czech. Chem. Commun., 1957, 22, 1052.
13. F. Krleža and N. Vuletič, Croat. Chem. Acta, 1970, 42, 75.
14. I.A. Crişan and V. Boloş, Revta. Chim., 1967, 18, 307; Anal. Abstr., 1968, 15, 4633.
15. L.J. Ottendorfer, Chemist-Analyst, 1958, 47, 96.
16. V. Chromý and J. Vřešt'ál, Chem. Listy, 1966, 60, 1537; Anal. Abstr., 1968, 15, 800.
17. S. Konishi, J. Metal Finish. Japan, 1962, 13, 159; Anal. Abstr., 1963, 10, 4104.
18. Y.C. Chen, C.Y. Hsiao and C.G. Fang, Acta Chim. Sinica 1964, 30, 330; Anal. Abstr., 1965, 12, 5091.
19. K.L. Cheng, Chemist-Analyst, 1963, 52, 12.
20. S. Wakamatsu, Japan Analyst, 1962, 11, 1151; Anal. Abstr., 1964, 11, 1745.
21. V.P. Zhivopistsev and V.S. Minina, Uch. Zap. Perm. Gos. Univ., 1968, No. 178, 196; Anal. Abstr., 1970, 18, 2323.
22. M. Furuya and M. Tajiri, Japan Analyst, 1963, 12, 59; Anal. Abstr., 1964, 11, 4735.

6.1.14   Mercury
$\overline{(\log K_{Hg(II)Y}} = 21.8)$

Although the stability constant of the HgY^{2-} complex is quite large, Ringbom [1] states that the conditional stability constant is never larger than 10^{12}, and then only in media at pH 3 - 5.5.   This large difference between the conditional and thermodynamic constants is largely a result of the formation of hydroxo-complexes.   The mixed complex $Hg(NH_3)Y$ is formed in ammoniacal media.   At pH 10 the apparent constant for HgY^{2-} is only $10^{8.6}$, but is increased to $10^{11.1}$ at the same pH in the presence of ammonia.   The formation of the $Hg(NH_3)Y$ complex was employed in the first published determination of mercury in ammoniacal medium [2,3].

EDTA also reacts with mercury(I), even with the solid salt, causing disproportionation according to the equation

$$Hg_2^{2+} + H_2Y = Hg + HgY^{2-} + 2H^+$$

This reaction has been used as a sensitive detection method for mercury(I) salts [4].

Direct determination

The use of Cu - EDTA - PAN [5] as indicator (PAR was used later [6]) was rapidly superseded by use of the newer indicators, especially Xylenol Orange [7 - 9] and Methylthymol Blue [10].   Diphenylcarbazide [11] has also been recommended for the direct determination of mercury, but the reaction between the indicator complex and EDTA is very slow, which the authors attributed to colloid formation. 1,10-Phenanthroline strongly increases the rate of this reaction but the authors were unable to explain this "catalytic" effect. It seems likely, however, that the mechanism is the same as that for the similar reaction already observed for the·interaction of copper with Xylenol Orange: the indicator is normally blocked, but traces of 1,10-phenanthroline enable direct complexometric determination of copper (see Section 6.2.1). The phenanthroline will form the $Hg(phen)_2$ complex, which will be decomposed by EDTA added after all the free mercury(II) has been titrated:

$$Hg(phen)_2 + H_2Y^{2-} = HgY^{2-} + 2phen + 2H^+$$

and the phenanthroline liberated then presumably displaces the indicator from its mercury complex. Alizarin Complexone [12] and the metallofluorescent indicator, o-dianisidinetetra-acetic acid [13] have also been recommended.

Interferences

Chloride in concentrations up to twice that of the mercury does not interfere in acid media, but higher concentrations make the colour transition, e.g. for Xylenol Orange, less sharp and it becomes difficult to determine the end-point accurately. Chloride interferes far less in alkaline media. In a study of the effect of chloride and thiocyanate, Matteucci [14] found that a chloride concentration of 10^{-3}M can be tolerated in direct titrations of mercury, with Xylenol Orange or Methylthymol Blue as indicator. He later recommended using a substitution reaction between the Zn-EDTA complex and mercury(I), increasing the tolerable chloride concentration to 10^{-2}M [15]. The chloro-complexes of mercury can be demasked by precipitation with silver nitrate, and the mercury can be determined directly without filtering off the AgCl precipitate [7]. It is, however, necessary to carry out the titration rapidly, as the AgCl soon darkens and makes evaluation of the colour transition difficult. Lead or zinc nitrate has been recommended for back-titration of mercury in weakly acid medium [16].

Bromide, iodide, thiocyanate, cyanide and thiosulphate also interfere, but fluoride does not and can be used to mask aluminium etc.

Masking of mercury

Ueno [17] masked mercury with potassium iodide in the analysis of mixtures of copper and mercury. The sum was determined by back-titration of excess of EDTA with copper sulphate, with murexide as indicator. After addition of potassium iodide, the EDTA liberated was determined by further titration with copper. Similarly, Kiss [18] masked mercury with tartrate, citrate or oxalate in the determination of copper in ammoniacal medium, using murexide, PAN, PAR or Chromazurol S as indicator. Up to 30 mg of calcium or magnesium did not interfere.

Barcza and Körös [19] masked mercury with potassium thiocyanate in the determination of bismuth, using Methylthymol Blue indicator at pH 0.7 - 1.2. After the titration, mercury could be demasked from the $[Hg(SCN)_4]^{2-}$ complex by means of silver nitrate and titrated after pH adjustment with hexamine.

Some thio-compounds have been recommended for fairly selective masking of mercury. Thiosemicarbazide [16] demasks mercury from its EDTA complex and does not react with a number of other elements. Copper interferes, however, as it forms an intensely blue thiosemicarbazide complex. Volf [20,21] recommended Unithiol for masking mercury.

The reaction of the Hg-EDTA complex with thiosulphate according to the reaction

$$HgY^{2-} + 2S_2O_3^{2-} + 2H_2O = Hg(S_2O_3)_2^{2-} + H_2Y^{2-} + 2OH^-$$

is interesting. A neutral solution of the HgY^{2-} complex becomes alkaline after addition of thiosulphate, as the Y^{4-} liberated is immediately hydrolysed:

$$Y^{4-} + 2H_2O = H_2Y^{2-} + 2OH^-$$

Sierra and Asensi [22] used this reaction for the acidimetric determination of mercury, using 0.1M hydrochloric acid to titrate the hydroxide produced; Methyl Red and Bromocresol Blue were used as indicators. The method can be used for the determination of mercury in mercury(II) chloride or bromide.

Among reducing agents, ascorbic acid has been used for reduction of mercury in the complexometric determination of bismuth [23].

The Hg-EDTA complex is an oxidant and can be used for the indirect determination of a number of strong reductants, e.g. hydroxylamine, hydrazine, phenylhydrazine, semicarbazide, the hydrazide of isonicotinic acid, etc. Budešínský employed a method [24] involving release of EDTA by reduction of the mercury, and its titration with lead nitrate in hexamine-buffered medium, with Methylthymol Blue as indicator. Specific optimal conditions are required for oxidation of each reductant.

Separation methods

Most methods are concerned with simple pharmaceuticals, in combination with other substances. Mercury can be removed from ores by distillation. Michal et al. [25] distilled mercury from a porcelain crucible in the presence of iron filings and zinc oxide. The mercury vapour was condensed on a water-cooled gold lid, dissolved in nitric acid and determined complexometrically.

Recommended procedures

Xylenol Orange as indicator. The pH of a weakly acid solution of mercury(II) is carefully adjusted to 5 - 5.5 with hexamine solution. Several drops of 0.1% indicator solution are added and the titration is done with 0.02 - 0.05M EDTA, depending on the mercury content, from red to clear yellow.

Alternatively, excess of EDTA is added to a solution containing up to 100 mg of Hg, along with several drops of indicator solution, and the pH is adjusted to 5 - 5.5 with hexamine solution. The surplus EDTA is titrated with zinc or lead nitrate from yellow to red-purple. If certain bivalent metals are present (Pb, Zn, Cd), a sufficient amount of 0.1M thiosemicarbazide is added to the solution after completion of the titration and the liberated EDTA, equivalent to the mercury content, is titrated [16].

Alizarin Complexone as indicator. A weakly acid solution containing 5 - 100 mg of Hg is mixed with 5 ml of acetate buffer (pH 4.3), 5 or 6 drops of indicator and 2 or 3 drops of Xylene Cyanol FF. The solution is then titrated with 0.02M EDTA until green.

PAR as indicator. The mercury(II) solution (5 - 100 mg of Hg) is mixed with 3 ml of acetate buffer (pH 4.3) and 5 or 6 drops of 1% PAR solution, and then titrated slowly with 0.02M EDTA from orange to yellow.

REFERENCES

1.   A. Ringbom, Complexation in Analytical Chemistry, Interscience, New York 1963, pp. 52 - 53.
2.   W. Biedermann and G. Schwarzenbach, Chimia, 1948, 2, 1.
3.   H. Flaschka, Mikrochim. Acta, 1952, 39, 38.
4.   R. Přibil, Coll. Czech. Chem. Commun., 1951, 16, 542.
5.   H. Flaschka and H. Abdine, Chemist-Analyst, 1956, 45, 58.
6.   P. Wehber, Z. Anal. Chem., 1959, 166, 186.
7.   R. Přibil and E. Körös, Acta Pharm. Hung., 1957, 27, 1; Anal. Abstr., 1960, 7, 48, I.
8.   R. Přibil, E. Körös and L. Barcza, Acta Pharm. Hung., 1957, 27, 145; Anal. Abstr., 1960, 7, 48, II.
9.   R. Přibil, E. Körös and L. Barcza, Acta Pharm. Hung., 1957, 27, 243; Anal. Abstr., 1960, 7, 48, III.
10. R. Přibil, E. Körös and L. Barcza, Pharm. Zentralhalle, 1961, 100, 252; Anal. Abstr., 1962, 9, 2237.
11. E. Banyai, E. B.-Gere and L. Erdey, Talanta, 1960, 4, 133.
12. R. Belcher, M.A. Leonard and T.S. West, J. Chem. Soc., 1958, 2390.
13. R. Belcher, D.E. Rees and W.I. Stephen, Talanta, 1960, 4, 78.
14. E. Matteucci, Chim. Ind. Milan, 1966, 48, 829; Anal. Abstr., 1967, 14, 7380.
15. E. Matteucci, Chim. Ind. Milan, 1967, 49, 474; Anal. Abstr., 1968, 15, 4587.
16. J. Körbl and R. Přibil, Coll. Czech. Chem. Commun., 1957, 22, 1771.
17. K. Ueno, Anal. Chem., 1957, 29, 1668.
18. T.A. Kiss, T.M. Suranyi and F.F. Gaal, Mikrochim. Acta, 1969, 266; Anal. Abstr., 1970, 18, 3747.
19. L. Barcza and E. Körös, Chemist-Analyst, 1959, 48, 94.
20. I.A. Volf, Zavod. Lab., 1959, 25, 1438; Anal. Abstr., 1960, 7, 3148.
21. Yu. V. Morachevskii and A.L. Volf, Zh. Analit. Khim., 1960, 15, 69.
22. F. Sierra and G. Asensi, An. Soc. Espan. Fis. Quim. Madrid, Ser. B, 1961, 57, 535.
23. J. Cífka, M. Malát and V. Suk, Coll. Czech. Chem. Commun., 1956, 21, 412.
24. B. Buděšínský, Coll. Czech. Chem. Commun., 1961, 26, 781.
25. J. Michal, J. Janovský and E. Pavlíková, Z. Anal. Chem., 1956, 153, 83.

6.2   COMPLEXOMETRY OF GROUP II CATIONS

This group includes all the other bivalent cations except beryllium and the alkaline earths, the tervalent cations aluminium, yttrium and the rare earths, and molybdenum and vanadium.

Except for manganese(II) and iron(II), these cations give EDTA complexes having stability constants differing by less than three orders of magnitude.   The copper complex is the most stable (log K = 18.8), and the aluminium complex is the least (log K = 16.1).   Thus all these elements can be determined at approximately the same acidity, at pH >4.   This group contains some pairs of elements with very similar behaviour, such as zinc and cadmium or cobalt and nickel. A number of problems arise in the determination of these elements in mixtures, especially at unfavourable concentration ratios, necessitating a search for suitable masking agents and rapid extraction or other separation methods.

Most of these elements are important in the metallurgical industry and their complexometric determination has been studied in some detail.   For example, the determination of copper has become very popular, with at least 100 publications dealing with the determination of copper with a wide variety of indicators, as well as determination by spectrophotometric, amperometric and similar titrations. A great many works have also been devoted to practical applications. The situation is similar for other members of this group and the number of publications is almost too great for complete survey.

The scope of this book permits only mention of the most important data, but these elements will be discussed in sufficient detail in the sections on practical applications.

6.2.1   Copper
        $\overline{(\log K_{CuY}} = 18.80)$

In acid medium, copper(II) forms an intensely blue complex with EDTA: in ammoniacal medium the complex is far less intensely coloured.   The blue colour of the Cu-EDTA complex affects the indicator colour transition, e.g., if the free indicator is yellow at the equivalence point (XO), the resultant colour ranges from green to purple, depending on the pH and the indicator concentration.   Free copper(II) also blocks a number of indicators in both alkaline (Erio T, etc.) and acid (XO, MTB) media.

Copper(I) forms a very weak complex with EDTA (log K = 7) and is not suitable for complexometric determination.   Traces of copper(I) produced in solutions of copper(II), e.g., by addition of potassium thiocyanate, form a redox system which can be titrated potentiometrically with EDTA [1].   Belcher et al. [2] titrated copper in a similar manner in a broad concentration range in ammonium acetate medium without addition of thiocyanate.   Iron(III) and copper can be determined stepwise [3].

Direct determination

The oldest determination procedure for copper still in use, that described by Schwarzenbach [4,5], employs murexide as indicator in

weakly ammoniacal medium and has also been used on the micro-scale
[6]. The method is poorly selective and the ammonia concentration
is a critical factor. Dyes of the Erio T type are not suitable for
the direct determination of copper, as they form strong complexes
with it and are blocked. This blocking can be prevented by addition
of a large amount of ethanol, methanol or propanol to the titration
solution [7].

At least twenty indicators have been recommended for the determination of copper. PAN [8] has become very popular; it forms
insoluble red complexes with copper and other cations in water, so
the reaction with EDTA at the equivalence point is very slow.
Alcohol must be added and the temperature raised. PAR is preferable;
it is soluble in water and its complexes are also soluble [9 - 11].
The colour transition in the titration of copper is from wine red to
blue-green.

Busev et al. [12] recommend using 7-(2-pyridylazo)-8-hydroxyquinoline,
permitting the quite selective determination of copper at pH 2.8.
This is the lowest pH value attained in determinations of copper.

Among other indicators, Pyrocatechol Violet [13] should be mentioned.
Methylthymol Blue and Xylenol Orange also undergo very sensitive
reactions with copper, but both are blocked. This blockage can be
effectively removed by addition of a small amount of 1,10-phenanthro-
line [14]. The reactions occurring near the equivalence point are
similar to those described for mercury (Section 6.1.14). Glycine in
large concentrations has a similar effect [15].

Glycinethymol Blue [16] is a quite selective indicator for copper.
High selectivity, or even specificity, of an indicator reaction for a
particular cation is of little importance in complexometry unless an
absolutely specific titrant is available. Triethylenetetramine
(trien) [17], for example, is a very selective titrant for copper.
Ethylenediamine-$\underline{N},\underline{N},\underline{N}',\underline{N}'$-tetrapropionic acid in combination with
Glycinethymol Blue has been used for the titration of copper [18].
This exceptional reaction will be described at the end of the copper
section.

Among metallofluorescent indicators, Calcein [19] has been recommended for the titration of copper. West states that the best indicator
for copper is o-dianisidinetetra-acetic acid at pH 6 - 7. The
titration is carried out under a mercury lamp and the end-point is
indicated by appearance of a bright blue fluorescence [20].

Indirect determination

This can be done with indicators which are otherwise blocked by
copper. Copper can be determined with a very sharp colour transition by back-titration with lead nitrate solution (Xylenol Orange as
indicator). A further advantage is that the masking action for
copper and other elements is very reliable. Other elements, such as
thorium (pH 4 and XO) [21] or cobalt (KSCN) [22] can be used for the
back-titration. Cameron and Gibson [23] also used a solution of
cobalt and potassium thiocyanate combined with triphenylarsonium
chloride. They extracted the blue ternary complex, formed at the
equivalence point, with chloroform.

Other titrants

Copper has been determined complexometrically with NTA [24] or DCTA, which forms a much stronger complex with nickel than with copper. In ammoniacal medium, the Ni-DCTA complex does not react with potassium cyanide, or does so only slowly, while the Cu-DCTA complex is readily demasked with cyanide. This difference has been employed in the stepwise determination of copper and nickel [25].

Interference

All elements in groups I and II interfere in the determination of copper. In the analysis of complex materials, such as alloys, the copper must be separated, or masked for a differential analysis. There are almost no methods for the determination of copper with masking of accompanying elements.

Separation of copper

In addition to electrolysis, a number of methods have been described for selective isolation of copper from solution. Davis [26] recommends precipitation of copper(I) tetraphenylborate in the presence of ascorbic acid. Only mercury interferes. Isagai [27] precipitates copper with an ethanol solution of salicylidene-2-hydroxybenzimine. Copper is determined complexometrically after isolation of the precipitate and its dissolution in hydrochloric acid. The method is suitable for 0.1 - 2.2 mg of copper. McCurdy et al. [28] studied the separation of copper from zinc and cadmium with thioacetamide. Copper can also be quantitatively precipitated from an acetic acid – tartaric acid solution with potassium ethylxanthate [29]. Mochalov and Bashkirova [30] reduced copper to the metal in ammoniacal medium with sodium borohydride solution. Silver is also reduced. The authors used this reaction for the determination of copper and zinc in silver solders. Karanow et al. [31] described a very effective method, involving the reduction of copper with a lead reductor. The precipitated copper is dissolved in sulphuric acid and determined complexometrically. Numerous elements with more negative standard potentials, such as zinc, nickel, cobalt, etc., are not precipitated by lead. Of the other elements, only silver and bismuth behave similarly to copper. The method has the further advantage that after removal of the copper, any iron in the solution is present as iron(II) and can be determined by permanganate titration. The method is recommended for the determination of copper in copper ores and concentrates.

Masking of copper

In alkaline media copper can be masked quite effectively with potassium cyanide. The advantages and disadvantages of this masking procedure have been discussed in detail in Section 3.1.1. Potassium cyanide is a poorly selective reagent as it binds other cations, such as silver, mercury, cobalt, cadmium, nickel and zinc, in strong cyanide complexes. Thioglycollic acid is also useful in alkaline medium [32], forming a colourless complex with copper(II) ions. It is, however, even less selective than potassium cyanide, as, in addition to the elements mentioned above, it also binds lead, tin, antimony, indium, etc. It is not suitable for masking free nickel and cobalt ions, as it forms coloured complexes with them. However, the EDTA complexes of nickel and cobalt are stable towards thioglycollic acid, which allows the stepwise determination of copper and cobalt or nickel. Iron(III) forms an intensely red complex with thioglycollic acid. In sodium hydroxide and triethanolamine medium iron is firmly bound as the TEA complex and does not undergo this colour reaction. Further details are given in Section 3.1.3.

In acid medium the masking of copper is accompanied by reduction to copper(I), so the solution becomes colourless, which is very useful in visual titrations. Cheng [33] suggested masking copper with sodium thiosulphate, which is quite selective (if present in not too great an excess it does not react with zinc etc.). Determination of the correct amount is quite simple: a fresh thiosulphate solution is added to the weakly acid solution in small amounts until the solution is decolorized. Other elements, such as nickel and cobalt, do not react with thiosulphate. Thiosulphate has been used, for example, in the determination of zinc in brass [34] or in the presence of nickel [35].

Thiourea is also quite selective for copper, binding it in a colourless complex and thus permitting the direct complexometric determination of some bivalent metals in this group, e.g. lead or zinc, with Xylenol Orange as indicator. Iron(III) interferes markedly, however, and must be masked with ammonium fluoride before the addition of thiourea [36]. This method has been used for the indirect determination of iron (pH 2 - 3) in the presence of large concentrations of copper. At pH 4 - 6 the Cu-EDTA complex does not react with thiourea. The reduction of copper in sufficiently acid solutions is, however, quantitative in a short time even in the presence of EDTA. Budevsky and Simova [37] used this reaction for the very selective determination of copper. The sum of all the components is determined by back-titration of EDTA with lead nitrate (Xylenol Orange as indicator). Then thiourea is added and the solution is acidified with nitric acid (1 + 1) to disappearance of the blue colour, the pH is adjusted to 5 - 5.5 with hexamine and the EDTA freed from the Cu-EDTA complex is titrated with lead nitrate. This procedure was employed for the determination of copper in alloys, minerals and copper concentrates.

Rao used a similar procedure [38]. All the elements present are determined by back-titration with lead nitrate, the solution is acidified to pH 0 - 1 and copper is reduced with hydrazine or ascorbic acid and precipitated as CuSCN. After pH adjustment to 5 - 5.5, the EDTA freed is determined as above. In agreement with the results of other authors, the method cannot be used in the presence of iron, manganese or cadmium. These two methods do not seem particularly suitable. The double pH adjustment leads to an accumulation of salts in solution, thiourea decomposes, and the precipitation of CuSCN seems unnecessary.

Körös and Remport-Horváth recommended that copper be reduced with ascorbic acid in the presence of iodide or thiocyanate, and precipitated as copper(I). This procedure has no advantages over those already described.

Recommended procedures

Murexide as indicator. An acid solution containing up to 20 mg of Cu per 100 ml is carefully neutralized with 1M ammonia solution to about pH 8. About 10 ml of ammonium chloride solution is added to weakly acid solutions.   The indicator is added (1:100 solid mixture with KNO_3) and titration is carried out with 0.02M EDTA from yellow to purple.   The colour transition is sufficiently sharp to permit titration with 0.001M EDTA.

Note.   Copper hydroxide is initially precipitated during the neutralization and then is readily dissolved by a small excess of ammonia. Large ammonia concentrations lead to formation of the tetra-ammine complex, $Cu(NH_3)_4^{2+}$, preventing the determination.   During the titration, the solution is neutralized by the hydrogen ions released and the pH can fall below the optimal value of $7 - 8$, this being shown by a clear change in the colour of the indicator.   Addition of 1 or 2 drops of ammonia just before the expected end of the titration considerably improves the colour transition.   The presence of ammonium chloride is necessary to give the requisite buffering capacity.   Triethanolamine cannot be used, as it forms a quite stable complex with copper.

Xylenol Orange as indicator.   An excess of $0.02 - 0.05M$ EDTA is added to an acid solution containing $2 - 25$ mg of Cu, and the solution is either partially neutralized with sodium hydroxide or the pH is adjusted to $5 - 5.5$ with solid hexamine.   The indicator is added and the solution diluted if necessary, and then titrated with $0.02 - 0.05M$ lead nitrate from a yellow or yellowish colour to red.

The Sanderson and West method [18].   This method is interesting in that it is an attempt to determine a single element very selectively by using a very selective titrant and indicator.   The titration is done with the sodium salt of ethylenediaminetetrapropionic acid (EDTPA), with Glycinethymol Blue as indicator.   The method is based on the very different properties of the EDTPA complexes of copper, zinc, cadmium, manganese, etc. from those of the EDTA complexes:

$\log K_{MY}$	Cu	Zn	Δ(Cu-Zn)	Cd	Δ(Cu-Cd)	Mn	Δ(Cu-Mn)
EDTA	18.8	16.5	2.3	16.5	2.3	14.0	4.8
EDPTA	15.4	7.8	7.6	6.0	10.4	4.7	10.7

Procedure.   A copper solution corresponding to 5 ml of 0.02M Cu is pipetted into a conical flask along with 5 ml of buffer (pH 5.8) and the solution is diluted to 100 ml.   Then 10 drops of freshly prepared 0.1% Glycinethymol Blue solution are added and the solution is titrated with 0.02M EDTPA from blue to yellow-green or green (depending on the copper content).   The colour transition is very sharp.   The pH 5.8 buffer is made by dissolving 150 g of sodium acetate and 4.6 ml of glacial acetic acid in water, and diluting to 1 litre.   The pH is checked with a pH-meter.

The authors carried out a number of determinations of copper in the presence of lead, zinc, cadmium, aluminium, magnesium and other alkaline earth metals with satisfactory results.   Of the common metals tested, only nickel interferes, giving a positive error. Iron(III) is expected to interfere, but could probably be masked with TEA.

This method, published twenty years ago, has not found practical

application.   One reason is that EDPTA is not yet produced commercially.   In addition, it does have limitations (iron and nickel must
not be present).

REFERENCES

1.   R. Přibil, Z. Koudela and B. Matyska, Coll. Czech. Chem. Commun.,
       1951, 16, 80.
2.   R. Belcher, D. Gibbons and T.S. West, Anal. Chim. Acta, 1955,
       13, 266.
3.   R. Přibil, Chimia, 1952, 6, 185.
4.   G. Schwarzenbach, Helv. Chim. Acta, 1946, 29, 1338.
5.   W. Biedermann and G. Schwarzenbach, Chimia, 1948, 2, 56.
6.   H. Flaschka, Mikrochim. Acta, 1952, 20, 324.
7.   T.A. Kiss and V.D. Canic, Glasnik Hem. Druzstva Beograd, 1962,
       27, 5;   Anal. Abstr., 1963, 10, 3091.
8.   K.L. Cheng and R.H. Bray, Anal. Chem., 1955, 27, 782.
9.   L. Sommer and M. Hniličková, Naturwissenschaften, 1958, 45, 544.
10.   P. Wehber, Z. Anal. Chem., 1959, 166, 186.
11.   F.J. Langmyhr and H. Kristiansen, Anal. Chim. Acta, 1959, 20,
       524.
12.   A.I. Busev, V.M. Ivanov and L.I. Talipova, Zh. Analit. Khim.,
       1963, 18, 33;   Anal. Abstr., 1963, 10, 5061.
13.   V. Suk, M. Malát and A. Jeníčková, Coll. Czech. Chem. Commun.,
       1955, 20, 158.
14.   R. Přibil, Talanta, 1959, 3, 91.
15.   W. Berndt and J. Šára, Talanta, 1960, 5, 281.
16.   J. Körbl, E. Kraus and R. Přibil, Coll. Czech. Chem. Commun.,
       1958, 23, 1219.
17.   H. Flaschka and S. Soliman, Z. Anal. Chem., 1957, 158, 253;
       1957, 159, 29.
18.   I.P. Sanderson and T.S. West, Talanta, 1962, 10, 247.
19.   F. Vydra, R. Přibil and J. Körbl, Coll. Czech. Chem. Commun.,
       1959, 24, 1623.
20.   R. Belcher, D.I. Rees and W.I. Stephen, Talanta, 1960, 4, 78.
21.   J. Kinnunen and B. Wennerstrand, Chemist-Analyst, 1957, 46, 92.
22.   S. Takamoto, J. Chem. Soc. Japan, Pure Chem. Sec., 1955, 76,
       1344.
23.   A.J. Cameron and N.A. Gibson, Anal. Chim. Acta, 1961, 25, 429.
24.   P. Wehber and W. Johannsen, Z. Anal. Chem., 1956, 153, 324.
25.   R. Přibil, Coll. Czech. Chem. Commun., 1955, 20, 162.
26.   D.G. Davis, Anal. Chem., 1960, 32, 1321.
27.   K. Isagai, J. Chem. Soc. Japan, Pure Chem. Sec., 1961, 82, 101;
       Anal. Abstr., 1962, 9, 4083.
28.   W.H. McCurdy, W.J.A. Vandenheuvel and A.R. Casazza, Anal. Chem.,
       1959, 31, 1413.
29.   M. Szelag and M. Kozlicka, Chem. Anal. (Warsaw), 1962, 7, 815.
30.   K.N. Mochalov and T.I. Bashkirova, Zavod. Lab., 1969, 35, 795;
       Anal. Abstr., 1970, 19, 2080.
31.   R. Karanow, A. Karolew and D. Toschewa, Talanta, 1962, 9, 409.
32.   R. Přibil and V. Veselý, Talanta, 1961, 8, 880.
33.   K.L. Cheng, Anal. Chem., 1958, 30, 1347.
34.   A. Karolev and O.B. Budevsky, Godichnik Nauch. Ins. Met., 1959,
       1, 163.
35.   G.G. Lystsova, Zavod. Lab., 1961, 27, 964;   Anal. Abstr., 1962,
       2, 971.
36.   R. Přibil and V. Veselý, Talanta, 1961, 8, 743.

37. O.B. Budevsky and L. Simova, Talanta, 1962, 9, 769.
38. B.V. Rao, Indian J. Technol., 1971, 9, 157; Anal. Abstr., 1972, 22, 625.
39. E. Köros and Zs. Remport-Horváth, Chemist-Analyst, 1957, 46, 91.

6.2.2   Lead
$\overline{(\text{Log } K_{PbY}} = 18.0)$

Lead can be readily determined directly in acid or alkaline medium.
No indicators are blocked by lead.

Direct determination

In alkaline media lead can be determined with Erio T as indicator in
solutions containing tartrate [1] or preferably triethanolamine [2].
Both these complexing agents prevent precipitation of lead hydroxide.
The colour transition is not as sharp as with Methylthymol Blue,
which is a better indicator.

At least a dozen indicators can be used in acid medium.   The following have been recommended: Pyrocatechol Violet [3], Alizarin
Complexone [4], Pyrogallol Red and Bromopyrogallol Red [5], and
especially Xylenol Orange [6] and Methylthymol Blue [7].   The colour
change of these last two indicators at pH 5 - 5.5 in solutions buffered with hexamine is the sharpest.

Kotrlý and Vytřas [8] studied these indicators in detail (except for
Alizarin Complexone) for the microdetermination of lead and found
that Xylenol Orange is the best, yielding even better results than
potentiometric or amperometric end-point detection.

Dithizone is a very sensitive indicator for lead and has been studied
in detail by Bovalini and Casini [9], Wänninen and Ringbom [10] and
Hara [11]. Kotrlý [12] used dithizone for the microdetermination of
up to 40 µg of lead.

In some communications ammonium acetate is recommended as a buffer
for the titration of lead, e.g. after its use for dissolving lead
sulphate, but the concentration of acetate must be as low as possible,
so as not to interfere in the colour transition of the indicator.
Ringbom [13] calculated that the error in the titration of 10^{-4}M lead
in a solution of 0.1M acetate (pH 5) is -10%, while the error at the
same pH in solutions buffered with hexamine is only -0.1%.

Indirect determination

Back-titrations of excess of EDTA with magnesium (Erio T as indicator)
or calcium (Methylthymol Blue as indicator) are primarily suitable
when masking agents such as potassium cyanide are used.

Separation of lead

The determination of lead is almost completely non-selective.   Only
in acid medium can lead be directly determined in the presence of
other metals giving EDTA complexes, namely the alkaline-earth metals,
and thus lead must generally be separated from interfering elements.
In the analysis of alloys, ores, etc., lead is usually separated as

the sulphate, and after this has been dissolved in ammonium acetate
solution, the lead is determined directly by titration with EDTA (see
the comments [13] above). Very good results can be obtained by
dissolving lead sulphate with EDTA, the excess of which is back-
titrated at pH 5 - 5.5, e.g. with zinc chloride (Xylenol Orange as
indicator).

Other separation methods are based on precipitation of lead as
sulphide with thioacetamide [14] or ammonium sulphide [15]. Small
amounts of lead can be separated by precipitation with sodium
diethyldithiocarbamate and extraction into chloroform [16,17]. Mahr
and Otterbein [18] proposed using thiourea (THU) for separation of
lead, which is precipitated from a small volume at 0°C as
$Pb(THU)_6(NO_3)_2$, and determined complexometrically after dissolution of
the precipitate. This method is especially recommended for
separation of lead from antimony and tin. Majumdar and Singh [19]
separated lead with Bismuthiol II in the presence of other complexing
agents (cyanide, tartrate and citrate) but gave results only for pure
lead solutions.

Ziegler [20] recommends ion-exchange separation of lead on Dowex-1
(sulphate form). None of these methods has found practical
application.

Masking of interferents

Many cations can be readily masked, and then lead can be determined
quite selectively. Generally potassium cyanide is used for certain
bivalent metals, fluoride for aluminium and tin, acetylacetone for
aluminium and traces of iron, and triethanolamine for aluminium, iron,
tin and antimony. Harzdorf [21] used combined masking with cyanide +
TEA + tartrate for the determination of lead in the presence of
copper, nickel, zinc, cadmium, aluminium, iron, tin(IV), antimony(V)
and bismuth. 1,10-Phenanthroline can be used in acid medium to mask
certain bivalent elements [22]. Thiourea is useful for masking
copper in weakly acid solutions.

On the other hand, the relatively low stability constant of the Pb-
EDTA complex allows stepwise titration of many binary mixtures, such
as bismuth and lead [23 - 25], or iron and lead [26]. By use of pH
control together with masking agents, such as 1,10-phenanthroline or
ammonium fluoride, ternary mixtures of the Bi-Pb-Cd or Bi-Pb-Sn type
can readily be analysed.

Masking of lead

No very selective masking agents for lead are known. In special
cases, 2,3-dimercaptopropanol (BAL) [27], thioglycollic acid [28],
β-mercaptopropionic acid [11] or Unithiol [29] can be used. Large
lead concentrations can be easily masked with sodium hydroxide,
allowing, for example, determination of cadmium in the presence of
lead and zinc by back-titration of excess of EDTA with calcium,
(Cresolphthalein Complexone as indicator) [30].

Recommended procedures

Determination in acidic solutions. Several drops of 0.2% Xylenol
Orange solution are added to an acid solution containing up to 50 mg
of lead, followed by solid hexamine in small amounts until a red-

purple colour appears.   Very acid solutions should be partially
neutralized with sodium hydroxide.   The solution is then titrated
with 0.02 - 0.05M EDTA, till lemon yellow.

Masking.   If copper is present, 10% thiourea solution is added to
the sample solution until it becomes colourless, and then 2 ml more.
If zinc, nickel, cobalt, etc. are present in addition to copper, but
in not too large concentrations, then to mask them a sufficient
amount of 0.15M 1,10-phenanthroline is added before the titration.

Determination in alkaline medium.   The lead solution is mixed with
10 ml of 10% triethanolamine solution and alkalized with concentrated
ammonia solution.   A small amount of Methylthymol Blue is added (a
1:100 mixture with KNO_3) and the solution is titrated with EDTA until
its colour turns from intense blue to smoke-grey.

Masking.   Iron and aluminium are masked directly with triethanolamine.   Copper, zinc, cadmium, nickel, cobalt and silver can be
masked by addition of potassium cyanide.   If alkaline-earth metals
are present, then the total sum is determined by back-titration with
calcium, then after addition of sufficient 10% thioglycollic acid
solution, the titration is continued, the additional calcium
consumption corresponding to the lead content.

Determination of lead in the presence of bismuth.   See Section 6.1.5.

REFERENCES

1.   H. Flaschka, Mikrochim. Acta, 1952, 22, 315.
2.   R. Přibil, Coll. Czech. Chem. Commun., 1954, 19, 64.
3.   J. Vřešťál and S. Kotrlý, Coll. Czech. Chem. Commun., 1957, 22,
      1775.
4.   R. Belcher, M.A. Leonard and T.S. West, J. Chem. Soc., 1958,
      2390.
5.   A. Jeníčková, M. Malát and V. Suk, Coll. Czech. Chem. Commun.,
      1956, 21, 1599.
6.   J. Kórbl, R. Přibil and A. Emr, Coll. Czech. Chem. Commun., 1957,
      22, 961.
7.   J. Kórbl and R. Přibil, Coll. Czech. Chem. Commun., 1958, 23,
      3726.
8.   S. Kotrlý and K. Vytřas, Coll. Czech. Chem. Commun., 1968, 33,
      287.
9.   S. Bovalini and A. Casini, Ann. Chim. Roma, 1953, 43, 287.
10.   E. Wänninen and A. Ringbom, Anal. Chim. Acta, 1955, 12, 308.
11.   S. Hara, Japan Analyst, 1961, 10, 633.
12.   S. Kotrlý, Coll. Czech. Chem. Commun., 1957, 22, 1765.
13.   A. Ringbom, Complexation in Analytical Chemistry, Interscience,
      New York, 1963, p. 93.
14.   A.A. Amin and M.Y. Farah, Chemist-Analyst, 1955, 44, 62.
15.   I. Weidenfeld, Israel J. Chem., 1966, 4, 247.
16.   J. Kinnunen and B. Wennerstrand, Chemist-Analyst, 1954, 43, 65.
17.   A. Hitchen, Mines Branch Report R 128, Dept. Mines Tech. Survey,
      Ottawa, Canada, 1965.
18.   C. Mahr and H. Otterbein, Z. Anal. Chem., 1955, 144, 28.
19.   A.K. Majumdar and B.R. Singh, Z. Anal. Chem., 1957, 156, 265.
20.   M. Ziegler, Z. Anal. Chem., 1961, 180, 1.
21.   C. Harzdorf, Z. Anal. Chem., 1964, 203, 101.

22. R. Přibil and F. Vydra, Coll. Czech. Chem. Commun., 1959, 24, 3103.
23. R. Přibil, Analyst, 1958, 83, 188.
24. R. Přibil and M. Kopanica, Chemist-Analyst, 1959, 48, 87.
25. S. Kotrlý and J. Vřešt'ál, Coll. Czech. Chem. Commun., 1960, 25, 1148.
26. K. Monoki and J. Sekino, Japan Analyst, 1963, 12, 1149.
27. R. Přibil and Z. Roubal, Coll. Czech. Chem. Commun., 1954, 19, 1162.
28. R. Přibil and V. Veselý, Talanta, 1961, 8, 880.
29. J.V. Morashevskii and A.L. Volf, Zh. Analit. Khim., 1960, 15, 69.
30. R. Přibil and V. Veselý, Chemist-Analyst, 1966, 55, 38.

6.2.3   Nickel
$\overline{(\log K_{NiY}} = 18.6)$

Direct determination

In 1948, Schwarzenbach described the determination of nickel in ammoniacal medium, with murexide as indicator; this was one of the first applications of titration with EDTA. In contrast to the similar determination of cobalt or copper, the concentration of ammonia is not very critical here; the colour transition is, in fact, strongest in concentrated ammonia medium (pH 11 - 12), with a sharp transition from yellow to a brilliant purple colour. The determination with murexide as indicator has also been studied on the micro-[2] and even the ultramicro-scale [3]. Pyrocatechol Violet [4] and Chrome Azurol S [5] have also been recommended as indicators for the determination of nickel in alkaline media. The determination in alkaline media is, however, completely non-selective, as all other metals titratable with EDTA (including the alkaline-earth metals) interfere.

At least ten indicators have been recommended for use in acid media, but have not found wide application. Xylenol Orange and Methyl-thymol Blue undergo sensitive reactions with nickel, but are blocked by it. In contrast to the case with copper, no means of breaking this blockage have been found. It must be borne in mind during direct titrations that the reaction of nickel with EDTA is rather slow and that the titration should be carried out slowly, especially near the end-point.

Indirect determination

Older methods based on back-titrations with zinc, magnesium or manganese solutions, with Erio T as indicator in ammoniacal media,are of little importance, as they are poorly selective, and even iron cannot be masked with triethanolamine. Back-titrations in alkaline media with calcium, with Thymolphthalexone, Methylthymol Blue or Calcein (Fluorexone) as indicators are very reliable, and back-titration with thorium nitrate at pH 2.8 (Alizarin S as indicator) is quite selective [6]. Flaschka exploited the strength of the Ni-EDTA complex and back-titrated excess of EDTA at pH 2 and 0°C, with Pyro-catechol Violet as indicator. Indirect titrations have the advantage that they can be done quickly.

Other titrants for nickel

DCTA forms a very strong complex with nickel (log K = 19.4) and has
been suggested for the determination of nickel (or cobalt) and
copper, zinc or cadmium [8]. In ammoniacal buffer medium the Ni-
DCTA complex does not react with potassium cyanide (or reacts only
very slowly), while copper, zinc and cadmium react quantitatively.
Magnesium chloride and Erio T are used for the back-titration. In
sodium hydroxide medium the Ni-DCTA complex does react with cyanide
[9]. Martínez and Castro [10] used EGTA to determine nickel. In
strongly ammoniacal medium nickel reacts with TTHA to form the Ni_2Y
binuclear complex. Back-titration of TTHA with a nickel solution
(murexide as indicator) yields a very sharp end-point.

Wilkins [11] recommended using 2-hydroxyethylenediaminetriacetic acid
(HEDTA) to determine nickel, aluminium and manganese in a single
solution. At pH 5, HEDTA bonds only nickel and aluminium in strong
complexes. The excess of HEDTA was determined by back-titration
with a copper solution with Methylcalcein or Methylcalcein Blue as
indicator. The end-point is shown by quenching of the fluorescence.
After demasking of the aluminium with fluoride, the titration with
copper solution is continued to determine aluminium. Then the pH is
adjusted to 9.5 and manganese is determined directly with an EDTA
solution. Good results are obtained for 1:1:1 mixtures of Ni, Al
and Mn.

Interferences

All the elements of groups I and II interfere in the determination of
nickel in weakly acid medium. Only rarely can this difficulty be
avoided by suitable masking, e.g. with fluoride (Al, etc.), thiourea
or sodium thiosulphate (Cu). Potassium fluoride can also be used to
mask iron as insoluble K_3FeF_6 [12]. Triethanolamine is useful in
alkaline media for masking iron, aluminium and traces of manganese
[13] and fluoride can be used to mask aluminium and calcium (CaF_2).
The Ni-EDTA complex is stable towards some thio-compounds, such as
BAL [14] and thioglycollic acid (TGA) [15]. Other EDTA complexes,
e.g. with copper, lead, zinc or cadmium are demasked by both these
substances, enabling stepwise determination of Ni-Cu, Ni-Zn or Ni-Pb
combinations. When potassium cyanide is also used, ternary mixtures
such as Ni-Pb-Zn or Ni-Pb-Cd can readily be analysed. Triethanolamine cannot be used in ammoniacal media, as the Fe-TEA complex
produced in masking of iron reacts with thioglycollic acid to produce
the intensely red Fe-TGA complex. This reaction does not occur in
sodium hydroxide medium and the Fe-TEA complex in solution remains
colourless after addition of TGA.

Masking of nickel

The only reliable masking agent for nickel in alkaline medium is
potassium cyanide. The $Ni(CN)_4^{2-}$ complex is stable towards formaldehyde or chloral hydrate, in contrast to the cyanide complexes of
cadmium and zinc, which are reaily demasked. Nickel can readily be
released from the cyanide complex by silver nitrate solution; the
cyanide complex of cobalt(III) is much more stable and does not undergo this reaction (see Section 6.2.4.3).

Only 1,10-phenanthroline is useful for masking small concentrations
of nickel in acid medium (but the reagent is expensive).

Separation of nickel

Dimethylglyoxime is most commonly used for separation of nickel;   it
has been recommended primarily for determining nickel in steels and
some alloys [16 - 20] and has been tested in various modifications by
different authors [21,22].   The precipitate is collected, and
decomposed in a suitable manner (hydrochloric or nitric acid) and the
nickel is determined complexometrically, with murexide as indicator.

Nickel can be extracted as the thioglycollate from alkaline medium,
with a chloroform solution of trioctylmethylammonium chloride
(Aliquat S-336).   Tervalent iron and cobalt are masked with
triethanolamine.   Chromium(VI) is extracted previously with an
Aliquat solution.   The nickel thioglycollate is transferred from the
chloroform phase into 0.2 - 0.3M nitric acid and determined complexometrically by back-titration of EDTA with a lead solution (Xylenol
Orange as indicator).   The method has been applied to the analysis
of alloys containing iron, cobalt, chromium, molybdenum and other
elements [23].

Extraction of interferents is used primarily when masking is not
suitable (e.g. for high concentrations).   For example, it is
recommended that iron be extracted with isobutyl methyl ketone from
6M hydrochloric acid [24].   This method is also suitable in determination of iron in the presence of large amounts of nickel.   Kurz
and Kober [25] developed an interesting method for the extraction
separation of iron, cobalt and nickel.   First iron is extracted from
hydrochloric acid medium with isobutyl methyl ketone and then cobalt
is extracted with the same reagent from ammonium thiocyanate medium.
Nickel remains in the aqueous phase and is determined complexometrically by back-titration with lead nitrate (Xylenol Orange as
indicator).   Cobalt can be determined directly in the organic phase.
Iron is determined complexometrically after stripping into water.
The authors state that the whole analysis takes an hour.

Iron and copper in large amounts (200 mg) in the presence of nickel
or cobalt in concentrations not exceeding 10^{-3}M can be selectively
extracted as the phenylacetates from hexamine medium [26].

A great deal of attention has been devoted to the determination of
nickel and cobalt, which will be discussed in the chapter on cobalt.

Recommended procedures

Two or three methods are of practical importance in complexometry.

Murexide as indicator.   A few drops of indicator are added to a
solution containing 2 - 15 mg of nickel in 100 ml, followed by dropwise addition of ammonia solution until the solution is a clear
yellow in colour.   The solution is titrated with 0.02 - 0.05M EDTA
until the colour begins to change, then 10 ml of concentrated ammonia
solution are added and the titration is continued until the colour
changes to blue-purple [2].

We have found that a nickel solution can be titrated directly in
strongly ammoniacal medium.   The colour transition is very sharp.
If the ammonia concentration is not high enough, the murexide colour
passes from yellow through orange-red to purple, but this method can
only be used for pure nickel solutions, e.g. after isolation by

precipitation with dimethylglyoxime.

Indirect titration, with Thymolphthalexone as indicator.   A small
excess of EDTA is added to a solution containing up to 25 mg of
nickel, along with 10 ml of concentrated ammonia solution;   the
solution is then diluted to 100 - 200 ml and titrated with 0.05M
calcium chloride to an intense blue colour.   Nickel can be determined in sodium hydroxide medium in the same manner;   accompanying
elements can be masked with a number of reagents (TEA, TGA, BAL, etc.).

Indirect titration, with Xylenol Orange as indicator.   Excess of
EDTA is added to an acid nickel solution and the pH is adjusted to
5 - 5.5;   after dilution, the solution is back-titrated with 0.02 -
0.05M lead nitrate till red.   Aluminium and other elements can be
masked with fluoride, and copper with thiourea.

PAR as indicator.   Two ml of 2M hydrochloric acid and about 1 g of
hexamine are added to a solution containing 5 - 25 mg of nickel in
100 ml.   When the buffer has dissolved, 5 - 8 drops of 0.1% indicator
solution are added and the solution is titrated with 0.02M EDTA from
red to yellow-green.

REFERENCES

1.   G. Schwarzenbach, Komplexon-Methoden, Siegfried, Zofingen, 1948.
2.   H. Flaschka, Mikrochem. ver. Mikrochim. Acta, 1952, 39, 38.
3.   H. Flaschka and F. Sadek, Mikrochim. Acta, 1957, 1.
4.   M. Malát and V. Suk, Chemist-Analyst, 1956, 45, 30.
5.   M. Malát and M. Tenorová, Coll. Czech. Chem. Commun., 1959, 24, 632.
6.   K. ter Haar and J. Bazen, Anal. Chim. Acta, 1956, 14, 209.
7.   H. Flaschka and R. Püschel, Z. Anal. Chem., 1955, 147, 353.
8.   R. Přibil, Coll. Czech. Chem. Commun., 1955, 20, 162.
9.   R. Přibil, unpublished results.
10.   F. Bermejo-Martínez and M. Paz Castro, An. Real. Soc. Espan. Fis. Quim. Madrid, Ser. B., 1959, 55, 803.
11.   D.H. Wilkins, Anal. Chim. Acta, 1960, 23, 309.
12.   R. Přibil and V. Veselý, Talanta, 1965, 12, 385.
13.   R. Přibil, Coll. Czech. Chem. Commun., 1954, 19, 58.
14.   R. Přibil and Z. Roubal, Coll. Czech. Chem. Commun., 1954, 19, 1162.
15.   R. Přibil and V. Veselý, Talanta, 1961, 8, 880.
16.   H. Flaschka, Chemist-Analyst, 1953, 42, 84.
17.   V.A. Nelson and L.J. Wrangell, Anal. Chem., 1960, 32, 99.
18.   L.L. Lewis and P.J. Straub, Anal. Chem., 1960, 32, 96.
19.   H. Green and P.J. Rickards, Brit. Cast Iron Steel Res. Assoc. J., 1964, 12, 578; Anal. Abstr., 1966, 13, 172.
20.   J. Iron Steel Inst., 1971, 209, 215; Anal. Abstr., 1972, 22, 154.
21.   M.A. Yagodnitsyn, Zavod. Lab., 1969, 35, 925; Anal. Abstr., 1970, 19, 2265.
22.   V.I. Bogdanova, Mater. Genet. Eksp. Mineral, 1971, 6, 260.
23.   R. Přibil and V. Veselý, Hutn. Listy, 1971, 26, 745.
24.   A.F. Bogenschütz and U. George, Metalloberflache, 1970, 24, 159; Anal. Abstr., 1971, 21, 3412.
25.   E. Kurz and G. Kober, Z. Anal. Chem., 1971, 254, 127.
26.   J. Adam, R. Přibil and V. Veselý, Talanta, 1972, 19, 825.

6.2.4   Cobalt
$$\frac{Cobalt}{(\log K_{CoY} = 16.31)}$$

Both oxidation states of cobalt are used in complexometry. The cobalt(II) complex is readily oxidized in alkaline medium to the blue $Co(OH)Y^{2-}$ complex, which, either by boiling or acidification, can be converted through the $Co(H_2O)Y^-$ complex into CoY^-. Both complexes have an intense red-purple colour. The oxidation can be done in cold solution with bromine and in weakly acid medium with cerium(IV) sulphate [1]. The CoY complex is very stable $(\log K = 36)$ and is stable towards substitution. Its intense colour has been used for the photometric determination of cobalt and has been variously modified [3]. The high stability, even in very acid solutions, permits very selective complexometric determination, which is, however, limited to only a few mg of Co because of the intense colour of the solution.

6.2.4.1   Determination of cobalt(II)

Direct determination

Complexometric methods for determining cobalt developed simultaneously with methods for determining nickel, and what was said of nickel in the previous section is also roughly true of cobalt. Only a few indicators have been recommended for the direct determination, e.g. Pyrocatechol Violet, Pyrogallol Red and Bromopyrogallol Red. Xylenol Orange can also be used, but only in hot solutions. A good deal of attention has been paid to the use of potassium thiocyanate as an indicator for cobalt, both in aqueous solution and in various mixtures with acetone and amyl alcohol [4,5].

Murexide has also been recommended for the direct determination of cobalt in alkaline media, but even under optimal conditions, the colour transition is far less sharp than that for nickel.

Indirect determination

Cobalt(II) can be determined in ammoniacal medium, e.g. by back-titration with zinc or magnesium solution, with Erio T as indicator. Calcium (Thymolphthalexone as indicator) is used in sodium hydroxide medium.   Copper and PAN [6,7] or zinc and dithizone can be used for the back-titration in weakly acid media [8].   Good transitions are obtained in solutions containing 50% alcohol.   Cameron and Gibson [9] used an extraction method, back-titrating with cobalt(II) sulphate solution and using the KSCN-triphenylarsonium chloride system as indicator (see Section 6.1.11).   Irving and Al-Jarrah [10] used a similar procedure and extracted the $Co(SCN)_4^{2-}$ produced at the equivalence point into a dichloroethane solution of tetrahexyl-ammonium chloride or trioctylmethylammonium chloride.   Back-titration can be reliably done with lead nitrate (Xylenol Orange as indicator).

6.2.4.2   Determination of cobalt(III)

Only indirect determinations can be used, after formation of the Co(III)-EDTA complex; the intense purple colour of the complex

limits the determination to a few mg of cobalt.   Kinnunen and
Wennerstrand [11] titrated excess of EDTA with thorium nitrate at pH
3, using Xylenol.Orange as indicator.   A calcium solution containing
Fluorexone (Calcein) indicator is used in alkaline media [12];   a
procedure is employed in which hydrogen peroxide is added to a cold
alkaline solution of the Co(II)-EDTA complex.   The intensely blue
Co(OH)Y^{2-} complex formed limits the amount of cobalt that can be
determined to 10 mg in 100 - 150 ml of solution.   In strongly
alkaline media, copper can be masked with thioglycollic acid, and
iron and aluminium with triethanolamine.[13].

Interferences

Elements that interfere in the determination of nickel also interfere
in the determination of cobalt.   Again, only a few elements can be
masked.   In alkaline medium, lead, copper, zinc, cadmium, etc. can
be masked with thioglycollic acid and aluminium with fluoride.
Triethanolamine can be used to mask iron only if the cobalt is
previously bonded as the EDTA complex.   Cobalt is partially oxidized
in ammoniacal TEA medium, forming the Co(III)-TEA complex which is
inert towards EDTA.   On the other hand, cobalt bonded as the EDTA
complex does not react with triethanolamine.

Masking of cobalt

Potassium cyanide is a common masking agent in alkaline medium,
enabling e.g., determination of manganese or the alkaline-earth
metals in the presence of cobalt, or determination of zinc and
cadmium in presence of cobalt after demasking from the cyanide
complex with formaldehyde.   Small amounts of cobalt (up to 3 mg in
100 ml) can be masked with triethanolamine and hydrogen peroxide [14].

Separation of cobalt

Takamoto [15] recommends that up to 5 mg of cobalt be separated by
precipitating as $K_2NaCo(NO_2)_6$ from lactic acid medium, which prevents
hydrolysis of tin, antimony, chromium and aluminium.   Titanium,
bismuth and copper are masked with dihydroxyethylglycine.   This
precipitation of cobalt is not disturbed by other cations in amounts
up to 100 mg.   More than 30 mg of magnesium and 50 mg of iron will
interfere.   Cobalt is determined complexometrically after centrifuging the precipitate.

Large amounts of cobalt - up to 1 g - can be removed from solution as
cobalt ammonium phosphate [16].   Small amounts of the cobalt remain
in solution but do not interfere in the determination of nickel or
copper.

Harris and Sweet [17] separated cobalt from nickel by precipitation
with 1-nitroso-2-naphthol and extraction of the complex into chloroform.

Chromatographic methods have also been developed for separation of
cobalt from nickel, copper, iron, etc.   For example, Wilkins [18]
analysed ferrites containing nickel, cobalt, iron and zinc by sorbing
all the anions on a strongly basic anion-exchanger in 9M hydrochloric
acid and eluting them stepwise: first nickel with 9M HCl, then
cobalt (4M HCl), iron (0.5M HCl) and finally zinc (3M HNO$_3$).   Wilkins
and Hibbs used similar methods for some magnetic alloys [19,20].

6.2.4.3   Determination of nickel and cobalt

The similarity in the behaviour of the two elements prevents their
complexometric determination in a single solution without prior
separation.   The oldest method is based on oxidation of cobalt with
hydrogen peroxide in the presence of EDTA.   The $Co(OH)Y^{2-}$ or CoY^-
complex formed does not react with potassium cyanide in ammoniacal
medium, whereas the EDTA complex of nickel reacts readily [21].   This
permits stepwise determination of the two elements.   The intense
colour of the cobalt complex prevents use of the method for cobalt
concentrations above 5 mg per 100 ml of solution.   The two elements
can be determined stepwise by back-titration with magnesium, Erio T
being used as indicator.   DCTA is also useful for this purpose [16].

A modification is based on oxidation of the Co-EDTA complex in
potassium hydroxide medium and back-titration of excess of EDTA with
calcium (Calcein as indicator);   the method is poorly selective,
though copper can be masked with thioglycollic acid and iron with
triethanolamine, as mentioned above [12,13].

Flaschka and Püschel [22] determined nickel in the presence of
cobalt(II) and other elements by a very interesting method based on
back-titration of excess of EDTA with bismuth nitrate, with Pyrocatechol Violet as indicator.   At room temperature, cobalt, nickel and a
number of other metals are replaced in the EDTA complex by bismuth.
The nickel complex is much more kinetically inert than the others,
however, and cooling the solution to close to 0°C "freezes" the
Ni-EDTA complex and prevents its replacement reaction with bismuth.
The other metals (e.g. Co, Zn, Cd, Cu, Pb, Hg) are readily replaced
from their EDTA complexes.

Procedure.   Solid ammonium nitrate (0.5g) is added to an acid
solution containing nickel, cobalt and other bivalent metals and the
pH is adjusted to 2 by careful addition of dilute ammonia solution.
Excess of 0.05M EDTA is added and (after pH adjustment if necessary)
pieces of metal-free ice are added until some stop melting and float
on the surface of the solution.   After addition of 2 or 3 drops of
Pyrocatechol Violet indicator the solution is titrated with 0.05M
bismuth nitrate from yellow to dark blue.

These authors also studied the differential determination of cobalt
and nickel, determining the sum of the metals in ammoniacal medium,
with murexide as indicator.   Then the solution was acidified to pH 2
and the indicator was decomposed with several crystals of ammonium
peroxodisulphate.   Pyrocatechol Violet was added, and the solution
cooled and titrated with bismuth nitrate.   The EDTA consumption
corresponded to the sum of Ni + Co and the consumption of bismuth
nitrate to the cobalt content.

The determination of cobalt through the Co(III)-EDTA complex in
strongly acid medium is also very selective.   This method was
developed by Flaschka and Ganchoff [23].

Procedure.   A solution containing nickel and cobalt is mixed with a
small excess of 0.1M EDTA and the pH is adjusted to 11 with 2M sodium
hydroxide.   Then 4 or 5 drops of 30% hydrogen peroxide are added and
the solution is acidified by dropwise addition of concentrated nitric
acid to a final pH of 0.6 - 1.5.   A small amount of urea is added,
along with several drops of Pyrocatechol Violet indicator and a

volume of 0.1M bismuth nitrate equivalent to the EDTA added.   Excess
of bismuth is determined by photometric titration with 0.1M EDTA [24].

Note.   Addition of the bismuth solution is necessary to ensure
replacement of all the other cations from their EDTA complexes.
Under these conditions, the presence of nickel, copper, zinc, cadmium,
lead and uranium does not interfere in the determination.   Aluminium
is replaced very slowly at pH 2, but much more rapidly at pH 0.5 - 0.7.
All elements which form particularly stable complexes with EDTA, such
as iron, thorium, zirconium, gallium, indium and bismuth, are co-
titrated.   Under the conditions used, manganese has a tendency to
form the $Mn(III)$-EDTA complex and must be reduced with ascorbic acid.
Chromium also interferes; in alkaline medium it is present as the
non-interfering chromate, but after acidification is partially
reduced by residual hydrogen peroxide and immediately forms the
$Cr(III)$-EDTA complex.   Mercury, silver and molybdenum also interfere.
Among anions, iodide, tartrate, citrate, fluoride and large concentrations of chloride interfere.   No indication of practical application of this procedure has been found in the literature.

A rather unselective but much simpler method is based on forming the
nickel and cobalt cyanide complexes and oxidizing the cobalt with
hydrogen peroxide to produce the $Co(CN)_6^{3-}$ complex.   The nickel
cyanide complex $Ni(CN)_4^{2-}$ is then selectively demasked with silver
nitrate and determined complexometrically.   With suitable dilution,
the method can be used to determine up to 50 mg of Ni and 30 mg of Co.
The sum of the two metals is determined in another aliquot of
solution, e.g. by back-titration with lead nitrate (Xylenol Orange as
indicator).   Two modified determinations of nickel, given below,
have also been developed [25].

Direct determination of nickel.   A solution containing nickel and
cobalt is mixed with 25 - 35 ml of concentrated ammonia solution and
diluted to 200 - 250 ml, then 10 - 15 ml of 5% potassium cyanide
solution are added along with 5 - 10 ml of 2% hydrogen peroxide and
the solution is mixed.   After 1 - 2 min the solution is titrated with
10% silver nitrate solution to the appearance of the first permanent
turbidity.   Then the nickel freed is titrated with 0.05M EDTA, with
murexide as indicator, from yellow or yellow-orange to a clear purple
colour.

Note.   A large excess of silver nitrate should not be added, as the
AgCN precipitated interferes in the colour transition.   The murexide
is decomposed by the hydrogen peroxide with 5 - 15 min, so the
titration should be carried out relatively quickly.

Indirect determination of nickel.   This method is based on back-
titration of excess of EDTA with calcium, with Methylthymol Blue as
indicator.   It has been found that a small amount of the cobalt is
co-titrated as a result of the replacement reaction between precipitated AgCN and $Co(CN)_6^{3-}$.   This reaction can be prevented by adding
potassium iodide to the solution, since AgI is precipitated instead
of AgCN.

Procedure.   Cobalt is oxidized as above.   The solution is mixed
with 5 ml of 5% potassium iodide solution and 10% silver nitrate
solution is added dropwise until a slight turbidity appears.   After
1 - 2 min, excess of 0.05M EDTA is added along with a small amount of
indicator and the solution is titrated with calcium from smoke-grey

to dark blue.

REFERENCES

1.   R. Přibil and V. Maličký, Coll. Czech. Chem. Commun., 1949, 14, 413.
2.   R. Přibil and J. Malík, Chem. Listy, 1951, 45, 237.
3.   R. Přibil, Analytical Applications of EDTA and Related Compounds, Pergamon Press, Oxford, 1972.
4.   S. Takamoto, Japan Analyst, 1959, 4, 178; J. Chem. Soc. Japan, Pure Chem. Sect., 1955, 76, 1339.
5.   R.N. Sen Sarma, Sci. Cult. (India), 1955, 20, 448.
6.   H. Flaschka and H. Abdine, Chemist-Analyst, 1956, 45, 2.
7.   H. Flaschka and H. Abdine, Chemist-Analyst, 1956, 45, 58.
8.   G.M. Lukaszewski, J.P. Redfern and J.E. Salmon, Lab. Practice, 1957, 6, 389.
9.   A.J. Cameron and N.A. Gibson, Anal. Chim. Acta, 1961, 25, 24.
10. H.M.N.H. Irving and R.H. Al-Jarrah, Chem. Anal. (Warsaw), 1972, 17, 779.
11. J. Kinnunen and B. Wennerstrand, Chemist-Analyst, 1957, 46, 92.
12. R. Přibil and V. Veselý, Chemist-Analyst, 1963, 52, 5.
13. R. Přibil and V. Veselý, Talanta, 1962, 9, 1053.
14. R. Přibil and V. Veselý, Chemist-Analyst, 1965, 54, 114.
15. S. Takamoto, J. Chem. Soc. Japan, Pure Chem. Sect., 1960, 81, 457; Anal. Abstr., 1961, 8, 2878.
16. R. Přibil and F. Vydra, Coll. Czech. Chem. Commun., 1956, 21, 1146.
17. W.F. Harris and T.R. Sweet, Anal. Chem., 1954, 26, 1648.
18. D.H. Wilkins, Anal. Chim. Acta, 1959, 20, 271.
19. D.H. Wilkins and L.E. Hibbs, Anal. Chim. Acta, 1958, 18, 372.
20. D.H. Wilkins and L.E. Hibbs, Anal. Chim. Acta, 1959, 19, 427.
21. R. Přibil, Coll. Czech. Chem. Commun., 1954, 19, 1174.
22. H. Flaschka and R. Püschel, Z. Anal. Chem., 1955, 147, 353.
23. H. Flaschka and J. Ganchoff, Talanta, 1961, 8, 885.
24. H. Flaschka and P. Sawyer, Talanta, 1961, 8, 521.
25. R. Přibil and V. Veselý, Talanta, 1966, 13, 515.

6.2.5   Zinc
$$\frac{\text{Zinc}}{(\log K_{ZnY} = 16.5)}$$

Zinc forms a sufficiently stable complex with EDTA even in weakly acid medium and can be determined by direct titration with EDTA at pH > 4.

In ammoniacal medium the Zn-EDTA complex does not react with either hydrogen sulphide or diethyldithiocarbamate, but the zinc can be freed from its EDTA complex by an excess of calcium and quantitative-ly precipitated, e.g. as ZnS [1].   This replacement reaction, in 4M ammonia solution, has been used for the indirect polarographic deter-mination of calcium in blood [2].   In sodium hydroxide medium the Zn-EDTA complex is converted into the $Zn(OH)_4^{2-}$ complex.

Direct determination

Many indicators have been recommended for the direct titration of

zinc in both alkaline and acid media.   In ammoniacal buffer (pH 10)
Erio T yields a very sharp colour transition from wine red to blue.
It was the first indicator proposed for zinc [3] and has also been
used for the micro [4] and ultramicro [5,6] determination of zinc.
It has the disadvantage that it is blocked by traces of copper or
aluminium.   Zincon [7], Naphthol Violet [8] and Pyrocatechol Violet
[9,10] have also been recommended for the determination in alkaline
media.   Murexide was recommended for the determination of zinc at pH
11 in ethanolamine medium[11].   Methylthymol Blue [12] has also been
recommended for the determination of zinc in alkaline and weakly acid
media.   In contrast to the determination of cadmium, Thymolphthalexone cannot be used for zinc, as the zinc blocks it, forming a pale
pink complex.

Titrations of zinc in alkaline media have poor selectivity, but this
can be improved by use of masking agents for some of the interfering
elements.   Generally, titration of zinc in weakly acid medium is
preferable.   PAN [13], the Cu–EDTA/PAN system [14] and especially
Xylenol Orange [15] and Methylthymol Blue [12] have been recommended
for this determination.   In solutions buffered with hexamine at pH
5 – 5.5 Xylenol Orange yields a very sharp colour transition from red-
purple to lemon yellow.   For this reason, standard zinc solutions
are often used for back-titration.   Dithizone is also excellent for
the determination of zinc at pH 4.5 [16,17] because of its high
sensitivity.   Alizarin Complexone [18] is also highly recommended
for the determination of zinc at pH 4.3, with a colour transition
from red to yellow.

Indirect determination

Back-titration with magnesium (Erio T as indicator) is frequently
used in combination with masking agents.   Various authors have
employed back-titration with zinc or lead solution at pH 5 – 5.5,
using Xylenol Orange or Methylthymol Blue as indicator.

Interferences

The selectivity of the determination of zinc is very low.   In weakly
acid medium, only alkaline-earth metals, with calcium in not much
more than 1:1 ratio to zinc, may be present.   On the other hand, the
large difference in the stability constants of the Zn–EDTA and Bi-
EDTA complexes permits their stepwise formation at pH 1 – 2 (Bi) and
pH 5 – 5.5 (Zn).

Masking of interferences

A few of the interferences can be removed by masking.   For example,
aluminium and calcium can be masked with fluoride, and copper with
thiosulphate, thiourea, or ascorbic acid and iodide.   Traces of iron
can be masked by extraction with acetylacetone [19].   In alkaline
media, iron and aluminium can be masked with triethanolamine.

Masking of zinc

In the determination of manganese and the alkaline-earth metals zinc
can be reliably masked with potassium cyanide.   Subsequent demasking
from the cyanide complex enables determination of zinc in the
presence of copper, nickel and cobalt.   Among other masking agents,
BAL [20], Unithiol [21] and thioglycollic acid [22] have been used.

Masking with sodium hydroxide (see below) is very simple.

Separation of zinc

There is as yet no known simple and highly selective separation
method for zinc which can be broadly applied. Separation by ion-
exchange is frequently useful, and extraction methods have been
employed. Onishi [23] determined zinc in aluminium alloys after
extraction with trioctylphosphine oxide and used the same method for
the determination of zinc in dust [24]. Zinc can be reliably
separated from manganese, nickel, cobalt, aluminium, copper, indium
and gallium by extraction from chloride medium with a chloroform
solution of trioctylmethylammonium chloride [25].

Other titrants for zinc

DCTA has been recommended for the indirect determination of zinc in
the presence of nickel. The Ni-DCTA complex does not react with
potassium cyanide in ammoniacal buffer medium, whereas the Zn-DCTA
complex does [26]. Zinc can also be titrated with a TTHA solution
to form the Zn_2-TTHA complex. The greater difference in the stability constants makes DTPA a much better reagent for the determination
of zinc in the presence of calcium [17]. EGTA forms a much more
stable complex with cadmium (log K_{CdY} = 16.1) than with zinc (log
K_{ZnY} = 13.0), permitting their stepwise determination (see Section
6.2.6).

Determination of zinc in the presence of cadmium

A great deal of attention has been paid to the complexometric determination of zinc at extreme Zn:Cd ratios. These methods are generally based on selective masking or separation of cadmium, or selective displacement reactions.

One of the first methods was based on determination of the sum of
Zn + Cd by direct titration with EDTA (Erio T as indicator), followed
by determination of the Cd. The cadmium was precipitated from its
EDTA complex by addition of fresh sodium diethyldithiocarbamate
solution and the EDTA freed was titrated with magnesium [27]. The
Schwarzenbach buffer concentration used is important in this titration. If more than 4 ml are added per 100 ml, the cadmium is masked
only partially. When the method is expanded by including cyanide or
thioglycollic acid as masking agents, more complicated mixtures can
be analysed, e.g. Cd-Zn-Ni-Mg or Cd-Zn-Pb.

Fabregas et al. [28] used an interesting reaction to mask cadmium,
based on replacement in the Pb-EGTA complex according to the equation

$$Cd^{2+} + Pb-EGTA^{2-} + SO_4^{2-} = Cd-EGTA^{2-} + PbSO_4$$

Zinc does not react with the Pb-EGTA complex. After the precipitated $PbSO_4$ has been filtered off, zinc is determined either at pH 5
with Xylenol Orange as indicator or at pH 10 with Erio T as indicator.
Satisfactory results are obtained for Cd:Zn ratios up to 17.

β-Mercaptopropionic acid (β-MCP) has been recommended for the selective masking of cadmium. In weakly acid medium, cadmium is masked
only with respect to Xylenol Orange, and direct titration with EDTA

yields high results because of partial reaction with the Cd-β-MCP
[29] complex.   Accurate results are obtained in titration with TTHA
solution, for Cd:Zn ratios up to 40.

Flaschka [30] used large concentrations of potassium iodide to mask
cadmium and titrated zinc with a 0.01M EDTA, using Xylenol Orange as
indicator.   Up to 50 g of KI per 100 ml of solution can be used to
determine zinc on the microscale at Cd:Zn ratios of up to 300.   Later,
Flaschka and Butcher [31] titrated zinc photometrically with 0.001M
DTPA, again using Xylenol Orange as indicator.   They state that
accurate results can be obtained for 0.1 - 0.3 mg of Zn at a Cd/Zn
ratio of 3300.   Very small amounts of some elements can be tolerated
if they are masked, e.g. copper, mercury and lead with iodide,
aluminium with tiron or sulphosalicylic acid, etc.   Small amounts of
calcium, bismuth and chromium do not disturb the determination.

Cadmium can also be separated from zinc by precipitating its 1,10-
phenanthroline complex with potassium iodide.   The insoluble ion-
association complex $Cd(phen)_2I_2$ is formed [32].   Zinc can be determined indirectly in the filtrate, in strongly ammoniacal medium, by
back-titration of DCTA with calcium, Methylthymol Blue being used as
indicator.   Small amounts of manganese, nickel and cobalt are not
precipitated under these conditions.   Cadmium can also be separated
from zinc in strongly alkaline medium by precipitation of its
sulphide with thiourea (see Section 6.2.6).

Recommended procedures

Erio T as indicator.   A solution containing up to 25 mg of Zn in
100 ml is first neutralized with sodium hydroxide and then 2 ml of
pH 10 buffer are added per 100 ml of solution.   After addition of
the indicator, the solution is titrated from wine red to clear blue.

Xylenol Orange as indicator.   Several drops of 0.1% Xylenol Orange
solution are added to the acid solution and sodium hydroxide
solution is added carefully until a red colour appears.   Then 1M
nitric acid is added dropwise until the red colour begins to change
to orange or yellow.   Solid hexamine is added to restore the red
colour and the solution is titrated with 0.02 - 0.05M EDTA until lemon
yellow.   Slightly acid solutions can be neutralized to pH 5 - 5.5
with hexamine solution.   It is essential to add excess of hexamine
to ensure adequate buffer capacity during the titration, and more
should be added after the end-point has been reached, to test whether
the pH has remained high enough (if the red colour reappears the
titration is continued).

Indirect determination.   If necessary, excess of EDTA can be back-
titrated with magnesium under the conditions given for Erio T as
indicator or with zinc or lead as described for Xylenol Orange as
indicator.

Determination in the presence of copper, nickel, cobalt and mercury
[33].

An amount of 2 - 3 ml of Schwarzenbach buffer (pH 10) is added per
20 ml of weakly acid solution containing the metals listed, followed
by solid potassium cyanide until the blue colour disappears.   In the
presence of cobalt and nickel, the final colour is yellow.   The
precipitate initially formed in the presence of mercury readily

dissolves.   On addition of Erio T, the solution turns blue.
Addition of several ml of 10% formaldehyde solution then produces a
wine-red colour and the solution is immediately titrated with 0.01 -
0.05M EDTA.   The colour transition from wine-red to steel-blue is
very sharp.   Solid chloral hydrate can be used instead of the formaldehyde, and slowly demasks zinc from the cyanide complex;   several
minutes should then be allowed for the demasking before titration
with EDTA.

Determination in the presence of copper.   Solid thiourea is added in
small portions to an acid solution until the blue colour disappears,
plus 20 - 30 mg extra.   After addition of Xylenol Orange, the pH is
adjusted with solid hexamine until a red-purple colour appears and
the solution is titrated with EDTA to a clear yellow colour.

Determination of zinc and lead.   Xylenol Orange is added to an acid
solution and the pH is adjusted to 5 - 5.5 with hexamine (red-purple
colour appears).   The sum of the two elements is determined by
titration with EDTA.   Then 0.1M 1,10-phenanthroline is added and the
solution is titrated with 0.05M lead nitrate to a red-purple colour.
Several more drops of masking agent are added and, if the red colour
disappears, the titration is continued.

Determination of bismuth and zinc.   The pH of the solution is
adjusted to 2 - 3 and bismuth is titrated, with Xylenol Orange as
indicator.   The pH is then adjusted to 5 - 5.5 with hexamine and the
titration is continued.

These methods can be combined, e.g. for the determination of Bi-Pb-Zn.

REFERENCES

1.   R. Přibil, Coll. Czech. Chem. Commun., 1951, 16, 86.
2.   R. Přibil and Z. Roubal, Coll. Czech. Chem. Commun., 1954, 19,
      252.
3.   W. Biedermann and G. Schwarzenbach, Chimia, 1948, 2, 1.
4.   H. Flaschka, Mikrochem. ver. Mikrochim. Acta, 1952, 39, 38.
5.   E.W. Debney, Nature, 1952, 169, 1104.
6.   H. Flaschka and F. Sadek, Mikrochim. Acta, 1957, 1.
7.   J. Kinnunen and B. Merikanto, Chemist-Analyst, 1955, 44, 50.
8.   B. Buděšínský, Coll. Czech. Chem. Commun., 1957, 22, 1579.
9.   M. Malát, V. Suk and A. Jeníčková, Coll. Czech. Chem. Commun.,
      1954, 19, 1156.
10.   V. Suk and M. Malát, Chemist-Analyst, 1956, 45, 30.
11.   A.A. Ashton, Chemist-Analyst, 1963, 52, 49.
12.   J. Körbl and R. Přibil, Coll. Czech. Chem. Commun., 1958, 22,
      873.
13.   K.L. Cheng and R.H. Bray, Anal. Chem., 1955, 27, 782.
14.   H. Flaschka and H. Abdine, Chemist-Analyst, 1956, 45, 58.
15.   J. Körbl, R. Přibil and A. Emr, Coll. Czech. Chem. Commun., 1957,
      22, 961.
16.   E. Wänninen and A. Ringbom, Anal. Chim. Acta, 1955, 12, 308.
17.   J.J. Hickey and J. Overbeck, Anal. Chem., 1966, 38, 932.
18.   R. Belcher, M.A. Leonard and T.S. West, J. Chem. Soc., 1958,
      2390.
19.   K. Yamaguchi and K. Ueno, Japan Analyst, 1963, 12, 55;   Anal.
      Abstr., 1964, 11, 4748.

20. R. Přibil and Z. Roubal, Coll. Czech. Chem. Commun., 1954, 19, 1162.
21. L.A. Volf, Zavod. Lab., 1960, 26, 271; Anal. Abstr., 1960, 7, 4679.
22. R. Přibil and V. Veselý, Talanta, 1961, 8, 880.
23. K. Onishi, Japan Analyst, 1969, 18, 1262.
24. K. Onishi, Japan Analyst, 1969, 18, 72.
25. R. Přibil and V. Veselý, Coll. Czech. Chem. Commun., 1972, 37, 13.
26. R. Přibil, Coll. Czech. Chem. Commun., 1955, 20, 162.
27. R. Přibil, Coll. Czech. Chem. Commun., 1953, 18, 783.
28. E. Fabregas, A. Prieto and C. Garcia, Chemist-Analyst, 1962, 51, 77.
29. R. Přibil and V. Veselý, Talanta, 1965, 12, 475.
30. H. Flaschka and J. Butcher, Microchem. J., 1963, 7, 407.
31. H. Flaschka and J. Butcher, Talanta, 1964, 12, 1067.
32. R. Přibil and V. Veselý, Talanta, 1964, 11, 1613.
33. H. Flaschka, Z. Anal. Chem., 1953, 138, 322.

6.2.6   Cadmium
         $\overline{(\log K_{CdY}} = 16.50)$

The complexometric behaviour of cadmium is the same as that of zinc. Everything that was said for the determination of zinc is also valid for cadmium.   Indicators, proposed for zinc, such as Xylenol Orange, Methylthymol Blue, PAN, Cu-EDTA/PAN, Pyrocatechol Violet, Naphthol Violet, etc. were tested for cadmium by the same authors and the determination procedures are described in the same publications.

Cadmium forms a somewhat weaker complex with the indicators than zinc does.   When Erio T is used as the indicator in determination of cadmium, care must be taken that the concentration of ammonia is as small as is compatible with the buffering requirements, to minimize the competitive formation of the $Cd(NH_3)_4^{2+}$ complex.   Similarly, when Xylenol Orange is used as the indicator care must be taken that the solution is well buffered with hexamine.   Several drops of pure pyridine should be added just before the expected end of the titration.   Thymolphthalexone has proved to be a very good indicator for cadmium: it forms an intensely blue complex with it in ammoniacal medium, in contrast to zinc, which forms a pale pink complex.

Interferences and masking

The interferences and masking agents are the same as those for zinc (Section 6.2.5).

Masking of cadmium

At one time, the masking of cadmium with potassium cyanide and demasking with formaldehyde or chloral hydrate was a common method for the determination of cadmium in the presence of nickel, cobalt, or copper. Other masking agents are BAL, Unithiol and thioglycollic acid.

Potassium iodide has also been used for masking cadmium, but the considerable dissociation of CdI_4^{2-} makes it necessary to use large concentrations (Section 6.2.5).   Cadmium can be masked in weakly acid medium with β-mercaptopropionic acid [1] and dimercaptosuccinic

acid [2]. 1,10-Phenanthroline is less selective for masking cadmium
but permits simultaneous determination of lead and cadmium [3].

Separation of cadmium

In addition to classical methods, such as precipitation of cadmium as
the sulphide in acid medium, several other precipitation methods have
been used. Horiuchi [4] recommends precipitating cadmium from
strongly alkaline solution, as the 2-(o-hydroxyphenyl)benzoxazole
complex, for the separation of cadmium from bismuth, tin and lead.
The method has been used for the determination of cadmium in Wood's
metal. Cadmium can be precipitated as the sulphide by boiling in
strongly alkaline medium with thiourea [5]. The cadmium sulphide
precipitated contains only a small amount of co-precipitated zinc.
Cadmium can be readily determined by the method described below. The
isolation of cadmium as $Cd(phen)_2I_2$ was mentioned on p. 167.

An extraction method has been developed which involves extraction of
cadmium as the CdI_4^{2-} complex by a chloroform solution of trioctylmethylammonium chloride [6]. Zinc is only slightly extracted from
weakly acid medium. Both elements are extracted quantitatively from
a chloride + iodide medium, cadmium as the CdI_4^{2-} complex and zinc as
the $ZnCl_4^{2-}$ complex. The zinc is readily stripped first with
potassium nitrate solution and then cadmium with an ammoniacal
solution of EDTA. Both cations are then determined complexometrically. The method can be used to separate cadmium and zinc from
manganese, nickel, cobalt, the alkaline-earth metals, gallium and
indium. Aluminium must be masked with fluoride before the extraction, and copper is masked with thiourea.

6.2.6.1   Determination of cadmium in the presence of zinc

In the Section 6.2.5, several methods were described for the determination of zinc in the presence of large concentrations of cadmium.
Attempts to determine small amounts of cadmium in the presence of
zinc have been less successful (because of the larger atomic weight
of cadmium). The methods which have so far been described depend on
the different stabilities of the Cd-EGTA and Zn-EGTA complexes. The
EGTA complex of cadmium is three orders of magnitude more stable than
that of zinc, but this is not quite sufficient for the stepwise
determination of cadmium and zinc without further adjustment of
conditions. Nevertheless, Flaschka and Carley [7] carried out a
stepwise determination of cadmium and zinc, using murexide as
indicator and 0.005M EDTA. Zincon can be used in a similar method
[8]. Both titrations were done spectrophotometrically. The two
breaks on the graph correspond to the consumptions for cadmium and
zinc. The method can be used only for microgram amounts of zinc.

The difference in the stabilities of the two EGTA complexes has been
used in a very simple procedure. In strongly alkaline medium, the
Zn-EGTA complex reacts with sodium hydroxide to produce tetrahydroxozincate, whereas the Cd-EGTA complex is non-reactive. The procedure
involves addition of 0.02 – 0.05M EGTA to a solution containing the
two elements and then strongly alkalizing the solution with 2M sodium
hydroxide. The solution then contains the Cd-EGTA complex and
$Zn(OH)_4^{2-}$. The excess of EGTA is determined by back-titration with a
calcium solution [5]. Visual detection of the end-point is
difficult. The high pH (13) of the solution makes use of Methyl-

thymol Blue or Thymolphthalexone impossible.   Fluorexone (Calcein)
exhibits strong fluorescence and cannot be used.   The only suitable
indicator is Cresolphthalexone, which yields a strongly purple colour
with calcium.   The indicator itself is also strongly coloured in
highly alkaline medium, but forms only a pink complex with zinc,
which is practically colourless at sufficient dilution, and is less
stable than the corresponding Ca complex.   In back-titrations, the
end-point is indicated by a change in colour from pink to intense
purple.   The method has been used for the determination of cadmium
in pigments of the lithopone type and for the determination of traces
of cadmium in raw zinc [5] after separation of the cadmium with
thiourea.   The method can be used in the presence of lead and
aluminium and in the presence of iron provided it is masked with
triethanolamine.

6.2.6.2   Determination of cadmium in the presence of copper

Böltz et al. described an interesting method which does not employ
demasking of the $Cd(CN)_4^{2-}$ complex with formaldehyde [9].   Potassium
cyanide is carefully added to an ammoniacal solution of cadmium and
copper until the blue colour disappears.   Then Mg-EDTA complex is
added to the solution.   According to the equation

$$Cd(CN)_4^{2-} + MgY^{2-} = CdY^{2-} + Mg^{2+} + 4 \ CN^-$$

magnesium is quantitatively replaced in the EDTA complex and can be
titrated directly with EDTA (Erio T as indicator) the amount being
equivalent to the cadmium present.   The authors recommend this
method for determination of 0.5 - 5% cadmium in copper alloys.

REFERENCES

1.   R. Přibil and V. Veselý, Talanta, 1965, 12, 475.
2.   T. Mekeda, K. Yamaguchi and K. Ueno, Talanta, 1964, 11, 1464.
3.   R. Přibil and V. Veselý, Chemist-Analyst, 1965, 54, 12.
4.   Y. Horiuchi, Rept. Technol. Iwate Univ., 1963, 16, 39.
5.   R. Přibil and V. Veselý, Chemist-Analyst, 1966, 55, 4.
6.   R. Přibil and V. Veselý, Coll. Czech. Chem. Commun., 1972, 37, 13.
7.   H. Flaschka and F.B. Carley, Talanta, 1964, 11, 423.
8.   H. Flaschka and J. Butcher, Mikrochim. Acta, 1964, 401.
9.   G. Böltz, H. Wiedmann and W. Kurella, Metall (Berlin), 1956, 10, 821.

6.2.7   Aluminium
$(\log K_{A1Y} = 16.1)$

The aluminium ion reacts very slowly with EDTA at room temperature in
weakly acid solutions.   The formation of the $A1Y^-$ complex is affected primarily by the acidity of the solution, the temperature and, to
a certain degree, by the concentration of neutral salts.   Ringbom
explains the slow formation of the complex at normal temperatures as
due to formation of polynuclear hydroxo-complexes during hydrolysis
of aluminium salts.   In his book [1] he states "Research on the
hydrolysis of aluminium(III) ions has shown that the formation of

polynuclear complexes occurs in a very narrow pH interval between pH
3.5 and 4.5 in 0.01M Al(III) salt solutions. Consequently, the
reaction between aluminium and EDTA at pH 3.5 is instantaneous and
above pH 3.5 the solution must be heated. Below pH 3.5, the
reaction equilibrium is unfavourable."

Consequently, attempts to determine aluminium directly have not been
very successful and have no importance in practical complexometry.
Nonetheless, a short description is given under "Determination of
iron and aluminium".

Indirect determination

The first indirect determination of aluminium was developed in 1949,
and was based on potentiometric back-titration of excess of EDTA with
ferric chloride in acetate medium (pH 4 - 6) [2]. The method is
poorly selective, like some proposed indirect titrations of aluminium
in alkaline medium, e.g. back-titration with zinc, using Erio T as
indicator [3]. Only back-titrations in acid medium are of any
importance; in this medium very sensitive metallochromic indicators
can be used.

Not all the possible variations of this determination will be
described here. It has been found that Xylenol Orange and dithizone
are the most important indicators in practice. Back-titration using
Xylenol Orange can be done with zinc chloride or lead nitrate in
solutions buffered with hexamine [4]. Thorium nitrate can also be
used in solutions which are well buffered with acetate (pH 3.6 - 4).
Back-titration with lead yields a very sharp colour transition, but
large concentrations of sulphate lead to precipitation of lead
sulphate, which dissolves very slowly. This difficulty is not
encountered in titrations with zinc salts (in well buffered solutions).

Wänninen and Ringbom [5] suggested using dithizone indicator in back-
titration with zinc. In a very detailed work based on theoretical
considerations, they recommend determining zinc by back-titration
with zinc sulphate at pH 4.0 - 4.5 in 40 - 50% ethanol medium. Their
work attracted a great deal of attention and was critically
evaluated by Nydahl [6] and modified by Gottschalk [7]. Brady and
Gwilt [8] consider both methods suitable for routine work, but feel
that the method using Xylenol Orange, in which the accuracy for
aluminium is 99.9 - 100.1%, is more "attractive".

Interferences

At the relatively high pH value (4 - 5.5) at which aluminium is usually determined, interferences are marked. All elements forming EDTA
complexes, except the alkaline-earth metals, interfere. Kiss [9]
states that even calcium interferes at Ca:Al ratios greater than 1:1
in the titration of aluminium with Xylenol Orange indicator. Nonetheless, a number of works have been published describing the determination of aluminium in the presence of a number of other elements.
The greatest attention has been paid to the Fe-Al-Ti, Fe-Al-Mn and
Fe-Al-Cr systems. The experience of various authors tends to vary
considerably (and will be discussed at the end of this section).

Masking of aluminium

Only a few reagents can be used to mask aluminium in practice.

Triethanolamine is well known as a masking agent for aluminium in alkaline medium [10]. The introduction of new metallochromic indicators has broadened the use of this reagent. Ammonium fluoride or sodium fluoride can be used to mask aluminium in acid medium [11 - 13]. The masking of aluminium is especially useful in the analysis of Fe + Al or Fe + Ti mixtures and for the differential determination of aluminium in the presence of all the bivalent metals except manganese. Other masking agents include salicylic acid, sulphosalicylic acid and p-aminosalicylic acid [12], acetylacetone, e.g. for the determination of zinc in the presence of aluminium [14,15] and tiron for the determination of manganese [16]. The last two masking agents cannot be used in the presence of iron without prior sample adjustment. With the first of these, the intensely red iron complex with acetylacetone must be extracted; with the second the iron must be converted into the cyanide complex. Halasz et al. [17] masked aluminium by precipitating it with sodium silicate in the determination of magnesium in the presence of aluminium.

Masking of interferents

Only a few masking agents can be used for this purpose. In practical applications, only masking of copper with thiourea (for aluminium bronzes) or of traces of cadmium with iodide can be used. Traces of heavy bivalent elements can be masked with 1,10-phenanthroline. The possibilities of selective determination of aluminium in alkaline media are broader and include masking with potassium cyanide or thioglycollic acid.

Separation methods

In addition to classical separation methods, a number of methods based on extraction, chromatographic separation, etc. have been proposed. Provided that manganese and titanium are not present in solution, the simplest method for separating aluminium is precipitation with ammonia, together with iron; the mixture of aluminium and iron is readily analysed complexometrically after dissolution. In special cases, aluminium can also be determined with ammonium benzoate [18]. Methods for separation of interferents have also been developed. Cupferron has frequently been recommended for precipitation of zirconium, iron and titanium. A more detailed description of these methods is given at the end of this section.

Other titrants for aluminium

Determination with DCTA. Contrary to expectation, it has been found that the complex between DCTA and aluminium is formed almost instantaneously at room temperature in the presence of large concentrations of neutral salts [19]. This fact has been employed in the determination of aluminium in the presence of both chromium(III) and chromate [20]. After determination of aluminium (by back-titration with lead solution, Xylenol Orange as indicator), chromium(III) is similarly determined after heating with a further portion of DCTA. Chromate does not interfere, as it does not oxidize DCTA even at increased temperatures. Burke and Davis [21] applied this method for the determination of aluminium in various alloys, Pritchard [22] used it for the analysis of silicates and Bennett and Reed [23] used it in the analysis of ceramic materials. Smithies and Wolstenholme [24] used titration with DCTA in the determination of aluminium in chrometanned leather.

Determination with HEDTA.   This reagent was used by Wilkins for the
determination of aluminium and nickel in the presence of manganese
(Section 6.2.3).

Determination with TTHA.   TTHA forms the Al_2-TTHA complex with
aluminium at room temperature and has been found to be a useful
reagent for the determination of aluminium in the presence of large
concentrations of manganese [25].

Recommended procedures for indirect determination

Xylenol Orange as indicator.   An excess of 0.05M EDTA is added to a
slightly acid solution containing up to 30 mg of aluminium and the
mixture is boiled gently for 1 - 2 min.   After cooling and dilution
of the solution, if necessary, to 100 ml, 1 - 3 drops of 0.5% Xylenol
Orange solution are added, followed by solid hexamine in small
amounts until the bright yellow colour pales slightly.   Then the
solution is titrated with 0.02 - 0.05M lead nitrate to a red-purple
colour.

Note.   The colour change is reversible.   After about 10 min the
red-purple colour changes to intense pink-red as a result of the
metathesis

$$PbIn + AlY = AlIn + PbY$$

(In is the indicator ion).   This colour generally does not change on
addition of 1 or 2 drops of EDTA, as the indicator is blocked by the
aluminium.   Aluminium can be determined in the same manner by back-
titration with zinc sulphate or acetate.   The metathetic reaction is
then much slower.   Aluminium can be determined with DCTA in the same
manner, but the solution must be heated, and the titration carried
out slowly because of the slow formation of the Pb-DCTA complex.

Dithizone as indicator [5].   A 40 - 50% excess of 0.1 - 0.05M EDTA and
10 ml of acetate buffer, pH 4.5, are added to a weakly acid solution
containing 5 - 30 mg Al in 50 - 100 ml of solution.   The solution is
boiled for 1 - 2 min, cooled, and diluted with an equal volume of
ethanol.   Each 100 ml of solution is mixed with 2 ml of freshly
prepared 0.3% ethanol solution of dithizone and the mixture is
titrated with 0.1 - 0.05M zinc sulphate, from green-purple to red.
The reagents must be added in the order stated.   If the buffer is
added before the EDTA, the results are somewhat lower, as polynuclear
complexes are formed.

Note.   Wänninen and Ringbom [5] state that heating of the aluminium
solution after addition of EDTA is necessary.   In contrast, Elo and
Polky [26] recommend adjusting the solution to pH 3 before addition
of the EDTA.   Al-EDTA is formed instantaneously and the solution
need not be heated.   After pH adjustment to 4.5, the excess of EDTA
is determined as described.

Recommended procedures for direct determination

Direct determination of aluminium is still generally considered
insufficiently reliable and is also rather demanding.   Two examples
will be given and the reader can judge their usefulness for himself.

Chrome Azurol S indicator. A weakly acidic solution containing 5 -
20 mg of Al in 100 ml is mixed with 3 - 5 drops of 0.4% indicator
solution and the pH is adjusted to 4 with acetate buffer.   The
solution is heated to 80°C and titrated with 0.05M EDTA to a colour
change from purple to red-orange.   Then 3 - 5 ml of buffer are added
to increase the pH.   If the purple colour appears again, the titration is continued to the formation of a yellow-orange colour.

Cu-EDTA/PAN system as indicator.   An acid solution containing 2 -
10 mg of Al in 20 - 30 ml is neutralized to pH 1 - 2 with dilute
ammonia solution.   Several drops of Bromophenol Blue indicator are
added and then 10% ammonium acetate solution until the colour is
blue-green.   Then 5 ml of glacial acetic acid are added and several
drops of the Cu-EDTA complex along with a sufficient amount of PAN
indicator.   The solution turns purple (formation of the Cu-PAN
complex).   The solution is heated to boiling and the temperature is
kept at 95°C during titration with 0.01M EDTA from purple-red to
yellow.   The colour transition is premature and the purple colour
rapidly reappears.   The titration is continued until the yellow
colour does not change after boiling for 30 sec.   Flaschka
recommends boiling the solution for another minute to confirm that
the titration is complete.   Sometimes a pale red colour appears,
which disappears after addition of another drop of EDTA.

Note.   Iwamoto [27] titrated aluminium in similar manner at pH 4 in
potassium hydrogen tartrate medium.   Addition of glycerol permits
increasing the temperature to 120°C, where the formation of the Al-
EDTA complex is much faster.

Some separation determinations

Iron and aluminium.   The large difference in the stability constants
of the Fe-EDTA and Al-EDTA complexes has led many authors to attempt
to develop a procedure for the successive determination of iron and
aluminium.   For example, even large concentrations of aluminium
should not interfere in the determination of iron at pH 2, as the
conditional stability constants at this temperature are $\log K'_{FeY}$ =
11.5 and $\log K'_{AlY}$ = 3.0.   It has, however, been found that the
reliability of the determination of iron is very dependent on both
the titration procedure and on the acidity conditions during the
titration.   Cheng et al. [28] state that only "large" concentrations
of aluminium interfere in the titration of iron, when tiron or
salicylic acid is used as indicator at pH 2 - 3.   Sweetser and
Bricker [29] found in the spectrophotometric determination of iron at
pH 1.7 - 2.3 that aluminium concentrations of 20 mg in 100 ml of
solution interfere in the measurement of the absorbance of the Fe-
EDTA complex at 525 nm.   In contrast, Davis and Jacobsen [30] determined iron by spectrophotometric titration, with sulphosalicylic acid
as indicator, at pH 1 with a precision of 0.3% in the presence of
100 - 150 mg of Al in 80 ml of solution at an Fe:Al ratio of 1:2.
These authors state that the positive error increases with decreasing
amount of iron, even when the 1:2 ratio is maintained.   For example
the error in the determination of 5.59 mg of Fe in the presence of
5 mg of Al was 2.7% (i.e. 0.15 mg of Fe).

Begelfer et al. [31] published a very interesting work, using sulphosalicylic acid in large concentrations both as an indicator for iron
and as a "temporary masking agent" for aluminium.   Iron is determined at pH 2 - 3 at 60 - 70°C, by direct titration, with sulpho-

salicylic acid as indicator (2 ml of 10% solution). Then an
additional equivalent amount of EDTA is added, the solution is heated
to the boiling point, and neutralized with 5% ammonia solution (Congo
Red paper). After addition of acetate buffer, the excess of EDTA is
titrated with 0.1M ferric chloride.

3-Hydroxy-2-naphtholic acid [32] has also been recommended as an
indicator for the stepwise determination of iron and aluminium
(Chrome Azurol S can also be used in this titration). The indicator
forms a blue complex with iron(III) at pH 2, and fluoresces at pH 3
in ultraviolet light; in the presence of aluminium, the fluorescence
is blue. The reaction of EDTA with aluminium is slow and can be
made even slower by cooling the solution. The principle of the
method is clear from the following procedure.

Procedure. An acid solution (pH 2) containing 12 – 62 mg of Fe and
25 – 30 mg of Al in 100 ml is stirred with a piece of ice to complete
the cooling. After addition of several drops of 0.1% indicator the
solution is titrated with 0.05M EDTA to a colour change from blue to
yellow. After completion of the titration the solution is heated to
50°C and the pH is adjusted to 3 by addition of 10 ml of 10% glycine
solution. Aluminium is titrated with EDTA in the dark under ultraviolet light to a change of the blue fluorescence to green. The
yellow colour of the Fe-EDTA complex somewhat decreases the sharpness
of the fluorescence change. The precision can be increased by
comparing the fluorescence of the titrated solution with the fluorescence of the indicator solution.

Indirect determination of iron and aluminium. The examples of the
successive determination of iron and aluminium given so far are more
or less of academic interest, as they can only be used for solutions
containing no other ions that form EDTA complexes. The determination of iron is quite selective, but the subsequent determination of
aluminium is disturbed by all the cations of group II. Indirect
determination is much better and is more frequently used, involving
determination of the sum of Fe + Al by back-titration of EDTA with
lead or zinc nitrate, with Xylenol Orange as indicator, at pH 5 – 5.5.
After complexation of the aluminium with ammonium fluoride (the
solution must be heated and cooled), the EDTA released is determined
in the same manner. The iron content is found from the difference
between the two titrations. The method is quite selective for
aluminium. Most bivalent cations present are co-titrated, but do
not interfere in the determination provided they do not react with
fluoride. Titanium and other elements forming stable fluoride
complexes interfere. The method is very useful for the analysis of
simple silicates containing calcium, magnesium or traces of manganese
or titanium in addition to aluminium and iron. This method will be
encountered later in slightly modified form.

Separational methods. The methods just described cannot be used for
the analysis of more complicated materials, containing, for example,
titanium, zirconium, niobium and the rare-earth elements. The cupferron method for separation of these elements is very popular, as
all of them, including iron, can be precipitated as insoluble
cupferronates and extracted with chloroform. Milner and Woodhead
[33] employed the cupferron extraction of iron, titanium, and zirconium in the determination of aluminium in silicates. Since that time,
the method has been variously modified. After extraction of the
elements mentioned, Weibel [34] determined aluminium by back-

titration of EDTA with zinc, using dithizone indicator as described
by Wanninen and Ringbom [5]. The method can be used to determine
small amounts of aluminium and the author states that it is preferable to the photometric determination of aluminium with 8-hydroxy-
quinoline. Thielicke [35] also extracted interferents into chloroform as their cupferronates, from dilute hydrochloric acid medium and
determined aluminium indirectly by potentiometric back-titration with
0.01M zinc sulphate at pH 4.5, using the redox indicator $Fe(CN)_6^{4-}/$
$Fe(CN)_6^{3-}$. This author states that this method is preferable to the
ordinary zinc sulphate back-titration with Xylenol Orange as
indicator.

Kraft and Dosch [36] have studied the cupferronate extraction in some
detail. They also extracted titanium, iron and other elements from
hydrochloric acid medium and then extracted aluminium from acetate
medium. In this way, a quite pure aluminium cupferronate solution
is obtained; after fuming with nitric acid and perchloric acid, the
aluminium is determined indirectly by voltammetric back-titration
with 0.025M zinc sulphate. The indicator system is a polarized
platinum cathode and a polarized platinum anode covered with a thin
layer of manganese dioxide [37,38]. The authors state that the
method is universal for aluminium and very precise. It has been
applied to a broad range of materials, both natural and technical
(silicates, iron ores, limestone, nickel and titanium alloys, etc.).
The standard deviation is 0.03% aluminium at the 10% level. The
method can be used for determining 0.005% Al in limestone and up to
45% in bauxite. It is technically quite complicated, so the reader
should refer to the original literature.

Note. Bartura and Bodenheimer [39] published some very surprising
results. They obtained low results for aluminium after removal of
iron etc. by extraction of the cupferronates into chloroform. They
attribute the loss to solubility of aluminium chloride in chloroform.
In the determination of 10 - 25 mg of aluminium in solutions which had
been extracted with pure chloroform, losses of 3 - 8% were experienced.
The authors thus regard the cupferron extraction of iron, titanium,
etc. as unsuitable. These results warrant careful testing in comparison with previous work. It is known, for example, that aluminium cupferronate is partially extracted into chloroform from 1M
hydrochloric acid [40], and that o-dichlorobenzene is a much better
solvent to use (practically zero extraction of aluminium).

Determination of titanium and aluminium. The principles of the
determination of aluminium and titanium were discussed in Section
6.1.12. The cupferron method has been repeatedly recommended for
separating titanium from aluminium. In addition to the method
already described, involving separation of titanium as the hydroxide
in triethanolamine medium, Pritchard [22] recommends precipitating
titanium and iron with sodium hydroxide and boiling the alkaline
solution with DCTA, which binds even any adsorbed aluminium. After
digestion for an hour and collection of the precipitated hydroxide,
aluminium is determined indirectly in the filtrate at pH 5 - 5.5 by
back-titration with zinc (Xylenol Orange as indicator). As DCTA
decomposes slightly in boiling solution, a blank must be run in
parallel. The digestion with DCTA must be done in polypropylene
vessels; if glass is used, the results are less consistent.

Most natural materials contain small amounts of titanium, which is
best determined spectrophotometrically. As titanium is always co-

titrated with aluminium, the aluminium content must be corrected for the titanium content.   Gorcey de Languyon [41] considers the interference from titanium in amounts below 4% to be negligible, provided the sample is decomposed by fusion with potassium hydroxide.   After dissolution and acidification of the cooled melt, K_2TiCl_6 is formed, which does not react with EDTA.   For titanium contents above 4%, however, a correction must be made.   According to this author the error for an aluminium content of 20 - 45% at a Ti content of 4% is 0.35% Al, at 6% Ti the error is 1.2% al, and at 10% Ti, 5.4% Al. The author considers the method suitable only at very low titanium contents.   Paniti [42] recommends neutralizing the titanium-containing aluminium solution to pH 2.5 - 3.0, at which titanium is quantitatively precipitated after 1 - 2 min.   After addition of 0.05M EDTA, the excess of reagent is back-titrated with ferric chloride, using sulphosalicylic acid indicator.   Results for aluminium in the presence of titanium are not given.

REFERENCES

1.   A. Ringbom, Complexation in Analytical Chemistry, Interscience, New York, 1963.   The passage quoted refers to C. Brosset, G. Biedermann and L.G. Sillén, Acta Chem. Scand., 1954, 8, 1917.
2.   R. Přibil, Z. Koudela and B. Matyska, Coll. Czech. Chem. Commun., 1951, 16, 69.
3.   R. Přibil, J. Čihalík, J. Doležal, V. Simon and J. Zýka, Cesk. Farm., 1953, 2, 223.
4.   M. Houda, J. Körbl, V. Bažant and R. Přibil, Coll. Czech. Chem. Commun., 1959, 24, 700.
5.   E. Wänninen and A. Ringbom, Anal. Chim. Acta, 1955, 12, 308.
6.   F. Nydahl, Talanta, 1960, 4, 141.
7.   G. Gottschalk, Z. Anal. Chem., 1960, 172, 192.
8.   G.W.C. Brady and J.R. Gwilt, J. Appl. Chem., 1962, 12, 75.
9.   E. Kiss, Australian Natl. Univ., private communication.
10.   R. Přibil, Coll. Czech. Chem. Commun., 1954, 19, 58.
11.   R. Přibil, Coll. Czech. Chem. Commun., 1954, 19, 64.
12.   R. Přibil and V. Veselý, Talanta, 1962, 9, 23.
13.   I. Sajó, Acta Chim. Acad. Sci. Hung., 1955, 6, 251.
14.   W.Z. Jablonski and E.A. Johnson, Analyst, 1960, 85, 297.
15.   K. Yamaguchi and K. Ueno, Japan Analyst, 1963, 12, 55.
16.   H. Flaschka and R. Puschel, Chemist-Analyst, 1955, 44, 71.
17.   A. Halasz, A. Janosi and K. Villanyi, Vesz. Vegyip. Egyet. Kozlemen., 1961, 5, 151.
18.   G.W.C. Milner and J.L. Woodhead, Analyst, 1954, 79, 363.
19.   R. Přibil and V. Veselý, Talanta, 1962, 9, 23.
20.   R. Přibil and V. Veselý, Talanta, 1963, 10, 1287.
21.   K.E. Burke and C.M. Davis, Anal. Chem., 1964, 36, 172.
22.   D.T. Pritchard, Anal. Chim. Acta, 1965, 32, 184.
23.   H. Bennett and R.A. Reed, Analyst, 1970, 95, 541.
24.   B. Smithies and S. Wolstenholme, J. Soc. Leather Trades Chemists, 1967, 51, 126.
25.   R. Přibil and V. Veselý, Talanta, 1971, 18, 395.
26.   A. Elo, Jr. and J.R. Polky, Anal. Chem., 1960, 32, 294.
27.   T. Iwamato, Japan Analyst, 1961, 10, 190; Anal. Abstr., 1963, 10, 2271.
28.   K.L. Cheng, R.H. Bray and T. Kurtz, Anal. Chem., 1953, 25, 347.
29.   P.B. Sweetser and C.E. Bricker, Anal. Chem., 1953, 25, 253.
30.   D.G. Davis and W.R. Jacobsen, Anal. Chem., 1960, 32, 215.

31. K.I. Begelfer, P.A. Sazonova and K.M. Funtikova, Steklo Keram.,
     1962, 30, 19.
32. H. Kristiansen, Anal. Chim. Acta, 1961, 25, 513.
33. G.W.C. Milner and J.L. Woodhead, Anal. Chim. Acta, 1955, 12, 127.
34. M. Weibel, Z. Anal. Chem., 1961, 184, 322.
35. G. Thielicke, Z. Anal. Chem., 1969, 246, 118.
36. G. Kraft and H. Dosch, Z. Anal. Chem., 1970, 249, 93.
37. G. Kraft, Z. Anal. Chem., 1968, 238, 321.
38. G. Kraft and H. Dosch, Erzmetall, 1968, 21, 308.
39. J. Bartura and W. Bodenheimer, Israel J. Chem., 1968, 6, 61.
40. C.C. Miller and R.A. Chalmers, Analyst, 1953, 78, 686.
41. G.L. de Languyon, Ber. Deutsch. Keram. Ges., 1958, 35, 155;
     Anal. Abstr., 1969, 16, 2947.
42. M. Paniti, Metalurgia (Bucharest), 1968, 20, 601.

6.2.8   Manganese
$$(\log K_{MnY} = 13.8)$$

Manganese(II) forms an analytically useful complex with EDTA in
solutions with pH greater than 6.   The Mn-EDTA and Mn-DCTA complexes
are readily oxidized, e.g. with lead dioxide or sodium bismuthate, to
yield the intensely purple complexes Mn(III)-EDTA and Mn(III)-DCTA,
both of which have been used for the spectrophotometric determination
of manganese [1,2].   They are readily reduced in acid medium by
iodide.   This reaction was used for the iodometric determination of
manganese in the presence of copper, iron, lead and other metals [3].

Direct determination

Erio T was the first indicator used for determining manganese in
ammoniacal medium (pH 10) [4].   As an auxiliary complexing agent to
prevent precipitation of manganese and its oxidation, tartrate was
first recommended [5].   The titration had to be carried out in hot
solution, as the Mn-tartrate complex reacts slowly with EDTA.   Large
tartrate concentrations have an unfavourable affect on the sharpness
of the colour transition.   The titration was improved by using
triethanolamine and hydroxylammonium chloride or ascorbic acid [6].
In the presence of these substances, the manganese solution is clear
and colourless.   The colour transition of Erio T from wine-red to
blue is very sharp.

Later, Erio T and its derivatives were replaced by Thymolphthalexone
[7] and Methylthymol Blue [8], which are much more suitable,
especially as they permit simultaneous masking of iron and aluminium
with triethanolamine.

A number of other indicators have been recommended for alkaline
medium, of which Naphthol Violet [9] and Acid Alizarin Black SN [10]
are the most important.

Manganese can be determined in very weakly acid medium (pH 6 - 7) with
Calcein (Fluorexone) [11], Methylthymol Blue [12], Methylcalcein or
Methylcalcein Blue [13] as indicator.

Indirect determination

All indirect determinations of manganese are carried out in ammonia

or ammonium chloride medium.   Manganese is usually determined by
back-titration with magnesium salts, Erio T, Thymolphthalexone or
Methylthymol Blue being used as indicator.   The last two indicators
can also be used in back-titration with calcium and also to determine
manganese in sodium hydroxide medium.   In special cases, the first
two can be used for back-titration with manganese sulphate containing
a small amount of hydroxylammonium chloride.

Interferences

The selectivity of the determination of manganese is very low, as all
cations forming EDTA complexes, including the alkaline-earth metals,
interfere.   In the titration of manganese at pH 6 - 6.5, with Methylthymol Blue as indicator, Köros [12] found that small amounts of
calcium and magnesium (4 - 6 mg) and up to 50 - 100 mg of barium do not
interfere in the determination of manganese.

Masking of interferents

Manganese(II) forms a very weak complex with cyanide, so all cations
of the "cyanide group" can be masked with potassium cyanide.   Iron
and aluminium can be masked with triethanolamine in the presence of
reducing agents, but only if either Thymolphthalexone or Methylthymol
Blue is used as indicator.   Thioglycollic acid does not react with
the Mn-EDTA complex and can thus be used to mask zinc, copper,
bismuth and indium.   As the Co-EDTA and Ni-EDTA complexes do not
react with thioglycollic acid, this can be combined with potassium
cyanide for stepwise determination of copper, nickel, manganese, etc.
BAL can be used similarly.

The determination of manganese in the presence of calcium and
magnesium is a very important problem.   Originally [14] ammonium
fluoride was proposed for this purpose, but can be used only for
small amounts of magnesium (10 mg).   Amounts of 15 - 20 mg of calcium
cause low results as a result of adsorption of manganese on the
calcium fluoride precipitate.   Povondra and Přibil [15] improved the
method by adding EDTA first, and then ammonium fluoride to the
solution.   Excess of EDTA is titrated with manganese sulphate, which
replaces all the magnesium and calcium during the titration;   the
latter is precipitated as CaF_2 in pure form without any adsorbed
manganese.   The method is suitable for analysis of materials rich in
manganese.

Masking of manganese

Manganese(III) has a marked tendency to form strong complexes and
this characteristic must be considered in masking this element.
Practically, only tervalent manganese can be masked.   A small amount
(4 - 5 mg in 100 ml of solution) of manganese(II) can be masked in
strongly alkaline triethanolamine medium [16].   An intensely green
complex is formed by aerial oxidation to Mn(III)-TEA, which prevents
visual titration of calcium when the manganese is present in high
concentration.   At sufficient dilution, the visual titration can be
carried out with Calcein as indicator.   The ready formation of the
Mn(III)-TEA complex can be used, as already mentioned, in the iodometric determination of manganese [3] or for its photometric determination [17].   The Mn(III)-TEA complex can readily be reduced by
hydroxylamine or ascorbic acid in alkaline medium, permitting the
selective determination of small amounts in the presence of calcium.

The procedure involves masking manganese in sodium hydroxide medium
and determining of calcium by direct titration, using Thymolphthalexone as indicator.   After addition of further EDTA and hydroxylamine,
the green colour of the Mn-TEA complex disappears and the excess of
EDTA is titrated with calcium.   Matsui [18] uses a procedure
involving the determination of the sum of Mn + Ca + Mg in one portion
of sample in the presence of buffer and hydroxylamine, and then
determining the sum of Ca + Mg in a second portion after masking
manganese with triethanolamine.

Large amounts of manganese - up to 80 mg - can be converted into the
Mn(III)-TEA complex by aerial oxidation at pH 13.   After addition of
potassium cyanide, the pH is adjusted to 10 - 11 by careful addition
of dilute acetic acid, the Mn(III)-TEA complex being converted into
the pale yellow $Mn(CN)_6^{3-}$ complex, which is completely inactive
towards EDTA [19].   Yotsuyanagi et al. [20] oxidized manganese at
pH 10 in the presence of TEA and cyanide in ammoniacal medium and
determined the sum of Ca + Mg by direct titration with EDTA, using
Thymolphthalexone as indicator.   The $Mn(CN)_6^{3-}$ formed in solution is
reduced at 60 - 70°C by ascorbic acid in the presence of EDTA.   Excess
of EDTA is determined by titration with 0.05M magnesium sulphate.
Any zinc present is finally demasked with formaldehyde and determined
by direct titration.

Note.   Asaoka [21] studied the reactions of the Mn(III)-TEA complex
polarographically.   He found that, in the presence of large concentrations of cyanide, the Mn(III)-TEA complex is gradually reduced to
the colourless Mn(II)-TEA complex, which is reoxidized by air to the
Mn(III)-TEA complex.   If the system is isolated from the air,
reoxidation does not occur, which the author explains as due to slow
formation of the ternary Mn(II)-TEA-CN⁻ complex, which is inactive to
oxidation by air.   The reduction of the original complex takes many
hours and does not affect the polarographic determination of manganese.   This effect was not found in the complexometric determination
of manganese on the principle described [19], probably as a result of
different conditions.

A very interesting reaction is observed in solutions containing iron
and manganese, in the masking of iron with triethanolamine or of
other cations with potassium cyanide.   The Fe(III)-TEA complex formed is converted in strongly alkaline medium in the presence of
cyanide into the cyanide complex of iron, which immediately oxidizes
manganese according to the equation

$$Fe(CN)_6^{3-} + Mn(CN)_6^{4-} = Fe(CN)_6^{4-} + Mn(CN)_6^{3-}$$

Thus both iron and manganese are masked simultaneously.   The reaction
is quantitative if iron is present in an amount equivalent to the
manganese.   Unfortunately, it cannot be used in determining calcium,
which forms a rather insoluble hexacyanomanganate complex [22].

Separation methods

A number of the methods proposed deal primarily with isolation of
manganese from samples containing calcium, magnesium, iron, aluminium,
titanium, etc.   Compared with use of masking agents, these methods
are rather complicated and can be used only in special cases.
Manganese can be precipitated from alkaline medium as the sulphide

with thioacetamide [23] or as the hydrated oxide by addition of chlorate [24,25] or from acid medium with ammonium peroxodisulphate [26]. The complications involved in these procedures are well known. The danger of loss of small amounts of manganese is always present and traces of calcium can be adsorbed on the precipitate.

A very interesting method [27] is based on the Volhard-Wolf reaction. The authors oxidized manganese with permanganate in acetate medium in the presence of zinc sulphate. The following reaction occurs:

$$3Mn^{2+} + 2MnO_4^- + 2H_2O = 5MnO_2 + 4H^+$$

The MnO_2 is isolated and dissolved and the manganese determined complexometrically. In the ideal case, three atoms of manganese in the original solution yield five manganese atoms in the precipitated MnO_2, giving a multiplication factor of 1.666. The experimentally determined factor is 1.63, which the authors explain as due to the side-reaction

$$4Mn^{2+} + 2MnO_4^- + 3H_2O = 4MnO_2 + Mn^{II}Mn^{IV}O_3 + 6H^+$$

(which would have a factor of 1.50). After dissolution of the MnO_2, the manganese must be determined in the presence of potassium cyanide (to bind any traces of zinc present). The authors applied this method to the determination of manganese in the presence of large amounts of the other cations of the cyanide group, such as zinc (250 mg), mercury (5 g), cadmium (1.1 g) etc., where direct masking with cyanide is practically impossible. After modification, the method can also be used for separation of manganese from iron and aluminium. When the correct conditions are used, very pure manganese dioxide is precipitated, containing only traces of other elements, which can readily be masked. This principle has not yet been used by other authors.

Manganese can be separated from the alkaline-earth metals, or from iron and aluminium, by a method based on using ion-exchange resins and exploiting the different stability of the EDTA and DCTA chelates [28,29]; this factor has been used in practice by other authors [30 - 33] and the methods are especially useful for routine series of analyses with a large number of samples.

The extraction of the thiocyanate complex $Mn(SCN)_6^{4-}$ with a benzene solution of trioctylmethylammonium chloride (Aliquat 336-S) has been described [34]. In a very simple extraction manganese and iron are separated from calcium and magnesium, which are determined complexometrically in the aqueous phase. The organic phase is stripped with ammonia, containing TEA (to mask iron) and hydroxylamine (as a reducing agent). The manganese is determined complexometrically, with Thymolphthalexone as indicator. The method has been applied to the determination of manganese, calcium and magnesium in manganese ores.

Other titrants for manganese

DCTA has been used for the stepwise determination of iron and manganese [35]. Iron was determined at pH 2, with salicylic acid as indicator, and manganese at pH 10, with Erio T. DCTA is preferable because the Fe-EDTA complex reacts with the indicator. Erio T has

been replaced by Thymolphthalexone in the determination with EDTA.
As already mentioned, Körös [12] used DCTA for the determination of
manganese at pH 6 - 6.5.   Other titrants can be used for manganese,
such as NTA [36] or EGTA [37], but have no advantages.   It is interesting that TTHA also forms a 1:1 complex with manganese, in contrast
to other bivalent metals.   This reaction has been used in the determination of aluminium in the presence of large concentrations of
manganese [38].

Recommended procedures

Thymolphthalexone as indicator.   Several crystals of hydroxylammonium chloride, 10 - 20 ml of concentrated ammonia solution and a small
amount of indicator are added to a weakly acid solution, and the
mixture is titrated with 0.02 - 0.05M EDTA from blue to smoke-grey.
For higher manganese concentrations, 10% triethanolamine solution
should also be added.

Erio T as indicator.   The procedure is the same as for Thymolphthalexone as indicator.

Note.   With Thymolphthalexone as indicator, iron and aluminium can
be masked with triethanolamine.   With Erio T, cooling to 5°C allows
masking of aluminium alone.   Other cations can be masked with
cyanide in both cases.

Methylthymol Blue as indicator (pH 6).   About 1 g of hexamine is
added to a weakly acid solution containing up to 130 mg of Mn in
100 ml, along with 8 - 10 drops of freshly prepared 0.1% indicator
solution, and the mixture is titrated with 0.1M EDTA from blue to
clear yellow.

REFERENCES

1.   R. Přibil and E. Hornychová, Coll. Czech. Chem. Commun., 1950,
      15, 456.
2.   R. Přibil and J. Vulterin, Coll. Czech. Chem. Commun., 1954, 19,
      1150.
3.   R. Přibil and V. Veselý, Chemist-Analyst, 1963, 52, 16.
4.   H. Flaschka and A.M. Amin, Mikrochim. Acta, 1953, 414.
5.   H. Flaschka, Chemist-Analyst, 1953, 42, 44.
6.   R. Přibil, Coll. Czech. Chem. Commun., 1954, 19, 58.
7.   J. Körbl and R. Přibil, Coll. Czech. Chem. Commun., 1958, 23,
      1213.
8.   J. Körbl and R. Přibil, Coll. Czech. Chem. Commun., 1958, 23,
      973.
9.   B. Buděšínský, Chem. Listy, 1957, 51, 726; Anal. Abstr., 1957,
      4, 726.
10.   R. Belcher, R.A. Close and T.S. West, Chemist-Analyst, 1958, 47,
      2.
11.   F. Vydra, R. Přibil and J. Körbl, Coll. Czech. Chem. Commun.,
      1959, 24, 2623.
12.   E. Körös, Proc. Intern. Symp. Microchem. Birmingham 1958,
      Pergamon Press, Oxford, 1959, p. 474.
13.   D.H. Wilkins, Anal. Chim. Acta, 1960, 23, 309.
14.   R. Přibil, Coll. Czech. Chem. Commun., 1954, 19, 64.

15. P. Povondra and R. Přibil, Coll. Czech. Chem. Commun., 1961, 26, 2164.
16. R. Přibil, Coll. Czech. Chem. Commun., 1954, 19, 465.
17. J. Adam, J. Jirovec and R. Přibil, Hutn. Listy, 1969, 24, 739.
18. Y. Matsui, Japan Analyst, 1961, 10, 183; Anal. Abstr., 1963, 10, 2206.
19. P. Povondra and R. Přibil, Coll. Czech. Chem. Commun., 1961, 26, 311.
20. T. Yotsuyanagi, T. Yamaguchi, K. Goto and M. Nagayama, Japan Analyst, 1967, 16, 1056; Anal. Abstr., 1969, 16, 1013.
21. H. Asaoka, Japan Analyst, 1963, 13, 1144; Anal. Abstr., 1965, 12, 1173.
22. R. Přibil and V. Veselý, Talanta, 1961, 8, 271.
23. H. Flaschka and H. Abdine, Chemist-Analyst, 1955, 44, 8.
24. H. Flaschka and R. Puschel, Chemist-Analyst, 1955, 44, 71.
25. A.G.G. Morris, Anal. Chem., 1961, 33, 509.
26. G. Staats and H. Bruck, Z. Anal. Chem., 1966, 223, 185.
27. L. Szekeres, E. Kardos and G.L. Szekeres, Mikrochim. J., 1966, 11, 476.
28. P. Povondra and Z. Šulcek, Coll. Czech. Chem. Commun., 1959, 24, 2398.
29. P. Povondra, R. Přibil and Z. Šulcek, Talanta, 1960, 5, 86.
30. G. Friese, Z. Angew. Geol., 1960, 6, 279.
31. G. Babachev and A. Nikolova, Stroit. Mater. Silikat. Prom., 1966, 7, 18; Anal. Abstr., 1967, 14, 3171.
32. Gy. Grasselly, Acta Univ. Szeged. Acta Mineral. Petrog., 1962, 15, 7; Chem. Abstr., 1963, 59, 3310.
33. B. Zagorchev, E. Danova and L. Dodova, God. Vissh. Khim. i Khim. Tekhnol. Inst., 1965, 12, 93; Chem. Abstr., 1967, 67, 104797.
34. R. Přibil and J. Adam, Talanta, 1973, 20, 49.
35. R. Přibil, Coll. Czech. Chem. Commun., 1955, 20, 162.
36. P. Wehber, Z. Anal. Chem., 1957, 154, 122.
37. F. Bermejo-Martínez and M. Paz Castro, An. Soc. Esp. Fis. Quim. Ser. B, 1960, 56, 27.
38. R. Přibil and V. Veselý, Talanta, 1971, 18, 385.

6.2.9 Molybdenum

$$\left(\log K_{(MoO_2)_2Y} = 27\right)$$

The complexometric behaviour of molybdenum is rather similar to that of uranium. In larger amounts it forms the relatively weak $(MoO_3)_2Y^{4-}$ complex, the stability constant of which has not yet been reliably determined. Molybdenum(V) also forms the $(MoO_2)_2Y^{2-}$ complex, which is very stable. Hruškova et al. [1] give the stability constant as > 2.5 x 10²⁷ and consider this one of the most stable complexes. Klygin et al. [2], however, give a value of (1.75 ± 0.52) x 10¹¹, obtained from measurements based on back-titration with zirconium. The conditions employed by the various authors are so different that the values can hardly be compared. The poorly stable Mo(VI)-EDTA complex has been variously employed in analytical chemistry. The selective precipitation of molybdenum with oxine in the presence of EDTA, which masks most cations, is interesting. This reaction has been used for the bromometric determination of molybdenum [3] and can still be considered competitive with the methods described below.

6.2.9.1   Determination of molybdenum(VI)

All attempts to determine molybdenum(VI) in weakly acid medium by
back-titration, e.g., with zinc and use of Xylenol Orange [4],
Pyrocatechol Violet [5] or the $Fe(CN)_6^{3-}/Fe(CN)_6^{4-}$ system [6] as
indicator were not very successful, because of their low selectivity
or unsatisfactory colour transition.

A group of indirect methods is based on precipitation of insoluble
molybdates.   De Sousa [7] precipitated molybdate with calcium and
after filtration determined the calcium complexometrically.   The
method was modified by Lassner and Schlesinger [8], but even this
improved procedure has not found practical use.   Similarly, molybdate can be precipitated with lead and the excess of lead determined
in the filtrate by a complexometric method [9].   This method has
been used in principle in the analysis of ferromolybdenum [10 - 12],
molybdenum disilicide [11] and permalloys [10].   Prior separation of
iron with sodium hydroxide is, however, necessary.

6.2.9.2   Determination of molybdenum(V)

Molybdenum(V) is always determined by back-titration of excess of
EDTA, after reduction of molybdenum(VI) with hydrazine in strongly
acid medium in the presence of EDTA.   The reaction proceeds readily
after short boiling.   The first determination was described simultaneously by Busev and Chan [13] and by Lassner and Scharf [14].   The
first of these authors titrated the excess of EDTA in ammoniacal
medium with zinc, using Erio T, and the others back-titrated with
copper, using PAN indicator at pH 4 in hot solutions containing 30%
ethanol.   Later [15] they determined a small amount of molybdenum
(0.5 - 5.0 mg) with simultaneous masking of tungsten with tartaric
acid, using PAN and 0.005M copper sulphate for the back-titration.

As higher concentrations of the Mo(V)-EDTA complex are strongly yellow,
Lassner and Scharf [16] recommend using Calcein (Fluorexone)
indicator and titrating the excess of EDTA in ultraviolet light with
0.05M copper solution.   Klygin et al. [2] determined molybdenum by
back-titration with zirconium sulphate (in 1M sulphuric aicd), using
Xylenol Orange as indicator.   The EDTA can also be back-titrated
with lead nitrate [17] or zinc sulphate [10].   Headridge [18]
obtained results that were 3 - 5% low in back-titration with zinc,
using Xylenol Orange as indicator, depending on the concentration of
hydrazine sulphate used.   He suggests that, even in weakly acid
medium (pH 5.3), oxygen reacts with hydrazine to yield hydrogen
peroxide, which reoxidizes molybdenum(V).   This suggestion is based
on the fact that molybdenum solutions reduced with a silver reductor
yielded the same results as complexometric titration with zinc.
Consequently, he recommends back-titration with zinc at a pH below 5.
Neither Xylenol Orange nor Alizarine Complexone can be used at this
pH.   Headridge used spectrophotometric titration in this work.

Note.   The value of 10^{27} for the stability constant of the Mo(V)-
EDTA complex seems too high.   Otherwise it would be expected that
molybdenum could be determined by back-titration very selectively
even at low pH (e.g., 1 - 2).   Attempts to determine molybdenum(V) by
back-titration with bismuth nitrate (Xylenol Orange as indicator)
were not successful because of the very vague colour transition,
although the high stability of the Mo(V)-EDTA complex ensures that

the molybdenum is not replaced by the bismuth [14].

6.2.9.3   Determination of molybdenum as the ternary complex with
          hydroxylamine

Methods used so far, in which molybdenum is determined as the
$(MoO_2)_2Y^{2-}$ complex, have a number of disadvantages.   In addition to
low selectivity, the colour of the complex at high molybdenum
concentrations lessens the sharpness of the colour transition.   The
2:1 Mo:EDTA ratio is also a disadvantage.

As mentioned above, all the methods used at present are based on
reduction of molybdenum with hydrazine in the presence of
EDTA.   No other reducing agents for molybdenum have been found
suitable.   Headridge [18] mentions using a silver reductor, but does
not consider sulphur dioxide, ascorbic acid and hydroxylammonium
chloride as suitable as none of them yields quantitative reduction of
molybdenum(VI) to molybdenum(V).   Hydroxylammonium chloride has,
however, since played an interesting role in the complexometry of
molybdenum.   Yaguchi and Kajiwara [19] used this reagent for reducing molybdenum and assumed formation of a simple 1:1 complex of
molybdenum(V).   Lassner [20], however, demonstrated that there is no
reduction and that a complex of molybdenum(VI) is formed, with
simultaneous depolymerization of the isopolymolybdates:

$$
\begin{array}{c}
O\text{----}NH_2 \\
Mo \\
OH_2
\end{array}
$$

He explained the yellow colour of the complex as due to deformation
of the molybdenum atom in formation of the three-membered ring.   The
Mo-HyOH complex (HyOH = hydroxylamine) is very stable, and molybdenum
sulphide is not precipitated from it in acid solutions.   EDTA
converts this complex into the ternary 1:1:1 Mo-EDTA-HyOH complex,
but the reaction is very slow and the solution must be maintained at
the boiling point for 15 min.   Lassner and Schedle [21] later gave
the conditions for forming this complex.   They added excess of EDTA
and back-titrated the surplus with either bismuth nitrate at pH 2 or
copper sulphate at pH 4.5.   Both titrations must be done at 80°C,
the former with Xylenol Orange as indicator and the latter with PAN.
In both cases, slight overtitration leads to replacement of molybdenum in the complex.   This method has two principal advantages:
the colour of the complex is less intense than that of the Mo(V)-EDTA
complex, and the stoichiometry is more favourable (by a factor of 2).
There are, however, great disadvantages: (a) the factor is empirical;
(b) the selectivity is low;   (c) tungsten interferes in the reaction
between molybdate and hydroxylammonium chloride, by formation of an
"isopoly blue".

Other titrants for molybdenum

Bermejo-Martínez and Gimenez Saiz [22] used DTPA for determination of
molybdenum after its reduction with hydrazine or hydroxylammonium
chloride (which they considered to be a reducing agent for molybdenum).
Excess of DTPA was titrated with copper solution at pH 4.5, with PAN

as indicator. Our results on the back-titration with zinc, using
Xylenol Orange as indicator, were unsatisfactory because of the poor
colour transition [23].   DCTA has also been used to determine molybdenum as the ternary complex [23].   Large amounts of hydroxylammonium chloride (up to 10 g) mask up to 300 mg of tungsten.   This
work is discussed by Lassner [24] in the light of later research.

Masking of interferents

The selectivity of the determination of molybdenum (usually at pH 4.5)
is very low.   Of the elements titratable with EDTA, only the
alkaline-earth metals and manganese do not interfere.   Titanium,
niobium, tantalum and tungsten can be masked with tartaric acid, and
thorium, aluminium, lanthanum and uranium are usually masked with
fluoride.

Masking of molybdenum

Molybdenum in weakly acid medium interferes in other titrations both
through partial formation of a complex with EDTA and through blocking
a number of indicators.   In acid medium molybdenum can be masked,
but not very selectively, with acetylacetone [25].   In alkaline
media, molybdenum generally interferes except when it forms an
insoluble molybdate.   Back-titrations can be used to analyse various
insoluble molybdates, e.g. those of barium or strontium.

Separation methods

The complexometric determination of molybdenum is limited to a rather
narrow range of materials.   Among alloys, primarily ferromolybdenum
and some ternary alloys, special catalysts, ores and concentrates can
be analysed.   It is recommended that iron be separated from molybdenum by the old method of precipitation with sodium hydroxide.
Manganese, copper or cobalt [as Co(OH)$_3$] can similarly be separated
in alkaline medium by use of hydrogen peroxide [27].   Acetylacetone
has been recommended for the extraction of iron in alkaline medium.

The separation of molybdenum from some metals has already been
mentioned;   it can be precipitated from solution as CaMoO$_4$ [7,8] or
as PbMoO$_4$ [9 - 12].   The oxine method is, however, much more
selective [3] and could be modified for complexometric purposes.

1,1-Diantipyrinyl-3-methylbutane [29] in chloroform has been recommended for the extraction of the peroxomolybdate complex from 3 - 5N
sulphuric acid containing 1 - 2% hydrogen peroxide.   After stripping
with concentrated ammonia solution, the molybdenum is reduced and
determined indirectly, by back-titration with zinc (Xylenol Orange or
Methylthymol Blue as indicator).   The authors state that this method
can be used to separate molybdenum from tungsten, the rare-earth
elements and other transition elements.

Recommended procedures

The Lassner and Scharf method [14].   A solution containing up to
30 mg of Mo in 100 ml is roughly neutralized and a measured amount of
0.05M EDTA is added along with 5 g of tartaric acid and 2 - 5 g of
hydrazine sulphate.   The mixture is acidified with 2 ml of 9M
sulphuric acid and boiled for 5 min.   The pH is adjusted to 4 with
ammonia, and enough ethanol is added to give a final concentration of

30%.   After addition of 5 - 8 drops of 0.1% ethanol solution of PAN, the solution is titrated with 0.05M copper sulphate from yellow to purple, and then back again with EDTA to pure yellow.

Note.   The molybdenum is determined as the $(MoO_2)_2Y$ complex and thus 1 ml of 0.05M EDTA = 9.595 mg of Mo.

Alizarin Complexone as indicator.   One ml of concentrated hydro-chloric acid and 1 ml of saturated hydrazine sulphate solution are added to 10 ml of solution containing 5 - 15 mg of MoO_4^{2-}.   The solution is heated to 95°C, kept at this temperature for 5 min, and cooled;   10 ml of 0.02M EDTA are then added, the solution is diluted to 50 - 100 ml, and 5 ml of 50% sodium acetate solution and 6 - 8 drops of 0.5% indicator solution are added.   The solution is titrated with 0.02M zinc sulphate from yellow to orange-red.

The yellow colour of the complex somewhat decreases the sharpness of the colour transition.   It is best to titrate to the first reddish colour.   Formation of the 2:1 complex results in 1 ml of 0.02M EDTA being equal to 3.838 mg of Mo.

Determination with DCTA [23].   Hydroxylammonium chloride (1 - 2 g) is added to an almost neutral solution containing 5 - 40 mg of molybdenum, and the solution is diluted to 150 - 200 ml and heated to the boiling point.   The solution is now yellow to light blue-green in colour, depending on the molybdenum concentration.   Excess of 0.05M DCTA (pH 4.5) is added and the solution is heated for a further 15 min. After cooling, the solution is adjusted with solid hexamine to pH 5 - 5.5, several drops of Xylenol Orange indicator are added and the solution is slowly titrated with 0.05M zinc sulphate to a purple-red colour, which disappears after about 3 min because of the slow replacement reaction between zinc and the Mo-EDTA complex.   In this determination 1 ml of 0.05M DCTA corresponds to 4.798 mg of Mo.

Note.   I later attempted a repetition of this procedure, but on boil-ing with hydroxylammonium chloride the molybdenum solutions rapidly became blue and could not be used further, as they practically did not change colour on addition of DCTA or EDTA.   In these experiments, demineralized water was used.   Use of water doubly distilled from an all-glass apparatus gave the expected results.   Traces of silicic acid, which were suddenly freed from the rather exhausted mixed-bed ion-exchanger, resulted in the earlier failure.   As traces of silicic acid are frequently present in normal analyses, the method was modified so that the excess of DCTA was added to the boiling solution first, followed by the hydroxylammonium chloride.   This change in the procedure prevented formation of the readily reducible silicomolybdic acid.

REFERENCES

1.   D. Hrušková, J. Podlahová and J. Podlaha, Coll. Czech. Chem. Commun., 1970, 35, 2738.
2.   A.E. Klygin, N.S. Kolyada and D.M. Zavrazhnova, Zh. Analit. Khim., 1961, 16, 442.
3.   R. Přibil and M. Malát, Coll. Czech. Chem. Commun., 1950, 15, 120.
4.   J. Kinnunen and B. Wennerstrand, Chemist-Analyst, 1958, 47, 38.

5. A.I. Busev and F. Chan, Vest. Mosk. Univ. Ser. Mat. Mech. Astron. Fiz. Khim., 1959, 14, 203.
6. I. Sajó, Mag. Kem. Foly., 1956, 62, 56.
7. A. de Sousa, Anal. Chim. Acta, 1955, 12, 215.
8. E. Lassner and H. Schlesinger, Z. Anal. Chem., 1957, 158, 195.
9. M. Umdea, Japan Analyst, 1960, 9, 172.
10. V.D. Konkin and V.I. Zhichareva, Zavod. Lab., 1963, 29, 791; Anal. Abstr., 1964, 11, 3680.
11. E.I. Nikitina and N.N. Adrianova, Zavod. Lab., 1965, 31, 654; Anal. Abstr., 1966, 13, 5517.
12. L. Šucha, Chem. Prum., 1967, 17, 324; Anal. Abstr., 1968, 15, 5355.
13. A.I. Busev and F. Chan, Zh. Analit. Khim., 1959, 14, 445; Anal. Abstr., 1960, 7, 2199.
14. E. Lassner and R. Scharf, Z. Anal. Chem., 1959, 167, 114.
15. E. Lassner and R. Scharf, Z. Anal. Chem., 1959, 168, 429.
16. E. Lassner and R. Scharf, Z. Anal. Chem., 1963, 183, 187.
17. Y. Endo and T. Higashimori, Japan Analyst, 1962, 11, 1310; Anal. Abstr., 1964, 11, 1712.
18. J.B. Headridge, Analyst, 1960, 85, 379.
19. H. Yaguchi and T. Kajiwara, Japan Analyst, 1965, 14, 785.
20. E. Lassner, J. Less-Common Metals, 1968, 15, 143.
21. E. Lassner and H. Schedle, Talanta, 1968, 15, 623.
22. F. Bermejo-Martínez and M.I. Gimenez Saiz, Quim. Ind. (Bilbao), 1968, 15, 147; Chem. Abstr., 1969, 71, 56322.
23. R. Přibil and V. Veselý, Talanta, 1970, 17, 170.
24. E. Lassner, Talanta, 1972, 19, 1121.
25. W.Z. Jablonski and E.A. Johnson, Analyst, 1960, 85, 297.
26. A.N. Zobnina and I.P. Kislyakov, Zavod. Lab., 1966, 32, 147; Anal. Abstr., 1967, 14, 3157.
27. E.I. Uvarova and V.N. Rik, Khim. Tekhnol. Topliv. Masel, 1964, 67; Anal. Abstr., 1965, 12, 6559.
28. R. Přibil and V. Veselý, Hutn. Listy, 1972, 27, 439.
29. V.P. Zhivopistsev, I.N. Ponosov and Z.I. Plyuta, U'chen. Zap. Perm. Gos. Univ., 1970, 229, 191; Anal. Abstr., 1972, 23, 2439.

6.2.10   Vanadium
$$(\log K_{VOY} = 18.8; \quad \log K_{VO_2Y} = 18.1)$$

All four oxidation states of vanadium react with EDTA. Schwarzenbach and Šandera [1] studied mainly the lower oxidation states and found their stability constants. Only the complexes of vanadium(IV) and (V) are of analytical importance. The vanadyl cation VO^{2+} forms the very stable VOY^{2-} complex $(\log K = 18.8)$ and the protonated complex $VOHY^-$. Ringbom et al. [2] studied the complexes of vanadium(V) in detail. They state that the VO_2Y^{3-} complex is formed in the titration of metavanadate with EDTA according to the equation

$$VO_3^- + H_2Y^{2-} = VO_2Y^{3-} + H_2O$$

In titration with EDTA at pH 6.5, no pH change occurs, as follows from this equation. In more acid solutions (below pH 3), the protonated complex is also formed:

$$VO_2Y^{3-} + H^+ = VO_2HY^{2-}$$

The existence of both complexes was demonstrated by spectrophotometric measurements on solutions of VO_3^- + EDTA, as a function of pH.
The VO_2Y^{3-} complex is very stable $(\log K = 18.1)$ but only over a
rather narrow pH interval.   At pH $10 - 12$ divanadate is formed, and
orthovanadate VO_4^{3-} appears only in even more alkaline solutions.
It follows from the stability constants that the titration of VO_3^-
should not be difficult. This is not true, however, as solutions of
vanadium(V) readily polymerize and the highly polymerized forms react
with EDTA quite differently, as will be described below.

6.2.10.1   Determination of vanadium(V)

Direct determination

Attempts to titrate vanadium(V) directly have not been very successful, primarily because of the very vague colour transition of the
indicator.   This difficulty results from that fact that EDTA can
react only with the VO_2^+ cation according to the equation

$$VO_2^+ + H_2Y^{2-} = VO_2Y^{3-} + 2H^+ \qquad (1)$$

If it is assumed that metavanadate solutions contain only VO_3^- anions,
which are converted into VO_2^+ cations in acid medium:

$$VO_3^- + 2H^+ = VO_2^+ + H_2O \qquad (2)$$

then formation of the simple cation should be favoured by increasing
the hydrogen-ion concentration.   However, increasing the acidity
also forces reaction (1) to the left towards decomposition of the
VO_2Y^{3-} complex.   Kakabadse and Wilson [3] tried direct titration of
vanadium(V) with EDTA, using Xylenol Orange as indicator and found
that the determination could be done according to equation (1) only
when the ratio of the vanadium concentration to the hydrogen-ion
concentration was greater than 5 and the pH above 1.8.

In practice, the composition of metavanadate solutions is much more
complicated.   Conductometric measurements have revealed that tri-
vanadate ions $V_3O_9^-$ predominate in solution instead of VO_3^- ions.
(In acid solutions, the hexavanadates $M_4^I[V_6O_{17}]$ can be prepared.)
All these polymeric forms react very slowly with EDTA.   Sajó [4]
encountered similar difficulties in attempts to determine vanadium
directly, with diphenylcarbazone as indicator.   The determination
can be carried out only over a very narrow pH range $(6.7 - 6.9)$, where
the author states that monovanadate, divanadate and tetravanadate
react instantly with EDTA.   At pH values below 6.7 polymerization to
higher polymers is rapid (pentavanadate or octavanadate) and these
species do not react with EDTA.   Their formation can be prevented at
pH $4.8 - 6.8$ by addition of glycerol or mannitol as auxiliary complexing agents.

None of these methods has been used in practice and they are mentioned only for the sake of interest.

Tanaka and Ishida [5] reduced vanadium(V) at pH $1.7 - 2$ with excess of
ferrous sulphate and titrated the resultant iron(III) with EDTA
either potentiometrically or with Variamine Blue B as redox indicator.

6.2.10.2   Determination of vanadium(IV)

In acid medium vanadium(V) is readily reduced with ascorbic acid or
hydroxylammonium chloride to vanadium(IV); this form yields the blue
vanadyl complex VOY^{2-} with EDTA. The reaction of the VO^{2+} with EDTA
is not instantaneous, but can nonetheless be used for the direct
determination of vanadium(IV).

Direct determination

The first direct determination of vanadium was described by Flaschka
and Abdine [6]. A vanadium solution at pH 3.5 (acetate buffer) is
reduced by boiling with ascorbic acid and the VO^{2+} formed is titrated
with EDTA (CuY-PAN system as indicator). The titration must be done
in hot solution. Arsenic, phosphate and tungsten do not interfere.
Molybdenum prevents the titration because of reduction to the quin-
quevalent form, which colours the solution intensely blue.

Other indicators, such as N-benzoyl-N-phenylanthranilic acid [7],
dithizone [8] and haematoxylin [9], have been recommended for these
direct titrations.

Indirect determination

The very sensitive metallochromic indicators can be used in the
indirect determinations. The high stability of the V(IV) - EDTA
complex (log K = 18.8) permits determination of vanadium rather
selectively, eg. at pH 2 - 3 by back-titration with thorium nitrate
[10], with Xylenol Orange as indicator. This indicator can also be
used for back-titrations with zinc or lead at pH 5 - 5.5. Manganese
(Erio T as indicator) [10] or copper (o-dianisidinetetra-acetic acid
as indicator) [11] can also be used for the back-titration. Similarly, vanadium can be determined indirectly with Calcein as indicator
[12]. Bromopyrogallol Red has also been found useful for back-
titrations with lead nitrate at pH 5 [13].

Interferences

Practically all elements titratable with EDTA except the alkaline-
earth metals interfere in the complexometric determination of
vanadium. Back-titrations at lower pH values are more selective,
e.g. with thorium nitrate [10]. Interferents have been very little
studied because of the comparative rarity of occurrence of vanadium.

Masking

Fluoride can be used to mask titanium, aluminium, etc. [14]. Lassner
and Scharf [15] masked tungsten with large concentrations of tartaric
acid in the direct determination, using the Cu-EDTA-PAN system as
indicator, and recommended their procedure for the analysis of
tungsten-vanadium alloys.

Masking of vanadium

Only hydrogen peroxide has been found useful for masking vanadium,
and has been used in the determination of aluminium [16] or titanium
in titanium-vanadium alloys [14].

Separation methods

It is best to precipitate iron and the heavy metals as the hydroxides
before the determination of vanadium [14]. On the other hand,
vanadium can be very selectively separated on Dowex 50W X8 cation-
exchanger, as described by Fritz and Abbink [17]. The vanadium is
selectively eluted with 0.01M sulphuric acid or perchloric acid
containing 1% hydrogen peroxide. After boiling, the vanadium is
reduced with ascorbic acid and determined indirectly by back-
titration of excess EDTA with zinc solution, Naphthyl Azoxine S being
used as indicator. In this way vanadium can be separated from 25
common cations, including iron, present in ratios of up to 100 to the
vanadium.

High molecular-weight amines, such as trioctylmethylammonium chloride
[18] can be used for selective extraction.

Recommended procedures

Direct determination of vanadium(IV). The pH of 50 ml of solution
containing 5 - 20 mg of vanadium is adjusted to 4 - 4.3 and the
solution is diluted with acetate buffer (pH 4.3). After addition of
0.5 g of ascorbic acid, the solution is heated to the boiling point,
5 or 6 drops of 0.005M Cu-EDTA complex are added along with 6 drops
of an ethanol solution of PAN indicator (0.05%), and the solution is
titrated (while hot) with 0.01M EDTA from red-purple to yellow-green.

Note. The pH of the solution must not fall below 3.5 during the
titration.

Indirect determination of vanadium(IV). Ascorbic acid (0.5 g) is
added to a weakly acid solution containing 5 - 20 mg of vanadium. The
pH is adjusted to 2 - 3 with ammonia or 1M nitric acid, excess of
0.01M EDTA is added (5 - 50 ml) along with several drops of Xylenol
Orange indicator, and the solution is titrated with 0.01M thorium
nitrate from yellow-green to red.

Alternatively, a measured amount of 0.05M EDTA (or DCTA) and 1 g of
ascorbic acid are added to a weakly acid solution containing 5 - 50 mg
of vanadium in 50 - 100 ml of solution. The original yellow colour
changes to intense blue, which gradually fades. The final solution
is pale blue. The reduction is carried out for about 10 min, with
occasional mixing. The pH is adjusted to 5 - 5.5 with solid hexamine
and the excess of EDTA (or DCTA) is titrated with 0.05M zinc chloride,
with Xylenol Orange as indicator, from yellow to intense purple.

In another method, after reduction of the vanadium by the method just
given, excess of 0.05M DCTA, 50 ml of concentrated ammonia solution
and a small amount of Methylthymol Blue indicator are added and the
solution is titrated with 0.05M magnesium chloride to a clear blue
colour.

Note. An EDTA solution cannot be used here as magnesium displaces
vanadium from the V(IV)-EDTA complex. Similar replacement of
vanadium in the V(IV)-DCTA complex is so slow that it does not affect
the end-point. Molybdenum(VI) does not interfere in alkaline medium.
Aluminium can be masked with fluoride.

REFERENCES

1.   G. Schwarzenbach and J. Šandera, Helv. Chim. Acta, 1953, $\underline{36}$, 1089.
2.   A. Ringbom, S. Siitonen and B. Skrifvars, Acta Chem. Scand., 1957, $\underline{11}$, 551.
3.   G. Kakabadse and H.J. Wilson, Analyst, 1961, $\underline{86}$, 402.
4.   I. Sajó, Z. Anal. Chem., 1962, $\underline{188}$, 594.
5.   M. Tanaka and A. Ishada, Anal. Chim. Acta, 1966, $\underline{36}$, 515.
6.   H. Flaschka and H. Abdine, Chemist-Analyst, 1956, $\underline{45}$, 58.
7.   V.R.M. Kaimal and C.C. Shome, Anal. Chim. Acta, 1962, $\underline{27}$, 594.
8.   S. Hara, Japan Analyst, 1961, $\underline{10}$, 629; Anal. Abstr., 1963, $\underline{10}$, 3043.
9.   G.K. Singhal and K.N. Tandon, Chemist-Analyst, 1967, $\underline{56}$, 60.
10. J. Kinnunen and B. Wennerstrand, Chemist-Analyst, 1957, $\underline{46}$, 92.
11. R. Belcher, D.I. Rees and W.I. Stephen, Talanta, 1960, $\underline{4}$, 78.
12. D.H. Wilkins and L.E. Hibbs, Anal. Chim. Acta, 1959, $\underline{20}$, 427.
13. M. Malát, V. Suk and M. Tenorová, Coll. Czech. Chem. Commun., 1959, $\underline{24}$, 2815.
14. R. Přibil and V. Veselý, Hutn. Listy, 1970, $\underline{11}$, 813.
15. E. Lassner and V. Scharf, Planseeberichte, 1961, $\underline{9}$, 51.
16. D. Filipov and N. Kirtcheva, Compt. Rend. Acad. Bulgare. Sci., 1964, $\underline{17}$, 467; Chem. Abstr., 1964, $\underline{61}$, 7686.
17. J.S. Fritz and J.E. Abbink, Anal. Chem., 1962, $\underline{34}$, 1080.
18. R. Přibil and V. Veselý, unpublished results.

6.2.11   The Rare-Earth Elements (RE) - The Lanthanides
$$(\log K_{LaY} = 15.50 \qquad \log K_{LuY} = 19.83)$$

Vickery [1] has suggested that scandium (at. no. 21) and yttrium (at. no. 39) can also be included among the rare-earth elements (at. no. 57 - 71) because of the similarity to lanthanum in electronic structure.   Both these elements also accompany the rare-earth elements in minerals.

The increasing industrial importance of the rare-earth elements is also reflected in the development of analytical methods for them. Before the introduction of EDTA into analytical chemistry, practically no titrimetric methods for direct determination of these elements were known.   A great deal of attention has been paid to the chemical properties of these RE complexes, especially their stability constants (Table 6.2), which are important for chromatographic separations as well as for complexometric determinations.   The very pure individual RE elements now available commercially are prepared chromatographically with NTA, EDTA and HEDTA and similar substances as elution reagents.   It is highly probable that details of the chromatographic preparation have been kept secret or only partially published.

Direct determination

The great chemical similarity of the rare-earth elements is also reflected in their complexometric behaviour.   The sum of these elements is always determined.   In aqueous medium their tervalent cations yield a colour reaction with a number of indicators.   Flaschka [2] first proposed determination of the rare-earth elements by titration at pH 8 - 9, in a tartrate or citrate acid medium to prevent hydrolysis, with Erio T as indicator. The titration must be done in boiling solution. At pH above 9, the complexes of these elements with the indicator are too stable and do not react with EDTA. Malát and Tenorová [3] proposed using Chrome Azurol S as indicator for titration of RE, in pyridine medium. The stability of the RE-EDTA complexes is sufficiently high to allow their formation in weakly acid medium. A number of more or less successful indicators have been proposed. Xylenol Orange [4] was first used in 1956 for the determination of lanthanum, and was also used for determining other rare-earth elements [5]. At present it is the usual indicator for this determination. Lyle and Rahman [6] and other authors consider it the most reliable of all indicators. Only arsenazo III is considered as more suitable for determining the rare-earth elements (in solutions with large sulphate contents) [7].

TABLE 6.2   Logarithms of the stability constants
of the ML complexes of the rare-earth elements
with various complexing agents

Metal	NTA[a]	EDTA[b]	EGTA	HEDTA[c]	DTPA[b]	DCTA	TTHA
La	10.37	15.50	15.55	13.82	19.43	16.3	22.22
Ce	10.83	15.98	15.70	14.45	20.50	16.76	19.20
Pr	11.07	16.40	16.05	14.96	21.07	17.23	
Nd	11.25	16.61	16.28	15.16	21.60	17.69	22.82
Sm	11.51	17.14	16.88	15.64	22.34	18.63	24.30
Eu	11.49	17.35	17.10	15.62	22.39	18.77	
Gd	11.54	17.37	16.94	15.44	22.46	18.80	
Tb	11.58	17.90	17.27	15.55	22.71	19.30	
Dy	11.71	18.30	17.42	15.51	22.82	19.69	
Ho	11.85	18.74	17.38	15.55	22.78	19.89	
Er	12.00	18.85	17.40	15.61	22.74	20.20	23.19
Tm	12.20	19.32	17.48	16.00	22.72	20.46	
Yb	12.37	19.51	17.78	16.17	22.62		
Lu	12.47	19.83	17.81	16.25	22.44	20.91	

The stability constants for complex ML_2 for a), complex MHL for b) and complex MOHY for c) are not given here.

Indirect determination

If it is necessary to carry out indirect determination, a zinc solution and Xylenol Orange are used for the back-titration (pH 5 - 5.5).

Interferences

All elements forming EDTA complexes with stability constants greater

than 10^{11} interfere or are co-titrated.   Among anions, fluoride
masks the rare-earth elements and phosphate precipitates them.   Large
concentrations of sulphate decrease the sharpness of the colour
transition.   Cerium(IV) also prevents determination of the sum of
these elements and must be reduced to cerium(III) with ascorbic acid.

Masking of interfering elements

There are only a very few cases where the rare-earth elements can be
determined without preliminary masking.   The alkaline-earth elements
do not generally interfere in the determination.   On the other hand,
elements forming EDTA complexes with a high conditional stability
constant at low pH can be determined without interference from the
rare-earth elements.   Milner and Edwards [8] described stepwise
determination of bismuth at pH 2 and neodymium or praseodymium at
pH 5 - 6, with Xylenol Orange as indicator.   Hung Shiu-Chieh and
Liang Shu-Chuan [9] described the microdetermination of first scandium and then the rare-earth elements at pH 1.8 - 2.2, with Xylenol
Orange as indicator.   They state that the determination is possible
up to a ratio of $Ln_2O_3:Sc_2O_3$ = 1000.   Zirconium can be determined by
EDTA titration in 0.4 - 0.6M sulphuric acid in the presence of all the
rare earths [10].

Flaschka recommends that copper, zinc and cadmium be masked with
potassium cyanide in alkaline medium.   Copper is easily masked with
thiourea.   A great deal of attention has been paid to the masking of
aluminium.   Chernikov et al. [11] determined 10 mg of rare-earth
elements and masked up to 3 mg of aluminium with sulphosalicylic acid.
Higher sulphosalicylic acid concentrations lead to loss of sharpness
of the colour transition of Xylenol Orange;   on the other hand, Brück
and Lauer determined lutetium [12] and dysprosium [13] in binary
alloys with masking of up to a 100-fold amount of aluminium with
sulphosalicylic acid.   They recommend neutralizing the solution with
ammonia alone and not with hexamine to maintain the pH at exactly
5.5 $\pm$ 0.1 and that the sulphosalicylic acid added should be exactly
equivalent to the aluminium content (0.5 ml of 20% solution per mg of
Al).

In the analysis of lanthanide zeolite catalysts for molecular sieves,
Marsh and Myers [7] recommend masking aluminium with sulphosalicylic
acid and direct determination of the lanthanides by titration with a
DTPA solution using arsenazo III indicator.   Sulphosalicylic acid
also masks iron(III), which in small (μg) amounts does not interfere
in the indicator colour transition.   In the analysis of Y-Al
granites, Wilke [14] first determined yttrium, using Xylenol Orange
as indicator, after masking aluminium, and then the sum Al + Y by
back-titration with zinc, using dithizone as indicator.

Acetylacetone has also been recommended for masking aluminium in
determination of the rare-earth elements.   Fritz et al. [15]
titrated these elements potentiometrically, using a mercury electrode,
and Milner and Gedansky [16] titrated them visually, using Xylenol
Orange at pH 5 - 5.5.   The optimal amount of acetylacetone is 3 - 5
times greater than that corresponding to the amount of aluminium.
The results are somewhat low when very large amounts of masking agent
are used.

Masking of the rare-earth elements

Fluoride reliably masks the rare-earth elements even when they are
bound in EDTA complexes, allowing stepwise determination of, for
example, iron and heavy metals and then of the rare-earth elements by
back-titration with zinc.   First the sum of all the elements is
determined and then, after addition of fluoride, the EDTA liberated
(corresponding to the sum of all the lanthanides) is determined.

Separation methods

Lanthanum can readily be separated from iron and aluminium by precipitation with sodium hydroxide from triethanolamine medium [17].   The
rare-earth elements are usually precipitated as the oxalates, which
are relatively insoluble (10^{-6} mole/l.).   However, the gravimetric
determination of the rare-earth elements as the oxides R_2O_3 has a
number of disadvantages.   Ignition of praseodymium and terbium
oxalates yields oxides with the approximate composition Pr_6O_{11} and
Tb_4O_7 respectively and cerium(III) oxalate yields CeO_2.   The first
two oxides can be reduced in a stream of hydrogen at 500 - 600°C to
Pr_2O_3 and Tb_2O_3, but ceric oxide is difficult to reduce.   If the
conditions are not optimal, the bioxalates $[R(C_2O_4H)_3]$ can also be
formed and, in the presence of sulphate, sulphate-oxalates [18].

In the separation of rare-earth elements from aluminium, Milner and
Gedansky [16] found not only co-precipitation of aluminium on the
precipitated oxalates, but also loss during filtration.   The content
in the filtrate was about 3 mg.

Gravimetric determination of the rare-earth elements was later
replaced by complexometric determination.   Misumi and Taketatsu [19]
isolated the oxalates of the rare-earth elements, dissolved them in
ammoniacal EDTA solution and determined the excess of EDTA by titration with magnesium, using Erio T as indicator.   Lyle and Rahman [20]
used a similar procedure, with 0.01M magnesium or zinc solution
(Erio T) or lanthanum and arsenazo III for the back-titration.   The
oxalates can also be decomposed with perchloric acid and the rare-
earth elements determined by direct titration, with Xylenol Orange as
indicator.

Complexometry is not limited to determination of the rare earths in
pure solutions.   They can be determined in the presence of thorium,
uranium, titanium and phosphate, as well as in the presence of EDTA,
various extraction agents, etc.   Some of these determinations will
be discussed below.

Recommended procedure

Several drops of Xylenol Orange indicator are added to an acid
solution containing up to 100 mg of the rare-earth element(s),
followed by portions of solid hexamine until the colour is purple-red.
The solution is titrated with 0.01 - 0.05M EDTA to a clear yellow.

Determination in the presence of phosphate

Rare-earth phosphates are precipitated in weakly acid solutions.
They cannot be determined indirectly, e.g. by back-titration of
surplus EDTA or DCTA, as they are readily replaced in the complexes
by the zinc used for the back-titration and precipitated as RPO_4.

Indirect titration can, however, readily be carried out with DTPA, which forms much more stable complexes with the rare-earth elements. Excess of 0.05M DTPA is added to a solution containing the lanthanide phosphates, the solution is neutralized with ammonia, a small amount of ascorbic acid is added to reduce the cerium, and the pH is adjusted with hexamine (and acid if necessary) to 5 - 5.5.   Surplus DTPA is then titrated with 0.05M zinc, with Xylenol Orange as indicator [21]. Scandium is determined simultaneously.   Thorium interferes, as it is precipitated as the phosphate during the neutralization, even in the presence of DTPA.   In the analysis of phosphate ores, Kinnunen and Wennerstrand [22] chose a somewhat different method.   They decomposed the ore with sodium peroxide, isolated the precipitated hydroxides of the rare-earth elements (plus Fe, etc.), dissolved them in nitric acid, precipitated and collected the rare-earth oxalates and ignited them to the oxides, which were then dissolved in nitric acid and determined complexometrically.

Determination in the presence of EDTA

The chromatographic separation of the rare-earth elements is based on separation of the RE-EDTA complexes at various pH values or elution with EDTA (and other compounds) at various acidities.   The eluate contains both EDTA and the rare-earth elements.   As this procedure is the basis of a technological process, the simultaneous determination of EDTA and these elements in solution is very important. Martynenko [23] recommends precipitating the rare-earth elements in one aliquot as the oxalates, and after ignition to the oxides, determining them complexometrically.   Free EDTA is determined in a second aliquot.   The overall EDTA content (bound and free) is found by adding the EDTA consumption in the two titrations.   A simpler procedure was proposed by Tereshin and Tananayev [24] who titrated all the EDTA at pH 2 with ferric chloride, using sulphosalicylic acid as indicator, and after adjusting the pH to 4.5 - 6 titrated the rare-earth elements with 0.04M EDTA, using Xylenol Orange as indicator.   The relative error in the individual determinations was $\pm 0.7\%$.   An alternative would be titration of free EDTA with zinc, and of the bound EDTA after demasking the rare-earth elements with fluoride.

Determination of thorium and the rare-earth elements

The stepwise determination of thorium at pH 2.5 - 3 and of the rare-earth elements with TTHA, DTPA or HEDTA was discussed in sufficient detail in Section 6.1.3, dealing with thorium.

Determination of scandium and the rare-earth elements

The stepwise determination of scandium and the rare-earth elements was mentioned in Section 6.1.4.   Combined titration with DCTA and TTHA allows determination both of thorium and scandium and of the rare-earth elements [25].   Details are given in the literature.

Analysis of a mixture of the rare-earth elements

Stepwise determination of the rare-earth elements, individually or in groups, is impossible, as their stability constants are so similar. From a theoretical point of view, a necessary condition for the stepwise titration of two elements is that their stability constants differ by a factor of at least 10^4.   This condition is not fulfilled even for the first and last members of the series (lanthanum and lutetium) and there is not much hope of finding a selective masking agent for individual lanthanides (or even groups of them).

Sometimes the mean relative atomic weight of the rare-earth elements is of interest.   This value can be found by a method described by Flaschka [2]: a weighed amount of the oxides is dissolved in nitric acid and the sum of the rare-earth elements is determined complexometrically.   The mean atomic weight is calculated according to the formula

$$A + 24 = \frac{100\ m}{V}$$

where A is the mean relative atomic weight, m is the weight of the oxides (in mg) and V is the consumption of 0.01M EDTA (in ml).

Patrovský [26] employed the large difference in the atomic weights of yttrium (88.92) and erbium (167.27) for their determination in mixtures.   He also dissolved a known weight of the mixture of the oxides of the two elements in nitric acid and determined them complexometrically, using Xylenol Orange as indicator.

The oxide contents (in mg), Y_2O_3, (x) and Er_2O_3, (y), are found from the equations

$$x = \frac{3.824V - m}{0.6935} \qquad y = \frac{2.582V - m'}{0.4094}$$

where V is the consumption of 0.02M EDTA (in ml), m and m' are the weights of the oxides dissolved (mg).   He states that this method can also be used to determine the sum of the yttrium earths if they are present in approximately equal amounts.   The atomic weight of terbium (158.93) is compensated by the atomic weight of lutetium (174.99) giving a numerical value of 166.96, which is very close to the atomic weight of erbium (167.26).

It seems suitable here to mention attempts to differentiate the rare-earth metals by complexometric titration.   The replacement reaction:

$$LnY^- + Zn^{2+} + PO_4^{3-} = ZnY^{2-} + LnPO_4$$

has been used [27].   First the sum of the rare-earth elements is determined by back-titration with zinc at pH 5 – 5.5, using Xylenol Orange as indicator.   After addition of 0.1M $NaHPO_4$, further titration with zinc gives replacement according to the equation above but at different rates, and the rare-earth elements can accordingly be separated into three groups:

(a)   La, Ce, Pr and Nd are replaced rapidly and quantitatively.
(b)   Sm, Eu, Gd, Tb, Dy, and Y are replaced very slowly and non-quantitatively.
(c)   Ho, Er, Tm, Yb and Lu are not replaced within 5 min after addition of the zinc.

An element in group (a) can be determined satisfactorily in the presence of an element in group (c), but the elements of group (b) obviously interfere.

Recommended procedure

Several drops of Xylenol Orange indicator are added to an acid
solution containing up to 100 mg of the rare-earth element(s),
followed by portions of solid hexamine until the colour is purple-
red. The solution is titrated with 0.01 - 0.05M EDTA to a clear
yellow.

REFERENCES

1.   R.C. Vickery, Chemistry of the Lanthanons, Academic Press, New
      York, 1963.
2.   H. Flaschka, Mikrochim. Acta, 1955, 55.
3.   M. Malát and M. Tenorová, Coll. Czech. Chem. Commun., 1959, 24,
      632.
4.   J. Korbl and R. Přibil, Chemist-Analyst, 1956, 45, 102.
5.   J. Kinnunen and B. Wennerstrand, Chemist-Analyst, 1957, 46, 92.
6.   S.J. Lyle and M. Rahman, Talanta, 1963, 10, 1177.
7.   W.W. Marsh Jr. and G. Myers Jr., Anal. Chim. Acta, 1968, 43,
      511.
8.   G.W.C. Milner and J.W. Edwards, Anal. Chim. Acta, 1958, 18, 513.
9.   Hung Shui-Chieh and Liang Shu-Chan, Scientia Sinica, 1964, 13,
      1619.
10.   R. Přibil, unpublished results.
11.   Yu. A. Chernikhov, R.S. Tramm and K.S. Pevzner, Zavod. Lab.,
      1960, 26, 921; Chem. Abstr., 1960, 54, 24124a.
12.   A. Brück and K.F. Lauer, Anal. Chim. Acta, 1965, 33, 338.
13.   A. Brück and K.F. Lauer, Anal. Chim. Acta, 1967, 39, 135.
14.   K.T. Wilke, Z. Anal. Chem., 1967, 232, 278.
15.   J.S. Fritz, M.J. Richard and S.K. Karraker, Anal. Chem., 1958,
      30, 1347.
16.   O.I. Milner and S.J. Gedansky, Anal. Chem., 1965, 37, 931.
17.   R. Přibil and V. Veselý, Chemist-Analyst, 1964, 53, 43.
18.   N.E. Topp, Chemistry of the Rare Earth Elements, Elsevier,
      Amsterdam, 1965.
19.   S. Misumi and T. Taketatsu, Bull. Chem. Soc. Japan, 1959, 32,
      873; Anal. Abstr., 1960, 7, 1691.
20.   S.J. Lyle and M. Rahman, Talanta, 1963, 10, 1183.
21.   R. Přibil and V. Veselý, Chemist-Analyst, 1967, 56, 23.
22.   J. Kinnunen and B. Wennerstrand, Chemist-Analyst, 1967, 56, 24.
23.   L.I. Martynenko, Nauch. Dokl. Vys. Shkol. Khim. i Khim.
      Tekhnol., 1958, 718.
24.   G.S. Tereshin and I.V. Tananayev, Zh. Analit. Khim., 1962, 17,
      526.
25.   R. Přibil and J. Horáček, Talanta, 1967, 14, 313.
26.   V. Patrovský, Coll. Czech. Chem. Commun., 1959, 24, 3305.
27.   R. Přibil and V. Veselý, Chemist-Analyst, 1965, 54, 100.

6.3   COMPLEXOMETRY OF GROUP III CATIONS

This group contains primarily the alkaline-earth metal cations.
These form relatively weak complexes with EDTA and can be determined
only in alkaline solutions; the determination is not selective. The
greatest attention has been paid to the complexometric determination
of magnesium and calcium, which is necessary in the analysis of raw

materials and of many products (glass, ceramics, cement, etc.). The
first publication on the determination of calcium and magnesium was
greatly welcomed by analysts, as up to that time titrimetric analysis
for these elements was practically restricted to the permanganate
titration of calcium oxalate.

6.3.1   Calcium and Magnesium
$(\log K_{CaY} = 10.7, \log K_{MgY} = 8.7)$

6.3.1.1   Direct determination of calcium

Calcium was among the elements which were first determined in 1948 by
Biedermann and Schwarzenbach [1] by titration with EDTA, using murexide as indicator; this method is still used to some extent. The
colour transition at pH 12 from red to blue-purple is not very satisfactory. For a number of years, however, no other indicator was
available. Several ways of improving the colour transition have
been proposed. Knight [2] recommended using a mixture of murexide
with Naphthol Green G in 2:5 ratio. Piglowski [3] used a mixture of
murexide, Naphthol Green B and Naphthidine Green. These modifications can improve the sharpness of the colour transition, but not the
sensitivity for calcium. The difficulty in determining the end-
point is completely removed by photometric titration, which gives
very good results [4], but it is always necessary to apply an indicator correction. A certain advantage of murexide - at least at the
beginning of complexometry - was its low sensitivity for magnesium.
In the presence of precipitated magnesium hydroxide murexide reacts
only with calcium, thus permitting a more or less satisfactory determination of calcium in the presence of magnesium.

Another classical indicator, Erio T, is even less suitable for direct
use in determining calcium in "pure solutions", as it forms a relatively weak calcium complex $(\log K_{CaIn} = 5.4)$ which is much less
strongly coloured than the magnesium complex. However, the Mg-Erio
T complex can be used as the indicator for direct titration of
calcium in solutions buffered at pH 10, as the calcium is preferentially complexed by the EDTA. For the same reason, the Mg-EDTA
complex can be added, with Erio T as indicator, for the analysis of
very pure calcium salts. As a result of the replacement reaction

$$MgY^{2-} + Ca^{2+} = CaY^{2-} + Mg^{2+}$$

the end-point is given by the final titration of the magnesium, with
a very sharp colour transition from wine-red to blue. Erio T is
still used mainly for determination of the sum of Ca + Mg.

A number of dihydroxyazo dyes have been recommended as indicators in
the determination of calcium. Many of them have a structure similar
to Erio T and their indicator function is similar. They differ in
their stability in aqueous solution or stability towards oxidants.
Analytical interest has centred primarily on substances which would
allow determination of calcium in the presence of magnesium or
precipitated magnesium hydroxide. They were discussed in Sections
2.5 and 2.7.

End-point detection in EDTA titrations of calcium improved consider-

ably on the introduction of metallochromic indicators of the phthalein and sulphophthalein series, especially Methylthymol Blue, Thymolphthalexone and Fluorexone (Calcein) and their derivatives.   The
first two of these indicators can be used both in ammoniacal medium
(Ca + Mg) and in sodium hydroxide medium (Ca), but only Fluorexone
can be used for the selective titration of calcium, similarly to
murexide.

6.3.1.2   Indirect determination of calcium

This determination is very useful in stepwise analysis for calcium
and other cations in a single solution by addition of suitable masking agents.   For example the sum Ca + Zn + Pb can be determined by
back-titration of excess of EDTA with magnesium, with Erio T or
another suitable indicator.   Further titration after adding cyanide
gives the content of zinc, and after addition of thioglycollic acid
yields the lead content.

6.3.1.3   Determination of magnesium

Because of the insolubility of magnesium hydroxide, magnesium can be
determined in ammoniacal medium only in the presence of ammonium
salts.   Erio T is quite suitable for the determination in solutions
at pH 10 (Schwarzenbach buffer); its derivatives can also be used.
The solution pH'is not as important in determinations using Methylthymol Blue or Thymolphthalexone.   It is sufficient to alkalize the
acidic solution of magnesium with a concentrated ammonia solution.

As with calcium, indirect determination is best done with Thymolphthalexone or Methylthymol Blue as indicator.   If masking agents
are used, several cations can be determined in a single solution.

6.3.1.4   Determination of calcium in the presence of magnesium

The stability constants of the Ca-EDTA and Mg-EDTA complexes differ
by only two orders of magnitude.   The simultaneous determination of
calcium and magnesium, in any ratio, is very precise, even on the
micro scale.   In analysis of a mixture of the two elements it is
frequently sufficient to determine calcium alone in an aliquot of the
solution and to calculate the magnesium content by difference from
the sum of the two.   Several hundred publications have dealt with
this problem.   Generally, one of four different procedures is
employed:

(a)   precipitation of magnesium as $Mg(OH)_2$;
(b)   masking of magnesium;
(c)   isolation of calcium;
(d)   determination of Ca and Mg with different complexing agents.

6.3.1.5   Determination of calcium in the presence of $Mg(OH)_2$

Calcium is determined in a strongly alkaline sodium hydroxide medium,
where magnesium is precipitated as the hydroxide, which interferes in
the determination not only by complicating the indicator colour
transition, but mainly by adsorbing calcium on its very active

surface.   With increasing magnesium concentration, i.e. with
increasing amounts of the $Mg(OH)_2$ precipitate, the negative error
becomes sufficiently large to make the determination of calcium useless.   The irregularity of the results obtained in the determination
of calcium with murexide as indicator has been explained as due to
unsuitable alkalinity of the solution, unsuitable choice of method
for precipitating magnesium hydroxide, etc.   In their study of the
microdetermination of calcium in blood serum, Kenny and Cohn [5]
found that the interference from magnesium decreases with increasing
murexide concentration, provided that a pH value of 12.4 - 12.5 is
maintained.   Van Schouwenburg [6] suggests that the $Mg(OH)_2$ precipitate does not adsorb calcium, but rather its complex with murexide.
No adsorption by $Mg(OH)_2$ was observed when Calcon and Eriochrome Blue
SE were used, in fact, the results for calcium increased with increasing magnesium concentration, as a result of partial co-titration of
magnesium.   Bond and Tucker [7] stated that the irreproducibility of
the results for calcium is a result of unsuitable solution alkalinity.
If the solution is insufficiently alkaline, magnesium is co-titrated;
however, with increasing alkalinity, the adsorption of calcium on
magnesium hydroxide increases.   They suggest adding sucrose to
prevent this adsorption [7,8].   Lott and Cheng recommend adding
gelatine [9] or poly(vinyl alcohol) [10] to suppress "adsorption"
during the determination of calcium with Calcon as indicator.   Van
Schouwenburg [6] found it impossible to reproduce the results of
these authors and recommended using Carbocel (sodium carboxymethyl-
cellulose).   Most authors feel that the results for calcium are very
dependent on the method of precipitating the magnesium hydroxide.
Slow dropwise addition of sodium hydroxide solution with thorough
mixing is recommended [11,12].   Flaschka and Huditz [13] and
Brunisholz et al. [14] recommend back-titration of EDTA with a calcium solution.   After strong alkalization of the solution all
the magnesium is replaced during the titration and the precipitated
$Mg(OH)_2$ should be very pure.   Brunisholz et al. state that the
results are satisfactory with up to 30 mg of magnesium present.   A
similar method was adopted by Lewis and Melnick [12].   They added up
to 95% of the required amount of EDTA and titrated the remaining
calcium after alkalization, using Calcon or Cal-Red as indicator.
Naidu and Sastry [15] first determined the sum of Ca + Mg with Erio T
as indicator and after precipitation of magnesium with sodium hydroxide determined the liberated EDTA with calcium chloride solution,
using Cal-Red as the indicator.   Lewis and Melnick [12] assumed
that EDTA is co-precipitated when $Mg(OH)_2$ is precipitated in the
presence of EDTA, yielding high results for calcium.   After several
minutes of thorough mixing, during which $Mg(OH)_2$ crystals are formed,
the EDTA is freed into solution and the results are improved.   This
co-precipitation of EDTA increases with increasing magnesium concentration.   Adsorption of both indicators can also be prevented by adding them after formation of the $Mg(OH)_2$ cyrstals.

Kodama [16] recommends a very interesting modification of the determination of calcium, in which the calcium is titrated directly with
EDTA (Cal-Red acid as indicator) 3 min after addition of the masking
agent (KCN, TEA).   The results are, understandably, low.   Then 6M
hydrochloric acid is carefully added dropwise to dissolve the $Mg(OH)_2$
precipitate, the solution is again alkalized and a further portion of
masking agent and indicator is added.   The calcium freed from the
$Mg(OH)_2$ precipitate is then titrated in the same manner.

It would be interesting to determine the degree to which the nature

of the indicator affects the results for calcium.   These effects can
be various and even mutually cancelling.   For example, in the titration of calcium, with Calcein as indicator, Diehl and Ellingboe [17]
did not observe any interference from magnesium hydroxide.   Socolar
and Salach [18] obtained good results in the microdetermination of
calcium (1 - 2 µg) with 0.002M EDTA, even in the presence of a 100-
fold molar ratio of magnesium.   Magnesium starts to interfere
seriously when the $Mg(OH)_2$ precipitate becomes visible in the
solution.   The results for calcium decrease in proportion to the
amount of precipitate.

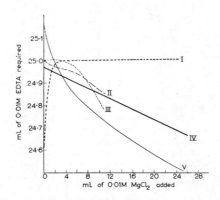

Fig. 6.1   The titration of 25 ml of 0.01M Ca in
the presence of an increasing amount of magnesium,
with various indicators [19].
I - Calcon, II - Methythymol Blue, III - Acid
Alizarin Black SN, IV - murexide, V - Calcein
(Fluorexone).   (By permission of Pergamon Press)

The results of various microdeterminations of calcium are difficult
to compare with results obtained on the macro scale.   In the micro
studies, the authors intentionally use microconcentrations of magnesium, where the possible effect of any precipitated $Mg(OH)_2$ would be
minimal.   The study published by Belcher et al. [19] is very important for practical complexometry.   These authors studied the behaviour of various indicators in the titration of calcium in the
presence of increasing concentration of magnesium and found that
Methylthymol Blue, Calcein and murexide are quite unsuitable, whereas
Acid Alizarin Black SN (C.I. 21725) yields good results with a sharp
colour transition up to an Mg:Ca ratio of 1:12.   Calcon is completely
unsuitable for the determination of calcium alone, but yields good
results in the presence of magnesium, even when the Mg:Ca ratio is
greater than 1:12.   This effect of magnesium on the titration of
calcium has not yet been satisfactorily explained.   Details of the
study are given in simplified form in Fig. 6.1.

This study was done at only one calcium concentration level, which
did not exceed 0.005M in the final solution, but nonetheless the
authors did not manage to avoid adsorption of calcium or of its

Calcon complex. At the higher concentrations of magnesium the
indicator colour transition was premature and the original colour
returned within 2 - 3 sec. However, even at the equivalence point,
the blue colour again changed to red within 10 sec. The authors
stated that it was possible to find the end of the titration reliably
as the point at which the colour stopped changing rapidly back to the
original red.

6.3.1.6   Masking of magnesium in the determination of calcium

It follows from the previous sections that the determination of
calcium in the presence of magnesium is not yet satisfactory. It
would be ideal if the magnesium could be selectively masked so that
it would not be precipitated as the hydroxide. Some additives, such
as sucrose [7,8], are intended to prevent precipitation of the hydroxide or at least to retard it. Tartaric acid has been recommended
[20,21] for masking magnesium, but Burg and Conaghan [21] feel that
it is unsuitable for EDTA titrations because of the partial co-
titration of magnesium, while Kosak and Ballczo [23] recommend using
it for the semimicrodetermination of calcium (0.5 - 7.5 mg) in the
presence of up to 5 mg of magnesium. It has also been recommended
that 1,10-phenanthroline be used for masking [24]. Complete masking
of magnesium in the presence of calcium remains an unsolved problem
and it is highly probable that it never will be solved.

6.3.1.7   Isolation of calcium or magnesium

The considerable difficulties encountered in the determination of
calcium in the presence of magnesium have led a number of authors to
attempt to separate calcium from magnesium by both "classical" and
new methods. Some years ago, Banevicz and Kenner [25] used the
oxalate method, separating iron and aluminium as the hydroxides and
then precipitating calcium with ammonium oxalate and determining
magnesium complexometrically in the filtrate. They stated that
excess of oxalate does not interfere in the determination of magnesium, with Erio T as indicator, provided that hydrogen peroxide is
added to the solution. This reagent does not react with oxalic acid
but ensures a sharp indicator transition. The slow colour transition in the titration of magnesium is a result of the presence of
the oxalato complex $Mg(ox)_2^{2-}$. It is best to determine the calcium
after collection as the oxalate, since the amount of adsorbed magnesium oxalate will be very low [26,17]. In addition, calcium can be
isolated from other metals if it is bound as the EDTA complex [28].
The calcium can also be determined by permanganate titration of the
oxalate. Gehrke et al. [29] recommend precipitating calcium together
with iron and aluminium at pH 5 - 7, with sodium sulphite. Ghosh
and Roy [30] precipitated calcium as the molybdate. Flaschka and
Huditz [13] precipitated calcium with napththalylhydroxylamine as a
brick-red precipitate which was then ignited to the oxide; this was
dissolved in hydrochloric acid and the solution titrated to determine
calcium.

Muraca and Reitz [31] proposed separating magnesium from calcium by
precipitating it as $Mg(OH)_2$ in the presence of mannitol, which
prevents co-precipitation of calcium. Štráfelda and Říhová [32]
precipitated magnesium in the usual manner as $MgNH_4PO_4 \cdot 6H_2O$ at pH 9
and determined calcium in the filtrate amperometrically, using a

mercury electrode. The conditions for the precipitation are not, however, simple and the reliability of the method is highly dependent on the amount of phosphate added. There is a constant danger of co-precipitation of calcium as $Ca_3(PO_4)_2$. None of these methods has so far been used in practice.

The proposed determination of calcium by flame photometry and of Ca + Mg complexometrically [33] should be mentioned.

6.3.1.8   Determination of calcium and magnesium with various complexing agents

Determination of calcium with EGTA

The stability constants of the Ca-EGTA and Mg-EGTA complexes are very different ($\log K_{CaY} = 11.0$, $\log K_{MgY} = 5.2$). This difference - the largest so far observed for calcium and magnesium complexes - should theoretically be sufficient for the stepwise determination of calcium and magnesium, assuming that suitable indicators are available. Unfortunately, no sufficiently sensitive indicator for calcium has been found that would enable determination in the presence of magnesium. Nonetheless, this problem was satisfactorily solved independently by Ringbom et al. [34] and Sadek et al. [35] by means of an indirect indicator reaction. They used zincon as an indicator in the titration of calcium (see Section 2.7.4). Zincon alone does not undergo a colour reaction with calcium and magnesium, but forms a blue complex with zinc. If the Zn-EGTA complex and zincon are added to a solution of calcium and magnesium, then only calcium replaces zinc in its EGTA complex, and the solution becomes blue:

$$Ca^{2+} + HIn^{3-} + Zn\text{-}EGTA^{2-} = Ca\text{-}EGTA^{2-} + ZnIn^{2-} + H^+$$

$$\phantom{Ca^{2+} +} \text{orange-red} \phantom{+ Zn\text{-}EGTA^{2-} = Ca\text{-}EGTA^{2-} +} \text{blue}$$

In the titration with EGTA, the zinc is displaced from its indicator complex at the equivalence point and the colour of the solution changes from blue to orange-red. This replacement reaction proceeds satisfactorily only under certain conditions, e.g. in the presence of auxiliary complexing agents which shift the reaction to the right. Ringbom et al. used a special buffer (pH 9.5 - 10) containing borax, ammonium chloride and sodium hydroxide. They stated that the Zn-EGTA content should be about a tenth of that of calcium. Similarly, the ammonium salt content is very important. Sadek et al. stated that the colour transition of the indicator is not affected by magnesium up to an Mg:Ca ratio of 20. With increasing magnesium concentration the colour transition becomes less marked. For the sake of illustration, both procedures will be given.

The Ringbom, Pensar and Wänninen method [34]

0.05M EGTA: 19.01 g of the acid dissolved in 100 ml of 1M NaOH and the solution diluted to 1 litre with distilled water.
0.025M Zn-EGTA: Equal volumes of 0.05M EGTA and 0.05M $ZnSO_4$ are mixed.
Buffer (pH 10): 25 g of borax $(Na_2B_4O_7.10H_2O)$ + 3.5 g of ammonium chloride + 5.7 g of sodium hydroxide dissolved in 1 litre of distilled water.
Indicator: 65 mg of zincon dissolved in 2 ml of 0.1M NaOH and

diluted with distilled water to 100 ml.

Procedure. The solution (20 - 50 ml) containing calcium and magnesium
is mixed with an equal volume of buffer, 2 ml of 0.025M Zn-EGTA and 3
drops of indicator, and titrated with 0.05M EGTA from blue to bright
orange. The colour transition is very sharp. The sum of Ca + Mg
can be determined similarly.

The Sadek, Schmidt and Reilley method [35]

0.01M Zn-EGTA: 10 ml of 0.1M $ZnSO_4$ mixed with the same volume of
0.1M EGTA, neutralized with sodium hydroxide and diluted to 100 ml
with distilled water.
Buffer (pH 9.5): 40 g of ammonium nitrate dissolved in 300 ml of
water, adjusted to pH 9.5 and diluted to 500 ml with water.
Indicator: 0.2 g of zincon dissolved in 5 ml of 0.1M NaOH and
diluted to 100 ml.

Procedure. First the solution containing calcium and magnesium is
adjusted to pH 5 with sodium hydroxide or hydrochloric acid, and the
smallest amount of buffer necessary to yield a pH of 9.5 is added.
Three drops of indicator are added, and the Zn-EGTA complex dropwise
until the solution turns blue. Then the solution is titrated with
0.01M EGTA to a sudden colour change. It is recommended that
several drops of indicator be added just before the end of the
titration, as the indicator fades quickly during the titration. At
a ratio of Mg:Ca = 20 the final colour at the equivalence point is
not clear orange-yellow, but has a slight blue tone. The titration
is complete when the colour no longer changes.

Not only have a number of modifications of this method of titrating
calcium with EGTA been recommended, but a number of methods have been
proposed for the stepwise determination of calcium with EGTA and then
magnesium with EDTA or DCTA. After titrating calcium by this method,
Sadek et al. [35] mask the zinc by addition of cyanide and titrate
magnesium with 0.05M EDTA, using Erio T as indicator. Here the
colour transition is not from wine-red to clear blue, but rather to
blue-grey. Traces of iron must be reduced with ascorbic acid. The
authors recommend that a small amount of cyanide and indicator be
added before the end of the titration. Flaschka and Ganchoff [36]
first titrate calcium with EGTA at pH 10, using murexide as indicator,
and then magnesium with EDTA, using Erio T. They studied both
titrations photometrically and state that calcium can be determined
with Mg:Ca ratios up to 100. Of the other possible indicator
systems, Zn-PAN has also been recommended for the determination of
calcium [37]. Burg and Conaghan [22] and Data and Toei [38]
recommend masking magnesium with tartaric acid during the titration
of calcium with EGTA. Tsunogai et al. [39] describe an interesting
extraction method for calcium. Glyoxal bis(2-hydroxyanil) (GHA)
forms a red complex with calcium, which can be readily extracted into
organic solvents. They titrate the extracted complex directly with
0.01M EGTA solution from red to colourless. The method has been
used for determining calcium in sea-water.

Determination of calcium in the presence of large amounts of magnesium

Nearly all published methods deal with the determination of calcium
in the presence of reasonable amounts of magnesium (e.g. in Ca:Mg
ratios from 1:10 to 10:1). Many of them are reliable only under

certain conditions, which have to be maintained in the solution and are often inconvenient for routine analysis.   The problem of how to determine calcium when the Mg:Ca ratio is more than 100 is still open.

Recently we have paid attention to use of the reaction of zinc ions with Thymolphthalexone [40].   In alkaline medium, this indicator forms intensely blue Me_2In^{2-} complexes with alkaline-earth metal cations.   The corresponding zinc complex, Zn_2In^{2-}, is practically colourless and only slightly pink in high concentration.   We have found that in borax solution (pH 9 - 10) only magnesium forms a colour-ed complex with the indicator, provided that any calcium (or stront-ium or barium) is bound as the EDTA complex.   Thus in the titration of EGTA with zinc in the presence of magnesium, at the end-point magnesium is displaced from its complex with the indicator:

$$Mg_2In^{2-} + 2Zn^{2+} = Zn_2In^{2-} + 2Mg^{2+} \qquad\qquad (1)$$

$$\text{blue} \qquad\qquad\qquad \text{colourless}$$

This reaction proceeds quantitatively even with very large amounts of magnesium (up to 200 mg), which in borax medium is not precipitated as hydroxide.   The course of the reaction is clear from Fig. 6.2 showing the spectra recorded during the titration of EGTA with zinc solution in the presence of a small amount of magnesium.   At the end-point the blue solution becomes colourless.   The colour change is also very sharp in visual titrations.   It should be mentioned that the magnesium must be present in at least 10^{-4}M concentration, because neither the Ca-EGTA complex nor free zinc gives a colour reaction with the indicator.

This reaction was made the basis for photometric and visual titra-tion of calcium in the presence of various amounts of magnesium.

Photometric titration of calcium.   A slightly acidic solution con-taining 0.5 - 10 mg of calcium and 1 - 200 mg of magnesium is pipetted into the cuvette, then 8 - 10 ml of 0.05M EGTA and 10 ml of saturated borax solution are added and the whole is diluted to about 60 ml; then 0.2 ml of 0.1% indicator solution is added.   The titration is performed by making successive additions of 0.05M zinc solution and after each addition measuring the absorbance at 640 nm.   The end-point is given by a sudden drop of the absorbance at 640 nm nearly to zero.

Note.   The order of addition of reagents must be strictly adhered to, because calcium (not bound with EGTA) is precipitated in borax solutions.   As already said, the titrations can be performed only in the presence of magnesium.   Very small amounts of magnesium are also nearly all bound with EGTA and the initial solutions after the addition of indicator are coloured only very slightly or not at all. During the titration with zinc, magnesium is displaced from its EGTA complex and reacts with the indicator:

$$2MgY^{2-} + 2Zn^{2+} + H_2In^{4-} = 2ZnY^{2-} + Mg_2In^{2-} + 2H^+ \qquad (2)$$

$$\text{colourless} \qquad\qquad\qquad\qquad \text{blue}$$

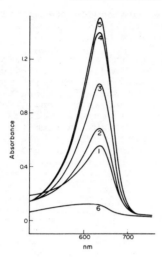

Fig. 6.2 Photometric
titration of EGTA with
zinc solution. Fraction
titrated, %: 1, 0; 2,
19; 3, 38; 4, 57; 5, 76;
6, 100.

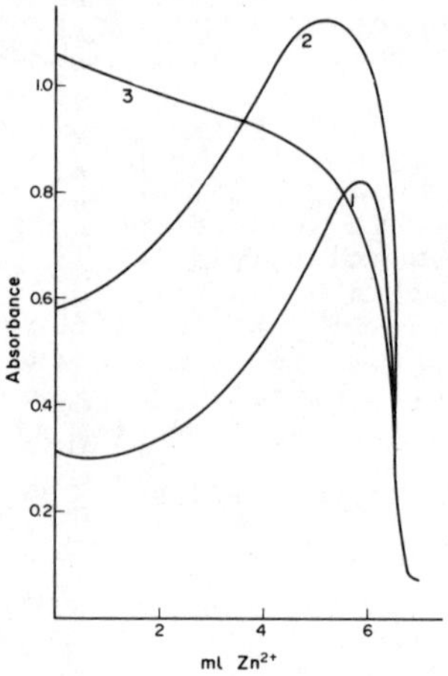

Fig. 6.3 Photometric
titration of EGTA with
zinc solution for various
Mg:Ca molar ratios: 1, 2:1;
2, 4:1; 3, 360:1.

(Reproduced by permission of the copyright
holder, Pergamon Press)

Therefore the blue colour becomes progressively deeper up to the end-
point, and disappears suddenly when excess of zinc is present, according to equation (1).   The photometric curves for small amounts of
magnesium are somewhat deformed sinusoids (Fig. 6.3, curves 1 and 2).
In the presence of large amounts of magnesium, only a relatively
small part of it is bound with EGTA and the curves are normal in
shape (Fig. 6.3, curve 3).   From Fig. 6.3 we also see that the
initial colour of the solution depends on the Mg:EGTA ratio and on
the concentration of indicator.   For a volume of 60 - 70 ml and a
10-cm cuvette, the most suitable volume of the indicator is 0.2 ml of
freshly prepared 0.1% solution.   A large excess of the indicator
causes a very deep blue colour which cannot be measured photometrically, but its disappearance can be followed visually.   By this
method we can determine as little as 1 mg of calcium in the presence
of 2 g of $MgSO_4.7H_2O$ (see Fig. 6.3, curve 3).

Visual titration of calcium.   The solution is prepared in a 500-ml
titration flask, as for the photometric titration, and diluted to
about 250 ml.   Then a small amount of indicator (1:100 solid mixture
with KCl) is added and the solution titrated with 0.05M (or 0.02M)
zinc until the solution is decolorized.   The solution must be
thoroughly mixed after each addition of titrant.

Note.   The method is applicable only to "pure" calcium-magnesium
solutions.   For the analysis of various geological materials (see
p. 299) it can be applied after separation of all interfering

elements.   First, iron and aluminium must be absent.   Triethanolamine cannot be used to mask them, because it needs a very high pH
(about 12), at which mangesium is precipitated as the hydroxide.
Precipitation with ammonia cannot be used either, because ammonium
salts interfere in the titration of EGTA with zinc.   It is convenient to use precipitation with hexamine.   Iron can be extracted with
a 50% solution of acetylacetone in chloroform.   Other elements such
as nickel, manganese and copper must also be removed before the
titration of calcium, or their sum with calcium determined and a
correction applied for the amounts of them present.

Determination of calcium with EGTA and of magnesium with DCTA

The difference in the stability constants of the Mg-DCTA and Mg-EGTA
complexes (log K values 10.3 and 5.2) is sufficient for titration of
magnesium with DCTA in the presence of EGTA, but a sufficiently
sensitive indicator must be available which can react in alkaline
medium with the Mg-EGTA complex.   Methylthymol Blue is suitable [41].
The method is based on masking calcium with EGTA, and titrating
magnesium with DCTA.   The reaction of magnesium with the indicator
in the presence of EGTA is satisfactory if the EGTA is not present in
too great an excess (not more than 3 ml of 0.05M EGTA in 100 ml).
Two portions of solution are required for determination of both
elements.   The principle of the method is clear from the procedures
given below.

(a)   Approximate determination of calcium with EGTA.   A weakly acid
solution of calcium and magnesium is treated with a tenth of its
volume of 1M potassium hydroxide and a small amount of Fluorexone is
added.   The solution is slowly titrated with EGTA until the green
fluorescence disappears.   Satisfactory results are obtained only at
low magnesium contents.   Otherwise, as might be expected, they are
low.

(b)   Determination of magnesium with DCTA.   The same amount of EGTA
as in the first titration and 1 - 2 ml in excess is added to a second
portion of the solution (equal in volume to the first).   After
addition of 20 ml of buffer (3 g of ammonium nitrate in 100 ml of
concentrated ammonia solution) and some Methylthymol Blue, the
solution is slowly titrated with 0.05M DCTA from blue to smoky-grey.

Note.   At higher magnesium concentrations, the magnesium is
partially precipitated as the hydroxide after addition of the buffer,
but dissolves again, especially if the solution is heated to 40°C.
Thus it is preferable to heat the solution after completion of the
titration and to observe whether the blue colour reappears.   If so,
the titration is continued until the blue colour again disappears.
Precipitation of $Mg(OH)_2$ can be prevented by a large addition of
ammonium salts, but the colour transition is less sharp.

The DCTA consumption corresponds sufficiently precisely to the
magnesium content, provided this is not greater than 25 - 30 mg (40 -
45 mg of MgO).

(c)   Determination of calcium after titration of magnesium.   With
high magnesium contents, the results for calcium obtained by procedure (a) are low.   It is then preferable to use the solution remaining after determination of magnesium according to procedure (b).
After the titration of magnesium, the solution at the equivalence

point contains Mg-DCTA, Ca-EGTA and free EGTA.   The overall EGTA
content is known, and thus the excess can be back-titrated with
calcium.   The results are satisfactory provided that the calcium
content is not too small (e.g. 2 mg), in which case they will be low.
If that happens, it is necessary to use a third portion to determine
the sum of Ca + Mg by adding excess of 0.05M DCTA and 10 ml of concentrated ammonia solution, and back-titrating with calcium chloride,
using Methylthymol Blue as indicator.

This method has the advantage that it can be used after masking of
iron, aluminium and traces of manganese with triethanolamine or of
other metals with potassium cyanide or thioglycollic acid.   The
method has been used in the analysis of silicates, magnesites, slags,
etc. [42].

Ishii [43] used a very similar procedure, titrating a mixture of
calcium and magnesium with an EGTA solution, using Calcichrome or
Patton-Reeder indicator. After addition of the same amount of EGTA
(+ 0.5 ml) to a second and equal portion of sample, the magnesium is
titrated at pH 9.5 - 10 with EDTA (Erio T as indicator).   The sum of
Ca + Mg is then titrated in a third portion.   Umemoto et al. [44]
used a procedure in which the excess of EGTA is determined by back-
titration in acetone medium with a calcium solution, at pH 13, using
BNP as indicator (see note below).   Magnesium is determined in the
same solution by direct titration with DCTA, with Methylthymol Blue
as indicator.   The method was used to determine calcium in organic
substances.

Note.   BNP is the abbreviation for 3,3-bis(2-hydroxy-3-carboxynaph-
thalenazo)phenolphthalein, which has been recommended as an indicator
for calcium [45].   It is prepared by condensation of diazophenol-
phthalein with 2-hydroxy-3-naphthoic acid [46].

Sato and Momoki [47] recommended using Phthaleincomplexone as an
indicator in the photometric titration of microgram amounts of
calcium and magnesium.   Calcium is first titrated in a glycine and
potassium hydroxide medium with EGTA, and then magnesium is titrated
with DCTA, the same indicator being used.   The method was applied to
the analysis of waters.

(d) Stepwise determination of calcium and magnesium [48].   The
indirect determination of calcium with EGTA by back-titration with
lead nitrate, and Thymolphthalexone as indicator, proceeds in the
same way as the method with zinc just described.   The Pb_2In^{2-}
complex is also colourless, so at the end-point the intensely blue
colour of the solution disappears on addition of the first excess of
lead.

Attempts to determine magnesium in the same solution with EDTA are
unsuccessful because the Pb-EGTA complex reacts with EDTA:

$$Pb\text{-}EGTA + EDTA = Pb\text{-}EDTA + EGTA \qquad\qquad (3)$$

$$\log K = 13 \qquad\qquad \log K = 18.0$$

The results are always high because of the co-titration of lead bound
by the EGTA.   In addition the EGTA liberated obscures the end-point
by forming a relatively weak complex with magnesium ($\log K = 5.2$).

We would expect that the reaction

$$Pb\text{-}EGTA + DCTA = Pb\text{-}DCTA + EGTA \tag{4}$$

$$\log K = 13 \qquad \log K = 19.7$$

would also interfere for the same reason, but we have found that this
reaction does not proceed at all, which is very surprising even
though formation of DCTA complexes usually proceeds a little more
slowly than formation of EDTA complexes.   This fact forms the basis
of a simple method for successive determination of calcium and
magnesium.

The calcium is determined as above, with lead nitrate for the back-
titration, then 1 or 2 drops of EGTA solution are added, and the blue
colour is restored after a few seconds.   Then 10 – 15 ml of concentrated ammonia solution are added and the magnesium is titrated with
0.05M DCTA, until the solution is colourless.   Alternatively, after
the calcium determination the indicator is left blocked, Erio T and
Schwarzenbach buffer are added, and the magnesium is titrated with
0.05M DCTA, from red to blue.

We suggest that the reason why the DCTA system works but the EDTA
system does not, is that the relative rigidity of the DCTA structure
(relative to EDTA) introduces an orientation factor into the kinetics
of the system, so that although DCTA should displace lead from the
Pb-EGTA complex, the kinetic factors prevent it from doing so in the
time needed for the titration.

(e) Potentiometric stepwise titration of calcium and magnesium.
EGTA cannot be considered a good titrant for magnesium, because it
forms too weak a complex for visual titration ($\log K_{MgY} = 5.2$).   However, it is well known that the stability of complexes can be considerably increased by addition of water-mixcible solvents such as
acetone, ethanol or methanol.   Wallén [49] used this to develop a
new method for the stepwise microdetermination of calcium and magnesium with EGTA.   Both titrations are followed potentiometrically,
with an amalgamated silver wire as indicator electrode and a mercury
sulphate reference electrode.

To the sample solution, which contains 0.005 – 0.1 mmole of each of
calcium and magnesium add 2 ml of 1M ammonium nitrate and concentrated ammonia solution dropwise until the pH is 8.9 – 9.0.   Add a few
drops of 0.002M Hg-EGTA solution and titrate with 0.01M EGTA.   Stop
the titration shortly after the first end-point break, which corresponds to the amount of calcium present, and make the solution 80%
(v/v) in ethanol.   Add a few drops of concentrated ammonia solution
to raise the pH to 10 and continue the titration until the potential
break is obtained.   The EGTA consumed between the two inflection
points corresponds to the amount of magnesium present.   According to
Wallén the method is suitable for Mg:Ca ratios between 10:1 and
1:10 (up to 2.3 mg of Mg and 3.6 mg of Ca).

Interference from cations

Of all complexometric determinations, that of calcium and magnesium
is the least selective.   Practically all cations interfere, but they
can generally be precipitated as the hydroxides in sodium hydroxide
medium.   They are co-titrated in ammonia medium if they form ammine

complexes.

At concentrations which do not greatly increase the ionic strength of
the solution, alkali metals do not interfere. The worsening of the
colour transition of the indicator can be avoided by sufficient
dilution. There are, however, exceptions. For example, sodium and
lithium salts prevent the determination of calcium with Fluorexone
(Calcein) as indicator. Interference from sodium was also observed
in the microdetermination of magnesium with DCTA, with Phthaleincomplexone as indicator [46], and potassium interferes even at the
0.02M level.

In the determination of calcium and magnesium, interfering cations
must be either separated or masked. Suitable masking agents (KCN,
TEA, BAL, TGA, etc.) have been described in previous sections and
will also be encountered in the practical section of this book.

Interference from anions

Anions interfere by forming precipitates in alkaline medium, e.g.
the carbonates, fluorides, phosphates, tungstates, molybdates, etc.
Carbonate precipitation becomes important only when hydroxide with a
high carbonate content is used. In normal analyses, any carbonate
is removed in the sample decomposition. Calcium tungstate and
molybdate dissolve readily in alkaline EDTA, so direct determination
of Ca and Mg is not difficult. Fluorides and phosphates accompany
calcium or magnesium in a number of raw materials (apatites) and
phosphates are found in plant materials, fertilizers, etc. The
interference from fluoride can generally be removed in the sample
decomposition. Only one work [50] describes the determination of
calcium (10 mg) in the presence of fluoride (20 mg), based on the
replacement reaction between the Zn-EDTA complex and calcium. The
zinc liberated is then determined by titration with 0.05M EDTA at pH
10, with Erio T as indicator. On the other hand, high concentrations of fluoride mask both calcium (formation of CaF_2) and magnesium
[51]. A great deal of attention has been paid to the determination
of calcium and magnesium in the presence of phosphate. Older methods
recommended separating phosphate from calcium and magnesium by ion-
exchange [52-55] or by precipitation as $FePO_4$ [56,57] or as phosphomolybdate [58]. Furuya [59] precipitated phosphate with thorium
nitrate solution, Kinnunen and Wennerstrand [60] precipitated phosphate with beryllium nitrate solution, etc. Collier [61] extracted
phosphomolybdic acid from a strongly acidic solution with a mixture
of butanol and chloroform (1:1).

Cimerman et al. [62] titrated calcium in the presence of phosphate
with 0.01M EDTA containing 0.05M Zn-EDTA, in Schwarzenbach buffer,
using Erio T as indicator. They stated that it is possible to
determine 2 - 10 mg of calcium in the presence of 60 mg of P_2O_5, or
1 - 2 mg of magnesium in the presence of 20 mg of P_2O_5. Ince and
Forster [63] carried out an indirect titration of calcium, using
Erio T and zinc solution, and used this method to analyse $Ca_3(PO_4)_2$
samples. Yalman et al. [64] determined calcium by back-titration of
EDTA with calcium solution, using Calcein as indicator, and followed
the titration photometrically. They obtained very good results for
10 - 20 mg of calcium in the presence of up to 30 mg of phosphate and
10 mg of magnesium. The method has been used to determine calcium
in bones, teeth and urine.

6.3.1.9   Masking calcium and magnesium

Though alkaline media are needed for the complexometric determination
of calcium or magnesium, these two elements sometimes need to be
masked in other analyses.   Fluoride is very useful [51] for this
purpose and has been used in the determination of manganese, zinc and
magnesium [65].

REFERENCES

1.   G. Schwarzenbach, Komplexon-Methoden, Siegfried, Zofingen, 1948.
2.   A.G. Knight, Chem. Ind. (London), 1951, 1141.
3.   J. Piglowski, Chem. Anal. (Warsaw), 1956, 1, 331.
4.   R.A. Chalmers, Analyst, 1964, 79, 519.
5.   A.D. Kenny and V.H. Cohn, Anal. Chem., 1958, 30, 1367.
6.   J.C. Van Schouwenburg, Anal. Chem., 1960, 32, 709.
7.   R.D. Bond and B.M. Tucker, Chem. Ind. (London), 1954, 1236.
8.   B.M. Tucker, Analyst, 1957, 82, 284.
9.   P.F. Lott and K.L. Cheng, Chemist-Analyst, 1957, 46, 30.
10. P.F. Lott and K.L. Cheng, Chemist-Analyst, 1959, 48, 13.
11. C.A. Baugh, K.H. Decker and J.W. Palmer, Anal. Chem., 1961, 33, 1804.
12. L.L. Lewis and L.M. Melnick, Anal. Chem., 1960, 32, 38.
13. H. Flaschka and H. Huditz, Radex-Rundschau, 1952, 181.
14. G. Brunisholz, M. Genton and E. Plattner, Helv. Chim. Acta, 1953, 36, 782.
15. P.P. Naidu and C.A. Sastry, Z. Anal. Chem., 1971, 253, 206.
16. K. Kodama, K. Mizuno, T. Oga and M. Matsunami, Japan Analyst, 1965, 14, 474; Chem. Abstr., 1965, 63, 6302d.
17. H. Diehl and J.L. Ellingboe, Anal. Chem., 1956, 28, 882.
18. S.J. Socolar and J.I. Salach, Anal. Chem., 1959, 31, 473.
19. R. Belcher, R.A. Close and T.S. West, Talanta, 1958, 1, 238.
20. V. Tichomírová and O. Šimáčková, Coll. Czech. Chem. Commun., 1957, 22, 982.
21. B. Bieber and Z. Večeřa, Coll. Czech. Chem. Commun., 1960, 26, 59.
22. R.A. Burg and H.F. Conaghan, Chemist-Analyst, 1960, 49, 100.
23. A. Kosak and H. Ballczo, Z. Anal. Chem., 1971, 263, 262.
24. V. Skřivánek and P. Klein, Chem. Listy, 1962, 56, 1110; Anal. Abstr., 1963, 10, 2171.
25. J.J. Banewicz and C.T. Kenner, Anal. Chem., 1952, 24, 1186.
26. R.C. Nigam and D. Prakash, Lab. Practice, 1970, 19, 605; Chem. Abstr., 1970, 73, 105167V.
27. P.B. Dunnill, P.H. Scholes and W. Tomlinson, B.I.S.R.A. Open Rept. MG/D/677/70, 1970; Anal. Abstr., 1971, 21, 1822.
28. R. Přibil and L. Fiala, Coll. Czech. Chem. Commun., 1953, 18, 301.
29. C.V. Gehrke, H.E. Affsprung and C.C. Lee, Anal. Chem., 1954, 26, 1944.
30. A.K. Ghosh and K.L. Roy, Anal. Chim. Acta, 1954, 14, 504.
31. R.F. Muraca and M.T. Reitz, Chemist-Analyst, 1954, 43, 73.
32. F. Štráfelda and J. Řihová, Coll. Czech. Chem. Commun., 1960, 25, 144.
33. A. Gee, L.P. Domingues and V.R. Deitz, Anal. Chem., 1954, 26, 1487.
34. A. Ringbom, G. Pensar and E. Wänninen, Anal. Chim. Acta, 1958, 19, 525.

35. F.S. Sadek, R.W. Schmidt and C.N. Reilley, Talanta, 1959, 2, 38.
36. H. Flaschka and J. Ganchoff, Talanta, 1961, 8, 720.
37. G. Nakagawa, H. Wada and M. Tanaka, Talanta, 1963, 10, 325.
38. Y. Data and K. Toei, Bull. Chem. Soc. Japan, 1963, 36, 5.
39. S. Tsunogai, M. Nishimura and S. Nakaya, Talanta, 1968, 15, 385.
40. R. Přibil and J. Adam, Talanta, 1977, 24, 117.
41. R. Přibil and V. Veselý, Talanta, 1966, 13, 233.
42. R. Přibil and V. Veselý, Chemist-Analyst, 1966, 55, 82.
43. H. Ishii, Japan Analyst, 1966, 15, 972; Chem. Abstr., 1967, 66, 34507y.
44. K. Umemoto, S. Hirose, K. Sakamoto, T. Kouri and K. Hozumi, Japan Analyst, 1970, 19, 191; Anal. Abstr., 1971, 21, 1165.
45. K. Toei and T. Kobatake, Talanta, 1967, 14, 1354.
46. T. Kobotake, T. Iwachido and K. Toei, Talanta, 1967, 14, 607.
47. H. Sato and K. Momoki, Anal. Chem., 1972, 44, 1778.
48. R. Přibil and J. Adam, Talanta, 1979, 26, 154.
49. B. Wallén, Anal. Chem., 1974, 46, 304.
50. J. Mashall and L. Geyer, Bull. Res. Council Israel, 1957, 6, 74.
51. R. Přibil, Coll. Czech. Chem. Commun., 1954, 19, 64.
52. G. Brunisholz, M. Genton and E. Plattner, Helv. Chim. Acta, 1953, 34, 782.
53. A.C. Mason, Analyst, 1950, 77, 529.
54. R. Jennes, Anal. Chem., 1953, 25, 966.
55. W.E. Schilz and G.N. Krynauw, Anal. Chem., 1956, 28, 1759.
56. W.A.C. Campen and L.H.J. Nijst, Chem. Weekbl., 1955, 51, 186; Chem. Abstr., 1955, 49, 8034g.
57. K. Lasiewicz, B. Byczynska and H. Zawadzka, Chem. Anal. (Warsaw), 1958, 5, 1041.
58. L. Kalman and A. Vagó, Mag. Kem. Foly., 1955, 61, 416.
59. M. Furuya and M. Tajiri, Japan Analyst, 1963, 12, 1139.
60. J. Kinnunen and B. Wennerstrand, Chemist-Analyst, 1955, 44, 41.
61. R.E. Collier, Chemist-Analyst, 1954, 43, 41.
62. Kh. Cimerman, A. Alon and J. Mashall, Anal. Chim. Acta, 1958, 19, 461.
63. A.D. Ince and W.A. Forster, Analyst, 1960, 85, 608.
64. R.G. Yalman, W. Bruegemann, P.T. Baker and S.M. Garn, Anal. Chem., 1959, 31, 1230.
65. W.G. Scribner, Anal. Chem., 1959, 31, 273.

6.3.2 $\underline{\text{Strontium}}$
$$\overline{(\log K_{SrY}} = 8.63)$$

$\underline{\text{Direct determination}}$

Strontium reacts with Erio T to produce a rather feeble colour, and thus older methods employed the substitution reaction:

$$MY^{2-} + Sr^{2+} = SrY^{2-} + M^{2+}$$

where MY is either the Mg-EDTA or the Zn-EDTA complex. The colour change in the replacement reaction with the Mg-EDTA complex is sufficiently clear, even though the actual reaction is not quantitative [1].

The same result is obtained if a known amount of magnesium is added to a strontium solution and their sum is determined. This reaction proceeds satisfactorily at pH 11 with the Zn-EDTA complex ($\log K_{ZnY} =$

16.5) and is again quite dependent on the concentration of ammonium
salts.   Budĕšínský [2] recommends Eriochrome Violet R as the
indicator.   West [3] considers Acid Alizarin Black SN as the best
indicator for titration of strontium, in diethylamine medium at pH
12.5.   The colour transition is from red to blue.   Cresolphthalein
Complexone is less suitable [4]; Ogawa and Musha [5] used it in
photometric titrations and determined 0.1 - 6 mg of strontium with a
precision of 1%.   The solution must be fairly alkaline (pH 10 - 11.5).
At pH below 10 the colour transition is not clear and at pH above
11.5 the indicator itself is strongly coloured.   Methylthymol Blue
and Thymolphthalexone are very useful indicators for the titration
and can be used in sodium hydroxide medium;   ammoniacal medium is not
suitable.

Indirect determination

Back-titrations of excess of EDTA with magnesium (Erio T as indicator)
or calcium (Thymolphthalexone as indicator) are the most important
methods.

Separation methods

Most attention has been paid to the separation of strontium on ion-
exchangers, especially its separation from other alkaline-earth
metals.   Because of the importance of strontium-90, many works have
employed radiometric detection.

Compared with other methods at present available, the complexometric
determination of strontium has little practical importance.

REFERENCES

1.   G. Schwarzenbach, Helv. Chim. Acta, 1956, 29, 1338.
2.   B. Budĕšínský, Coll. Czech. Chem. Commun., 1958, 23, 895.
3.   R.A. Close and T.S. West, Anal. Chim. Acta, 1960, 23, 261.
4.   G. Anderegg, H. Flaschka, R. Sallman and G. Schwarzenbach, Helv.
      Chim. Acta, 1953, 37, 113.
5.   K. Ogawa and S. Musha, Bull. Univ. Osaka Prefect Ser. A., 1960,
      8, 63.

6.3.3   Barium
$$(\log K_{BaY} = 7.76)$$

The barium complex with EDTA is one of the weakest known.   The first
complexometric determination was based on the replacement reaction
described for strontium in the previous section.   The Erio T colour
transition is even less marked than for strontium [1] and the results
are very dependent on the concentrations of buffer and ammonium salts
[2].   Schwarzenbach [3] used Cresolphthalein Complexone to determine
barium.   A 50% ethanol medium must be used to suppress the colour of
the indicator itself;   there is a marked danger that barium will be
precipitated as the carbonate in this medium.   The solution must be
titrated immediately after preparation.   Methylthymol Blue and
Thymolphthalexone are very suitable indicators for the determination
of barium.   The colour transition in sodium hydroxde and 20%

triethanolamine solutions [4] is very sharp.   Thymolphthalexone has
been used, for example, in control of the purity of technical barium
salts.   Traces of iron and copper can be masked with triethanolamine.

Combs and Grove [6] carried out a spectrophotometric titration of
microgram amounts of calcium, using Erio T as indicator, and of Ca +
Ba in a further aliquot of the solution, with Methylthymol Blue
indicator.   Magnesium must be absent in this titration.

Wänninen [7] proposed determining barium with diethyltriaminepenta-
acetic acid (DTPA);   barium is titrated in the presence of the Mg-
DTPA complex and Erio T indicator.   The Ba-DTPA complex is more
stable than the Ba-EDTA complex.   The method was used for the
indirect determination of sulphate [7], and the theory has been
discussed [8].

Olsen and Novak [9] studied this reaction in detail, in view of the
fact that the concentration of the Mg-DTPA complex must affect the
sharpness of the Erio T colour transition and thus the reliability of
the determination.   The optimal amount of Mg-DTPA complex is dependent on the concentration of barium present.   Consequently, Olsen and
Novak recommend using a mixture of DTPA and the Mg-DTPA complex as
the reagent.   They consider a 0.01M DTPA/0.015M Mg-DTPA solution as
optimal.   The colour transitions of the indicator are then very
sharp and the relative standard deviation is 0.3% or less.   Wänninen
states that increasing the temperature to 40°C has no effect.   If
the indirect determination of sulphate is performed with 0.01M DTPA +
0.005M Mg-DTPA as titrant, then ethanol must be present during the
titration of excess of barium in the presence of precipitated $BaSO_4$.
The authors state that this method is also useful for the determination of strontium or calcium.

Complexometric determination of barium has been used in only a few
cases, in analysis of waters and of barytes, ferrites, silicates and
special glasses.   Much more attention has been paid to the determination of sulphate, discussed in Chapter 7.

Recommended procedures

A weakly acid or neutral solution of barium (100 ml) is mixed with
3 - 10 ml of 1M sodium hydroxide and a small amount of indicator
(1:100 solid mixture with KNO_3) and titrated with 0.01 - 0.05M EDTA
from blue to colourless or smoky-grey.

Alternatively excess of EDTA and some indicator are added to a
solution containing barium, and the solution is strongly alkalized
with ammonia solution.   The excess of EDTA is titrated with 0.05M
magnesium from smoke-grey to bright blue.

REFERENCES

1.   T.J. Manns, M.U. Reschowsky and A.J. Certa, Anal. Chem., 1952,
      24, 908.
2.   R. Sijderius, Anal. Chim. Acta, 1954, 10, 517.
3.   G. Anderegg, H. Flaschka, R. Sallman and G. Schwarzenbach, Helv.
      Chim. Acta, 1953, 37, 113.
4.   R. Přibil and V. Veselý, Chemist-Analyst, 1964, 53, 43.

5.   J. Stabryn, Chem. Prum., 1970, 20, 436; Anal. Abstr., 1971, 21,
     929.
6.   H.F. Combs and E.L. Grove, Anal. Chem., 1964, 36, 400.
7.   E. Wänninen, Suom. Kemistilehti, 1956, 29B, 184; Anal. Abstr.,
     1957, 4, 2582.
8.   E. Wänninen, Acta Acad. Aboensis Mat. Phys., 1960, 21, 1; Anal.
     Abstr., 1960, 7, 5134.
9.   E.D. Olsen and R.J. Novak, Anal. Chem., 1966, 38, 152.

6.3.4   Beryllium
$$(\log K_{BeY} = 9.8 \ [1]$$

It would appear from the log K values for beryllium that indirect
determination, e.g. by back-titration with magnesium should not be
difficult, but this is not the case; the Be-EDTA complex readily
hydrolyses and beryllium is quantitatively precipitated by ammonia [2]
or phosphate [3] even in the presence of excess of EDTA.   Both
reactions have been used for the gravimetric determination of beryllium.   A number of photometric methods have been developed for the
determination of beryllium with masking of interferents with EDTA.
It follows from the data published by various authors that the Be-
EDTA complex occurs only in a narrow pH range.   Flaschka [4] concluded from the neutralization titration curves of a Be-EDTA solution
that the complex exists at pH 4 - 5 with a probable value of log K =
9.2 for the stability constant. Starostin et al. [5] assume that the
complex exists at pH 7 - 7.5, with log K = 10.2.   The BeOHY^{3-} complex
is assumed to exist at pH above 7.5.   The conditions necessary for
maintaining beryllium in solution in media which are only slightly
alkaline are so sensitive that the complexometric determination of
beryllium is of practically no importance.   Nevertheless, Takamoto
[6] proposed an indirect determination of beryllium by back-titration
of EDTA with cobalt in 50% acetone solution, using thiocyanate as
indicator.

Pirtea et al. [7 - 9] recommended precipitation of the complex salt
$[Co(NH_3)_6][(H_2O)_2Be_2(CO_3)_2(OH)_3].3H_2O$ for the gravimetric determination of beryllium;   it is produced by precipitation in carbonate
medium with hexa-amminecobalt(III) chloride.   The method is useful
for both micro and macro determinations of beryllium.   Misumi and
Taketatsu [10] made a detailed study of the conditions for precipitating this compound, for subsequent titration of the cobalt with EDTA.
They determined small amounts of beryllium (1.40 mg) in the presence
of iron, aluminium and magnesium.   The beryllium must first be
separated as the hydroxide from EDTA medium.   Monk and Exelby [11]
improved the method by dissolving the precipitate in sulphuric acid
and determining the cobalt by direct titration with EDTA at pH 8.5 - 9
in the presence of thiocyanate, using tetraphenylarsonium chloride
in chloroform as the indicator (Section 6.2.4).

In a study of the composition of the compound, Sengupta [12] found
three different crystal forms,   the composition of all three being
$[Co(NH_3)_6]_2[Be_4O(CO_3)_6]$ with 10 or 11 molecules of water.   Monk and
Exelby [13] found the same composition but demonstrated that the
number of water molecules in the crystalline precipitate is dependent
on the atmospheric humidity and is between 10.8 and 11.1 at relative
humidity from 32 to 80%.   After vacuum drying over phosphorus pent-
oxide, the number of water molecules is reduced to 3.   This is very

important for the gravimetric determination of beryllium, but the
ratio Co:Be = 1:2 is decisive for complexometry, regardless of the
water content.   This method has not yet found practical use, as it
is tedious and rather difficult.

REFERENCES

1.   J.R. Merrill, M. Honda and J.R. Arnold, Anal. Chem., 1960, $\underline{32}$,
      1420.
2.   R. Přibil and J. Kucharský, Coll. Czech. Chem. Commun., 1950,
      $\underline{15}$, 132.
3.   J. Huré, M. Kremer and F. le Berquier, Anal. Chim. Acta, 1952,
      $\underline{5}$, 37.
4.   H. Flaschka, unpublished results.
5.   V.V. Starostin, V.I. Spitsyn and G.F. Silina, Zh. Neorgan.
      Khim., 1963, $\underline{8}$, 660.
6.   S. Takamoto, J. Chem. Soc. Japan, Pure Chem. Sect., 1955, $\underline{76}$,
      1344.
7.   T.I. Pirtea, Rev. Chim. (Bucharest), 1956, $\underline{7}$, 427.
8.   T.I. Pirtea and G. Mihail, Z. Anal. Chem., 1958, $\underline{159}$, 205.
9.   T.I. Pirtea and V. Constantinescu, Z. Anal. Chem., 1959, $\underline{165}$,
      183.
10.   S. Misumi and T. Taketatsu, Bull. Chem. Soc. Japan., 1959, $\underline{32}$,
      593.
11.   R.G. Monk and K.A. Exelby, Talanta, 1965, $\underline{12}$, 91.
12.   A.K. Sengupta, J. Inorg. Nucl. Chem., 1964, $\underline{26}$, 1823.
13.   G.R. Monk and K.A. exelby, Talanta, 1964, $\underline{11}$, 1633.

6.3.5   Silver
         $\overline{(\log K}_{AgY} = 7.3)$ [1]

EDTA forms the AgY^{3-} complex with silver.   This complex is so
unstable that silver reacts with all important reagents for it even
in the presence of excess of EDTA.   This permits a number of gravimetric and photometric determinations of silver which are quite
selective, with masking of interferents with EDTA [2].   This
"passive" role of EDTA is more important than all the published
complexometric determinations of silver.   They will be mentioned
only briefly.

Direct determination

Cresolphthalexone [3] and Thymolphthalexone [4] have been recommended
for the direct determination of silver, but can be used only for pure
solutions of silver nitrate.   Štráfelda [5] titrated silver
potentiometrically in borate buffer at pH 9, using a silver electrode.
The method is completely non-selective and not as satisfactory as
ordinary potentiometric titration with sodium chloride.

Indirect determination

Flaschka and Huditz [6] suggested the replacement reaction

$$Ni(CN_4^{2-} + 2Ag^+ = 2Ag(CN)_4^{2-} + Ni^{2+}$$

for the determination of silver.   The freed nickel is determined by
direct titration with EDTA, with murexide as indicator.   Solid
$K_2Ni(CN)_4$, which can readily be prepared, is used as reagent.   The
method can also be used for insoluble silver compounds such as AgCl,
AgBr and AgSCN, but not for AgI (which is too stable).   Among other
disadvantages, the stoichiometry is unfavourable (Ni = 2Ag).

Liteanu [7] used a similar procedure, but replaced the $Ni(CN)_4^{2-}$
complex with the $Cd(CN)_4^{2-}$ complex, which is prepared directly in the
solution by mixing a cadmium solution with cyanide so that cadmium is
in a slight excess.   This is first titrated with EDTA and then the
silver solution is added.   The displaced cadmium is determined
directly by titration with EDTA, with Erio T as indicator.

Flaschka's original method was used by Gedansky and Gordon [8] for
the indirect determination of silver by photometric titration of the
replaced nickel with EDTA.

Another replacement reaction employs reaction with copper diethyldithiocarbamate solution in carbon tetrachloride.   The replacement
reaction

$$Cu(DDC)_2 + 2Ag^+ = 2AgDDC + Cu^{2+}$$

liberates an equivalent amount of copper into the aqueous phase,
where it is determined complexometrically in weakly acidic medium [9].
The method is more selective than the preceding one, but still has
unfavourable stoichiometry (Cu = 2Ag).

Grdinić and Gertner [10] have described an ultramicrodetermination of
silver (1 - 40 μg) by titration with 0.001M EDTA.   This method is
based on the exchange reaction between cupric sulphide and silver.
The copper sulphide is fixed on filter paper as a round spot
(prepared by the "ring-oven" technique).   After reaction with silver
the replaced copper is washed into the outer zone and is determined
complexometrically.   The silver must first be separated as the
bromide by a similar technique.

Separation and masking of silver

Silver separates automatically when alloys are decomposed with aqua
regia and generally does not interfere in other determinations.   It
can readily be masked in alkaline medium with potassium cyanide or
thioglycollic acid.

REFERENCES

1.   A. Ringbom and E. Linko, Anal. Chim. Acta, 1953, 9, 80.
2.   R. Přibil, Analytical Applications of EDTA and Related Compounds,
      Pergamon Press, Oxford, 1972.
3.   R. Belcher, M.A. Leonard and T.S. West, Chem. Ind. (London),
      1958, 128.
4.   P. Klein and V. Skřivánek, Chem. Prum., 1962, 12, 359; Anal.
      Abstr., 1962, 10, 543.
5.   F. Štráfelda, Coll. Czech. Chem. Commun., 1962, 27, 343.
6.   H. Flaschka and F. Huditz, Z. Anal. Chem., 1952, 137, 104.

7. C. Liteanu, I.A. Crişan and G. Theiss, Rev. Roum. Chim., 1967,
    12, 567; Anal. Abstr., 1968, 15, 4568.
8. J.S. Gedanski and L. Gordon, Anal. Chem., 1957, 29, 566.
9. R. Přibil, unpublished results.
10. V. Grdinić and A. Gertner, Mikrochim. Acta, 1969, 800.

6.3.6 Alkali Metals

Of the alkali metals, only lithium (log K = 2.8) and sodium (log K =
1.7) form demonstrable complexes with EDTA.   Formation of complexes
of the others is rather improbable.

In principle, the alkali metals could be determined after isolation
as insoluble salts containing another complexometrically determinable
cation.   The precipitation must, however, be quantitative and the
precipitated compound must have a precisely defined composition.
Experience gained in the gravimetric determination of these cations
is extensively employed in this area, but most instrumental methods
(e.g. flame photometry) are superior.   The authors of the complexometric methods described below often chose unnecessarily difficult
procedures.   For example, de Sousa [1] first isolates lithium
chloride, and then instead of using "complicated" drying and weighing,
he precipitates the chloride as AgCl, to which he applies the
exchange reaction with Ni(CN)$_4^{2-}$.   The displaced nickel is then
determined complexometrically, with murexide as indicator (see
Section 6.3.5).   The argentometric determination of chloride is much
simpler.   De Sousa also converts potassium [2], rubidium [3] and
caesium [4] perchlorates into the corresponding chlorides by heating
with ammonium chloride and then determines these compounds by the
same procedure as that used for lithium.

A great deal more attention has been paid to the complexometric
determination of sodium and potassium, which can be of practical
importance in a number of cases.   Flaschka [5] recommends that
sodium be determined by precipitation of NaZn(UO$_2$)$_3$.(CH$_3$COO)$_9$ followed by EDTA titration of the zinc in the precipitate, in ammonia +
ammonium carbonate medium (to mask the uranium).   In this way 20 -
250 µg of sodium can be determined.   The zinc in this precipitate
can better be determined in 50% ethanol medium, with dithizone as
indicator, where uranium does not interfere [6].   This method has
been used in the determination of sodium in blood serum [7].   Other
"triple acetates" can also be used.   Gagliardi and Reimers [8]
studied these in detail and state that the nickel or cobalt triple
acetate is the most suitable for both gravimetric and complexometric
determinations.   Sodium can be determined in the presence of the
other alkali metals, calcium, barium, manganese, cadmium, zinc,
copper and cobalt.   Doležal et al. [9] used the cobalt triple
acetate and determined the uranium in it by reduction with ascorbic
acid, and back-titration of excess of EDTA with thorium nitrate,
using Xylenol Orange as indicator.   The zinc triple acetate method
has been critically re-examined [10].

The low solubility of sodium potassium cobaltinitrite can be used for
the determination of potassium.   Flaschka and Holasek [11], and
Taketatsu [12], determined the cobalt in the precipitate complexo-
mtrically, using murexide as indicator.   Sodium in amounts up to
five times that of potassium does not interfere.   The determination
of cobalt in the precipitate has been variously modified [13 - 15].

The greatest difficulties arise from supersaturation effects, the solubility of the precipitate during wahsing, and the unfavourable stoichiometry $(Co = 2K)$. The method has been applied mainly to the determination of potassium in blood serum [11,13,15,16], erythrocytes [17], urine [18], etc.

Another gravimetric reagent, sodium tetraphenylborate, $NaB(C_6H_5)_4$, ("Kalignost"), has been recommended for the determination of potassium. Rubidium, caesium, thallium(I), ammonium and a number of nitrogen-containing organic substances are also precipitated. Flaschka and Sadek [19] dissolve the $KB(C_6H_5)_4$ precipitate in acetone and react it with the Hg-EDTA complex:

$$KB(C_6H_5)_4 + 4HgY^{2-} + 4HCl + 3H_2O = 4Hg(C_6H_5)Cl + 4H_2Y^{2-} + KH_2BO_3$$

The EDTA freed is titrated with zinc, with PAN as indicator. An advantage of the method is the very favourable stoichiometry $(1K = 4Hg = 4H_2Y^{2-})$.

Sen [20] recommended determining the alkali metals (except lithium) by precipitation as $M_2Ag[Co(NO_2)_6]$. He determines cobalt in the isolated precipitate by direct titration with EDTA, using KSCN as indicator in aqueous acetone medium.

There are few methods for the simultaneous determination of sodium and potassium, since there are no suitable gravimetric determinations available. Gagliardi and Reimers [21] published a very interesting paper, describing the determination of sodium or potassium on the basis of exchange of these cations for magnesium on an ion-exchanger. They used Lewatit-S 100 in a column (20 x 250 mm), saturated (in a carbon dioxide or nitrogen atmosphere) with 5% magnesium chloride solution. After passage of the sodium or potassium solution (ammonium salts must be absent) through the column, the magnesium in the eluate is determined directly by EDTA titration with Erio T as indicator. The sum of the two metals can be determined (or either of them alone if the other is absent). The determination takes only 30 min and can be complemented by determination of potassium with Kalignost. This work was published in 1958 but has not yet been used in practice.

REFERENCES

1. A. de Sousa, Mikrochim. Acta, 1961, 732.
2. A. de Sousa, Mikrochim. Acta, 1961, 644; Anal. Chim. Acta, 1960, 22, 522.
3. A. de Sousa, Mikrochim. Acta, 1961, 729.
4. A. de Sousa, Talanta, 1961, 8, 686.
5. H. Flaschka, Mikrochem. ver. Mikrochim. Acta, 1952, 39, 391.
6. A. Holasek and M. Dugandzić, Mikrochim. Acta, 1959, 488.
7. M. Dugandzić, H. Flaschka and A. Holasek, Clin. Chim. Acta, 1959, 4, 819.
8. E. Gagliardi and H. Reimers, Mikrochim. Acta, 1957, 784.
9. J. Doležal, L. Novozámský and J. Zýka, Coll. Czech. Chem. Commun., 1962, 27, 1830.
10. B.C. Sinha and S.K. Roy, Talanta, 1979, 26, 596.
11. H. Flaschka and A. Holasek, Z. Physiol. Chem., 1956, 303, 9.

12. T. Taketatsu, J. Chem. Soc. Japan, Pure Chem. Sect., 1957, 78,
     640; Anal. Abstr., 1958, 5, 361.
13. H. Flaschka, A. Holasek and M. Rosenthal, Z. Physiol. Chem.,
     1957, 308, 183.
14. R. Pileri, Z. Anal. Chem., 1957, 157, 1.
15. W. Herbirger and W. Hubmayer, Klin. Wochenschr., 1960, 38, 822.
16. M. Pećar, Mikrochim. Acta, 1960, 567.
17. A. Herbirger and W. Hubmayer, Klin. Wochenschr., 1960, 38, 1064.
18. A. Holasek and M. Pećar, Clin. Chim. Acta, 1961, 6, 125.
19. H. Flaschka and F. Sadek, Chemist-Analyst, 1958, 47, 30.
20. B. Sen, Anal. Chim. Acta, 1958, 19, 320.
21. E. Gagliardi and H. Reimers, Z. Anal. Chem., 1958, 160, 1.

6.3.7   Gold and the Platinum Metals

These elements are generally of less interest for analysts. Complexometry has made no great advances in this area and will thus be
discussed only briefly.   Very little is known about the formation of
complexes of these metals with EDTA.   For example, gold(III) is
reduced by EDTA to colloidal gold;   this reaction has been used for
the microdetection of gold [1].   The formation of the very stable
cyanide complexes has been used for the determination of gold and
palladium.   Both elements react similarly to silver with $K_2Ni(CN)_4$
solution, according to the equations

$$Ni(CN)_4^{2-} + Au^{3+} = Au(CN)_4^- + Ni^{2+}$$

$$Ni(CN)_4^{2-} + Pd^{2+} = Pd(CN)_4^{2-} + Ni^{2+}$$

The displaced nickel is then titrated with EDTA, with murexide as
indicator [2,3].   The determination of palladium is not disturbed by
the presence of ruthenium, osmium, rhodium and iridium, but platinum
interferes.   It appears that only palladium forms an analytically
useful complex with EDTA.   Potentiometric measurements have yielded
a value of $\log K = 18.5$ for the Pd-EDTA complex [4].   Kinnunen and
Merikanto [5] recommend determining palladium by back-titration of
excess of EDTA with thorium (pH 3) or thallium (pH 4), Xylenol Orange
being used as indicator.   Kragten gives $\log K = 26.4$ [6] or 25.8 [7].

REFERENCES

1.   W.A. Hynes, L.K. Yanovski and J.E. Ransford, Mikrochim. Acta,
     1950, 35, 160.
2.   J. Kinnunen and B. Merikanto, Chemist-Analyst, 1955, 44, 11.
3.   H. Flaschka, Mikrochim. Acta, 1953, 226.
4.   W.M. MacNevin and O.H. Kriege, J. Am. Chem. Soc., 1955, 77, 6149.
5.   J. Kinnunen and B. Merikanto, Chemist-Analyst, 1958, 47, 11.
6.   J. Kragten, Talanta, 1978, 25, 239.
7.   J. Kragten, Personal communication.

CHAPTER 7

The Complexometry of Anions

The discussion of the complexometric determination of alkali metals
(Section 6.3.6) also holds for the complexometric determination of
anions. They can be determined only after conversion into insoluble
salts containing complexometrically titratable metals. The method
is based either on EDTA titration of the metal in the precipitate, or
on determination of the excess of metal in the filtrate. The
complexometric properties of the anions themselves (cyanide, fluoride)
are used in only a few cases, and lead to an unfavourable stoichiometry.

Methods have been proposed for determining almost all anions,
frequently without consideration of any practical importance. Many
useful methods are available, however, and the determinations of
fluoride, sulphate and phosphate are especially important, as the
usual gravimetric methods are replaced by rapid complexometric
procedures. Greater attention will be paid to these determinations.

7.1 FLUORIDE

Fluoride can be determined either by utilizing the insolubility of
some of its compounds, such as CaF_2, $PbClF$ or $PbBrF$, or by using its
ability to form complexes with some cations. Metallochromic
indicators are found to be very useful for this purpose, e.g. Xylenol
Orange or Methylthymol Blue, which allow the direct determination of
the fluoride of a particular cation, e.g. thallium.

7.1.1 Determination of Fluoride as CaF_2

Belcher and Clark [1] determined 5 - 65 mg of fluoride by precipitation with calcium at pH 4.5. The excess of calcium was titrated
with EDTA, Erio T being used as indicator. The time required for
quantitative precipitation of calcium fluoride increases with
decreasing amount of fluoride, and consequently the authors recommend
that the solution be left to stand overnight for complete precipitation. The determination of fluoride is not disturbed by the
presence of chloride, bromide, iodide and carbonate, and excess of

calcium can be determined directly in the presence of the precipitate.
The authors observed an interesting phenomenon: in the presence of
phosphate, sulphate or arsenate, the CaF_2 dissolved slowly during
titration with 0.1M EDTA and the indicator colour continuously
returned, rendering the determination impossible. This "induced
dissolution" could only be eliminated by filtering off the precipitate before the titration. The presence of these anions in the
filtrate also decreases the sharpness of the colour transition of
Erio T.

Among other authors, Okada and Sugiyama [2] used a similar procedure,
precipitating the fluoride in 75% ethanol solution. Wilska [3]
recommends 60% ethanol for more rapid precipitation of CaF_2.

7.1.2   Determination of Fluoride as PbClF

Laszlovsky [4] precipitated fluoride from neutral solution with a
0.75% solution of lead chloride. The isolated PbClF precipitate was
dissolved in hot 2M nitric acid and the lead was titrated with EDTA
in ammoniacal tartrate medium, with Erio T as indicator. Traces of
copper were masked with potassium cyanide.

Vřešt'ál et al. [5] also determined fluoride by precipitating with
lead chloride solution and titrating the excess of lead with EDTA,
but used Xylenol Orange as indicator. The method can be used to
determine 1 - 60 mg of fluoride in a volume of 50 - 60 ml. Up to 2%
of sodium sulphate does not interfere; at higher concentrations it
causes precipitation of $PbSO_4$.

Sakharova and Shishkina [6] precipitated fluoride in the presence of
sodium chloride by reaction with lead acetate and then used the
procedure given by Laszlovszky, but with Chromogene Black ET00.

Leonard [7] carried out a detailed study of the conditions for the
precipitation of PbClF. He precipitated fluoride with lead nitrate
in the presence of sodium chloride and, after filtration, titrated
the excess of lead with EDTA, using Xylenol Orange as indicator
(hexamine as buffer). The optimum pH for the precipitation was
found to be 4.5 - 10, but it should not be adjusted with a weak-acid
buffer. Chloride to fluoride ratios up to 4:1 do not affect the
composition of the precipitate provided they exceed 1:1. The
solubility of PbClF is much less in 20% ethanol medium. Temperatures
up to 50°C do not affect the determination if the final solution
volume does not exceed 70 ml.

Note. Erdey et al. [8] state that instead of complexometric
titration of the lead, the precipitate can be dissolved in excess of
EDTA and the chloride titrated with silver nitrate. They used
Variamine Blue B as indicator, and recommended the method for determining fluorine in organic materials.

7.1.3   Determination of Fluoride as PbBrF

Harzdorf and Steinhauser [9] made a detailed study of the precipitation conditions for both PbClF and PbBrF. The pH is particularly
important, and should be 3.6 for the former and 3.2 for the latter
compound. The precipitate should be left to stand for at least 5 hr,
and preferably overnight. Both methods are equally useful for

determination of fluoride in pure solutions.   The PbClF method
yields high results if silicon is present.   Though the PbBrF method
normally yields very good results, the results are about 1% low for
solutions with large silicon contents, but can be improved by
increasing the bromide concentration.

The final procedure given by the authors is as follows.   A solution
containing 0.5 - 6 mmole of fluoride is mixed with 50 ml of 0.25M
potassium bromide (or chloride), diluted to 250 ml, and acidified
with 4M nitric acid to Dimethyl Yellow indicator.   An extra 2 or 3
drops of acid are added and the solution is heated to 90 - 95°C to
expel carbon dioxide, but must not be boiled because of the risk of
loss of hydrogen fluoride.   After cooling to 60 - 70°C the pH is
adjusted to 3.5 - 3.7 with 1M sodium acetate (yellow colour of the
indicator), then 50 ml of 0.25M lead nitrate are slowly added and the
solution is left to cool.   The final pH is precisely adjusted (pH-
meter) with sodium acetate (to 3.3 for PbBrF, 3.6 for PbClF).   Two
methods can be used for the final determination: either lead is
photometrically titrated with EDTA, using Xylenol Orange as indicator,
or the bromide or chloride is titrated potentiometrically with silver
nitrate.   The authors used the bromide method to determine fluorine
in aluminium fluoride and cryolite.   The use of acetate buffer
should be compared with Leonard's recommendations [7].

7.1.4   Determination of Fluoride as CeF$_3$

Yamamura [10] quantitatively precipitated fluoride by reaction with
cerous nitrate at pH 1.75.   After centrifuging the precipitate, he
titrated the excess of cerium in an aliquot of the solution with
EDTA, using arsenazo as indicator.   The effect of common anions was
considered.   Hydrogen fluoride is liberated pyrolytically from
organic or inorganic samples by heating with tungstic oxide.   The
method can be used for samples containing beryllium, aluminium or
zirconium.

7.1.5   Determination of Fluoride as LaF$_3$(H$_2$O)$_3$

Mixuno and Miyaji [11] recommended precipitating fluoride with
lanthanum and titrating the excess of lanthanum with EDTA, using
Methylthymol Blue as indicator in hexamine-buffered medium.   The
method was applied to the determination of fluorine in organic
compounds, which were decomposed by the Schöniger method in quartz
flasks, with 5 ml of 0.01M lanthanum chloride as absorption solution.
The authors studied the effect of the carbon dioxide formed during
the combustion, which could interfere in the determination of
fluoride in weakly acid medium.   They recommend heating the
absorption solution (containing the fluoride) to 70°C, with magnetic
stirring to ensure complete removal of carbon dioxide without loss of
hydrogen fluoride.

7.1.6   Direct Determination of Fluoride by Titration with Thorium
        Nitrate

The very low solubility of thorium fluoride has often been used for
the precipitation titration of fluoride.   Schmidt and Ortloff [12]
titrated fluoride in glycine-perchlorate buffer directly with 0.02M

thorium nitrate, using Xylenol Orange as indicator. They state that
the method can be used for 1 - 10 mg of fluorine. Körös [13]
titrated fluoride at pH 5 - 6.5 directly with thorium nitrate, using
Pyrocatechol Violet as indicator. Hlucháň et al. [14] titrated up
to 100 μg of fluoride at pH 2,2, using Methylthymol Blue as indicator,
but even when all the conditions listed by these authors are
maintained, the colour transition is not very sharp.

An official analytical methods committee has studied and recommended
[15] the titration of fluoride with 0.02M thorium nitrate, with use
of Methylthymol Blue as indicator. The final procedure is as
follows. A solution containing 2.0 - 15.0 mg of fluoride is diluted
to 50 ml with water and adjusted to pH 3.4 ± 0.1 (pH-meter or pH-
paper) with 1M perchloric acid or sodium hydroxide. After addition
of indicator the solution is titrated with 0.02M thorium nitrate from
a microburette with constant stirring, until a permanent blue colour
is formed. Iron present (up to 5 mg) must be reduced before the
titration with 1 ml of 2% ascorbic acid solution.

7.1.7   <u>Direct Determination of Fluoride with $U(SO_4)_2$ Solution</u>

Harzdorf [16] used the very low solubility of "fluorouranate" for
determination of fluoride by direct titration with uranium(IV)
sulphate, the indicator being Xylenol Orange or Methylthymol Blue.
Turbidimetric titration can also be used. X-Ray studies revealed
that the simple compound KUF_5 is not formed in the titration, but
rather $7KF.6UF_4$. The author used a procedure in which the pH of a
solution containing 4 - 50 mg of fluoride was adjusted to 3 - 5 with
chloroacetate buffer, 1 g of potassium chloride was added and the
solution was titrated with $U(SO_4)_2$ (in 2N H_2SO_4), with either
indicator. A strong side-light should be used because of the marked
turbidity of the titrated solution.

The last two methods are not really complexometric titrations. Most
of the methods given require "pure" fluoride solutions, necessitating
separation of the fluoride. The distillation method of Willard and
Winters is still used for this purpose. A pyrolytic method was
mentioned in Section 7.1.4. Ion-exchange can sometimes be used.
The extraction method described by Chermette et al. [17] should also
be mentioned; it is based on extraction of the fluoride complex with
triphenylantimony(V) chloride, $(C_6H_5)_3SbCl_2$, which can be extracted
with carbon tetrachloride or benzene. The authors state that the
fluoride can be extracted in the presence of aluminium.

REFERENCES

1. R. Belcher and S.J. Clark, Anal. Chim. Acta, 1953, <u>8</u>, 222.
2. K. Okada and T. Sugiyama, Ann. Rept. Takamine Lab., 1957, <u>9</u>,
    185; Anal. Abstr., 1958, <u>5</u>, 4105.
3. S. Wilska, Suom. Kemistilehti, 1958, <u>B31</u>, 400; Anal. Abstr.,
    1959, <u>6</u>, 3493.
4. J. Laszlovszky, Mag. Kem. Foly., 1954, <u>60</u>, 210.
5. J. Vřešťál, J. Havíř, J. Brandštetr, and S. Kotrlý, Coll.
    Czech. Chem. Commun., 1958, <u>23</u>, 886.
6. Ya. G. Sakharova and N.I. Shishkina, Zavod. Lab., 1959, <u>25</u>,
    1442; Anal. Abstr., 1960, <u>7</u>, 3239.

7.   M.A. Leonard, Analyst, 1963, 88, 404.
8.   L. Erdey, L. Mázor and M. Pápay, Mikrochim. Acta, 1958, 482.
9.   C. Harzdorf and O. Steinhauser, Z. Anal. Chem., 1966, 218, 96.
10. S.S. Yamamura, M.E. Kussy and J.E. Rein, Anal. Chem., 1961, 33,
     1655.
11. K. Mizuno and N. Miyaji, Japan Analyst, 1972, 21, 631.
12. J. Schmidt and R. Ortloff, Z. Chem., 1965, 5, 236.
13. E. Körös, Acta Pharm. Hung., 1957, 27, 1.
14. E. Hluchań, M. Sedlák and M. Jeník, Chem. Zvesti, 1969, 23, 219;
     Anal. Abstr., 1970, 19, 203.
15. Analytical Methods Committee, Analyst, 1972, 97, 734.
16. C. Harzdorf, Z. Anal. Chem., 1965, 214, 345.
17. H. Chermette, C. Martelet, D. Sandino and J. Tousset, Compt.
     Rend., 1971, 273C, 543; Chem. Abstr., 1972, 76, 30316u.

7.2 DETERMINATION OF SULPHATE

Sulphate is determined complexometrically by precipitation as either
barium sulphate or lead sulphate. Both methods, each of which has
advantages and disadvantages, have been studied in a large number of
publications, which frequently differ only in small details.   Consequently, only the most important characteristics will be mentioned.
Methods based on other principles will also be discussed.

7.2.1   Determination of Sulphate as BaSO$_4$

All the methods described involve determination either of excess of
barium after precipitation of the sulphate or of the barium in the
precipitate.   Most authors use the precipitation conditions regarded
as optimal for the gravimetric determination of barium or sulphate.
Various indicators have been used.   In the older literature,
Phthaleincomplexone [1,2] was recommended, sometimes in conjunction
with other dyes, such as Naphthol Green, Methyl Red, etc., as well as
Calcein [3] or sodium rhodizonate.   Erio T is not suitable because
the colour transition is not sharp.   Addition of the Zn-EDTA complex,
from which barium partially displaces zinc in ammoniacal medium,
considerably increases the sharpness of the transition [4].   Methylthymol Blue is also useful.   The filtration step in methods based on
titration of the barium in the precipitate has received a good deal
of attention.   Belcher et al. [5] recommended using a crucible
filled with crushed paper, whereas Jackson [6] used a Gooch crucible
with a rolled up filter paper.   West [7], however, stated that this
method may lead to loss of substance and is not to be recommended.
A sintered-glass crucible is unsuitable because of the difficulty in
dissolving all the barium sulphate in alkaline EDTA solution.   There
are various views on the dissolution of barium sulphate in EDTA; it
is very dependent, for example, on the grain size and on the aging of
the precipitate.   Belcher et al. recommend dissolving in hot ammoniacal EDTA, whereas Rumler et al. [8] consider this method unreliable
and recommend 5% triethanolamine solution and a 200 - 300% excess of
EDTA.   Barium sulphate readily dissolves in this mixture on brief
heating.   It dissolves even better in an EDTA solution containing
sodium hydroxide (pH 12).   Ballczo and Doppler [9] recommend
ignition of the filter paper and barium sulphate in a platinum
crucible and fusion with a small amount of sodium metaphosphate at
1000°C.   Decomposition proceeds according to the equation

$$BaSO_4 + NaPO_3 = BaNaPO_4 + SO_3$$

The cooled melt is dissolved in 10 ml of 1M hydrochloric by heating
on a water-bath; excess of metaphosphate is converted into orthophosphate. The authors determined barium indirectly by back-
titration of excess of EDTA with 0.01 - 0.001M magnesium chloride.
A solid mixture of Erio T (2 parts) and Tropaeolin OO (1 part) in 500
parts of sodium chloride was used as indicator. The method is useful for solutions containing 0.05 - 10 mg of sulphate.

A great deal of attention has also been paid to precipitation of the
sulphate. Most authors use the usual method involving precipitation
by dropwise addition of barium chloride solution to a boiling acid
sulphate solution. Iritani and Tanaka [2] precipitated barium
sulphate from homogeneous solution by adding a solution of the Ba-
EDTA complex to a neutral sulphate solution, adjusting the pH with
ammonia to 9 - 10 and boiling, with dropwise addition of dilute hydrochloric acid to adjust the pH to 2.5 - 3. After filtration, the EDTA
liberated was titrated at pH with barium chloride, Phthaleincomplexone
being used as indicator. In this manner 10 mg of sulphate could be
determined with a precision of 0.1 mg.

A much faster method involves titration of excess of barium without
separation of the precipitate, usually with addition of ethanol to
decrease the solubility. This method was used by Wilson et al. [10],
Sijderius [11], Ueno and Yamaguchi [12], Légrádi and Molnár [13],
Tettweiler and Piltz [4] and many others. Lounamaa and Fugmann [14]
state that determination of the end-point is unreliable in these
methods.

The complexometric determination of sulphate is much faster than the
gravimetric determination, if the aging period is ignored. The time
required for quantitative precipitation of sulphate has received
renewed attention in this connection. Belcher et al. [5] for
example, found that the precipitation of sulphate (3 - 6 mg) is
quantitative after 1 - 2 hr. Standing for 10 - 15 min before
filtration is sufficient for solutions containing 30 - 50 mg of
sulphate. Similarly, Wilson et al. [10] titrated the excess of
barium practically immediately after precipitation of the sulphate.
Munger et al. [15], Faber [16] and Gwilt [17] also used only a short
standing period before the titration, Sijderius [11], however,
states that it is not possible to precipitate barium sulphate quantitatively in a period shorter than 24 hr.

Styunkel' and Yakimets [18] feel that the complexometric determination of sulphate is not reliable and that it can only be used semi-
quantitatively. They also consider that agreement between complexometric and gravimetric results is quite accidental and independent
of the precision of the method, but rather dependent on suitable
choice of conditions, which differ for individual cases. They
further think that barium sulphate is always partially soluble in
EDTA, which must be important in indirect determinations, and that
use of large amounts of barium sulphate chloride leads to adsorption
of barium. On the other hand, they do not recommend precipitating
with very small excesses of barium, because of the low sensitivity of
the indicators used. These opinions are unique. If optimal
conditions are maintained, the complexometric determination of
sulphate can be considered as sufficiently precise and quite satis-

factory for series analyses.

Interferences

Theoretically, cations should not interfere in the sulphate methods
based on determination of the barium in a well-washed barium sulphate
precipitate, especially if the sulphate is precipitated in the
presence of EDTA, which binds practically all cations in complexes.
This procedure was first proposed for the gravimetric determination
of sulphate by Přibil and Maričová [19]. Under these conditions, a
pure BaSO$_4$ precipitate is obtained, even in the presence of large
amounts of lead. Other considerations prevail, however, if the
sulphate is determined indirectly by titration of excess of barium
either in the filtrate or in the presence of the precipitate, and
this determination becomes impossible in the presence of any cations
which form complexes with EDTA. Some anions, such as chromate or
phosphate, also interfere. Only in a few cases is this determination
possible without masking or prior separation. For example, small
amounts of iron or aluminium can be masked with triethanolamine, and
copper and some other cations with potassium cyanide. Generally,
separation of cations by ion-exchange is recommended, especially in
water analysis. Sijderius [11] used Dowex-50 ion-exchanger; Hunt
[20] used Amberlite IR-112. Page and Spurlock [21] used the same
procedure.

In water analysis, a very interesting problem arises in the determination of sulphate and water hardness in a single sample. Published
methods are based on determination of the sum of Ca + Mg by direct
titration with EDTA. After acidification, the sulphate is precipitated with barium chloride, the excess of which is determined by one
of the methods described above, after alkalization of the solution.
The two neutralization steps cause accumulation of salts, especially
ammonium salts, which can interfere strongly in the indicator colour
transition. A method involving determination of Ca + Mg in one
portion of the water sample and barium (in the precipitated BaSO$_4$) or
the sum Ca + Mg + excess of barium in a second portion is preferable.

Lounamaa and Fugmann [14] described an interesting method for determining sulphate. In the determination of sulphur in steels, they
collected the sulphate on aluminium oxide (by the Brockmann method)
and thus separated it from the large amounts of iron. The sulphate
was eluted with ammonia, which was then removed by boiling, and the
sulphate was precipitated with barium chloride. The precipitate was
filtered off after 12 hr and the excess of barium titrated with EGTA
in the presence of triethanolamine, with Erio T as indicator. In
this way the authors determined up to 0.0085% of sulphur in a 5 g
sample. It is not clear how competitive this method is with
combustion methods for determining sulphur in steels.

7.2.2   Determination of Sulphate as PbSO$_4$

Indirect determination of sulphate after precipitation as PbSO$_4$ has
certain advantages over the barium chloride method. Lead sulphate
can be obtained in much purer form, as a coarsely crystalline
precipitate. A disadvantage is its greater solubility, but this can
be decreased by addition of various alcohols.

Sporek [22] precipitated lead sulphate quantitatively from a solution

containing 50% isopropyl alcohol.   The precipitation is quantitative
in 10% nitric acid medium.   The lead sulphate is washed with
isopropyl alcohol and dissolved in an ammoniacal solution of EDTA;
the excess of EDTA is titrated with 0.05M zinc chloride, with Erio T
as indicator.   The determination is not disturbed by uranium, iron
or arsenate.   High concentrations of phosphate lead to a positive
error.   The authors used this method to determine sulphate in
extracts of uranium ores.   The determination takes 30-45 min and is
much more reliable than the barium sulphate method.

Other authors also recommend addition of alcohol.   Ashbrook and
Ritcey [23] recommend 50% ethanol and standing for 3 hr, Odler and
Gebauer [24] precipitate sulphate from a 20% methanol solution.

Several methods have been recommended for determination of the lead
(in either the lead sulphate or the filtrate).   Iritani et al. [25]
precipitated lead sulphate in 30% ethanol solution at pH 3.8-4.2, in
a volume less than 150 ml.   The excess of lead in the filtrate was
titrated with EDTA, Cu-PAN being used as indicator.   As the lead is
determined in weakly acid media, alkaline-earth metals and carbon
dioxide do not interfere.   Later, Iritani and Tanaka [26] described
the determination of phosphate and sulphate in a single solution:
phosphate was determined as $MgNH_4PO_4.6H_2O$ (magnesium titrated
complexometrically) and sulphate was then precipitated with a known
amount of lead nitrate, the excess of which was titrated in acetate
medium with EDTA (Xylenol Orange as indicator).   Buffering with
hexamine would be better (cf. p. 153).

In addition to the Sporek method [22], other procedures have been
used for dissolving the lead sulphate.   Galleja and Fernandez-Paris
[27] dissolved it in a tartrate solution and titrated the lead
directly at pH 10, using Erio T indicator.   Tanaka and Tanabe [28]
studied the dissolution of lead sulphate in ammonium acetate solution,
taking into account the interference from large amounts of acetate
(see p. 153), and titrated the lead with 0.01M EDTA, using Xylenol
Orange as indicator.   Odler and Gebauer [24] used two methods for
determining sulphate.   In the first, 2-20 mg of sulphate were
precipitated in 50 ml with excess of lead acetate.   After addition
of 10 ml of methanol, the lead sulphate was filtered off, dissolved
in 50 ml of 10% ammonium acetate solution and titrated with 0.02M
EDTA (Xylenol Orange as indicator).   As might be expected, the
results were low (by up to 5%, depending on the lead content).   In
the second method, the lead sulphate was filtered off and the excess
of lead in the filtrate was titrated in hexamine-buffered medium,
again with Xylenol Orange as indicator.   The results were then up to
3.7% high.   The authors apparently ignored earlier experience with
the effect of acetate, which strongly competes with formation of the
Pb-EDTA complex.   Consequently, the negative errors were larger in
the first method (larger acetate contents) than the positive errors
in the second, where the acetate was only that added as lead acetate
and the actual determination was carried out in hexamine medium.
Precipitation of sulphate with lead nitrate is much better.   After
collection, the lead sulphate can readily be dissolved in excess of
EDTA, the surplus of which can be back-titrated with zinc chloride at
pH 5-5.5, with Xylenol Orange as indicator.

Interferences

Large concentrations of chloride or iodide and traces of hydrogen

sulphide interfere in precipitation of the lead sulphate.   Also, all
elements which are precipitated as hydroxides at the acidity used and
are collected along with the lead sulphate will interfere.   Barium
and strontium are precipitated as the sulphates, but these are not
soluble in ammonium acetate.   They do not interfere if the dissolution in EDTA is used, provided the lead is determined in acid medium.
Sporek [22] states that phosphate interferes.   In the determination
of sulphate in detergents, Taminori et al. [29] hydrolysed poly-
phosphates and diphosphates present, by boiling with nitric acid to
obtain orthophosphate.   This was then separated as the silver salt
before the actual determination of sulphate.   The authors state that
borate must also be separated as the methyl ether if the BO_3^{3-}/SO_4^{2-}
ratio is greater than 0.6.

REFERENCES

1.   G. Anderegg, H. Flaschka, R. Sallman and G. Schwarzenbach, Helv.
     Chim. Acta, 1953, 37, 113.
2.   N. Iritani and T. Tanaka, Japan Analyst, 1958, 7, 42.
3.   M. Effenberger, Chem. Listy, 1958, 52, 1501.
4.   K. Tettweiler and W. Pilz, Naturwissenschaften, 1954, 41, 332.
5.   R. Belcher, D. Gibbons and T.S. West, Chem. Ind. (London), 1954,
     127.
6.   P.J. Jackson, Chem. Ind. (London), 1954, 435.
7.   T.S. West, Chem. Ind. (London), 1954, 608.
8.   F. Rumler, R. Herbosheimer and G. Wolf, Z. Anal. Chem., 1959,
     166, 23.
9.   H. Ballczo and G. Doppler, Mikrochim. Acta, 1954, 403.
10. H.N. Wilson, R.M. Pearson and D.M. Fitzgerald, J. Appl. Chem.,
     1954, 4, 488.
11. R. Sijderius, Anal. Chim. Acta, 1954, 11, 28.
12. K. Ueno and Y. Yamaguchi, Japan Analyst, 1954, 3, 331.
13. L. Légrádi and A.F. Molnár, Mag. Kem. Foly., 1958, 64, 29.
14. N. Lounamaa and W. Fugmann, Z. Anal. Chem., 1963, 199, 98.
15. J.R. Munger, R.W. Nippler and R.S. Ingols, Anal. Chem., 1950,
     22, 1455.
16. J.S. Faber, Pharm. Weekbl., 1954, 89, 705.
17. J.R. Gwilt, Chem. Ind. (London), 1954, 309.
18. T.B. Styunkel' and E.M. Yakimets, Zavod. Lab., 1956, 22, 653.
19. R. Přibil and D. Maričová, Chem. Listy, 1952, 46, 542.
20. W.G. Hunt, J. Am. Water Works Assoc., 1953, 45, 535.
21. J.O. Page and W.W. Spurlock, Anal. Chim. Acta, 1965, 32, 593.
22. K. Sporek, Anal. Chem., 1958, 30, 1032.
23. A.W. Ashbrook and G.M. Ritcey, Analyst, 1961, 86, 740.
24. I. Odler and J. Gebauer, Chem. Zvesti, 1961, 15, 563; Anal.
     Abstr., 1962, 9, 1450.
25. N. Iritani, T. Tanaka and H. Oishi, Japan Analyst, 1959, 8, 30;
     Anal. Abstr., 1960, 7, 102.
26. N. Iritani and T. Tanaka, Japan Analyst, 1960, 9, 1; Anal.
     Abstr., 1961, 8, 4603.
27. J. Galleja and J.M. Fernández Paris, Rev. Cienc. Appl., 1961,
     15, 120; Anal. Abstr., 1962, 9, 162.
28. T. Tanaka and H. Tanabe, Japan Analyst, 1959, 8, 826.
29. S. Tanimori, K. Abe and M. Tobari, J. Chem. Soc. Japan, Ind.
     Chem. Sec., 1965, 68, 273.

7.3   DETERMINATION   OF   PHOSPHATE

The indirect complexometric determination of phosphate is based on
the insolubility of the ammonium phosphates of magnesium and zinc.
Far less attention has been paid to the precipitation of phosphate
with thorium nitrate, bismuth perchlorate and solutions of some other
cations.   Several direct precipitation titrations of phosphate in
the absence of EDTA have also been described.

7.3.1   Determination of Phosphate as $MgNH_4PO_4 \cdot 6H_2O$

The classical gravimetric method, based on precipitation of phosphate
as $MgNH_4PO_4 \cdot 6H_2O$ from a medium containing ammonia and ammonium
chloride is completely non-selective and can practically be used only
in the absence of cations other than sodium and potassium.   The
precipitation of the double phosphate in the presence of EDTA is
quite selective, however;   the EDTA reliably binds the bivalent
metals, including the alkaline-earth metal ions.   Contrary to expectation, a gravimetric study [1] has indicated that the excess of
magnesium slowly replaces iron and aluminium in the Fe-EDTA and Al-
EDTA complexes and that these metals are precipitated as the hydroxides together with the magnesium ammonium phosphate.   This co-
precipitation of small concentrations of iron is slow and, on standing overnight, the ferric hydroxide forms a layer on the magnesium
ammonium phosphate precipitate.   This can be prevented by adding a
further complexing agent - tiron.   This method is useful for the
gravimetric determination of magnesium, but cannot be used in
complexometric determinations, because of the intense red colour of
the Fe-tiron complex.

Huditz et al. [2] developed the first complexometric titration of
phosphate, based on this principle.   Magnesium ammonium phosphate is
precipitated from a medium containing EDTA and citric acid.   The
precipitate is collected, and dissolved in hydrochloric acid, excess
of EDTA is added and the surplus is titrated with magnesium   (Erio T
as indicator).   Flaschka and Holasek [3] used a simpler procedure in
the microdetermination of phosphate; after dissolving the precipitate in hydrochloric acid, they strongly alkalized the solution with
ammonia and titrated the magnesium directly with EDTA.   The method
can be used only for amounts of phosphorus below 1 mg, as the
magnesium ammonium phosphate is reprecipitated at once at higher
phosphate concentrations.   The method was applied to the determination of phosphate in blood serum [4].

A great deal of attention has been paid to the selective precipitation of phosphate.   As mentioned above, EDTA does not bind all the
elements present.   In addition to the use of citric acid to mask
iron, aluminium and titanium, lactic acid has been recommended for
traces of these elements [5].   A mixture of EDTA and triethanolamine
has been found very useful, but cannot be used to mask titanium [6].
Beryllium can be masked with sulphosalicylic acid [2].

If phosphate is to be precipitated as $MgNH_4PO_4 \cdot 6H_2O$, the conditions
used must be cafefully controlled, or compounds with different
composition from this will be precipitated.   In gravimetric work it
is common practice to make a preliminary separation, followed by
reprecipitation under the proper conditions.

St. Chamant and Vigier [7] recommended gravimetric determination of
phosphate by neutralizing an acid solution of the phosphate to pH 5
with ammonia, and then adding magnesium chloride solution and monoethanolamine. Well-developed crystals with exactly defined
composition are precipitated. Eschmann and Brochon [5] used a
modification of this precipitation for the complexometric determination by the procedure of Huditz et al. [2] and applied it to determine phosphate in various foodstuffs [8,9]. Further applications
will be described where appropriate.

7.3.2   Determination of Phosphate as $ZnNH_4PO_4.6H_2O$

This procedure has two advantages over the magnesium method: no
complications in the composition of the precipitate occur during the
precipitation of phosphate, and the zinc ammonium phosphate is
readily soluble in ammonia buffer at pH 10, allowing the direct
determination of zinc with EDTA (Erio T as indicator). Optimal
conditions for the precipitation were worked out by Ishibashi and
Tabushi [10]. The phosphate must be precipitated with a 2 - 5-fold
amount of zinc chloride at pH 5.8 - 7.5 (p-nitrophenol as indicator)
and the final concentration of ammonium chloride must be 2 - 3M. The
precipitate is filtered off after 3 hr, dissolved in hydrochloric
acid, and neutralized. After addition of Schwarzenbach buffer the
zinc is titrated directly with 0.01 - 0.1M EDTA, Erio T being used as
indicator. The authors state that the error is 0.3%. The results
are high in the presence of molybdate and one reprecipitation is
necessary. Citrate interferes. The authors used this method for
the determination of phosphate in ores.

Buss et al. [11] found the same conditions. They precipitated the
double phosphate at pH 6.7 - 7.0 by heating an acid solution of zinc
and phosphate and adding a solution of monoethanolamine dropwise till
the required pH was reached. The precipitate was dissolved in
hydrochloric acid, excess of EDTA was added, the pH adjusted to 10,
and the surplus EDTA back-titrated with zinc (Erio T as indicator).
(The results of the Japanese authors suggest that back-titration is
unnecessary [10]). The authors developed three modifications: for
macrodetermination (25 - 200 mg of phosphorus), semimicro determination (2.5 - 10 mg) and for micro determination (0.1 - 1.0 mg). Some
applications and further modification of this method can be found in
further publications by these authors [12,13].

7.3.3   Determination of Phosphate as $BiPO_4$

The precipitation of phosphate as $BiPO_4$ is very limited in scope.
Chloride and sulphate interfere and the precipitation must be done
with bismuth nitrate or perchlorate. The acidity of the solution is
very important. If it is too high, the precipitation is incomplete,
and if it is too low the bismuth is hydrolysed. Ueda et al. [14]
precipitate bismuth phosphate from ∼0.1M nitric acid, dissolve the
precipitate in hot 8M nitric acid, add excess of EDTA, adjust to
pH 2 - 2.5 and titrate with EDTA, using Xylenol Orange as indicator.

Riedel [15] solved the problem of the optimal acidity in an unusual
manner: the solution was first evaporated to dryness with nitric
acid (for removal of chloride), 13 ml of perchloric acid were added,
and the solution was evaporated to fumes of perchloric acid. After

cooling, bismuth solution was added, the solution was heated to the boiling point and 80 ml of boiling water were added. The decrease in the acidity led to precipitation of a coarsely crystalline precipitate of BiPO . The filtrate was diluted to 300 ml and the excess of bismuth titration with EDTA (Pyrocatechol Violet indicator). The method can be used for the determination of down to 5 mg of phosphate. Indium, gallium, zirconium, thorium, mercury, antimony and iron(III) interfered. Mercury and antimony can usually be removed during preparation of the solution, and Riedel recommended reducing iron(III) with a dilute hydrazine solution neutralized with perchloric acid.

7.3.4   Determination of Phosphate as $Th_3(PO_4)_4$

Kinnunen and Wennerstrand [16] used the very low solubility of thorium phosphate for the indirect determination of phosphate, which they precipitated from a boiling weakly acid solution with thorium nitrate. The mixture was cooled and diluted to known volume in a standard flask, the precipitate was collected on a dry filter, and thorium was determined in an aliquot of filtrate by EDTA titration, with Xylenol Orange as indicator. Furuya and Tajiri [17] similarly determined phosphorus in alloys containing copper, which they masked with thiourea. Kirkbright et al. [18] determined micro-amounts of phosphate (10 - 50 µg) by precipitation from a volume of 4 ml at pH 3 with a dilute thorium solution; excess of thorium was determined, without filtration, by titration with EDTA, using Xylenol Orange as indicator. The end-point was indicated by a change in the colour of the precipitate from red to yellow. The authors state that there is no interference from a 100-fold amount of chloride, bromide, nitrate, sulphate and perchlorate. Fluoride, tellurium, selenium and molybdenum interfere. Other interfering elements can be removed by ion-exchange [19]. The precision was 1 µg.

7.3.5   Determination of Phosphate as $Zr(HPO_4)_2$ or $ZrO(HPO_4)$

The insolubility of zirconium phosphate in strongly acidic sulphuric or nitric acid media led to the development of a method for the indirect determination of phosphate, based on EDTA titration of excess of zirconium in strongly acid medium, with Xylenol Orange as indicator. Ideally, this method should be better than all other methods, because both the precipitation of phosphate and the determination of zirconium are very selective; however, this is true only under certain conditions. The main difficulties lie in the behaviour of zirconium solutions, which have a tendency not only to hydrolyse, but also to form polynuclear complexes. The behaviour of zirconium solutions was described in detail in Section 6.1.1.

Budevsky et al. [20] developed a method based on precipitation of phosphate with a special zirconyl sulphate solution containing $ZrO(SO_4)_2^{2-}$ ions. These authors state that precipitation from 15 - 25% sulphuric acid solution leads to a constant precipitate composition, $ZrO_2:P_2O_5$ = 1:1, i.e. to $Zr(HPO_4)_2$, independent of time and of the excess of zirconium (20 - 300%). The determination is, however, strongly limited by the phosphate concentration, and can be used only for the narrow range of 35 - 55 mg of P_2O_5. Outside this range, there is a systematic error. The authors did not succeed in determining phosphorus in nitric acid media, because of the ready polymerization of zirconyl nitrate solutions.

The preparation of the special precipitant is rather tedious.　A
suitable zirconium salt is fused with potassium hydrogen sulphate at
$600 - 700°C$, then the cooled melt is dissolved in concentrated
sulphuric acid.　Next day water is added and the solution is heated
until completely clear $(2 - 3 hr)$, and diluted to the required volume.
No information is given on the stability of this solution;　the
concentration can readily be checked by EDTA titration.

Sinha et al. [21] came to completely different conclusions.　They
stated that only nitric acid can be used for the precipitation of
phosphate (with zirconyl nitrate) and for titration of excess of
zirconium with EDTA.　When the phosphate is precipitated from $2 -
3.5M$ nitric acid and the solution is heated, a readily filterable
precipitate is obtained, containing zirconium and phosphorus in 1:1
ratio.　Probably the compound $ZrO(HPO_4)$ is formed.　The precipitate
is filtered off, and zirconium determined in the filtrate (diluted to
give an acidity of $1 - 1.25M$ HNO_3) by EDTA titration at a temperature
of at least 90°C.　On the basis of earlier work [22], they criticize
the Budevsky method.　The basic difference between the two methods
lies in the fact that in the Budevsky method the compound $Zr(HPO_4)_2$
is formed, i.e. with twice the phosphorus content of the precipitate
in the Indian method.　According to Slumper and Metlelock this
compound is formed only when zirconium is precipitated in presence of
excess of phosphate [23].　However, the work by Pilkington and
Wilson (Section 6.1.1) favours the Budevsky method;　these authors
found that only sulphuric acid medium is suitable for depolymerization of zirconyl sulphate solutions.　Sulphuric acid also prevents
repolymerization of zirconyl sulphate solutions.　It must be pointed
out, however, that the conflict between the two methods is more
apparent than real, since Sinha et al. do not seem to have tested the
Budevsky method, and the Sinha and Das Gupta work [22] did not use
the Budevsky conditions either, and Budevsky et al. did not use the
nitric acid concentrations later recommended by Sinha et al.　Further,
the quotation of the Slumper and Metlelocke work is irrelevant, since
they used 6M hydrochloric medium.　It seems evident that the pecul-
iarities of the zirconium polymerization systems are such that all
these authors could be correct.　For the sake of completeness, both
methods are given below.

The Budevsky method [20].　A $5 - 7$ ml volume of a solution containing
$35 - 55$ mg of P_2O_5 is mixed with 8 ml of 9M sulphuric acid and 12 ml
of the special 0.5M zirconium sulphate solution.　The walls of the
beaker are washed with $1 - 2$ ml of distilled water and the solution is
heated for 10 min on a water-bath.　It is then transferred to a 100-
ml standard flask and diluted to the mark.　The solution is filtered
through a dry filter, the first part of the filtrate being discarded,
and 50 ml of the remainder are pipetted into a titration flask and
diluted to 150 ml.　The solution is heated to the boiling point and
titrated with 0.025M EDTA (Xylenol Orange indicator) from pink to
lemon yellow.　One ml of 0.025M EDTA corresponds to 7.10 mg of P_2O_5.

The Sinha method [21].　A solution containing $0.7 - 80$ mg of P_2O_5 is
pipetted into a 250-ml standard flask, mixed with 125 ml of 4M nitric
acid and diluted to about 200 ml.　Then $10 - 30$ ml of 0.05M zirconyl
nitrate are added and the solution is heated on a water-bath for 30
min.　After cooling, the solution is diluted to the mark with 1M

nitric acid.   The precipitate is filtered off on a dry filter and
100 ml of the filtrate are pipetted into a 500-ml titration flask,
diluted to 200 ml and boiled for 1 min to remove nitrogen oxides.
The temperature is kept at 90°C or above during titration of the
zirconium with 0.05M EDTA, added very slowly towards the end of the
titration.   The end-point is given by the very sharp transition of
the Xylenol Orange from pink-red to lemon yellow.   One ml of 0.05M
EDTA corresponds to 7.10 mg of P_2O_5.

The second method seems preferable because the equivalent is half as
large and the determinable phosphorus range is much broader.   A great
advantage of both methods is their selectivity.   For example, titanium, iron, thorium, bismuth, aluminium, etc. do not interfere, but
large concentrations of fluoride do.

The complexometry of phosphate can be completed by further methods
based on the same principle.   Phosphate can be quantitatively
precipitated as $FePO_4$ [24,25] or as $AlPO_4$ [26] and the excess of
cation determined complexometrically.   Selig [27] determined
phosphorus (up to 6 mg) in organic compounds after combustion by the
Schöniger method, by precipitating phosphate with 0.02M cerous
nitrate (in 0.01M nitric acid) and titrating the excess of reagent at
pH 5 - 6 with 0.02M EDTA, using Xylenol Orange or Methylthymol Blue as
indicator.   The standard deviation was 0.1% for a phosphorus content
of 10%.

7.3.6   Determination of Phosphate by Precipitation Titration

The excellent indication properties of the common metallochromic
indicators also permit some direct precipitation titrations of
phosphate with magnesium or another suitable cation.   These are not
complexometric titrations, but are often used in conjunction with
them.   For example, Bakács-Polgár [28] titrated phosphate with
magnesium solution in an ammonia buffer solution (pH 10), with Erio T
as indicator, but the determination could be used only for pure
phosphate solutions.   Later [29], the same author used a combination
of DCTA and EDTA for binding calcium, magnesium, aluminium and iron
in presence of phosphate.   Excess of the complexing agents was
determined in the usual manner with magnesium solution and Erio T.
After addition of ammonium chloride and ethanol, the phosphate
present was titrated by precipitation with magnesium, the end-point
being indicated by the red-purple colour of the Mg-Erio T complex.
In a similar manner, Szekeres et al. [30] first determined calcium
(or Ca + Mg) by back-titration of excess of EDTA with magnesium
chloride (Thymolphthalexone as indicator).   After addition of
ethanol, the precipitation titration of phosphorus was carried out.
The end-point was indicated by a permanent blue colour.   The authors
masked iron and aluminium with triethanolamine.   The method is useful for simultaneous determination of calcium and phosphate in
natural phosphates.

Conclusion

This chapter on the complexometry of anions has been concerned with
three anions most frequently encountered in practical determinations:
fluoride, sulphate and phosphate.   Where determinations of other
anions are necessary, they will be mentioned in the appropriate
context.

REFERENCES

1. R. Přibil and V. Jelínková, Coll. Czech. Chem. Commun., 1953, 18, 307.
2. F. Huditz, H. Flaschka and I. Petzold, Z. Anal. Chem., 1952, 135, 333.
3. H. Flaschka and A. Holasek, Mikrochem. ver. Mikrochim. Acta, 1952, 39, 101.
4. H. Flaschka and A. Holasek, Z. Phys. Chem. (Leipzig), 1952, 289, 279.
5. H. Eschmann and R. Brochon, Chemist-Analyst, 1956, 45, 38.
6. R. Přibil, unpublished results.
7. H. Saint-Chamant and R. Vigier, Bull. Soc. Chim. France, 1954, 180.
8. H. Eschmann and R. Brochon, Chimia, 1956, 10, 58.
9. R. Brochon and H. Eschmann, Mitt. Lebensmitt. Hygiene, 1956, 47, 155.
10. M. Ishibashi and M. Tabushi, Japan Analyst, 1958, 7, 376.
11. H. Buss, H.W. Kohlschütter and M. Preiss, Z. Anal. Chem., 1963, 193, 264.
12. H. Buss, H.W. Kohlschütter and M. Preiss, Z. Anal. Chem., 1963, 193, 326.
13. H. Buss, H.W. Kohlschütter and M. Preiss, Z. Anal. Chem., 1964, 204, 97.
14. S. Ueda, Y. Yamamoto and H. Wakizaka, J. Chem. Soc. Japan, Pure Chem. Sect., 1961, 82, 873.
15. K. Riedel, Z. Anal. Chem., 1959, 168, 106.
16. J. Kinnunen and B. Wennerstrand, Chemist-Analyst, 1957, 46, 92.
17. M. Furuya and M. Tajiri, Japan Analyst, 1963, 12, 288; Anal. Abstr., 1964, 11, 4737.
18. G.F. Kirkbright, A.M. Smith and T.S. West, Analyst, 1969, 94, 321.
19. R.M. Dagnall, K.C. Thompson and T.S. West, Analyst, 19 8, 93, 72.
20. O. Budevsky, L. Pencheva, R. Russinova and E. Russeva, Talanta, 1964, 11, 1225.
21. B.C. Sinha, S. Das Gupta and S. Kumar, Analyst, 1968, 93, 409.
22. B.C. Sinha and S. Das Gupta, Analyst, 1967, 92, 558.
23. R. Slumper and P. Metlelock, Compt. Rend. 1947, 224, 122.
24. C. Liteanu, I. Crişan and L. Truta, Stud. Univ. Babes-Bolyai Ser. Chim., 1963, 2, 31.
25. C. Liteanu and I. Crişan, Stud. Univ. Babes-Bolyai Ser. Chim., 1963, 2, 39.
26. C. Liteanu and I. Crişan, Stud. Univ. Babes-Bolyai Ser. Chim., 1963, 2, 45.
27. W. Selig, Z. Anal. Chem., 1969, 247, 294.
28. E. Bakács-Polgár, Mag. Kem. Foly., 1954, 60, 229; Anal. Abstr., 1955, 2, 1498.
29. E. Bakács-Polgár, Z. Anal. Chem., 1959, 167, 353.
30. L. Szekeres, E. Kardos and G.L. Szekeres, J. Prakt. Chem., 1965, 28, 113.

CHAPTER 8

Analytical Applications

The following chapters are intended to give a wide-ranging review of
the use of complexometry in practice, but the task is not an easy one,
for complexometry has penetrated all fields of inorganic analysis,
and to satisfy all the informational requirements of specialists
would be very difficult.   Complexometry has developed in different
fields with uneven speed, depending both on interest in this titrimetric method and on the demands for particular types of analysis.

Complexometry is well worked out for analysis of alloys, silicates,
slags, inorganic raw materials and industrial products such as
cements, glasses, ceramic materials, ferrites, etc.   In other fields,
e.g. in analysis of electroplating solutions, paint pigments, fuels
or foodstuffs, interest is concentrated on the determination of a
limited number of elements.   A still narrower spectrum of elements
is encountered in clinical analysis, where only calcium and occasion-
ally magnesium and phosphorus are determined complexometrically.   The
application of complexometry in pharmaceutical or organic analysis is
limited to the compounds which either contain a metal or form
insoluble metal compounds.   Complex-forming or reductive properties
can also be utilized for determination of certain other compounds.

As in every new field, we find attempts made to use complexometry for
the sake of doing so, even when other methods are available and more
suitable.   Such applications will either not be presented at all, or
only in abbreviated form for the reader's information.

Each chapter will deal with a particular field of application, rather
than with determination of one particular element.   This division is
considered advantageous for it enables the specialist to select
quickly the section of interest.   The extent of individual chapters
varies considerably, of course, according to the importance of
complexometry in the field dealt with.

CHAPTER 9

Analysis of Alloys

Innumerable alloys are produced in the world today. They are used by almost all industries, ranging from the manufacture of parts for micro-electronics to nuclear power plant construction.

The requirements concerning the quality of alloys are increasing all the time and include not only their physical properties - mechanical strength, flexibility, metallurgical properties, thermal and electrical conductivity, fire resistance - but also their chemical properties, such as their resistance towards water, the atmosphere, chemicals, corrosive gases at high temperature, etc. Analytical testing of pure metals and their alloys is therefore necessary. Analytical chemistry has kept abreast with the development of technology and exploits a whole range of modern physical techniques requiring costly instruments that can sometimes solve in a matter of seconds difficult analytical problems, such as direct control of the alloy composition during the smelting process. Nevertheless, these instrumental methods are still far from completely replacing chemical procedures, to which complexometry makes an important contribution, and chemical procedures will always be needed for analysis of the standards used for calibrating the instruments.

Demands for the analysis of alloys are very diverse. When there is only one principal component (binary alloys, ferro-alloys), it is frequently the only component that needs to be determined. For other alloys the analysis is often limited to determination of a few components or of trace elements which affect adversely the properties of the particular alloy. The combination of complexometry with spectrophotometric, spectrographic, polarographic and other methods can make the complete analysis of alloys substantially faster than can be achieved by the use of classical methods alone.

9.1 ALUMINIUM AND ALUMINIUM ALLOYS

In the complexometric analysis of aluminium alloys great interest has arisen in the determination of magnesium and zinc, even in low concentrations. Determination of lead, bismuth and manganese has also be reported. In all these alloys aluminium is the principal

239

component and is not determined.   Analysis of binary alloys of
aluminium with titanium, lanthanum, lutetium etc. has been reported.

In some types of alloys, the aluminium is present in minor concentration, and sometimes even in traces.   Analysis of such materials will
be discussed in the appropriate sections.

9.1.1   Determination of Magnesium

Aluminium alloys are often decomposed with sodium hydroxide solution.
During the dissolution aluminium, zinc and also silicon and traces of
calcium are separated from iron, nickel, cobalt, copper, titanium,
manganese and magnesium, which remain undissolved.   Because the
amount of magnesium in aluminium alloys varies widely, the weight of
sample (and the amount of alkali used for dissolution) should be
chosen according to the type of material.   Useful data on the
dissolution of Al-Mg alloys are summarized in Table 9.1.

TABLE 9.1   Dissolution of aluminium alloys

Reference	Mg (%)	Sample weight (g)	30% NaOH (ml)	40% NaOH (ml)
Sergeant [1]	0.5	2	50	
	0.5 - 1.0	1	35	
	1 - 3	0.5	25	
	3 - 12	0.2	15	
Študlar and Janoušek [2]	0.1	5	50	
	0.1 - 1.0	1 - 2	30	
	1 - 3	0.5	30	
	3 - 6	0.25	15	
	6	0.2	10	
Jackson et al. [3]	1	2		25
	1 - 5	1		25
	5 - 15	0.5 - 0.2		25 - 20

An interesting method for dissolution of Al-Mg alloys was proposed by
Flaschka and Myers [4], with the object of avoiding contamination or
losses during the dissolution process.   The sample is placed in a
vacuum desiccator together with a beaker containing concentrated
hydrochloric acid, and the desiccator is partly evacuated.   The
hydrochloric acid vapour will then dissolve the sample.   Copper
alloys can be dissolved in a similar way with nitric acid vapour.
According to the authors this procedure can also be applied in
silicate analysis, for dissolving the "cake" from fusion of the
sample with sodium carbonate or other suitable flux.

The further procedure also depends on the composition of the alloy.
According to Jackson et al. [3], alloys containing iron and manganese
(less than 0.25%) should be dissolved with sodium hydroxide to which
iron(III) sulphate has been added (5 ml of 0.05% solution).   The

addition of iron facilitates the retention of manganese during the
dissolution process. When the alloy contains more than 4% silicon
the dissolution should be carried out without heating [3].   On the
other hand, Boden [5], and Matelli and Vicentini [6] recommend
dissolving the alloys with acids, and separating the SiO_2 produced.
After the alkaline dissolution of the alloy the heavy metals and
magnesium are filtered off, washed and dissolved in acid, and further
separation is then usually necessary. This is done by precipitation
with ammonia or zinc oxide. Some difficulties may arise in quantitative precipitation of manganese. A series of separation procedures
described in the literature was studied by Študlar and Janoušek [2].
For the separation of manganese from magnesium they do not recommend
the precipitation of manganese with hydrogen peroxide or ammonium
sulphide in alkaline medium, because magnesium is easily adsorbed on
the precipitate. The reaction of manganese with permanganate
recommended by Sajó [7] was found quantitative but is not suitable
for higher amounts of manganese, since the large volume of precipitate formed has a strong adsorption ability. The oxidation of
manganese with bromine water was recommended instead [2].

The procedures discussed above are very time-consuming and several
filtrations are usually necessary. A very promising method is based
on the masking of aluminium and other metals. Berák [8] recommends
masking iron and aluminium with triethanolamine, and other elements
(including up to 1 mg of manganese) with a large amount of potassium
cyanide (1 g), after dissolution of the alloy in hydrochloric acid.
However, Berák did not notice the interesting reaction between the
Mn - TEA complex and potassium cyanide in alkaline solution, resulting
in formation of the very stable cyanide complex of tervalent manganese (see p. 63). In this way even large amounts of manganese can be
masked, and this has been used in the analysis of aluminium alloys,
by Burke [9] and Fog [10].

Usually the presence of calcium in aluminium alloys is ignored
because it is generally present only in traces. If necessary it can
be separated by precipitation with oxalate [11].

The magnesium is finally titrated with EDTA, with Eriochrome Black T
as indicator. According to Boden [5], who checked the method
recommended by the American Society for Testing Materials for
complexometric determination of magnesium in aluminium alloys, a
sharper end-point is obtained by DCTA titration, with Methylthymol
Blue as indicator. A similar procedure was adopted by Matelli and
Vicentini [6]. Smart [12] recommends the use of Calmagite instead
of Erio T, especially when chromium is present as chromate.

Some detailed procedures are given below.

Determination of magnesium according to Sergeant [1]. Dissolve a
suitable weight of sample (see Table 9.1) in sodium hydroxide. When
digestion is complete, dilute to about 200 ml with water, add a small
amount of filtration aid or paper pulp, and bring to the boil.
Filter off on a sintered-glass funnel of medium porosity, with a pad
of paper pulp on top of the fritted disc. Wash 4 or 5 times with
hot water made just alkaline to litmus with sodium hydroxide, to
ensure maximal removal of aluminium, silicon and zinc. Moisten the
pad with water and with a thin glass rod transfer it to the original
beaker. Rinse the funnel and glass rod into the beaker with hot
water, add 15 ml of 5% sulphuric acid and simmer for 1 - 2 min.

Filter the solution through the original funnel, and wash four times with a small amount of hot water; transfer the filtrate to the original beaker. Add 6 - 8 g of ammonium chloride and 5 ml of saturated bromine water, boil for 1 - 2 min, and add sufficient concentrated ammonia solution to decolorize the solution. Finally add 20 ml of bromine water, 2 - 3 g of sodium acetate and a little paper pulp, and boil for 2 min.

Filter off on a pulp pad, wash four times with a small amount of hot water, and transfer the filtrate to the original beaker. Add 5 ml of concentrated ammonia solution and sufficient potassium cyanide to complex any copper and nickel, and titrate with 0.05M EDTA to the blue end-point, using Eriochrome Black T as indicator.

Remark. Essentially the same method was described by Študlar and Janoušek [2].

Determination of magnesium according to Wakamatsu [11]. Dissolve 1 g of sample in 15 ml of 20% sodium hydroxide solution and 15 ml of 20% potassium cyanide solution. Add 2 g of sodium carbonate, filter, and wash the precipitate with water. Dissolve the residue in 20 ml of hydrochloric acid (1 + 1), add 2 ml of concentrated nitric acid and boil for 2 min. Make ammoniacal with ammonia solution (1 + 1), add 5 ml of saturated ammonium oxalate solution, adjust to pH 5.5 - 6.5 and boil for 2 min. Add 15 ml of 20% potassium cyanide solution, let stand for 2 min, adjust the pH to 10 with ammonia and titrate with 0.05M EDTA, using Erio T as indicator.

Determination of magnesium according to Berák [8]. Dissolve 1 g of sample in 20 ml of hydrochloric acid (1 + 1), evaporate to dryness, and take up the residue in 200 ml of water (the dissolution can be done directly in a 200-ml standard volumetric flask). Dilute a 20-ml aliquot with 80 ml of water, add a few crystals of hydroxylamine hydrochloride and cool to about 15°C. Add 10 ml of pure triethanol-amine and immediately 10 ml of buffer solution (11 g of NH_4Cl + 70 ml of concentrated NH_3 in 200 ml of water), then cool again in a water-bath till the solution becomes almost clear. Add 10 ml of freshly prepared 10% potassium cyanide solution and titrate very slowly with 0.05M EDTA, using Erio T as indicator. Near the end-point wait awhile after every drop, because the formation of the Mg-EDTA complex is a little sluggish in the cold.

Determination of magnesium according to Burke [9]. Mix about 2 g of the sample with 1 g of sodium cyanide, 10 - 20 ml of triethanolamine, 5 - 6 g of sodium hydroxide and 10 - 20 ml of water. After the reaction has subsided, heat until dissolution is complete, then add dropwise 5 ml of 30% hydrogen peroxide, dilute with 150 ml of water and digest for 10 - 15 min. Filter off the precipitate of magnesium hydroxide and wash it with hot water. Dissolve the precipitate in 40 ml of hot hydrochloric acid (1 + 1), add a few drops of hydrogen peroxide, boil off the excess of acid (but do not evaporate to dry-ness) and dilute with water to standard volume. To an aliquot containing 5 mg of magnesium add 10 ml of 10% triethanolamine solution, 0.2 g of ascorbic acid, 20 ml of ammonia buffer solution (pH 10), and 1 g of potassium cyanide. Titrate with 0.05M EDTA, with Erio T as indicator.

Determination of magnesium according to Fog [10]. The method is a

combination of those just described.   It is based on decomposition
of the sample with sodium hydroxide solution.   The residue is
dissolved in acid and aluminium, iron and manganese are masked with
triethanolamine, potassium cyanide and hydrogen peroxide, according
to the method of Povondra and Přibil [13].   The titration is done
with 0.01M EDTA and Erio T as indicator.

The British Standards Institution has published a slightly different
method for 0.1 - 5.0% of Mg in Al-alloys [14] based on collection of
the magnesium with ferric hydroxide.   The precipitate, containing
iron, aluminium, manganese and titanium, is dissolved and iron and
aluminium are separated by the zinc oxide method.   The final
solution is treated with KCN, ammonia and EGTA (for masking of Mn)
and titrated with DCTA, Methylthymol Blue being used as indicator.

The Burke and Fog methods [9,10] are faster than the others, and with
use of DCTA as titrant [14] (see p. 209) they could be the most
convenient for the determination of magnesium in aluminium alloys.
The principle described by Fog seems the simplest, because it needs
only one filtration.

9.1.2   Determination of Zinc in Aluminium Alloys

According to Faller [15], when an aluminium alloy is decomposed with
sodium hydroxide solution the residue cannot be satisfactorily washed
and always retains 0.1 - 0.3 mg of zinc.   On the other hand, with
alloys containing manganese, 0.2 - 0.3 mg of this element is always
found in the alkaline solution.   Alloys containing 2% of copper are
difficult to dissolve in potassium hydroxide solution and therefore
Faller recommended the use of potassium hydroxide solution containing
potassium sulphide.   The precipitated sulphides of zinc, iron,
copper and manganese are then dissolved and the metals determined.
Faller's procedures are rather tedious, and rather faster is the
procedure of Sergeant [16], who dissolves the alloy in hydrochloric-
nitric acid mixture and precipitates the sulphides of zinc and copper
at pH 5 - 7 in the presence of tartaric acid.   After dissolution of
the precipitate, iron and aluminium are precipitated with ammonia and
zinc is determined complexometrically in the filtrate.   These
classical methods are of course tedious and laborious.   A number of
other methods have been proposed for the separation of zinc.   Hunter
and Miller [17,18] separated zinc by ion-exchange on Amberlite IRA-
400 from 2M hydrochloric acid medium.   A similar separation using
Amberlite IRA-410 (Cl-form) and Amberlite Ir-120 $(NH_4^+$ form) was
worked out by Kadama and Kenie [19].   Freegarde [20] used a De-
Acidite FF column to separate zinc from 0.12M hydrochloric acid
medium containing 10% of sodium chloride.   The strong base type
exchanger retains only zinc and copper.   After elution of copper
with 2M hydrochloric acid, zinc can be eluted with 1M nitric acid and
determined complexometrically.   Magnesium, aluminium, iron,
manganese, nickel, titanium and silicon do not interfere, but tin
does and must be separated.   According to the author, the method is
suitable for the determination of 0.1 - 6% of zinc in aluminium alloys
and its precision is the same as that of classical methods.   The
method was worked out in 1958 and much later was accepted (with some
minor modifications) as a British standard procedure [21].   According
to the modified procedure zinc is separated on De-Acidite FF from 2M
hydrochloric acid and eluted with 0.001M hydrochloric acid.   The
method is recommended for complexometric determination of 0.1 - 12% of

zinc (for 0.01 - 0.25% of zinc a polarographic determination is recommended).

In the last few years extraction methods have enjoyed increasing popularity since under optimal conditions they are fast and have good selectivity.   Several procedures have been proposed for the extraction of zinc in analysis of Al-alloys.   Ishibashi and Komaki [22] use a 15% solution of Amberlite LA-1 or Amberlite LA-2 in kerosene for extraction of zinc from 3M hydrochloric acid.   The organic phase is stripped with 0.3M nitric acid.   The co-extracted traces of aluminium and copper are also stripped and must be masked with triethanolamine and potassium cyanide, which also masks zinc.   After demasking with 4% formaldehyde solution, the zinc is titrated with EDTA at pH 9 - 10, with Erio T as indicator.   Iron interferes and must be extracted beforehand with methyl isobutyl ketone (MIBK).

Onishi [23] recommends extraction of zinc from 1.5 - 3.0M hydrochloric acid with 0.1M trioctylphosphine oxide (TOPO) solution in benzene. The organic phase is washed with dilute perchloric acid (1 + 15). After addition of ammonium fluoride and sodium thiosulphate (to masking aluminium and copper) and pH adjustment to 5.5, zinc is titrated with 0.01 - 0.005M EDTA, with Xylenol Orange as indicator.

Graffmann and Jackwerth [24] extract zinc as its thiocyanate complex into MIKB.   The organic phase also contains traces of aluminium and other metals, but these can be masked with triethanolamine and potassium cyanide.   Demasking of the zinc from its cyanide complex is necessary.   Cadmium, if present, is titrated together with zinc. The final step is titration with 0.1M EDTA at pH 9, zincon being used as indicator.   A determination takes 15 min and 5 or 6 can be done per hour.   The method was modified by Naumann and Schröder [25] who extracted zinc into MIKB and then stripped it with hydrochloric acid (1 + 1).   After evaporating the bulk of the resulting aqueous solution, they adjusted the pH to 5, masked copper with 2-mercapto-propionic acid, and titrated the zinc with 0.05M EDTA, using Xylenol Orange as indicator.

Zinc can also be extracted from weakly acid chloride solution (pH 5) with a chloroform solution of trioctylmethylammonium chloride (Aliquat S-336), aluminium being masked with fluoride and copper with thiourea [26].

In the older methods the zinc had to be titrated in alkaline solution, with Erio T as indicator.   The presence of copper (even in traces) required its masking with potassium cyanide and demasking of zinc with formaldehyde.   The newer methods using weakly acid medium in the presence of masking agents [NH_4F, $Na_2S_2O_3$, $SC(NH_2)_2$ etc.] are more selective and do not use potassium cyanide.

Several procedures for zinc determination are given below in chronological order of development.

Determination of zinc according to Sergeant [16].   After dissolution of the sample, zinc and heavy metals are precipitated as the sulphides;   these are dissolved in acid and aluminium and iron are separated as the hydroxides.   After masking of copper, zinc is titrated with 0.05M EDTA, with Erio T as indicator.

Add to the sample (in a 400-ml tall beaker), a few drops at a time,

5 ml of concentrated hydrochloric acid for every 0.5 g of sample,
followed by an equal amount of concentrated nitric acid.   When the
violent reaction has subsided wash down the cover glass and the
beaker wall with a small amount of water, place the beaker on the
hot-plate and boil for 1 - 2 min.   Then dilute to 50 - 60 ml with cold
water, add 3 g of tartaric acid per 0.5 g of sample, and neutralize
with ammonia, using Methyl Red as indicator.   Re-acidify the
solution with 10 ml of nitric acid (1 + 10), add sufficient 5% sodium
carbonate/10% sodium sulphide solution and adjust the pH to 5 - 7 if
necessary.   Add a little paper pulp and boil for 1 - 2 min on the
hot-plate.   Filter off on a pulp pad, and wash 4 or 5 times with
dilute carbonate/sulphide solution.   Dissolve the precipitate off
the pad with 15 ml of hot dilute hydrochloric acid, and wash through
with hot water.   Transfer the filtrate to the original beaker, add
2 - 3 g of ammonium chloride and sufficient bromine water to colour
the solution, bring to the boil to oxidize sulphur, and make just
ammoniacal to precipitate the iron and aluminium.   Filter off on a
pulp pad and wash with ammonium chloride solution.   Transfer the
filtrate to the titration beaker, add 4 ml of 20% potassium cyanide
solution to mask copper, and Erio T as indicator.   If the solution
turns blue add 10 ml of 20% formaldehyde solution and then titrate
with EDTA to a slightly blue or greenish end-point.   The whole
procedure takes 20 - 25 min for one sample.

Determination of zinc according to Freegarde [20].   The sample
solution is passed through a column of De-Acidite FF, aluminium and
any copper are eluted, and then zinc is eluted with 1M nitric acid.

The columns used are 1 cm in diameter, and about 25 cm long.   A
slurry of the resin (15 - 16 mesh) in 0.5M hydrochloric acid is poured
into the column to give a resin bed about 15 cm long, which is
successively washed with 100 ml of 1M nitric acid and then 100 ml of
0.12M hydrochloric acid containing 150 g of sodium chloride per litre
(solution A).   A suitable weight of sample, containing up to 40 mg
of zinc, is dissolved in a mixture of 10 ml each of concentrated
hydrochloric acid and nitric acid with 20 ml of water (with warming
if necessary) and the solution is then evaporated to dryness.   The
residue is taken up in solution A and any insoluble material filtered
off.   The solution is run through the column at a flow-rate of 6 ml/
min.   The column is eluted with 100 ml of solution A to remove
aluminium and then with 2M hydrochloric acid to remove any copper
(20 ml are sufficient for aluminium alloys, and 100 ml for copper-
base alloys).   Finally zinc is eluted with 250 ml of 1M nitric acid
and titrated with EDTA (PAN as indicator).

Determination of zinc according to Graffmann and Jackwerth [24].
After dissolution of the sample, zinc is extracted as the $Zn(SCN)_4^{2-}$
complex (together with traces of iron and copper) into MIBK.   After
stripping, zinc is determined by EDTA titration, with zincon as
indicator.

Dissolve 1 g of the alloy, containing 1 - 6% of zinc, in 20 ml of
hydrochloric acid (1 + 1) and at the end add a few drops of 30%
hydrogen peroxide.   Boil the solution until it is clear and
evaporate most of the acid, cool, adjust the volume to 60 - 70 ml, add
10 ml of 8M ammonium or sodium thiocyanate and extract with 50 ml of
MIBK.   Discard the aqueous phase, wash the organic phase with 25 ml
of 1M thiocyanate solution and transfer the organic phase to the
titration flask, diluting it with 100 ml of water.   Add a few drops

of triethanolamine, a little ascorbic acid and 25 ml of pH-9 buffer
solution (see below) and 5 ml of 10% potassium cyanide solution. Add
zincon as indicator; the colour must be orange. Add 5 - 10 ml of
30% formaldehyde solution (the colour changes to blue). Titrate
with 0.1M EDTA to the orange end-point.

The buffer is a 2.2% H_3BO_3, 2.6% KCl, 0.6% NaOH solution, the
indicator is a 1:100 mixture of solid zincon and sodium chloride.

For analysis of alloys containing 0.1 - 1.0% Zn a 5-g sample is used
and the quantities of the reagents are modified accordingly.

A single determination takes 15 min. In routine analysis 5 or 6
samples can be analysed per hour.

Determination of zinc according to Naumann and Schröder [25]. Zinc
is extracted as $Zn(SCN)_4^{2-}$ as in the method above, but the zinc is
stripped without the use of potassium cyanide.

Dissolve 4 g of sample in 80 ml of hydrochloric acid (1 + 1) and 1 - 3
ml of hydrogen peroxide. Add 50 ml of water, filter off the residue
and dilute the filtrate to volume with water in a 200-ml standard
flask. Take a 50-ml aliquot (for 1 - 5% zinc) or 25 ml (for 2.5 - 10%),
add 30 ml of 50% ammonium thiocyanate solution and extract successively with 50 and 30 ml of MIBK. Wash the organic phase with 100 ml
of a 17:3 v/v mixture of 5% NH_4SCN and 20% NH_4F solutions and finally
with 30 ml of 5% NH_4SCN solution. Strip zinc from the organic
phase with two 50-ml portions of hydrochloric acid (1 + 1) and
evaporate the combined extract to 10 ml. Dilute with 50 ml of water,
adjust the pH to 5 with hexamine and add 3-mercaptopropionic acid
dropwise till the disappearance of any blue colour and then three
more drops. Titrate the zinc with 0.05M EDTA, using Xylenol Orange
as indicator. The error is 2% for a zinc content of 2 - 5%.

The older methods for the determination of zinc in aluminium alloys
(Sergeant, Faller) have been critically evaluated by Bem and Rubel
[27] who also devised several modifications of the zinc titration with
EDTA, using Xylenol Orange, Methylthymol Blue, and Eriochrome Black T
as indicators.

9.1.3 Determination of Aluminium in Binary Al-Zn Alloys (ALZINOY)

These alloys are frequently used for the preparation of galvanizing
baths. The aluminium content is very important, because it affects
the surface quality of the zinc coating. Bhargava [28] does not
recommend methods based on the determination of zinc and subtraction
from 100% to obtain the aluminium content, because of the presence of a
small amount of iron and lead (0.5% of each). He recommends the
method described earlier [29], based on indirect determination of
aluminium after its masking with ammonium fluoride. The method
presented below has been in use for many years in his laboratory.

Weigh a 0.2 g sample (filings or turnings) into a 150-ml beaker.
Add 5 ml of hydrochloric acid (1 + 1) and warm to dissolve. Clear
the solution with a few drops of concentrated nitric acid and boil
for 1 - 2 min. Cool and dilute to volume in a 100-ml standard flask.
Pipette a 10-ml aliquot into 250-ml beaker, and add from the burette
40.00 ml of 0.01M EDTA. Adjust the pH to 4 with 2M sodium hydroxide.
Boil gently for 5 min and cool to room temperature. Rinse the wall

of the beaker with water and add 1 g of hexamine. When this has
dissolved add 7 drops of Xylenol Orange solution and titrate with
0.01M lead solution. Add 10 ml of 10% ammonium fluoride solution,
boil for 2 - 3 min, cool and repeat the titration with 0.01M lead
solution. The difference between the two titration volumes gives
the amount of aluminium.

9.1.4   Determination of Lead and Bismuth in Aluminium Alloys

The reason given by the authors for working out this method was that
lead and bismuth, as components of various alloys, may occur in
aluminium alloys made by reprocessing of waste metals.

Bismuth and lead are extracted as the diethyldithiocarbamates from an
ammoniacal tartrate solution. Interfering elements are masked with
potassium cyanide (and manganese with EGTA).

Dissolve 2 g of sample in 40 ml of acid mixture (700 ml of water +
730 ml of conc. HCl + 500 ml of conc. HNO$_3$). After filtration (if
necessary) add 15 g of tartaric acid and 75 of concentrated solution.
Cool, then add 2 g of potassium cyanide and 5 ml of 0.05M EGTA.
Transfer the solution into a 250 - 400 ml separating funnel and
precipitate lead and bismuth with 5 ml of 2% sodium diethylthiocarbamate solution. Extract with four 10-ml portions of chloroform. Mix
the combined extracts with 10 ml of nitric acid (1 + 1) and evaporate
to 3 ml. In the solution titrate bismuth at pH 1.5 directly with
EDTA and Xylenol Orange. To determine lead add 1 g of ammonium
fluoride, 1 g of thiourea and excess of EDTA solution, adjust the pH
to 5.5 - 6 with hexamine and titrate the surplus EDTA with lead
solution.

9.1.5   Analysis of Binary Alloys of Aluminium and Titanium

A few methods have been proposed for the analysis of these alloys,
covering a very wide range of Ti:Al ratios.

Determination of titanium in Al-Ti alloys according to Přibil and
Veselý [31].   Dissolve 0.5 - 1.0 g of the alloy in 40 ml of sulphuric
acid (1 + 4) and a few drops of nitric acid (1 + 1) by boiling. Cool
and transfer to a 500-ml standard flask and dilute to the mark with
water.

To an aliquot of sample solution (containing up to 110 mg of Al and
40 mg of Ti) add 10 - 20 ml of sodium salicylate solution, 20 - 40 ml
of 20% triethanolamine solution (depending on the amount of Al
present), 2 drops of phenolphthalein and 2M sodium hydroxide until a
violet colour appears. Adjust the solution to pH 1 - 2 with nitric
acid (1 + 1), using indicator paper, and cool to below 20°C. Add
10 ml of hydrogen peroxide, a measured volume of 0.05M EDTA and a few
drops of Xylenol Orange, and back-titrate with 0.05M bismuth nitrate
to a red colour. The alkalization of the solution is necessary,
because the very stable salicylate complex of aluminium is not otherwise formed; this complex remains stable after the acidification
[32].   At the maximum levels of titanium (40 mg) and aluminium
(150 mg) the solution must be diluted to 300 - 400 ml before titration
because of the intense yellow colour of the peroxo-titanium-EDTA
complex.

For the determination of less than 0.1% of titanium a larger sample
must be used. The resulting large amount of aluminium cannot be
masked properly and the titanium must be separated by precipitation
with sodium hydroxide, in the presence of triethanolamine, which also
reliably masks any iron in the sample.

To an aliquot containing 2 - 40 mg of titanium add 20 - 50 ml of 20%
triethanolamine solution. If some precipitate appears, dissolve it
in dilute nitric acid. Add an excess of 2M sodium hydroxide. If
the sample contains a large amount of iron, the solution is initially
yellow but this colour disappears after a while (formation of the
colourless Fe-TEA complex). Boil for 1 min and let stand on a hot-
plate for 10 - 15 min, then filter off and wash 5 times with 1% TEA
solution and twice with hot water. Dissolve the precipitate in 25 -
30 ml of hot nitric acid (1 + 3) and wash the filter thoroughly with
hot water. The titanium solution thus obtained may contain traces
of aluminium, so continue with the determination as in the first
procedure. This method is also suitable for the separation of
titanium from aluminium and iron.

Analysis of binary Al-Ti alloys according to Culp [33]. The sum of
aluminium and titanium is determined by back-titration of excess of
EDTA with 0.05M bismuth nitrate at pH 5 - 5.5, with Xylenol Orange as
indicator. Aluminium is determined in another sample after masking
of titanium with lactic acid.

Weigh 0.5 g of powdered alloy and transfer it to a 400-ml beaker.
Add 40 ml of hydrochloric acid (1 + 1), cover and let stand until
vigorous action stops, then warm gently until all action ceases. Add
5 ml of concentrated nitric acid to destroy carbides and heat
gradually to boiling to expel brown fumes. If any insoluble
material remains, filter off on Whatman 41H paper, ignite the washed
paper and precipitate and fuse the residue with potassium pyrosulphate in a platinum crucible. Dissolve the cooled melt in water and
add to the filtrate in a 250-ml standard flask and make up to volume.

To determine Al + Ti, to a suitable aliquot of this solution, in a
500-ml conical flask, add a measured volume of 0.05M EDTA and dilute
with water to about 200 ml. Boil the solution for about 10 min and
cool nearly to room temperature. Adjust to pH 5.0 - 5.5 with solid
hexamine and cool to below 15°C. Add 6 or 7 drops of 0.5% Xylenol
Orange solution and titrate the excess of EDTA slowly with 0.05M
bismuth nitrate to a "pink-orange end-point", adding the titrant
dropwise and swirling for several seconds after each drop, as the
end-point is approached.

To determine the Al, to a suitable aliquot of the sample solution in
a 500-ml conical flask add 5 ml of 80% lactic acid solution (for up
to 35 mg of Ti) and a measured volume of 0.05M EDTA, and dilute to
200 ml. Boil for 10 min, then cool to room temperature. Adjust
the pH, cool and titrate as above, to a pink-orange colour which
fades rapidly (10 - 15 sec) on standing.

The fading of the indicator colour can be attributed either to
displacement of aluminium from its EDTA-complex by bismuth (the
aluminium released reacts very slowly with the indicator or to
hydrolysis of the bismuth indicator complex.

Determination of aluminium in Al-Ti alloys according to Lazarev [34].
Titanium is masked with tartaric acid and aluminium is determined
indirectly by back-titration of excess of EDTA with 0.05M zinc
acetate (Xylenol Orange as indicator).

Dissolve 0.2 g of alloy in hot sulphuric acid (1 + 3) and a few drops
of nitric acid.   Dilute to 60 - 70 ml with water, add 25 ml of 0.05M
EDTA (accurately measured) and 10 ml of 20% tartaric acid solution.
Add the indicator, followed by ammonia solution (1 + 1) until a lilac
colour appears, then just sufficient hydrochloric acid (1 + 1) to
make the colour yellow.   Add 8 ml of 50% sodium acetate solution and
titrate the excess of EDTA with 0.05M zinc acetate.

9.1.6   Analysis of Binary Alloys of Aluminium and Lanthanides

Brück and Lauer worked out two methods for the determination of
lutetium [35] and dysprosium [36] in amounts exceeding 1% in the
experimental alloys with aluminium.   Both methods are based on
direct titration of the lanthanide with EDTA, Xylenol Orange being
used as indicator.   Aluminium must be masked with sulphosalicylic
acid.   The conditions for the titration are citical.   The pH at the
beginning of the titration should be 5.5 $\pm$ 0.1 for lutetium and 6.0 $\pm$
0.3 for dysprosium.   The sulphosalicylic acid added must also be
carefully measured.   However, Milner and Gedansky [37] do not
recommend masking with sulphosalicylic acid when Xylenol Orange is
used as the indicator for EDTA titration of rare-earth elements,
because the end-point colour change is poor.   They recommend acetylacetone instead, since the amount added (0.5 ml per 10 mg of Al) is
less critical and a 2 or 3-fold excess has no adverse effect.   However, sodium hydroxide (not ammonia) must be used for the preliminary
neutralization before pH adjustment with hexamine (otherwise a yellow
colour develops).   The titration is done directly with 0.05M DTPA.
According to these authors [37] this method is more reliable than the
classical gravimetric (oxalate) method, which always gives low
results due to losses during filtration.   Milner and Gedansky
frequently found up to 3 mg of lanthanide in the filtrate from the
oxalate precipitation.

9.1.7   Analysis of some Other Aluminium Alloys

A number of procedures can be found in the literature for complexometric analysis of various experimental alloys.   Tramm and Pevzner
[38], for example, analysed binary zirconium-aluminium alloys.
Since the complexometric behaviour of zirconium is somewhat different
from that of other metals, the analytical procedure is given below in
detail.

Fuse 0.1 g of Zr-Al alloy with 2 g of potassium pyrosulphate.
Dissolve the cooled melt in hydrochloric acid (1 + 9) and filter off
aluminium and zirconium hydroxides after precipitation with ammonia.
Dissolve the hydroxides in 60 ml of hot hydrochloric acid (1 + 5) and
boil for 5 min, keeping the original volume by adding water.   Add
1 ml of 0.1% Xylenol Orange solution and titrate immediately with
0.05M EDTA to a yellow colour.   If the pink colour reappears, warm
the solution again and finish the titration.   Cool the solution and
add a measured volume of 0.05M EDTA (2 - 5 ml more than expected for
complexing the Al present.   After 10 - 15 min adjust the pH to 4 with

ammonia, dilute to 100 ml and titrate with 0.05M thorium nitrate to a
red colour.

Note.   The method does not seem to be very reliable.   The presence
of hydrochloric acid is not favourable in the determination of
zirconium (see p. 93).   The formation of the Al-EDTA complex at room
temperature is very slow (optimal pH about 3.5), though it forms more
easily on boiling.   This may be why this problem was re-examined
nine years later by Tramm et al. [39], who produced the following
method, which seems more reliable.

Dissolve 0.1 g of the sample in 25 ml of concentrated sulphuric acid,
dilute with 100 ml of 2% ammonium sulphate solution and boil for 3
min.   Titrate the hot solution with 0.05M EDTA (Xylenol Orange as
indicator) from red to yellow.   For the determination of aluminium,
cool the solution and adjust the pH to 4 with ammonia, add 5 ml of 1M
acetic acid, 1 g of hexamine and a measured excessive volume of 0.05M
EDTA, and boil for 5 min.   After cooling, add a little more indicat-
or if necessary, and titrate the excess of EDTA with 0.05M zinc
chloride, from yellow to red.

According to the authors the average error is less than 0.5% for 15%
aluminium and 1.5% for 85% zirconium.   It is left to the reader to
judge the feasibility of this procedure, after reading the section on
zirconium (p. 92).

This account could be extended to include rather specialized analytical procedures.   Some alloys will undoubtedly be considered as
rarities, e.g. aluminium-plutonium or gallium-aluminium-antimony
alloys.   The first is analysed by separating plutonium by ion-
exchange and titrating aluminium complexometrically.   In the second,
the gallium is extracted from 6M hydrochloric acid with butyl acetate,
and the gallium and aluminium are determined by EDTA titration.
Antimony is determined by the usual bromate method.

Aluminium is also present in a large number of other alloys (aluminium bronze, magnetic alloys) where its determination is only a part
of the overall analysis.   In still other alloys aluminium occurs in
only minute quantities, and such determinations will be dealt with in
the appropriate sections.

REFERENCES

1.   J.C. Sergeant, Metallurgia, 1953, 48, 261.
2.   K. Studlar and I. Janoušek, Hutn. Listy, 1958, 13, 543; Anal.
      Abstr., 1959, 6, 1639.
3.   R.C. Jackson, W. Stross and L.H. Wadsworth, Metallurgia, 1958,
      58, 1.
4.   H. Flaschka and G. Myers, Z. Anal. Chem., 1975, 274, 279.
5.   H. Boden, Chemist-Analyst, 1966, 55, 75; Anal. Abstr., 1967,
      14, 6708.
6.   G. Matelli and V. Vicentini, Allum. Nuova Met., 1968, 37, 343;
      Chem. Abstr., 1969, 70, 73914.
7.   I. Sajó, Kohászati Lapok, 1954, 9, 322.
8.   L. Berák, Hutn. Listy, 1957, 12, 434.
9.   K.E. Burke, Anal. Chim. Acta, 1966, 34, 485.
10.   H.M. Fog, Acta Chem. Scand., 1968, 22, 791.

11. S. Wakamatsu, Japan Analyst, 1957, 6, 294; Anal. Abstr., 1958, 5, 382.
12. A. Smart, Metallurgia, 1964, 69, 245; Anal. Abstr., 1965, 12, 4452.
13. P. Povondra and R. Pribil, Coll. Czech. Chem. Commun., 1961, 26, 311.
14. British Standards Institution: BS 1728: Part 22 (1972); Anal. Abstr., 1973, 24, 1481.
15. F.E. Faller, Z. Anal. Chem., 1953, 139, 14.
16. J.C. Sergeant, Metallurgia, 1954, 50, 252; Anal. Abstr., 1955, 2, 875.
17. J.A. Hunter and C.C. Miller, Analyst, 1956, 81, 79.
18. C.C. Miller and J.A. Hunter, Analyst, 1954, 79, 583.
19. K. Kodama and T. Kanie, Japan Analyst, 1955, 4, 627; Anal. Abstr., 1956, 3, 2427.
20. M. Freegarde, Metallurgia, 1958, 58, 261; Anal. Abstr., 1959, 6, 4301.
21. British Standards Institution: BS 1728: Part 18 (1970); Anal. Abstr., 1970, 19, 2114.
22. M. Ishibashi and H. Komaki, Japan Analyst, 1962, 11, 43; Anal. Abstr., 1964, 11, 55.
23. K. Onishi, Japan Analyst, 1969, 18, 1262.
24. G. Graffmann and E. Jackwerth, Z. Anal. Chem., 1969, 244, 391.
25. H.C. Naumann and D. Schröder, Metall, 1970, 24, 942; Anal. Abstr., 1971, 21, 3284.
26. R. Přibil and V. Veselý, Coll. Czech. Chemm. Commun., 1972, 37, 13.
27. J. Bem and S. Rubel, Chem. Anal. (Warsaw), 1973, 18, 869; Anal. Abstr., 1974, 26, 1454.
28. O.P. Bhargava, Talanta, 1975, 22, 471.
29. R. Přibil, Talanta, 1965, 12, 925.
30. R. Crawford and D. Lush, Metallurgia, 1970, 81, 159; Anal. Abstr., 1971, 20, 2347.
31. R. Přibil and V. Veselý, Chemist-Analyst, 1963, 52, 43; Anal. Abstr., 1964, 11, 1670.
32. R. Přibil and V. Veselý, Talanta, 1963, 10, 233, 383.
33. S.L. Culp, Chemist-Analyst, 1967, 56, 29; Anal. Abstr., 1968, 15, 1944.
34. M.M. Lazarev, Tekhnol. Legk. Splavov. Nauchno.-Tekh. Byull. VILS, 1971, 105; Anal. Abstr., 1972, 23, 2405.
35. A. Brück and K.F. Lauer, Anal. Chim. Acta, 1965, 33, 338.
36. A. Brück and K.F. Lauer, Anal. Chim. Acta, 1967, 39, 136.
37. O.I. Milner and S.J. Gedansky, Anal. Chem., 1965, 37, 931.
38. R.S. Tramm and K.S. Pevzner, Zavod. Lab., 1965, 31, 163; Anal. Abstr., 1966, 13, 2921.
39. R.S. Tramm, L.V. Kharchenko and E.G. Namvrina, Nauch. Trudy Nauchno-Issled. Proekt. Inst. Redkometall Prom., 1973, No. 47, 36; Anal. Abstr., 1974, 27, 2528.

9.2   IRON AND ITS ALLOYS

Production of iron, cast iron, high-grade steels and other alloys is
undoubtedly one of the main industrial fields in economically
developed countries.   Analytical chemistry, especially the methods
applied to the steel industry, has a long tradition in this field.
Recently, however, there has been an increase in the use of industrial methods and a trend towards automation of the analytical
procedures.   Many such methods are recommended by the various
national standards institutions, but in some cases even old and more
or less empirical procedures still find use, for instance the determination of manganese in ferromanganese according to Volhard [1] and
Wolff [2].   The introduction of new master alloys, materials suitable for the construction of nuclear reactors (e.g. cobalt-free steel)
etc. requires new reliable and sufficiently fast analytical methods.
For instance the development of malleable alloys containing spheroidal graphite and a small amount of magnesium created a need for the
determination of the latter element, the content of which was not
previously of interest.   Similarly, new demands have arisen for
analytical control during the production of special alloys.   Classical methods are frequently too tedious for the purpose and do not
yield reliable results.   Because of the relatively high content of
the components to be determined, photometric or polarographic methods
cannot be recommended, whereas complexometry has proved to be very
useful, especially in the analysis of binary and magnetic alloys.
These methods will be dealt with in the following sections.

9.2.1   Analysis of Binary Master Alloys

Nowadays a considerable number of these binary alloys with iron are
produced.   They are priced not only according to their purity, but
also according to the content of the second component, the cost of
which many times exceeds that of iron (e.g. vanadium, molybdenum).
The iron is not usually determined.

Analysis of ferromanganese

A number of methods were suggested for the determination of manganese
in ferromanganese shortly after the appearance of complexometry, but
in the course of time they have been replaced by more advantageous
methods.   Some authors recommend removal of iron first, for example
by use of amyl acetate extraction [3], by precipitation as basic
benzoate [4] or by the classical zinc oxide method.   Flaschka and
Püschel [5] recommended masking the iron with potassium cyanide after
its reduction with ascorbic acid, and the masking of aluminium with
tiron.   Masking of both with triethanolamine is better [6].   The
simplest methods of all are based on the successive EDTA titration of
iron and manganese.   A selection of several procedures is given
below.   For determination of aluminium see p. 263.

One such method was described long ago by the author [7].   It is
based on the determination of iron by DCTA titration at pH 2 - 3
(salicylic acid as indicator), followed by addition of triethanolamine, hydroxylamine and ammonia (to give a pH of 10) and titration
of manganese with 0.05M DCTA (Erio T as indicator).   Titration with
EDTA was not possible because the Fe-EDTA complex reacts with Erio T
in ammonia solution.   At that time Thymolphthalexone was not known;
it is superior to Erio T, forming an intensely blue complex with

manganese(II) and giving no interference reaction with the Fe-EDTA
complex.   Rad'ko and Yakimets [8] have used EDTA for both titrations - of iron with sulphosalicylic acid as indicator, and of
manganese at pH 8 - 9 with Fast Mordant Blue B (C.I. 16680).   Sajó [9]
chose the same determination of iron but titrated the manganese at
pH 6 - 7 with the vanadium(V)-EDTA-diphenylcarbazide system as
indicator.

These analytical principles were applied to the determination of both
metals in ferromanganese by Chekhlarova et al. [10], who dissolved
0.5 g of the alloy in 30 ml of nitric acid (1 + 1), then made up the
solution to standard volume.   An aliquot was diluted, adjusted to
pH 2 and titrated with DCTA solution, with sulphosalicylic acid as
indicator.   After addition of potassium tartrate, ammonia and Thymolphthalexone, the manganese was titrated to a smoky-grey end-point.
The relative standard deviation was 0.16% at a manganese content of
60 - 80%.   Spauszus [11] applied the principle of the Sajó method
(using modified indicator preparation) with good results to the
determination of manganese in ferromanganese (75 - 80% Mn) and spiegel-
eisen (12.4%).

Direct determination of manganese in ferromanganese [6]

After dissolution of the alloy, iron(III) is masked with triethanolamine in ammoniacal medium, and after addition of hydroxylamine the
manganese is titrated with EDTA, Thymolphthalexone being used as
indicator.

Dissolve a weighed amount of the alloy (about 0.25 g) in concentrated
nitric acid and then add a small amount of hydrochloric acid.
Evaporate to small volume.   After further addition of the acids,
repeat the evaporation (but not to dryness).   Dilute with water,
transfer to a 250-ml standard flask without filtration of any silicic
acid precipitated, and dilute to volume with water.

To a 25 or 50 ml aliquot, in a 300-ml conical flask, add 5 ml of 10%
hydroxylamine hydrochloride solution, 10 ml of 20% triethanolamine
solution, 30 - 35 ml of concentrated ammonia solution, and dilute to
150 ml.   Add sufficient indicator to secure a clear blue colour and
titrate with 0.05M EDTA to colourless or a slight pink.   Near the
end-point add the EDTA only slowly.   When the end-point is reached,
wait 2 - 3 min to see whether the blue or violet colour reappears.
If it does, continue the titration until a permanent colour change is
clearly visible (usually 1 or 2 drops of EDTA will be required).

Aluminium is also masked with triethanolamine.   Small amounts of
copper and zinc can be masked with 50 mg of potassium cyanide.
Ammoniacal solutions of the Fe-TEA complex are almost colourless;
the pale yellow colour observed at high concentrations is caused by
traces of colloidal ferric hydroxide, but this does not influence the
colour change of the indicator.   If more than 25 mg of iron are
present in 100 ml of solution, the colour may be so intense that the
end-point is a little obscured.   Standing of the solution or slight
warming very often leads to colourless solutions.

This method was carefully tested by Morris [12] and Študlar [13].
Morris found that variation in the results was caused only by
incomplete dissolution of the sample because of careless heating,
undissolved particles of sample (probably carbides) adhering to the

beaker, at the top of the foam formed on the addition of hydrochloric
acid.   Further concentration to small volume causes low results,
possibly associated with the precipitation of silicic acid.   Hence
the details of the sample dissolution must be closely followed (see
below).

Študlar [13] studied the interference from other elements.   Copper
up to 2% does not interfere even without masking (according to my own
experience, copper is partly masked with triethanolamine).   Higher
amounts of zinc, nickel and lead decrease the sharpness of the end-
point colour change.   Zinc, nickel and copper can be masked with
potassium cyanide, but more than 50 mg of it will cause low results.
On the other hand, a small amount of cyanide improves the sharpness
of the Thymolphthalexone end-point.   Both authors [12,13] agree that
the method is precise, as proved by statistical treatment of the
results.   The dissolution procedures recommended by both authors are
given below.

Morris's method [12].   To 0.15 ($\pm$ 0.01) g of the sample (containing
75 - 80% of manganese) in a 250-ml beaker, add 10 ml of concentrated
nitric acid and boil gently until the sample has dissolved.   Continue
boiling until the volume is about 5 ml.   Without cooling, add 10 ml
of concentrated hydrochloric acid and evaporate again to a volume of
5 ml.   Repeat the acid addition once or twice more, boiling down to
5 ml each time, until the final solution is golden yellow.   Cool,
add 100 ml of cold water, 10 ml of 10% hydroxylamine solution, 20 ml
of 33% triethanolamine solution and adjust to pH 10 with aqueous
ammonia.   Add a pinch of 1:100 w/w Thymolphthalexone-KNO_3 mixture
and titrate with 0.05M EDTA to a colour change from green to brick-
red.

Študlar's method [13].   Dissolve a 0.25-g sample in 10 ml of nitric
acid (1 + 1) (dissolution is rapid and complete).   Add 5 ml of
concentrated hydrochloric acid and evaporate to small volume, add
1 ml of concentrated hydrochloric acid and 100 ml of water and bring
the solution to the boil.   Then cool, transfer the solution to a
250-ml standard flask and dilute to the mark.   Take 50 ml for the
determination of manganese according to the original procedure [6],
using 0.025M EDTA.

Bec and Kazior [14] proposed using the Přibil and Kopanica method [6]
for the routine analysis of ferromanganese.   They titrated manganese
with 0.03M EDTA, using Thymolphthalexone as indicator.   The absolute
error in the analysis of standard samples containing 50 - 80% of
manganese was said to be $\pm$0.3%.   The determination takes 45 min.

Jurczyk [4] rather unjustly considers these masking methods less
suitable, regarding the presence of the masking agents as responsible
for the sluggish indicator colour change.   He therefore recommends
precipitation of iron, aluminium, titanium, and chromium, if present,
as basic benzoates, filtration, and titration of the manganese in the
filtrate with EDTA (Erio T as indicator).   The use of masking is not
eliminated, however, for he recommends masking of copper, cobalt and
nickel with potassium cyanide [15].   The method is time-consuming,
mainly because a double benzoate precipitation is required.   The
aluminium content is always very low and triethanolamine should
suffice to mask it.   According to our experience only chromium
(3 - 5 mg) interferes but can be masked by boiling with triethanolamine or ascorbic acid [15].

The Karolev method [16].   This method seems to be very simple and
rapid for routine analysis.   The authors use a special masking
mixture of 25 g of hydroxylamine hydrochloride, 12.5 g of ammonium
fluoride and 6.5 g of sodium potassium tartrate in 500 ml of 10%
triethanolamine solution, adjusted to pH 6.0 with hydrochloric acid.

Dissolve 100 mg of the sample in 20 ml of nitric acid (1 + 1), and
evaporate the solution to 2 - 3 ml.   Add 5 ml of concentrated hydrochloric acid and heat for 2 - 3 min.   Dilute to 100 - 150 ml with
water, add 20 ml of masking mixture, adjust to pH 10 with ammonia and
titrate the manganese with 0.05M EDTA, with Methylthymol Blue as
indicator, from intense blue to smoky grey.

Analysis of ferromolybdenum

Complexometric determinations of molybdenum in ferromolybdenum are
all based on three principles.   The first is indirect determination
of molybdenum(V), which is easily formed by reduction of moly-
bdenum(VI) with hydrazine sulphate in the presence of EDTA (formation
of Mo_2-EDTA complex).   The second is based on formation of a ternary
complex by boiling molybdenum(VI) with hydroxylamine (HyOH) and EDTA.
For complexometric determination its composition Mo-HyOH-EDTA is more
advantageous than the binuclear complex in the first type of determination.

In this context we ought to mention a basic error which repeatedly
appears in the literature, though the whole problem has been properly
elucidated by Lassner and Schedle [17].   During the reaction of
molybdenum with hydroxylamine and EDTA, molybdenum is not reduced,
but a ternary 1:1:1 complex is formed instead (see p. 186).

The third group of methods is based on precipitation of lead molybdate and the subsequent complexometric determination of the excess
of lead added.

All three types of determination require separation of iron by
precipitation with sodium or potassium hydroxide [18,19], or ammonia
[20] or by directly fusing the sample with sodium peroxide and sodium
carbonate, followed by the extraction of the cooled melt with hot
water [21].   Another method uses extraction of iron with acetylacetone from alkaline triethanolamine medium [22].

Determination of molybdenum(V) as the $(MoO_2)_2$-EDTA complex [21].
Fuse the sample (0.5 g) with 4 g of sodium peroxide and sodium
carbonate at 550 - 600°, cool, and extract the molybdate with hot
water.   Take a fraction of the filtrate, containing 10 - 30 mg of
molybdenum.   Add an excess of 0.005M EDTA and 20 ml of 1% hydrazine
sulphate solution in 0.4M hydrochloric acid and heat to boiling.
The reduction of molybdenum is complete within 2 min.   Cool and
adjust the pH to 4.5 - 5.0 with ammonia, add PAN indicator and ethanol
(one-third of the volume of the neutralized solution) and titrate
with 0.005M copper sulphate until the colour changes to yellow (1 ml
of 0.005M EDTA corresponds to 0.9595 mg of Mo).   A similar method
was published by Endo and Higashimori [20].   After dissolution of
the alloy (0.1 g) in sulphuric acid and hydrogen peroxide they
precipitate iron with ammonia.   In a fraction of the filtrate they
reduce molybdenum with hydrazine sulphate in acidic medium in the
presence of 0.02M EDTA and mask copper with thiourea.   They back-
titrate the excess of EDTA with 0.02M lead nitrate (Xylenol Orange as

indicator). In this case 1 ml of 0.02M EDTA corresponds to 3.838 mg of molybdenum.

Determination of molybdenum(VI) as the Mo-HyOH-EDTA complex [18]. Dissolve 0.2 g of sample in 30 ml of nitric acid (1 + 3) and 10 ml of hydrochloric acid (1 + 1), dilute to 200 ml with hot water and precipitate iron, manganese and copper with an excess of 20% sodium hydroxide solution. Boil for 5 min, cool, dilute to volume in a 500-ml standard flask and filter. Acidify 250 ml of the filtrate (to Congo Red paper) with hydrochloric acid. Add 25 ml of 0.1M EDTA and 10 ml of 20% hydroxylamine solution, and boil. When the solution no longer gives a blue colour with Congo Red paper add 30 ml of 20% ammonium acetate solution and titrate the excess of EDTA with 0.1M zinc sulphate to the red end-point of Xylenol Orange. In this case 1 ml 0.1M EDTA corresponds to 9.595 mg of molybdenum. These authors still suppose that hydroxylamine reduces molybdenum (see remarks above). A similar method has been worked out by Poljaková and Čakon [19] who also think the Mo(VI) is reduced with hydroxylamine. They mask aluminium with ammonium fluoride. Vanadium(V) is co-titrated and must be determined separately by spectrophotometry. DCTA can be used instead of EDTA [22].

Determination of molybdenum as PbMoO₄ [23]. Dissolve 0.1 - 0.7 g of sample in 1 ml of concentrated hydrochloric acid with a few drops of nitric acid. Evaporate the acids and dissolve the residue in 10 ml of hydrochloric acid (1 + 1), dilute to 50 ml with water, heat the solution and precipitate iron with an excess of 10% sodium hydroxide solution. After 10 min filter, and dilute the filtrate to known volume. Dilute an aliquot to 100 ml, adjust the pH to about 5.5 with hexamine and 0.2M nitric acid, heat the solution, add a few drops of Xylenol Orange, then 0.02M lead nitrate dropwise till a pink colour appears, and then add an excess. Titrate the excess with 0.02M EDTA to lemon yellow.

Gusev and Nikolaeva [24] proceed similarly but use 2-ethylamino-5-(2-pyridylazo)-p-cresol as the indicator.

Popova and Seraya [25] determine the sum of W + Mo in the same way as for molybdenum. In another portion of the solution they determine molybdenum after reduction to molybdenum(V) with hydrazine as previously described, after masking of tungsten with tartaric acid. Tungsten is calculated by difference. Nikitina and Adrianova [26], after separation of iron, titrate molybdenum directly in hot solution with 1% lead acetate solution, with Xylenol Orange as indicator.

Analysis of ferrotungsten (determination of W and Al)

There is little possibility of determining tungsten complexometrically. One method is based on precipitation of lead tungstate in the same way as lead molybdate [25]. For other elements present in ferrotungsten, only one procedure (for the determination of aluminium) has been described [27], which may be of some importance and is given here.

Fuse 0.5 g of the sample for 30 min at 700 - 800° with 6 g of sodium peroxide. Dissolve the cooled melt in 100 ml of water, transfer to a 500-ml standard flask, dilute to the mark, mix and filter. To 25 ml of the filtrate add 15 ml of 0.01M DCTA and adjust the pH to 2.5 with hydrochloric acid. Boil the solution for 3 min and cool.

Add enough acetate buffer to give pH 5.5, and a few drops of Xylenol Orange indicator. Titrate with 0.01M zinc chloride to the red colour of the indicator. Then add 10 ml of 4% sodium fluoride solution, boil again for 3 min (formation of AlF_6^{3-}) and titrate the liberated DCTA with the zinc solution to the same end-point.

The procedure has the advantage that the aluminium is determined without separation of tungsten and according to the authors is suitable for an Al:W ratio of 1:50.

Analysis of ferrotitanium (determination of Ti and Al)

A great many methods for direct and indirect determination of titanium are described in the section on titanium (p. 134). The most convenient method is that proposed by Bieber and Večeřa [28]. It is based on formation of the very stable yellow $Ti-EDTA-H_2O_2$ peroxo-complex with excess of EDTA, the surplus being back-titrated with bismuth nitrate at pH 2 - 3 with Xylenol Orange as indicator. Because of the intense yellow colour of this peroxo-complex the method is only suitable for less than 25 mg of titanium in 100 ml. All elements - including iron - which form stable EDTA complexes at pH 2 - 3 interfere. Large amounts of aluminium, copper, nickel and cobalt also interfere. The negative error caused by chloride and sulphate can be reduced by dilution of the solution before titration.

For the analysis of ferrotitanium a reliable procedure is the separation of titanium as hydroxide in the presence of triethanol-amine (and EDTA) [29].

Dissolve the sample in concentrated nitric acid and dilute the solution to known volume. Pipette an aliquot containing not more than 25 mg of titanium into a large beaker and add 20 ml of 20% triethanolamine (TEA) solution. The original solution must be acid enough not to form a precipitate with the TEA added. Then add 2M sodium hydroxide with mixing until the solution becomes colourless (formation of Fe-TEA complex). Boil the solution for 1 min and let the precipitate settle. Filter off on a medium porosity filter paper and wash five times with 1% triethanolamine solution and twice with hot water. Dissolve the precipitate by adding 25 - 30 ml of hot 5M nitric acid to the filter and then wash through with hot water. Collect the filtrate in a 500-ml titration flask, cool, add 10 ml of hydrogen peroxide solution and a measured volume of 0.05M EDTA. Dilute to about 200 ml, adjust the pH to 2 - 3, and thoroughly cool to below 20°. Titrate with 0.05M bismuth nitrate to the red end-point of Xylenol Orange.

Triethanolamine has the advantage that it keeps bivalent metals in solution, so that they are only very slightly adsorbed on the precipitated titanium hydroxide. Very pure $Ti(OH)_4$ is obtained if 1 g of EDTA is added before the precipitation. Titanium hydroxide is then precipitated as dense flakes, which dissolve in nitric acid less easily. Magnesium is also precipitated but does not interfere in the titration. Thorium, zirconium and rare earths are also precipitated, but are not usually present in ferrotitanium.

The method has been applied for the determination of titanium in ilmenite $(FeTiO_3)$ and titanium slags [30], and Ti-Al alloys [31]. Titanium can also be selectively separated from iron by extraction of its ascorbate complex with trioctylmethylammonium chloride (TOMA,

Aliquat 336-S) [31].

Several methods have been proposed for the determination of iron,
aluminium and titanium in a single solution. With the use of
various masking agents all three elements can be determined sucess-
ively. These procedures have been applied mostly in the analysis of
ores, slags, cements, minerals etc. One method has been used by
Bruile [33] for the analysis of ferrotitanium.

Dissolve 0.3 - 0.5 g of sample in 60 ml of a 3:1 mixture of concentrated hydrochloric and nitric acids and 10 - 15 ml of concentrated
sulphuric acid. Evaporate to white fumes. Dissolve the residue in
100 ml of 5% sulphuric acid and filter. Ignite the filter and the
residue, cool, treat with H_2SO_4 + HF to remove silicic acid, and
evaporate to dryness. If some residue remains, fuse it with $K_2S_2O_7$,
dissolve the cooled melt in 5% sulphuric acid and combine with the
main filtrate. Dilute the filtrate to volume in a 250-ml standard
flask. Pipette 25 ml of this solution, add 50 ml of 0.05M EDTA and
some Xylenol Orange. Neutralize the solution with ammonia to
reddish-violet, then add 20% sulphuric acid until yellow. Add 20 ml
of 25% sodium acetate solution, adjust the pH as before with ammonia
and sulphuric acid. Boil the solution for 2 - 3 min. Cool, and
titrate the excess of EDTA with 0.05M zinc sulphate. (The consumption of EDTA corresponds to the sum of Fe + Ti + Al + trace metals
such as copper; therefore disregard the volume of zinc sulphate used
in the titration). Add to the same solution 10 - 15 ml of 20% tartaric acid solution to demask titanium from its Ti-EDTA complex and
set the solution aside for 5 min. Adjust the pH to 5 - 5.5 and
titrate again with zinc sulphate (the amount used corresponds to the
titanium present). Then add 10 - 15 ml of 5% potassium fluoride
solution to demask aluminium, acidify with 20% sulphuric acid until
yellow, and boil for 2 - 3 min. Cool, adjust the pH as before, and
titrate once more with zinc sulphate. This final consumption of
zinc sulphate corresponds to the amount of aluminium. This method
does not seem to be very suitable for the determination of small
amounts of aluminium, considering that about 20 ml of 0.05M EDTA will
be needed for the sum of Fe + Ti + Al in a 50-mg sample. The
consumption of EDTA for a 2% content of aluminium (1 mg Al in 50 mg)
would be only 0.74 ml. Moreover the masking of titanium with
tartaric acid could cause a poor indicator colour change.

Analysis of ferrovanadium

Complexometric analysis of this alloy is not very attractive in
comparison with the commonly used reductometric titration with ferrous
sulphate, but it is fairly selective for the determination of vanadium in steels. Only strong oxidants $(MnO_4^-, Cr_2O_7^{2-})$ interfere.
Nevertheless some methods have been proposed for the determination of
vanadium in steels and other alloys. Vanadium(IV) is the best
oxidation state for the complexometric determination. The V(IV) -
EDTA complex is easily formed from vanadate by reduction with
ascorbic acid in the presence of EDTA. The excess of EDTA is then
determined with zinc solution, with Xylenol Orange as indicator.
Iron(III) interferes even if simultaneously reduced with ascorbic
acid, and therefore must be removed with sodium hydroxide or by suitable extraction. Hydrogen peroxide is used for masking vanadium(V).
Similar methods are based on the determination of the sum of other
metals (e.g. Al, Fe etc.) in the presence of hydrogen peroxide. The
sum of all metals + vanadium is determined in a second aliquot after

reduction of vanadium(V) with ascorbic acid.   The details will be
found in the literature [33,34].

Analysis of ferrochromium

Complexometry has not so far found use in the determination of
chromium in ferrochromium.   It is much simpler to use the redox
titration of chromium(VI) with ferrous sulphate.   In the complexometric determination of chromium(III) many cations interfere.   Moreover, the $Cr(III)$-EDTA complex is formed only in boiling solution and
because of its intense violet colour only a few mg can be determined
by indirect titration with a visual end-point.   These problems can
be overcome, however (see Section 6.1.11, p. 129).

Chromate has been determined by precipitation with barium chloride
and back-titration of the excess of barium in the filtrate [35].
This method has been applied to determination of chromium in steels
after ion-exchange separation of iron and chromate.

Analysis of iron-zirconium alloys

Zirconium is not only a component of ferrozirconium but is also
present in other alloys, in which its content may vary from traces to
several per cent.   Complexometric determination of zirconium is very
advantageous, because it can be done very selectively in 1 - 2N acid
medium, with Xylenol Orange as indicator.   Xylenol Orange is regarded by many authors as the best indicator for zirconium titration,
even for photometric determination of traces [36 - 40].

Two examples will be given below of the analysis of an alloy containing some other elements besides zirconium and iron.

Wakamatsu [41], after dissolution of the sample, precipitates iron,
nickel and copper at pH 1.5 with cupferron in the presence of EDTA,
and extracts the cupferronates with diethyl ether.   He then adds
acetone to the aqueous phase and determines zirconium at pH 3 by adding excess of EDTA and back-titrating with copper sulphate (PAN as
indicator).   This method is a little time-consuming, and ignores the
possibility of titrating zirconium directly in strongly acidic medium
without separation of other components (see Section 6.1.1, p. 92).

Klygin and Kolyada [42] recommend 0.1 - 0.6N sulphuric acid medium for
direct determination of zirconium, with Xylenol Orange as indicator.
The method is applicable for 2 - 20 mg of zirconium when a boiling
solution is titrated with 0.02M EDTA.   Up to 10 mg of iron, indium,
scandium, thorium etc. will not interfere.   The method was mainly
applied by the authors for the analysis of Zr-Al and Zr-U alloys.

Musil [43] has paid attention to alloys containing 30% each of zirconium, iron and aluminium, and 5% of silicon.   The method for zirconium is as follows.

Dissolve 0.4 g of sample in a mixture of 25 ml of nitric acid (1 + 1),
5 ml of concentrated hydrochloric acid and 25 ml of sulphuric acid
(1 + 1).   Heat the solution to white fumes and continue with low
heating for 5 - 10 min.   Cool, add 50 - 100 ml of water and boil for a
while.   Filter through a dense filter, and treat the silicaceous
residue on the filter as usual in silicate analysis, combining any
resultant solution with the filtrate.   Dilute the total filtrate

accurately to 200 ml.   The acidity must be 2N in sulphuric acid.
Pipette 50 ml, dilute with 20 ml of 2N sulphuric acid, add 1 g of
ascorbic acid and titrate slowly with 0.01M EDTA after addition of
5 drops of 0.2% Xylenol Orange solution, till yellow (compare the
colour with that of a slightly over-titrated solution).

For very accurate determination of zirconium Musil recommended its
precipitation with mandelic acid and gravimetric determination as
ZrO_2.   He determined aluminium in the alloy complexometrically and
iron by titration with potassium dichromate.

Determination of nickel in binary iron alloys and steels

A simple method for the determination of nickel was proposed as early
as 1954, based on the masking of iron with triethanolamine in strongly ammoniacal solution and subsequent direct EDTA titration, with
murexide as indicator [44].   Only a relatively low concentration of
iron can be masked in ammoniacal medium because the solution is
yellow, but in sodium hydroxide medium the Fe-TEA complex is colourless.   In the presence of EDTA and TEA, any ferric hydroxide precipitated on alkalizing the solution is easily dissolved, yielding a
colourless solution.   Nickel is then determined by back-titration of
excess of EDTA with a calcium solution.   The best indicator is
Thymolphthalexone.   This method was applied for the determination of
nickel in Fe-Ni alloys and pellets [45].   Another but somewhat
tedious method is based on precipitation of nickel with dimethylglyoxime and on titration of the dissolved precipitate.   The method is
applicable to more complicated Fe-Ni alloys [46-49].

Determination of nickel in Fe-Ni alloys and Ni pellets [45].
Dissolve 0.5 g of sample in concentrated hydrochloric acid and nitric
acid and evaporate almost to dryness.   Dilute with water.   For
nickel pellets transfer the solution to a 250-ml standard flask and
dilute to volume with water and use 50 ml for the determination.   For
nickel alloys transfer the whole solution to a 500-ml titration flask.
Partially neutralize the solution with 1M sodium hydroxide, add 10 ml
of 0.05M EDTA and 10 ml of 20% triethanolamine solution and with
continuous stirring alkalize with 15-20 ml of 1M sodium hydroxide.
Dilute to about 250 ml, add the indicator (Thymolphthalexone) and
titrate with 0.05M calcium chloride to the blue colour.   After the
titration add one drop of thioglycollic acid and observe whether the
solution decolorizes owing to bonding of traces of copper and zinc.
If so, continue the titration with calcium chloride to the new end-
point.   Run a reagent blank in the same way.

The method gives good results for the analysis of pellets containing
5-15% of nickel.

The method originally described for the determination of nickel with
murexide as indicator [44] was applied by Khlystova [50] for ferrous
alloys containing 65-80% of nickel.   Zel'tser and Skorospelova [51]
extended the method to include the determination of iron after its
separation as hydroxide.   They also described the complexometric
analysis of Fe-Ni-Mo alloys.

Determination of nickel after separation with dimethylglyoxime.   These
procedures [46-49] are based on the well-known classical method.
They differ only in the preparation of the sample solution and the
dissolution of the nickel dimethylglyoximate, the nickel in which is

determined by EDTA titration, with murexide as indicator.   The
alloys are usually dissolved in hydrochloric and nitric acid.   The
nickel is precipitated in the presence of citric acid [46] or tartaric acid [48] to keep iron in solution.   The precipitate is dissolved
off the filter with 30 ml of 9M hydrochloric acid [46] or with a
mixture of hydrochloric acid and nitric acid [48].   In both cases
the solution is evaporated to dryness and nickel determined by EDTA
titration in ammoniacal solution.   A slightly modified procedure was
included in British Standards [49] and is recommended for a wide
range of nickel content in iron, steels and permanent magnet alloys
also containing aluminium and cobalt.

9.2.2   Analysis of Steel and Cast Iron

This section will deal with determination of certain elements in cast
iron and steel and also in some binary alloys.   Some of the binary-
alloy methods can be used for steel analysis, of course.

Determination of aluminium in steels

Several methods have been proposed for the determination of small
amounts of aluminium (0.2 - 0.5%) in steels and also in binary alloys
such as ferrosilicon and ferromanganese (see p. 263).   The iron must
first be separated.   Its precipitation with sodium hydroxide, though
commonly used, is not completely satisfactory because of the adsorption of aluminium on the precipitated iron hydroxide.   Spauszus [52]
long ago exploited the fact that the loss of aluminium is least when
the sample solution is gradually added to a hot solution of sodium
hydroxide.   He recommended for this purpose the use of a funnel with
the stem drawn out to a narrow tip, so that flow-rate does not exceed
10 ml/min.   Spauszus and Schwarz [53] later used the finding of Hill
[54] that the adsorption of aluminium on ferric hydroxide can be
completely prevented by addition of borate ions, which occupy all the
active adsorption sites on the precipitate.   It should be noted that
these adsorption phenomena are disregarded in many newer works.

Other methods used for the separation from iron include ion-exchange,
extraction, electrolysis at a mercury cathode, and the basic benzoate
method after reduction of iron to iron(II) with thioglycollic acid
[55].   The choice of final determination of aluminium is more or
less arbitrary.   Nowadays back-titration of excess of EDTA is commonly used, with a standard zinc or lead solution and Xylenol Orange as
indicator.   DCTA, which readily forms the aluminium complex at room
temperature, does not seem to have been used in this field so far.

Determination of aluminium according to Spauszus [52].   Dissolve 2 g
of sample in 30 ml of concentrated hydrochloric acid and add a few
drops of concentrated nitric acid.   Boil to expel nitrogen oxides
and dilute to 50 ml with water.   In a 1-litre polyethylene beaker
dissolve 40 g of sodium hydroxide and 3 g of boric acid in 100 ml of
water.   Add the hot sample solution to this with stirring.   Cool
the mixture, transfer it to a 500-ml standard flask and dilute to
volume.   Filter and collect 250 ml of filtrate in a standard flask.
Transfer this 250 ml to a 750-ml beaker and neutralize with hydrochloric acid to pH 7.   Add to the solution 1 ml of ammonium acetate
solution (1 kg/l.) and heat to boiling.   After addition of 3 drops
of Cu-EDTA complex and 5 drops of 0.1% ethanolic solution of PAN,
titrate with 0.05M EDTA to a yellow colour.   It is necessary to

titrate the boiling solution, otherwise the orange colour returns and
the titration must be continued; the titration is finished when the
yellow colour does not turn to orange within 30 sec.

Spauszus defends this Cu-EDTA-PAN titration as the only one suitable
for direct titration of aluminium. It has the disadvantage that a
boiling solution must be titrated, and the end-point is a little
sluggish. Moreover, for the declared minimum amount of 0.05% of
aluminium, the consumption for 1 g of alloy is only 0.371 ml of 0.05M
EDTA.

Determination of aluminium according to Jurczyk [55]. The author
recommends the following weights of sample and volume of acids according to the expected aluminium content.

Al, (%)	Weight of sample, (g)	Concentrated HCl, (ml)	Concentrated HNO$_3$, (ml)	H$_2$SO$_4$ (1 + 1), (ml)	HClO$_4$ (70%), (ml)
0.2 - 2.5	1	30	10	20*	-
2.5 - 5.0	0.5	20	5	20*	20
over 5	0.25	20	5	15†	20

* For difficultly soluble samples the H$_2$SO$_4$ is replaced by 20 ml
of HClO$_4$.

† In this case a further addition of 15 ml of HClO$_4$ is made.

Dissolve the sample in a 250-ml tall beaker in the mixture of acids
and evaporate to fumes but not to dryness. Then add 10 ml of
hydrochloric acid, boil for 2 - 3 min, add 75 ml of water and continue
boiling till all salts have dissolved. Separate the silicic acid
and treat it as usual (evaporation with HF and fusion of the residue
with KHSO$_4$). The combined solution after these operations should
have a maximum volume of 200 - 250 ml. Neutralize the solution with
ammonia (1 + 1) to first appearance of a precipitate, then make acid
with a few drops of hydrochloric acid. Add 2 g of ammonium chloride,
3 ml of glacial acetic acid, 5 ml of thioglycollic acid [to reduce
iron(III) and any chromium(VI)] and 20 ml of 10% ammonium benzoate
solution. Adjust the pH to 4 - 4.5 and boil for 5 min. After 10
min filter off the precipitated aluminium benzoate on paper, and wash
five times with hot 3% ammonium benzoate solution (adjusted to pH 4.5
with acetic acid). Rinse the precipitate into the original beaker
with hot water and wash the filter with 15 ml of hot hydrochloric
acid (1 + 1). Dilute the combined solution to 150 ml, heat to 40 -
50° (any benzoic acid precipitated dissolves), add 1 ml of 1%
hydrogen peroxide (sufficient for up to 10 mg of titanium) and 50 ml
of 0.025M EDTA. Adjust the pH to 6 - 6.5. Add 20 ml of acetate
buffer (pH 6.2) and Xylenol Orange as indicator, heat the solution
for 2 min, and titrate the solution with 0.025M zinc sulphate to the
red colour of the zinc-XO complex. The consumption of EDTA
corresponds to the sum of Al + Ti. Add 30 ml of saturated sodium
fluoride solution, boil for 2 min and titrate the liberated EDTA with

zinc sulphate.

Jurczyk supplements his procedure by a series of small remarks which
will not be given here since their content is trivial to an experienced analyst.   The method was said to give very good results even
in the analysis of very highly alloyed steels (18% Cr, 30% Ni).   An
analysis takes about 60 - 90 min.

However, the method requires considerable experience, for it has
several possible stumbling blocks.   The first is boiling the EDTA
solution in the presence of hydrogen peroxide (risk of oxidation) and
then the possible reaction between the Ti-EDTA-H_2O_2 complex and
fluoride.   Neither reaction can be ruled out.   Therefore a safer
procedure seems to be the masking of titanium with lactic acid as
proposed by Chen and Li (p. 136), which avoids the use of peroxide and
one boiling of the solution.

Some other methods for the determination of aluminium in steels and
iron alloys have been published.   They differ mostly in the preparation of the sample and will be mentioned only briefly.   Kurjaković
and Plepelić [56] fuse 1 g of alloy with 4 g of potassium pyrosulphate, treat the cooled melt with 15 ml of sulphuric acid, and filter
off and wash any residue.   They electrolyse the filtrate (volume not
more than 50 ml) at 4 A for 30 min at a mercury cathode.   Aluminium
is then determined by back-titration of excess of EDTA by the Cu-EDTA-
PAN method.   The method is applicable for steels containing at least
0.1% of aluminium.

Habercetl [57] proceeds very simply, dissolving the sample in acid,
and precipitating iron with 10M sodium hydroxide (30 ml).   In a
suitable fraction of the filtrate aluminium is determined indirectly
by back-titration of excess of 0.01M EDTA with 0.01M zinc solution
with Erio T as indicator.   The results are said to be about 0.03%
lower than gravimetric results.

Kalina-Zhikhareva [58] has worked out a similar method for steels
containing 2 - 15% of aluminium and 5 - 25% of chromium.   After
dissolution of 0.2 g of the steel in acid, chromium is oxidized with
persulphate without a catalyst.   The persulphate must be added in
portions until the boiling solution is orange (or violet if manganese
is present).   Aluminium, titanium, iron, phosphate and part of the
manganese are then precipitated with ammonia.   After dissolution of
the precipitate aluminium is isolated with sodium hydroxide and
finally determined indirectly by back-titration of excess of EDTA
with zinc solution, at pH 8.5, with Erio T as indicator.   According
to the author small amounts of phosphate and vanadium do not interfere.

Determination of aluminium in ferrosilicon and ferromanganese

Determination of aluminium in these alloys is relatively simple.
Silicon (45 - 90%) is separated during the dissolution of the sample
in nitric, perchloric and hydrofluoric acids.   Iron must also be
separated.   Brháček [59] removed most of the iron by electrolysis at
a mercury pool electrode.   With a 50-cm^2 electrode and a current
density of 0.4 A/cm^2 the Fe:Al ratio in the solution can be decreased
to <1 in 10 - 20 min of electrolysis.   The remaining iron is determined complexometrically at pH 2 - 2.5, with salicylic acid as indicator.   Aluminium is determined at pH 5 in the same solution by any

convenient method, e.g., back-titration of excess of 0.05M EDTA with
0.05M ferric chloride.

Maekawa et al. [60] described two methods for the determination of
aluminium in ferrosilicon and ferromanganese. In the analysis of
ferrosilicon the sample is dissolved in a mixture of acids (HF +
H_2SO_4 + $HClO_4$), most of the iron is extracted with pentyl acetate,
and manganese and titanium are separated by precipitation with sodium
hydroxide. Aluminium is then determined by the Cu–EDTA–PAN method.
In the analysis of ferromanganese, iron is also extracted with pentyl
acetate and the remainder, together with aluminium, is separated from
manganese by the basic acetate method.

Rybalov and Eshmukhambetova [61] described the complete analysis of
AMS-alloys containing at least 20% each of silicon, manganese and
iron (besides Al, Ca, Mg, F, C, S). For the complexometric determination of aluminium they used dissolution of the sample with hydrofluoric and nitric acid evaporation, fusion of the residue with
sodium peroxide at 800–1000°C and reduction of manganese with ethanol.
In the dissolved melt, aluminium was determined by EDTA titration by
the usual method. For the complete analysis the authors dissolved
another sample in the acid mixture, evaporated the solution to dryness, fused the residue with potassium pryosulphate at 800 – 1000°C,
dissolved the cooled melt in hydrochloric acid (1 + 1), then determined iron and calcium + magnesium complexometrically, and manganese
and phosphorus photometrically.

Determination of magnesium in cast iron and steels

The main problem in the determination of magnesium in malleable cast
iron and steels is the quantitative separation of iron and other
accompanying elements from magnesium. Because of the low content of
magnesium (0.01 – 1%) large samples must be taken for the analysis,
which causes some difficulties. Various methods of separation have
been published, exploiting the classical precipitation method as well
as more modern methods such as extraction and ion-exchange procedures.
The aim is to obtain a solution containing, if possible, only magnesium, which can then be determined by complexometry (or photometry).

As early as 1955 Green [62] published the first complexometric method
for the determination of magnesium in nodular cast iron. He
dissolved 10 g of sample in acid, then precipitated iron with zinc
oxide and manganese by boiling with 2 g of ammonium persulphate. The
filtrate contained zinc and a small amount of aluminium and magnesium.
The zinc was masked with potassium cyanide and aluminium with tiron.
Excess of 1% EDTA solution was added and back-titrated with 1%
magnesium solution (Erio T as indicator). According to the author
the method is not stoichiometric and an empirical factor must be
determined from the recovery of known amounts of magnesium. This
method did not seem to win much appreciation at the time of its
publication.

Green's method has been carefully re-examined by Kraft and Fischer
[63] and considerably improved by the use of previously published
information [64]. In this modification iron is precipitated with
zinc oxide and manganese with permanganate. The small excess of
permanganate is removed with ethanol. The oxidation of manganese
with bromine is not satisfactory because of the volatility of bromine.
In the permanganate reaction less than 10 µg of manganese escapes

into the filtrate.   For masking zinc tetraethylenepentamine (tetren)
is recommended.   The blue colour of the Cu-tetren complex does not
disturb the determination of magnesium with Erio T as indicator.   A
3.06-g/l. EDTA solution was used (1 ml corresponds to 0.2 mg of
magnesium, or 0.01% of magnesium in a 2-g sample).

Reichert [65,66] used a different method.   After dissolution of 1 -
10 g of alloy (0.01 - 0.001% Mg) all the heavy metals were precipitated in ammoniacal solution with hydrogen sulphide.   Traces of iron in
the filtrate were reduced with ascorbic acid and masked with
potassium cyanide.   After addition of triethanolamine magnesium was
determined as usual.   Later [67] he used ^{27}Mg to prove that
magnesium is not adsorbed on the sulphide precipitate.

By this method, magnesium in the range 0.025 - 0.428% can be determined in cast iron with satisfactory results.   Leo [68] came to the
same conclusions, whereas McLaren [69] obtained contradictory results.
According to his experience some magnesium is always adsorbed on the
sulphide precipitate, causing low results.

Iron can also be separated by electrolysis at the mercury cathode.
Leo [70] performed the electrolysis at 20-30°C with a current density
of 9 A/cm.   Manganese was then removed with permanganate.
Siekierska [71] proposed a similar method (see p. 266).

McLaren [69] determined magnesium after chromatographic separation on
a cellulose column, using a mixture of ethyl methyl ketone and 10M
hydrochloric acid (98:2 v/v) as eluent.   He tested his procedure on
synthetic solutions containing the equivalent of 0.02 - 0.06% magnesium and considered it rather empirical and not very precise ($\pm$10%),
but suitable for routine work because of its speed.

The extraction methods for iron are very useful.   A considerable
number of them can be found in the monograph by De et al. [72].
Most were originally devised for other purposes, e.g. for photometric
determination of extracted traces of iron.

The method of Rooney and Carter [73] is based entirely on the extraction of interfering components.   Iron is nearly all extracted with
isobutyl acetate and the rest, together with Al, Mn, Ni, is extracted
with cupferron and diethyldithiocarbamate into chloroform.   The
magnesium is then determined with 0.01M EDTA, with Erio T as indicat-
or.

Because of the very large extraction capacity of isobutyl acetate a
5-g sample could be used, allowing determination of 0.017 - 0.10% of
magnesium.   A single determination takes an hour, six samples can be
analysed in three hours.   For highest accuracy the authors recommend
running a reagent blank, which usually corresponds to 0.002 - 0.005%
of magnesium.

This method, which dates from 1958, is a little tedious.   It needs
several extractions and can be shortened.   For example, after the
extraction of iron, any traces left, together with aluminium and
manganese or zinc, can be masked with triethanolamine and potassium
cyanide.

Details of this method can be found in the literature [73].   Other
methods are much shorter and will be described here in detail.

Siekierska method [71]. Dissolve 3 g of sample in 40 ml of sulphuric acid (1 + 9) and evaporate to fumes. Dissolve the residue in
water and filter off silica and graphite. After electrolysis at a
mercury cathode, add ammonia and ammonium persulphate and heat to
boiling. Filter off and wash the hydroxide precipitate and dilute
the filtrate to volume in a 200-ml standard flask. To 100 ml add
0.5 g of hydroxylamine, 0.2 g of triethanolamine hydrochloride and
0.2 g of potassium cyanide. Titrate with 0.05M EDTA, with Erio T as
indicator. The relative error is $\pm$0.2%. Siekierska obtained very
good results for 0.0466 - 1.12% of magnesium. However, if a small
amount of manganese escapes to the final filtrate the original
procedure should not be used. For example, 0.337 mg of manganese
would be equivalent to the presence of 0.01% of magnesium. This
amount of manganese can be masked directly with triethanolamine as
the Mn(III)-TEA complex without addition of hydroxylamine and
potassium cyanide, both of which keep manganese in the bivalent form
which is normally co-titrated with magnesium unless there is a large
surplus of cyanide present. Presumably the original procedure was
carefully worked out so that no manganese escaped to the filtrate.

Method of Tikhonov et al. [74]. Dissolve 0.2 g of sample in 15 ml
of hydrochloric acid (1 + 1), then add 2 ml of concentrated nitric
acid. Evaporate the solution nearly to dryness, dissolve the
residue in 20 - 25 ml of water, and add 35 ml of 10% sodium diethyldithiocarbamate solution. Filter off the precipitate and wash it
with water. Add to the filtrate 2 ml of 0.005M EDTA, 10 ml of 1:1
v/v glycerol-water mixture to mask up to 0.5% of chromium (or 30 ml
for 1% of Cr), 5 ml of ammonia buffer (pH 10) and 1 ml of 0.1% Erio T
solution. Then add an excess of 0.005M EDTA and back-titrate with
0.005M magnesium chloride.

Determination of calcium (+ magnesium) in cast iron and steels

In all the procedures mentioned above for the determination of
magnesium in cast iron and steels, the presence of calcium was not
considered. The calcium is always co-titrated with the magnesium,
but its content is very low - only a few thousandths of 1%. In
addition, the ratio of the atomic weights of calcium and magnesium is
sufficiently favourable for the calcium content to be disregarded
(0.001% of Ca = 0.0006% of Mg).

Nevertheless some procedures for the determination of calcium have
been proposed. Bezuglii and Amsheeva [75] extract the iron from
hydrochloric acid solution with butyl acetate and precipitate calcium
(0.05 - 1.0 mg) as oxalate, then determine the calcium with 0.004M
EDTA, using Methylthymol Blue as indicator. The true amount of
calcium (x mg) must be calculated from the experimental value (y mg)
by means of the equation y = 0.0011 + 0.9974.

Maekawa et al. [76] described another method which allows the
separate determination of the calcium and magnesium.

Method of Maekawa et al. [76]. Heat 2 g of sample with 15 ml of
perchloric acid (60%) to white fumes, dissolve the residue in 60 ml of
water, filter and dilute to 100 ml. Electrolyse at a mercury cathode
at 15 A for 2 hr to remove most of the iron and other heavy metals.
Evaporate the solution again to white fumes, dissolve the residue in
10 ml of hydrochloric acid (1 + 1) and evaporate to dryness. Add a
further 5 ml of hydrochloric acid and 40 ml of water, and make

ammoniacal to precipitate all remaining heavy metals. Filter,
acidify the filtrate and make up to volume in a 100-ml standard flask.
Pipette 40 ml of this solution, add 50 mg of ascorbic acid, 3 ml of
50% v/v triethanolamine solution and 5 ml of 20% sodium cyanide
solution and adjust the pH to 10.   Titrate the sum of Ca + Mg with
0.02M EDTA, using Erio T as indicator.   For the determination of
calcium use another 40 ml of the solution.   Add 3 ml of 0.01%
magnesium chloride solution to ensure complete precipitation of
magnesium, then add TEA and sodium cyanide as above and 10 ml of 20%
sodium hydroxide solution.   Set aside for 3 min and then titrate the
calcium with 0.002M EDTA, using the Patton-Reeder indicator (p. 23).

Determination of calcium (magnesium) in ferrosilicon [77].   Dissolve
0.5 g of sample in nitric acid and hydrofluoric acid mixture, add
perchloric acid and evaporate nearly to dryness.   Take up in 20 ml
of 6M hydrochloric acid and extract iron with 20 ml of methyl
isobutyl ketone.   Precipitate residual iron together with manganese
with ammonia and bromine water.   Determine the calcium as just
described.

Determination of lead in leaded free-cutting steel

Wakamatsu [78] proposes the following method.   Dissolve 5 g of
sample in 60 ml of 6M hydrochloric acid, add 3 - 5 ml of concentrated
nitric acid and 50 ml of 60% perchloric acid.   Evaporate to white
fumes.   Add a little 6M hydrochloric acid, and dilute to volume in a
100-ml standard flask with this acid*.   Shake a 20-ml aliquot with two
20-ml portions of methyl isobutyl ketone for 30 sec each time, to
remove iron, molybdenum and chromium.   Evaporate the aqueous phase
nearly to dryness, add 3 ml of concentrated nitric acid and 2 ml of
60% perchloric acid and evaporate to white fumes.   Dissolve the
residue in 50 ml of water, add 5 ml of 20% triethanolamine solution
and an excess of 0.005M EDTA (2 - 6 ml).   Adjust the pH to 10 - 11
with aqueous ammonia.   Then add 5 ml of 10% potassium cyanide
solution and small amount of Thymolphthalexone and titrate the excess
of EDTA with 0.005M calcium to the intensely blue end-point colour.
The determination takes 50 min.

Determination of bismuth in cast iron and steels [79]

Dissolve 1 g of sample (containing 0.1 - 0.2% of Bi) in 15 ml of warm
sulphuric acid (1 + 5), and oxidize iron with a few drops of nitric
acid.   Filter off any graphite.   Dilute the filtrate to 100 - 150 ml
with water, add 3 ml of 5% manganese sulphate solution and 3 ml of 1N
potassium permanganate, boil the solution for 3 - 5 min and set aside
in a warm place for a further 5 - 10 min to co-precipitate the
bismuth with the manganese dioxide.   Filter off and wash the
precipitate 6 - 8 times with hot water and dissolve the bismuth in a
suitable amount of nitric acid and a little 30% hydrogen peroxide.
Adjust the pH to 2 - 3 and titrate with 0.005M EDTA (Xylenol Orange as
indicator).

The original procedure used sulphuric acid and hydrogen peroxide to
dissolve the bismuth off the precipitate but large amounts of
sulphate adversely affect the bismuth titration.   The author also
described the spectrophotometric determination of bismuth with
Xylenol Orange.

---

*N.B. Anal. Abstr., 1962, 9, 4704 erroneously says "dilute with water".

Determination of tin in modified cast iron [80]

Dissolve 0.5 g of sample with 20 ml of nitric acid (1 + 3) and
continue with the precipitation of manganese dioxide as described
above, to co-precipitate tin.   Dissolve the precipitate with 10 ml
of sulphuric acid (2 + 5) and a few drops of hydrogen peroxide (30%),
then add 20 ml of 5% beryllium sulphate solution and 20 ml of 0.1M
EDTA and dilute with water.   Heat the solution to 70 - 80° and co-
precipitate the tin with beryllium hydroxide by alkalizing to Phenol
Red with 10% ammonia solution and adding 1 - 2 ml in excess.   After
30 min filter off and wash the precipitate with 40 - 50 ml of 1%
ammonium sulphate solution and dissolve it in 11 ml of hot sulphuric
acid (2 + 5).   Add 3 ml of 30% hydrogen peroxide solution and
evaporate to fuming.   Dissolve the residue in water and make up to
volume in a 250-ml standard flask.   Take a suitable aliquot, adjust
to pH 2, and add a little ascorbic acid and 10 - 15 ml of 0.01M EDTA.
Titrate the excess with 0.01M bismuth nitrate, using Xylenol Orange
as indicator.

Determination of sodium in iron alloys [81]

Determination of sodium in such alloys is probably never needed.
Amsheeva proposes extracting iron with butyl acetate, then precipitating sodium as sodium magnesium uranyl acetate, dissolving the
precipitate, reducing the uranium with ascorbic acid and determining
it by addition of excess of EDTA and back-titration with 0.01M
bismuth nitrate (Xylenol Orange as indicator).   The error is said to
be about 7%.

Determination of phosphorus in ferrophosphorus [82]

Fuse 0.5 g of sample with sodium peroxide for 5 min, cool, dissolve
in 100 ml of water and boil for 2 min.   Then dilute to volume in a
500-ml standard flask.   To a 50-ml aliquot add 10 ml of concentrated
hydrochloric acid and 5 ml of 10% tartaric acid solution, neutralize
with ammonia and add 5 ml of 20% potassium cyanide solution.   Add
25 - 30 ml of magnesium solution (see below), and adjust to pH 10 with
ammonia.   After 10 min add 0.2% ethanolic Erio T solution and
titrate rapidly with 0.03M EDTA to the greenish-blue end-point.

Magnesia solution: dissolve 8.5 g of $MgCl_2.6H_2O$ and 17 g of NH_4Cl in
800 ml of water, neutralize with ammonia, filter, adjust the pH to
5 - 6 and dilute to 1000 ml.

According to the author at least 5 mg of phosphorus can be determined
by this method (1% of P in a 0.5-g sample), if the excess of magnesium
is at least 5 mg.   The titration must be carried out as fast as
possible, because the indicator (Erio T) reverts to its red colour
30 - 60 sec after termination of the titration.   A single determination takes 20 min and the method is suitable for up to 20% of
phosphorus.

Analysis of special steels and iron alloys

The preceding sections gave complexometric methods for the analysis
of binary alloys of iron, including the determination of some
elements that may be present in minor or trace amounts in cast iron
and steel.   From the analytical point of view steels represent a
very varied multicomponent system and the determination of individual

components is not always an easy task.   For this reason authors have
often limited the application of complexometry to the determination
of only certain elements.   The other components are either not
mentioned at all, or the possibility of using "other methods" is
suggested.   For instance Lewis and Straub [83] determine complexometrically in stainless steels only nickel and cobalt.   The various
metal ions are sorbed on Dowex 1-X8, and nickel, vanadium, chromium,
titanium and manganese are eluted with 9M hydrochloric acid.   Nickel
is precipitated with dimethylglyoxime and then determined by EDTA
titration.   Cobalt is eluted with 7M hydrochloric acid and determined complexometrically.

Separation on ion-exchangers is still very popular and alloys of all
sorts are very suitable for application of this technique.   Wilkins
and Hibbs published a series of works in which they dealt with the
analysis of some steels.   Most of the components were determined
complexometrically.

Separation on ion-exchangers is relatively time-consuming, but on the
other hand it can be done semiautomatically on a series of columns
simultaneously for a large number of samples.   The use of complexometric determination considerably speeds up the whole analysis.
Therefore some interesting procedures worked out by these authors
will be given here.

Analysis of ALNICO alloys [84]

Determination of copper.   Dissolve 1 g of sample in a 150-ml beaker
with aqua regia and evaporate to a syrupy consistency to remove
nitric acid (add a little water after the first evaporation, and
evaporate again).   Add 5 ml of concentrated hydrochloric acid and
50 ml of water, and electrolyse at 0.5 A with a divided cell (its
construction is described in the original paper) until the copper is
completely deposited.   This should take about 30 min.   Remove the
beaker from the apparatus and rinse the electrode into it with water.
The solution contains the iron, nickel, cobalt and aluminium.   Insert
the electrode assembly into a 250-ml beaker containing 10 ml of 20%
ammonium acetate solution and add sufficient water to cover the
copper deposit.   Reverse the current flow in the electrode assembly
and strip the copper into the buffer solution.   Add an excess of
0.01M EDTA, 4 or 5 drops of PAN (alcoholic solution) and back-titrate
with standard copper solution.

Determination of Al, Ni, Co, Fe.   Add to the copper-free solution
3 ml of 30% hydrogen peroxide and evaporate to 4 – 5 ml.   Transfer
the solution into a column (35 x 1 cm) of Dowex 1-X4 (100 – 150 mesh,
chloride form) which has been previously washed with 9M hydrochloric
acid.   Elute aluminium and nickel with two column volumes of 9M
hydrochloric acid.   Then begin elution of the cobalt with 4M hydrochloric acid.   The cobalt will move down the column in a blue band,
and when it is 5 cm from the bottom of the column, remove the beaker
containing the aluminium and nickel and receive the cobalt in another
250-ml beaker.   When the cobalt has been removed from the column,
elute the iron with 0.5M hydrochloric acid into another 250-ml beaker.
Evaporate each eluate to 10 ml and dilute it to volume in a 250-ml
standard flask with water.

All four elements are finally determined by essentially the same
method as follows.

Aluminium and nickel.   To a 25-ml aliquot add enough buffer (pH 4.8)
and excess of EDTA and back-titrate with standard copper solution,
using PAN as indicator.   Mask the aluminium with 1 g of ammonium
fluoride and titrate the liberated EDTA in the same way.

Cobalt and iron.   Use a 25-ml aliquot of solution for determination
of cobalt (or iron) by back-titration of excess of EDTA with standard
copper solution and PAN as indicator.   The authors recommended this
titration, because for all metals we need only one standard metal
solution (copper).

Analysis of soft magnetic alloys

In another work [85] Wilkins and Hibbs paid attention to the analysis
of alloys containing either manganese or vanadium besides iron and
cobalt.   These magnetic alloys are important for their high
saturation value.   Manganese or vanadium is added to facilitate
cold-working of the material.

Dissolve 1 g of alloy in 9M hydrochloric acid, add 5 drops of 30%
hydrogen peroxide solution and boil for 10 min.   Add 0.1 g of sodium
sulphite and boil for 10 min.   Transfer the solution to a 100-ml
standard flask and dilute to volume with 9M hydrochloric acid.
Transfer a 10-ml aliquot to a 35 x 1 cm column of Dowex-1 (100 - 150
mesh, chloride form) previously washed with 9M hydrochloric acid.
Elute the manganese (or vanadium) with 30 ml of 9M hydrochloric acid
into a 250-ml beaker.   Place another beaker under the column and
elute with 4M hydrochloric acid until the blue band of cobalt is
removed from the column.   Then elute the yellow band of iron into
another beaker with 0.5M hydrochloric acid.   Evaporate each fraction
to a few ml, add 50 ml of water and 10 ml of sodium acetate-acetic
acid buffer and determine the individual metals complexometrically.
The authors back-titrated excess of EDTA with copper sulphate
solution, the end-point being indicated by quenching of the fluorescence of Calcein.

Analysis of permanent magnet alloys (for Fe, Al, Cu, Ni, Co and Ti)

Hibbs and Wilkins [86] worked out the complete analysis of these
alloys by the ion-exchanger method.   It is more complicated than the
previous method and will be given here only in brief.

Dissolve a 0.5-g sample in aqua regia and evaporate to a syrupy
consistency (repeating this several times, with dilution between
evaporations) to remove nitric acid.   Transfer the solution to a
column containing 80 g of strongly basic anion-exchanger (Dowex 1 X8,
200 - 400 mesh, chloride form) with 7M hydrochloric acid.   Elute Al +
Ni + Ti with 100 ml of 7M hydrochloric acid (followed by 50 ml of 5M
hydrochloric acid).   Then elute cobalt with 5M hydrochloric acid
(complete removal of the blue cobalt band) and copper (the lower
yellow band) with 3M hydrochloric acid.   Finally elute iron with
0.5M hydrochloric acid.   Evaporate each fraction to about 5 ml.

By this method four solutions are obtained.   The authors determined
the individual metals complexometrically by a method similar to that
used for the soft magnetic alloys.   Cobalt, iron and copper were
determined by back-titration of excess of EDTA with copper sulphate,
with Calcein as indicator.

The fraction containing aluminium, nickel and titanium was treated chromatographically.   The details can be found in the original paper [86].

According to the authors the method is also suitable for other nickel-base alloys containing aluminium and titanium.

Wilkins [87] later extended this ion-exchange work to analysis of heat-resistant alloys containing iron, nickel, cobalt, chromium, titanium, tungsten, molybdenum, niobium and tantalum.

Determination of aluminium and iron in magnetic alloys [88]

Dissolve 0.2 g of sample in 30 ml of concentrated hydrochloric acid and 10 ml of nitric acid (1 + 1) and evaporate the solution with 15 - 20 ml of sulphuric acid (1 + 1) to fumes.   Cool and dilute to volume in a 100-ml standard flask with water.   For the determination of iron dilute a 25-ml aliquot to 50 ml and neutralize with ammonia. Add 2 g of ammonium chloride and 30% hexamine solution in excess and heat at 80° for 10 - 15 min.   Collect the precipitate and wash it with water, then dissolve it with hydrochloric acid.   Adjust the pH to 2, heat, add 10 - 15 drops of 25% sulphosalicylic acid solution and titrate with 0.1M EDTA to the transition from reddish violet to lemon-yellow.

To determine Al + Fe dilute another 25-ml aliquot to 50 ml and boil the solution after addition of 4 ml of 40% lactic acid solution (to mask Ti).   Add 20 ml of 0.05M EDTA, adjust to pH 5.5 with hexamine and back-titrate with 0.05M bismuth nitrate, using Xylenol Orange as indicator.

This method is a modification of the procedure of Culp [89].   It is only one more variation for the determination of iron and aluminium and does not consider the titanium, which can be present in some alloys in titratable amounts (see preceding section on permanent magnet alloys).   Perhaps more convenient is the determination of Fe and Al by the fluoride method (p. 114) (masking of Al + Ti) by back-titration of EDTA with zinc or lead solution with Xylenol Orange as indicator.   Titanium, if present, can be separated as the hydroxide from triethanolamine medium [29] (see p. 257).

Analysis of iron-base nickel-chromium alloys [90]

Iron alloys containing chromium and nickel are widely used as refractory and corrosive-resistant materials, as electrical resistors and as alloying additives in metallurgy.   Their composition is mostly suitable for complexometric analysis.

Use 0.5 g of sample, as drillings.   Dissolve alloys high in nickel in 30 - 50 ml of aqua regia, and alloys high in chromium and low in nickel in sulphuric acid (1 + 3) and oxidize with a few drops of nitric acid.   In both cases heat on a sand-bath to ensure complete dissolution.   If aqua regia is used add 10 - 15 ml of sulphuric acid (1 + 3) after complete dissolution and evaporate to white fumes (avoid losses by not overheating).   Add water carefully and re-evaporate.   Dissolve the residue in hot water.   Transfer the solution to a 250-ml standard flask, dilute to the mark with water, and mix.

Determination of Fe (and Ni) and Cr.   Pipette an aliquot of the
sample solution into a large conical flask, dilute to 100 ml with
water, and add enough 0.05M EDTA to complex all the iron and nickel.
Adjust to pH 5 - 6 with hexamine, add some Xylenol Orange and titrate
with 0.05M lead nitrate to a red-violet colour.   The result
corresponds to iron or nickel or their sum.   Now add enough 0.05M
EDTA to complex all the chromium, adjust to pH 1 - 2 with 1M nitric
acid and boil for 15 min.   Cool, dilute to 400 ml with water, adjust
to pH 5 - 6 with hexamine, and titrate (with vigorous swirling) with
0.05M lead nitrate to a pale rose to red-violet end-point.

The end-point in both titrations is very sharp and within 0.02 ml of
0.05M EDTA.   The violet Cr-EDTA complex is formed quantitatively by
sufficient boiling at pH 1 - 2.   At higher or lower pH, the results
are low.

Because of the intense colour of the Cr-EDTA complex, the solution
should be diluted before the second titration;   the end-point colour
change is sharp enough when the chromium content is < 2.5 mg/100 ml.

Determination of Cr and Ni in the presence of Fe.   Transfer a sample
aliquot containing not more than 25 mg of iron to an 800-ml tall-form
beaker.   Dilute to 100 ml with water.   Add a known excess of 0.05M
EDTA, adjust to pH 1 with hydrochloric or nitric acid and boil for 15
min.   Dilute the hot solution to 400 ml with water, add 20 ml of 20%
triethanolamine solution and make the warm solution alkaline with
30 - 50 ml of 1M potassium hydroxide.   (The colour changes from red-
violet to light blue).   Add Fluorexone (Calcein) until the solution
is wine red in colour (if a slight fluorescence appears after
addition of the indicator, the solution is insufficiently alkaline
and more potassium hydroxide must be added).   Titrate with 0.05M
calcium chloride to appearance of a distinct green fluorescence.
The result corresponds to the sum of chromium and nickel.   Now add
1 g of potassium cyanide, stir, and after 5 min titrate the liberated
EDTA with 0.05M calcium chloride to a green fluorescence.   The
result corresponds to nickel.

If too much iron is present in the sample, make the aliquot initially
6M in hydrochloric acid and extract with ether, and evaporate the
aqueous phase to about 10 ml.   Then continue as above.   The reason
for doing this is that not more than 25 mg of iron can be properly
masked with triethanolamine.

As the reaction of cyanide with the Ni-EDTA complex is not rapid, it
is best to add the cyanide to the solution warmed to 35 - 40°C and wait
for a few minutes before the titration.   If Thymolphthalexone is
used as indicator the end-point colour is deep blue.

Analysis of chromium-nickel alloys according to Crişan and Pfeffer
[91]

These authors proposed a simpler method than the one just given.
After the dissolution of 2 g of alloy in hydrochloric and nitric acid,
the solution is diluted to 1 litre.   In an aliquot containing 2 -
5 mg of nickel and 8 - 10 mg of chromium, iron is determined by
titration with 0.02M EDTA after addition of sulphosalicylic acid as
indicator (colour change from red-orange to yellow green).   Then
the solution is made alkaline to pH 8 with ammonia buffer and nickel
is titrated with 0.02M EDTA and murexide as indicator (colour change

from yellow-orange to violet).   In the same solution chromium also
can be determined as follows: after neutralization of the solution,
concentrated hydrochloric acid is added to destroy the murexide, a
known (and excessive) amount of 0.02M EDTA is added, the mixture is
boiled for 1 - 2 min (formation of the Cr-EDTA complex), then the pH is
adjusted to 4 - 5 and the surplus EDTA back-titrated with 0.05M ferric
chloride (colour change of sulphosalicylic acid from pale violet to
red violet).

This method probably has limited use, and is not applicable for
samples rich in nickel, which interferes in the determination of iron.
Also the amount of chromium present (8 - 15 mg) seems too high because
of the slow formation of the Cr-EDTA complex (boiling for at least
10 min is necessary) and because of the intense colour of the Cr-EDTA
complex.

Analysis of iron-nickel-cobalt-copper alloys [92]

Iron alloys containing moderate amounts of nickel and cobalt and
varied amounts of copper are widely used in electronics.   The rapid
analysis of such alloys is therefore of considerable interest.   The
alloys can be analysed completely by complexometric titrations,
provided that all four components are present in titratable amounts
(at least a few per cent).   Very low copper contents (< 2%) are
better determined photometrically.

Dissolve 0.25 - 0.5 g of sample in nitric acid (1 + 1) and dilute to
volume in a 250-ml or 500-ml standard flask.

Determination of Ni and Co.   To a portion of sample solution containing 2 - 50 mg of cobalt and 2 - 75 mg of nickel add sufficient 0.05M
EDTA to complex all the nickel and cobalt, 10 ml of 20% triethanolamine solution and 20 ml of 1M potassium hydroxide.   Stir and dilute
to 100 - 150 ml with water.   Add 1 - 5 ml of 3% hydrogen peroxide and
let stand for 5 min.   Add 5 ml of 10% thioglycollic acid solution
(to mask Cu) and some Fluorexone (Calcein) and titrate with 0.05M
calcium chloride solution to a green fluorescence.   The result
corresponds to the sum of cobalt and nickel.   Add 0.2 - 1.0 g of
solid potassium cyanide and after 5 - 10 min titrate again with 0.05M
calcium chloride.   The result corresponds to nickel.

Determination of copper.   Take an aliquot containing at least 15 mg
of copper, add 0.05M EDTA in excess, adjust the pH to 5 - 5.5 with
hexamine, add a few drops of Xylenol Orange and titrate the excess of
EDTA with 0.05M lead nitrate.   The results correspond to the sum of
Fe + Ni + Co + Cu.   Pipette another aliquot (of the same volume),
add 1 g of ammonium fluoride and 20% thiourea solution.   Add an
excess of 0.05M EDTA, adjust the pH to 5 - 5.5 and titrate again with
0.05M lead nitrate.   The result corresponds to the sum of Fe + Ni +
Co.   The difference between the two results corresponds to copper.
The iron can also be calculated by difference.

The determination of copper is only reliable if enough copper is
present in the alloy.   For low amounts of copper the photometric
determination described by Šedivec and Vašák [93] should be used.
It is based on the reaction of copper with diethyldithiocarbamate in
ammoniacal solution containing EDTA to mask iron, cobalt and nickel.
The yellow-brown copper complex is extracted with chloroform and
measured.   The method of Šedivec and Vašák can also be used for

AC - J

milligram amounts of copper as follows. After extraction of the
copper complex into chloroform, copper is displaced into the aqueous
phase by addition of mercuric nitrate dropwise until the organic
phase is colourless. The small excess of mercury in the aqueous
phase is masked with 5% thiosemicarbazide solution. After addition
of 0.05M EDTA and adjustment of the pH with hexamine, the copper is
determined by back-titration with lead nitrate as above.

REFERENCES

1.   I. Volhard, Liebigs Ann., 1879, 198, 318.
2.   N. Wolff, Stahl Eisen, 1891, 11, 377.
3.   J. Kinnunen and B. Merikanto, Chemist-Analyst, 1954, 43, 93.
4.   J. Jurczyk, Neue Hütte, 1968, 13, 58; Chem. Abstr., 1968, 68,
     119147x.
5.   H. Flaschka and R. Püschel, Chemist-Analyst, 1955, 44, 71.
6.   R. Přibil and M. Kopanica, Chemist-Analyst, 1959, 48, 35.
7.   R. Přibil, Coll. Czech. Chem. Commun., 1955, 20, 162.
8.   V.A. Rad'ko and E.M. Yakimets, Trudy Ural. Politekh. Inst.,
     1960, No. 96, 166; Anal. Abstr., 1961, 8, 4634.
9.   I. Sajó, Acta Chim. Acad. Sci. Hung., 1961, 28, 253.
10. I. Chekhlarova, I. Gabrovskii and P. Ganev, Rudodobiv. Metal,
     1970, 25, 45; Anal. Abstr., 1971, 21, 3402.
11. S. Spauszus, Neue Hütte, 1963, 8, 108; Chem. Abstr., 1965, 63,
     10672d.
12. A.G.C. Morris, Chemist-Analyst, 1960, 49, 105.
13. K. Študlar, Chemist-Analyst, 1960, 49, 106.
14. H. Bec and E. Kazior, Hutnik (Katovice), 1972, 19, 83; Anal.
     Abstr., 1973, 24, 1582.
15. R. Přibil and V. Veselý, Talanta, 1961, 8, 565.
16. A. Karolev, R. Popova and J. Daneva, Metalurgiya (Sofia), 1976,
     31, 18; Anal. Abstr., 1977, 33, 5B 173.
17. E. Lassner and H. Schedle, Talanta, 1968, 15, 623.
18. V.D. Konkin and V.I. Zhikhareva, Zavod. Lab., 1963, 29, 791;
     Anal. Abstr., 1964, 11, 3680.
19. K. Poljaková and M. Čakon, Hutn. Listy, 1972, 27, 743; Anal.
     Abstr., 1973, 24, 3473.
20. T. Endo and T. Higashimori, Japan Analyst, 1962, 11, 1310;
     Anal. Abstr., 1964, 11, 1712.
21. M. Kawahata, H. Mochizuki, R. Kajiyama and M. Ishii, Japan
     Analyst, 1962, 11, 748; Anal. Abstr., 1964, 11, 2616.
22. R. Přibil and V. Veselý, Hutn. Listy, 1972, 27, 439; Anal.
     Abstr., 1973, 11, 804.
23. L. Šucha, Chem. Prum., 1967, 17, 324; Anal. Abstr., 1968, 15,
     5355.
24. S.I. Gusev and E.M. Nikolaeva, Izv. Vyssh. Ucheb. Zaved. Khim.
     Khim. Tekhnol., 1969, 12, 98; Anal. Abstr., 1970, 18, 3118.
25. O.I. Popova and O.G. Seraya, Zh. Analit. Khim., 1968, 23, 791;
     Anal. Abstr., 1970, 18, 177.
26. E.I. Nikitina and N.N. Adrianova, Zavod. Lab., 1965, 31, 654;
     Anal. Abstr., 1966, 13, 5517.
27. V.N. Tikhonov and V.A. Budnichenko. Zavod. Lab., 1974, 40, 381;
     Anal. Abstr., 1974, 27, 3265.
28. B. Bieber and V. Večeřa, Coll. Czech. Chem. Commun., 1961, 26,
     2081.
29. R. Přibil and V. Veselý, Talanta, 1963, 10, 233.

30. R. Přibil and V. Veselý, Hutn. Listy, 1963, 18, 512; Anal. Abstr., 1964, 11, 4216.
31. R. Přibil and V. Veselý, Chemist-Analyst, 1963, 52, 43.
32. J. Adam and R. Přibil, Talanta, 1975, 22, 905.
33. E.S. Bruile, Zh. Prikl. Khim. Leningr., 1966, 39, 1192; Anal. Abstr., 1967, 14, 5408.
34. D. Filipov and N. Kirtcheva, Compt. Rend. Acad. Bulg. Sci., 1964, 17, 467; Anal. Abstr., 1965, 12, 4545.
35. K. Isagai and N. Takeshita, Japan Analyst, 1955, 4, 222; Anal. Abstr., 1956, 3, 98.
36. K.L. Cheng, Talanta, 1959, 2, 61, 186.
37. K. Hosohara, R. Kuroda and K. Kazuki, Japan Analyst, 1962, 11, 841.
38. D. Čechová, Chemist-Analyst, 1967, 56, 94.
39. J. Müllerová and J. Janošik, Hutn. Listy, 1969, 24, 448; Anal. Abstr., 1970, 19, 1279.
40. J. Sláma and J. Tůma, Hutn. Listy, 1971, 26, 137; Anal. Abstr., 1971, 21, 4094.
41. S. Wakamatsu, Japan Analyst, 1958, 7, 578; Anal. Abstr., 1959, 6, 3446.
42. A.E. Klygin and N.S. Kolyada, Zavod. Lab., 1961, 27, 23; Anal. Abstr., 1961, 8, 3224.
43. J. Musil, Hutn. Listy, 1970, 25, 194; Anal. Abstr., 1971, 20, 1617.
44. R. Přibil, Coll. Czech. Chem. Commun., 1954, 19, 58.
45. R. Přibil and M. Kopanica, Chemist-Analyst, 1959, 48, 66.
46. W.F. Harris and T.R. Sweet, Anal. Chem., 1952, 24, 1062.
47. H. Flaschka, Chemist-Analyst, 1953, 42, 84.
48. H. Green and P.J. Rickards, B.C.I.R.A.J., 1964, 12, 578; Anal. Abstr., 1966, 13, 172.
49. J. Iron Steel Inst., 1971, 209, 215; Anal. Abstr., 1972, 22, 154.
50. K.B. Khlystova, Uchen. Zap. Yaroslav Tekhnol. Inst., 1966, 9, 121; Anal. Abstr., 1968, 15, 3937.
51. E.Yu. Zel'tser and M.V. Skorospelova, Trudy Vses. Nauchno-Issled. Inst. Electromekh., 1971, 35, 178; Anal. Abstr., 1972, 23, 3176.
52. S. Spauszus, Neue Hütte, 1961, 6, 653; Chem. Abstr., 1962, 56, 6642h.
53. S. Spauszus and C. Schwarz, Neue Hütte, 1962, 7, 180; Chem. Abstr., 1962, 57, 4012a.
54. U.T. Hill, Anal. Chem., 1959, 21, 429.
55. J. Jurczyk, Chem. Anal. Warsaw, 1971, 18, 717; Anal. Abstr., 1972, 22, 3153.
56. M. Kurjaković and R. Plepelić, Kemija Ind., 1965, 14, 650; Anal. Abstr., 1966, 13, 6852.
57. H. Habercetl, Hutn. Listy, 1963, 18, 138.
58. V.I. Kalina-Zhikhareva, Trudy Ukr. Nauchno-Issled. Inst. Metal., 1964, No. 7, 285.
59. L. Brháček, Coll. Czech. Chem. Commun., 1959, 24, 2811.
60. S. Maekawa, Y. Yoneyama and T. Kamada, Japan Analyst, 1961, 10, 187; Anal. Abstr., 1963, 10, 2286.
61. A.B. Rybalov and M.S. Eshmukhambetova, Zavod. Lab., 1975, 41, 544; Anal. Abstr., 1976, 30, 1B 18.
62. H. Green, J. Brit. Cast Iron Res. Assoc., 1955, 6, 20; Anal. Abstr., 1956, 3, 353.
63. G. Kraft and J. Fischer, Giessereiforschung, 1968, 20, 131; Anal. Abstr., 1970, 18, 202.
64. J. Fischer and G. Kraft, Spheroidal Cast Iron Bull., 1957, 20, 39, 52.

65. R. Reichert, Z. Anal. Chem., 1956, 150, 250.
66. R. Reichert, Giesserei, 1957, 44, 51; Chem. Abstr., 1957, 51, 6431e.
67. H. Gamsjäger and R. Reichert, Z. Anal. Chem., 1957, 158, 356.
68. R. Leo, Wissen. Z. Hochschule Schwermaschinenbau, Magdeburg, 1958, 1, 237.
69. K.G. McLaren, Anal. Chim. Acta, 1959, 21, 165.
70. R. Leo, Wissen. Z. Hochschule Schwermaschinenbau, Magdeburg, 1957, 1, 121.
71. J. Siekierska, Hutn. Listy, 1967, 17, 667; Anal. Abstr., 1963, 10, 2284.
72. A.K. De, S.M. Khopkar and R.A. Chalmers, Solvent Extraction of Metals, Van Nostrand-Reinhold, London, 1970.
73. R.C. Rooney and R. Carter, J. Res. Brit. Cast Iron Assoc., 1958, 7, 442.
74. V.N. Tikhonov, N.A. Gordeeva and B.G. Karnaukhov, Zavod. Lab., 1974, 40, 940; Anal. Abstr., 1975, 28, 4B 146.
75. D.V. Bezuglii and A.A. Amsheeva, Zh. Analit. Khim., 1962, 17, 1045; Anal. Abstr., 1963, 10, 4674.
76. S. Maekawa, Y. Yoneyama and D. Morinaga, Japan Analyst, 1967, 16, 455; Anal. Abstr., 1968, 15, 6659.
77. N. Yamaguchi, A. Hata and M. Hasegawa, Japan Analyst, 1967, 16, 253; Chem. Abstr., 1967, 67, 39807b.
78. S. Wakamatsu, Trans. Japan Inst. Metals, 1962, 3, 27; Chem. Abstr., 1963, 58, 1905a.
79. A.A. Amsheeva and D.A. Bezuglii, Zh. Analit. Khim., 1964, 19, 97; Anal. Abstr., 1965, 12, 2229.
80. A.A. Amsheeva, Zavod. Lab., 1968, 34, 789; Anal. Abstr., 1969, 17, 2740.
81. A.A. Amsheeva, Vestn. Kharkov. Politekhn. Inst., 1968, No. 80, 45; Anal. Abstr., 1970, 18, 3987.
82. S. Wakamatsu, Japan Analyst, 1957, 6, 579; Anal. Abstr., 1958, 5, 2589.
83. L.L. Lewis and W.A. Straub, Anal. Chem., 1960, 32, 96.
84. D.H. Wilkins and L.E. Hibbs, Anal. Chim. Acta, 1958, 18, 372.
85. D.H. Wilkins and L.E. Hibbs, Anal. Chim. Acta, 1959, 20, 427.
86. L.E. Hibbs and D.H. Wilkins, Talanta, 1959, 2, 16.
87. D.H. Wilkins, Talanta, 1959, 2, 355.
88. N.F. Budyak and N.N. Nikolaeva, Zavod. Lab., 1970, 36, 1198;
89. S.L. Culp, Chemist-Analyst, 1967, 56, 29.
90. R. Přibil and V. Veselý, Chemist-Analyst, 1961, 50, 100.
91. I.A. Crişan and M.M. Pfeiffer, Revta. Chim. Bucharest, 1967, 18, 109; Anal. Abstr., 1968, 15, 2645.
92. R. Přibil and V. Veselý, Chemist-Analyst, 1963, 52, 5.
93. V. Šedivec and V. Vašák, Coll. Czech. Chem. Commun., 1950, 15, 260.

9.3   LEAD AND LEAD ALLOYS

Lead is a major component of many important alloys, which have wide
applicability, such as white metals, special low-melting alloys,
solders, bearing metal, type metals etc.   Lead itself is also very
much used and in some cases its purity is important (e.g. bismuth-
free lead for accumulators).   Many alloys used in electronics must
have a closely defined composition.

Analysis of such alloys by EDTA titration is very useful as it is
also possible to determine complexometrically further components such

as bismuth, tin, copper, cadmium, nickel, zinc and cobalt, though not
antimony, which is usually determined by bromate titration.   Because
of the diverse composition of the various alloys a uniform procedure
cannot be given.   Therefore determination of lead will be described
first and then analysis of simple alloys and special multicomponent
alloys.

9.3.1   Determination of Lead in the Presence of Other Metals

Lead is usually determined (gravimetrically or by EDTA titration)
after separation as lead sulphate, or after masking of interfering
metals.

Študlar and Janoušek [1] have studied the separation of lead as
sulphate.   They find that this method gives somewhat low results
owing to the solubility of lead sulphate in the sulphuric acid used
for washing.   They state that the optimal concentration of sulphuric
acid is 5% in agreement with the general opinion in the literature
(3 - 5%).   They also state that small amounts may pass into the
filtrate even if a fine-pore filter paper is used.   They prefer
direct EDTA titration of lead in ammoniacal tartrate solution after
masking of copper, cadmium, zinc, and nickel with potassium cyanide,
and recommend Methylthymol Blue as a superior indicator to Erio T.

The problem of interfering metals was studied in detail by Harzdorf
[2].   He masks iron, aluminium, bismuth, tin and antimony with
triethanolamine, and bivalent metals with potassium cyanide.

It is well known that the best determination of lead is its EDTA
titration at pH 5 - 5.5 in hexamine-buffered solution, with Xylenol
Orange as indicator.   The selectivity is very low, however, because
all elements present in lead alloys interfere.   Masking is usually
limited to ammonium fluoride for aluminium and thiourea for copper.
Very suitable, but not much used, is phenanthroline for masking
copper, nickel, cadmium, zinc.   Titration in acid medium also has
the advantage that we can determined bismuth and lead successively,
and, with the aid of phenanthroline, also cadmium in a single sample
solution.

9.3.2   Determination of Lead in Lead Drosses

Pinkston and Kenner [3] proposed a simple method for determination of
total lead, based on separation of lead as the sulphate, its
dissolution in ammonium acetate solution, and titration in ammoniacal
tartrate medium.

Dissolve (in a conical flask) a suitable weight of sample containing
0.40 - 0.48 g of lead, in 20 ml of concentrated sulphuric acid with
boiling.   Add a small piece of filter paper to reduce antimony.
After the dissolution add a few drops of concentrated nitric acid to
destroy the filter paper and other organic matter.   Cool the
solution, dilute with 200 ml of water, add approximately 2 g of
tartaric acid and boil the solution to facilitate complexation of
iron and antimony.   Cool the solution and filter off the sulphate on
paper of medium porosity and wash with 2M sulphuric acid.   Return
the precipitate to the original flask, boil for 30 min with 30 ml of
4M ammonium acetate, add 200 ml of water, 2 g of tartaric acid and

25 ml of concentrated ammonia solution and titrate with EDTA at 70-
80°C, with Erio T as indicator.

We may expect the results to be low, as explained above (p. 277).
For the influence of ammonium acetate on the accuracy of the lead
titration see p. 153.

9.3.3   Determination of Lead in Accumulator Plates

Filippova and Dubrovskaya [4] described a method for the determination of lead oxide, lead sulphate and lead or lead dioxide in
charled and uncharged lead plates. First, lead oxide is selectively
extracted with acetic acid, then lead sulphate with sodium chloride.
The lead in the extracts is determined complexometrically.   The
residue contains the lead and lead dioxide.

Add to an appropriate amount of sample 100 ml of 5% acetic acid and
stir for 15 min.   Collect the residue and wash it with 5% acetic
acid and then with 40 - 50% aqueous ethanol to remove soluble lead.
Dilute the filtrate to volume in a 50-ml standard flask with water
and determine the lead with EDTA in a 10 or 20 ml portion after
addition of ammonia and tartaric acid buffer as described above.   The
consumption of EDTA corresponds to the amount of lead oxide.

Place the filter and residue in the original beaker, add 100 ml of
20% sodium chloride solution and stir for 1 hr.   Filter, dilute the
filtrate to volume in a 250-standard flask with water, determine lead
in an aliquot with EDTA and calculate the amount of lead sulphate.
The residue on the filter contains metallic lead and lead dioxide.
Dissolve it with 25 ml of nitric acid (1 + 1) and a few drops of
hydrogen peroxide, dilute to 250 ml (standard flask) with water and
determine lead in an aliquot in the usual way.

The weight of sample should be 1 g for a discharged plate and 2 - 2.5 g
for a charged one.   The extraction of lead sulphate with sodium
chloride is incomplete in the presence of lead oxide, which reduces
the solubility of the sulphate.   Therefore the oxide is extracted
first with 5% acetic acid as described above.

In a further paper [5] the authors described analysis of a mixture of
lead and lead dioxide by dissolution in a mixture of EDTA, acetic
acid and potassium iodide.   The liberated iodine is determined with
sodium thiosulphate and total lead in the same solution with EDTA.

9.3.4   Analysis of Simple Pb-Sn Alloys

The analysis of such alloys has been made very fast by use of EDTA
titration.

Determination of tin [6].   In a 50-ml conical flask dissolve the
sample (containing 10 - 12 mg of tin) in 4 ml of sulphuric acid (1 + 1),
on a hot-plate at 200 - 225°C.   The tin is dissolved and lead remains
in the solution as the sulphate.   Cool the mixture, add 20 ml of
0.005M EDTA, and adjust the pH to 2 - 2.5 with 70% ammonium acetate
solution (pH-meter).   If copper is present in small amount add some
thiourea and allow to stand for several minutes.   By pipette add
5.00 ml of 0.005M bismuth nitrate, then one drop of Xylenol Orange,

and titrate the excess of bismuth with 0.005M EDTA from red to yellow. The determination takes only 10 min. According to the authors it is possible to determine as little as 4 mg of tin with excellent precision and accuracy.

Determination of lead [7].   Dissolve 0.3 - 0.5 g of the alloy in 10 - 20 ml of nitric acid (1 + 1). Dilute to 200 ml (standard flask) and filter off the metastannic acid from part of the solution (dry filter, funnel and receiver). To 25 ml of the filtrate add 0.5 - 1.0 ml of 20% tartaric acid solution, a small amount of indicator (1% solid mixture of Methylthymol Blue with potassium nitrate) and 10% ammonia solution until the solution is blue. Add 10% NH_4Cl/NH_3 buffer and titrate with 0.025M EDTA from blue to grey-brown. According to the authors metastannic acid does not adsorb any lead if the solution is not evaporated to dryness.

The determination of lead in alkaline medium can be improved by replacement of tartaric acid with triethanolamine and addition of only ammonia to adjust the pH. A large amount of buffer causes an unsharp end-point, which might be from pure blue to smoky grey. The brownish colour at the end-point in the original method was probably caused by traces of iron, which would be masked if triethanolamine were used.

The determination of lead in acid medium is more convenient, however. The filtrate could be diluted and buffered with hexamine to pH 5 - 5.5 and titrated directly with EDTA (Xylenol Orange as indicator).

Determination of lead [8]. For the batch analysis of a large number of samples Tschetter and Bachman [8] recommended a special method based on modified dissolution of samples in a mixture of sodium fluoride and nitric acid, prepared by dissolving (in a 250-ml propylene beaker) 10 g of sodium fluoride in 200 ml of nitric acid (1 + 1).

Dissolve up to 750 mg of the alloy, in a suitable size of propylene beaker (30 - 100 ml), in 1 ml of dissolution mixture per 20 mg of sample (to obtain the optimal concentration of NaF for masking tin - higher concentrations precipitate PbF_2). Cover the beaker with plastic film to avoid evaporation and let stand at room temperature until dissolution is complete. The further steps depend on the amount of lead and the weight of sample. For sample weights containing less than 15 mg of lead, transfer the whole solution to a 150-ml glass beaker, add 300 - 600 mg of tartaric acid to prevent precipitation of lead, add 0.5 - 1.0 g of hexamine and adjust to pH 5 with nitric acid. Titrate with 0.02M EDTA, using Xylenol Orange as indicator. For larger weights of lead use an aliquot containing 15 - 35 mg and titrate as described. According to the authors this method is suitable for Pb-Sn alloys containing 20 - 80% lead. The original paper gives numerous useful practical hints.

9.3.5   Analysis of White Metals (Pb-Sn-Sb-Cu)

These alloys are normally analysed gravimetrically for Pb, titrimetrically for Sn and Sb, and electrolytically for Cu. The gravimetric determination of lead as lead sulphate is tedious. When other elements such as cadmium are also to be determined, the whole analysis is rather time-consuming.

Ottendorfer [9] proposed complexometric determination of lead and
copper, in alkaline tartrate medium to prevent interference by tin
and antimony.   Copper is selectively masked with potassium cyanide.
He extended the complexometry to determination of the tin [10].
Because these procedures are fast and give good results, they will be
described in detail.

Dissolve 0.5 g of sample in 10 ml of concentrated hydrochloric acid
and 2 ml of concentrated nitric acid, avoiding unnecessary heating.
Add 25 ml of 4% potassium chloride solution and 25 ml of 0.2M EDTA.
Boil gently for 1 min, cool, and dilute to the mark in a 250-ml
standard flask.

Determination of lead + tin.   Transfer a 25-ml portion to a 250-ml
conical flask, add a saturated solution of thiourea until the blue
colour of copper disappears, then 0.5 ml in excess.   Add 15 ml of
30% hexamine solution, dilute to about 150 ml, add a few drops of
Xylenol Orange solution and titrate the excess of EDTA with 0.05M
lead nitrate to a red end-point.

Determination of tin.   To the same solution after the titration, add
2 g of ammonium fluoride (the solution becomes yellow as EDTA is
freed from the tin complex) and titrate slowly with lead nitrate to
an intense red that remains for more than 1 min.

Determination of lead + tin + copper.   Transfer another portion of
original solution to a 250-ml conical flask, add 15 ml of 30%
hexamine solution, dilute to 150 ml with water, add a few drops of
Xylenol Orange and titrate the excess of EDTA with 0.05M lead nitrate
to the red colour.

The author obtained very good results for alloys of various composition, with 4 – 68% tin, 72 – 14% lead and 2 – 8% copper.   Antimony
(10 – 20%) was determined in a separate sample by dissolving it in
sulphuric acid and titrating with potassium bromate.

Teperek [11] developed a relatively fast method for determination of
copper and less than 1% of lead in white metals.   After dissolution
of the sample (2 g) in hydrochloric acid (+ hydrogen peroxide) he
reduced the copper with sodium metabisulphite and added pyridine and
sodium hydroxide followed by potassium thiocyanate to co-precipitate
lead with the copper.   He dissolved the precipitate in nitric acid
and diluted the solution to volume in a 100-ml standard flask.   In
one aliquot he determined the sum Cu + Pb by titration with 0.01M
EDTA (PAN as indicator).   In a second aliquot he masked copper with
sodium thiosulphate and titrated lead with 0.01M EDTA, using Xylenol
Orange as indicator.

9.3.6   Analysis of Low-Melting Alloys [12]

Low-melting Bi-Pb-Cd-Sn alloys of the type of Wood's and Rose's metal
contain all these metals in substantial amounts and appear ideal
material for EDTA titration methods.   Successive determination of
bismuth, lead and cadmium in the same solution offers no difficulties
but tin, in the dissolution of the sample, is separated as insoluble
metastannic acid, which adsorbs other heavy metals (these can be
recovered by a sulphide separation).   On the other hand tin can be
removed by evaporation with nitric - hydrobromic acid mixture and

determined in a separate sample by bromate titration after reduction
to tin(II). If metastannic acid is precipitated it can be purified
by washing with an ammoniacal solution of EDTA [13] and determined
gravimetrically. This last method, combined with complexometry, is
described in Procedure A, below. Separation of tin by volatilization as tin bromide will be described in procedure B. A third
possibility for avoiding interference by tin is to mask it with
fluoride, and will be described in the next section (Bi-Pb-Sn alloys).

Procedure A. Dissolve 0.5 - 1.0 g of alloy in concentrated nitric
acid and evaporate to small volume. Dilute to 150 ml with water and
boil briefly. Filter off on fine-pore filter paper. Into a small
beaker pipette a known amount of 0.05M EDTA and make slightly alkaline with aqueous ammonia. Wash the metastannic acid on the filter
with this solution and then with about 50 ml of hot water (the final
washings should give no reaction with 5% sodium sulphide solution).
Char the filter and after ignition weigh the tin(IV) oxide.

Transfer the combined filtrate and washings to a 250-ml standard
flask and dilute to the mark with water.

Determination of bismuth. Into a conical flask pipette a known
volume (50 - 100 ml) of the solution thus prepared, dilute it to
150 ml and adjust to pH 1 - 2 with nitric acid or ammonia (pH-paper).
Add five drops of 0.1% Xylenol Orange solution and titrate with 0.05M
EDTA to a lemon-yellow end-point. The EDTA consumed plus the EDTA
originally present in the portion taken corresponds to the amount of
bismuth present.

Determination of lead and cadmium. After titration of the bismuth,
to the same solution add small amounts of hexamine until an intensely
red-violet colour persists. Again titrate with EDTA. The consumption of EDTA corresponds to the lead plus cadmium content. Add 5 -
10 ml of 0.05M 1,10-phenanthroline (Phen) and with 0.05M lead nitrate
slowly titrate the EDTA thus freed, to an intensely red-violet end-
point. Add 1 - 2 ml more Phen solution and note whether the
indicator colour changes. If it does, continue with the titration
with lead nitrate; the consumption of lead nitrate correspond to the
cadmium content.

It is necessary to wash the metastannic acid thoroughly with the EDTA
solution. It has been found that this requires at least 60% of the
amount of EDTA necessary to bind all bismuth present in the sample.
Under these conditions, the volume of EDTA used in the titration of
the bismuth in an aliquot is small (2 - 6 ml), but the end-point
colour change is so sharp that the duplicate titrations agree to
within 1%.

If only small amounts of bismuth are present it may happen that the
solution contains excess of EDTA. In this case the excess of EDTA
can be back-titrated at pH 1 - 2 with bismuth solution to the red
colour of the Bi-XO complex and then brought back to the yellow end-
point with EDTA.

In the absence of bismuth, the pH can be adjusted directly to 5 - 6
and total lead plus cadmium determined with a lead solution and then
cadmium as described above.

Procedure B. This is based on volatilization of the tin as bromide.

AC - J*

Decompose 0.2 - 0.4 g of sample with 20 ml of concentrated nitric acid
and evaporate almost to dryness. Evaporate three times, adding
10 ml of 40% hydrobromic acid before each evaporation. Dissolve the
final residue in 5 ml of concentrated nitric acid, transfer to a
100-ml standard flask and dilute to the mark with water. Pipette
10 ml of this solution into a conical flask, dilute with water and
determine bismuth, lead and cadmium as before.

In titration of the bismuth, the solution may become turbid owing to
the hydrolysis of BiOBr, but the turbidity disappears during the
titration, and does not influence the results.

The method is applicable to simple low-melting alloys including Pb-Cd,
Bi-Cd-Sn, Bi-Pb-Zn, Pb-Ni-Sn. Tin must be determined in a separate
sample or calculated by difference.

The method can be modified by using one aliquot for determining
bismuth by EDTA titration at pH 1 - 2 and then the sum of Pb + Cd at
pH 5 - 6, and then a second aliquot for bismuth again and (after
addition of Phen) the lead at pH 5 - 6.

9.3.7   Analysis of Bi-Pb-Sn Alloys

Such alloys can be analysed by the two methods just described. In
both, separation of tin is necessary, because tin(IV) forms a stable
complex with EDTA in relatively acid solution and interferes in all
determinations. A superior alternative is to mask tin(IV) with
fluoride or tartaric acid, as below.

Determination of Bi and Pb according to Shteiman et al. [14].
Dissolve 0.2 g of the alloy in a mixture of 25 ml of 4% sodium
fluoride solution and 10 ml of concentrated nitric acid. Add to the
solution 20 - 30 ml of hot water (60-70°C) and dilute with 100 ml of
cold water. After addition of Xylenol Orange adjust to pH 1.5 - 2.0
with 10% ammonia solution and titrate the bismuth with 0.05M EDTA.
Then neutralize the solution with 10% ammonia to a red-violet colour
(formation of Pb-XO complex). After addition of 10 ml of acetate
buffer (500 g of sodium acetate and 10 ml of glacial acetic acid per
litre), titrate again with 0.05M EDTA to a "pale pink colour".

The authors' description of the colour at the end-point proves once
again how acetate interferes (5 g in 100 ml of titrated solution!).
Hexamine is far better (see p. 153).

Determination of Bi, Pb and Sn according to Shakhova [15]. Dissolve
0.5 g of the sample with 50 ml of concentrated hydrochloric acid by
heating. After dissolution add 30% hydrogen peroxide, boil to expel
all chlorine, cool and dilute to volume in a 100-ml standard flask.
The solution must contain at least 15% v/v hydrochloric acid. To a
10-ml aliquot add 1.5 - 3.0 ml of 20% tartaric acid solution, a few
drops of Xylenol Orange and 20% ammonium acetate to give a crimson
colour. Titrate bismuth with 0.025M EDTA to a yellow end-point.
After adjusting to pH 5 - 5.5 with ammonium acetate and hexamine,
titrate the lead with 0.025M EDTA.

In a further 10-ml aliquot determine the sum Bi + Pb + Sn in the same
way, but omitting the addition of tartaric acid.

Hydrochloric acid is not the best medium for the titration of bismuth, which easily hydrolyses to insoluble BiOCl, which dissolves only slowly in EDTA. According to our experience small amounts of tartaric acid prevent the precipitation of BiOCl, but large amounts make the end-point indistinct.

A similar analysis of ternary Pb-Sn alloys (containing, for example, cadmium, zinc or nickel) was described by Yurist and Korotkova [16]. It is based on a set of differential determinations. First the sum of all the metals is found in one aliquot by back-titration of EDTA with ferric chloride at pH 5, using sulphosalicylic acid as indicator. In a second aliquot the total amount of metals other than tin is determined (after addition of tartaric acid) in ammonia solution (pH 10), by back-titration of excess of EDTA with magnesium solution (Erio T indicator). In other aliquots they determine successively lead and zinc, using diethyldithiocarbamate for precipitation of lead. Cadmium can be determined after precipitation of lead and tin with ammonia, by the usual titration with Erio T as indicator.

9.3.8   Analysis of Low-Melting Pb-In Alloys [17]

The presence of indium complicates the bismuth-lead-cadmium determination described in the previous section, because indium forms an extremely stable complex with EDTA (log K = 24.9). It is not possible to determine all four metals without the use of masking agents. The analysis of the alloys in the absence of tin is based on the following procedures.

Determination of bismuth and indium. To an acidic aliquot of sample solution add an excess of EDTA. Adjust to pH 1-2 with ammonia or nitric acid and then to pH 2.2-2.5 with a few ml of monochloroacetate buffer (prepared from 1M monochloroacetic acid by adding sodium hydroxide until the desired pH is obtained). Heat to 50-70°C, add some Xylenol Orange and titrate with 0.05M bismuth nitrate to a red-violet colour. If turbidity appears during the titration, swirl until it disappears, and continue. The consumption of EDTA (A) corresponds to Bi + In.

The direct determination of indium is possible at pH above 3, where bismuth hydrolyses, but back-titration is more reliable. Moderate amounts of lead and cadmium do not interfere.

Determination of the sum Bi + In + Pb + Cd, and of Cd. To another aliquot add an excess of EDTA. Adjust to pH 5-5.5 with hexamine, add some Xylenol Orange and titrate with 0.05M lead nitrate. Consumption of EDTA = B. Add 0.1M 1,10-phenanthroline (2-3 ml for 5 mg of Cd) and titrate the liberated EDTA with 0.05M lead nitrate. Consumption of EDTA = C.

Determination of In + Pb + Cd. Dilute an acidic aliquot to 100-150 ml, add an excess of 0.05M EDTA and 3-5 ml of 16% mercaptoacetic acid solution, and heat the solution to boiling (for masking of bismuth). Add a few drops of Xylenol Orange and titrate slowly with 0.05M zinc nitrate to a red-violet colour. Consumption of EDTA = D. From all four titrations A, B, C, D, the amount of the metals is calculated as follows:

$$B - D = Bi, \quad C = Cd, \quad B - A - C = Pb, \quad A + D - B = In$$

Determination of tin.   For more complicated alloys containing all
four metals and tin we must separate the tin as metastannic acid.
After its purification with EDTA as described above (Procedure A) we
determine tin gravimetrically as SnO_2. The filtrate is then
analysed as already described, but it must be remembered that it
already contains some EDTA, and its amount must be taken into account
in all titrations and calculations.

Direct determination of indium in simpler alloys (e.g. Pb-Sn-In-Ag)
was described by Shakhova [15].   After dissolution of the sample
(0.2 - 0.5 g) in 15 ml of sulphuric acid and cooling, she dilutes the
solution (together with precipitated $PbSO_4$) to 250 ml (in a standard
flask and then filters (dry filter paper).   To a 50-ml aliquot of
filtrate she adds tartaric acid, adjusts to pH 4.5 - 5, and titrates
indium with 0.025M EDTA, using Xylenol Orange as indicator.

Yurist [18] developed a similar method for the analysis of In-Pb-Sn
solder.   According to his procedures, after masking of tin with
tartaric acid, indium is determined with EDTA at pH 3 in boiling
solution and then lead at pH 4.   Xylenol Orange is used as indicator
for both titrations.   He gives a new determination of tin in tartaric
acid solution.   Because tin does not react with Xylenol Orange in
tartaric acid medium at pH 1.8 he adds 2 ml of 0.01M bismuth nitrate
and titrates both tin and bismuth with 0.01M EDTA, and corrects for
the bismuth added.   Biechler [19] has also described a method for
the analysis of In-Pb-Sn alloys.   After the dissolution of 0.25 g of
sample in 25 ml of concentrated sulphuric acid he adds 50 ml of
concentrated hydrochloric acid and dilutes with water to volume in a
250-ml standard flask.   He uses a 50-ml aliquot to determine the sum
of In + Zn by back-titration of excess of EDTA at pH 8.5 with 0.01M
indium sulphate, using Methylthymol Blue as indicator.   After
addition of 5 g of potassium cyanide he titrates the freed EDTA
corresponding to the zinc.   In both titrations fluoride must be
added to mask tin, for the determination of which the author
recommends reduction in another aliquot with metallic lead and
titration of tin(II) with iodate, with starch-iodide as indicator.
Both the Yurist and the Biechler methods could be used for analysis of
the corresponding lead-containing alloys after separation of the lead
as sulphate.

9.3.9   Analysis of Lead-Silver Alloys

Besides simple binary Pb-Ag alloys, several more complicated alloys
are used in electronics.   They contain a certain amount of copper,
zinc and cadmium.   Silver is the main component in many of the
alloys.   Silver in large amounts interferes in the other determinations and must be removed by precipitation or reduction.   In some
cases silver can be masked with potassium cyanide.

Binary Pb-Ag alloys.   For the determination of 2% or more of lead in
silver alloys Erdey et al. [20] describe a very simple method based
on direct titration of lead with 0.01M EDTA, with Xylenol Orange as
indicator, without prior separation of silver.   The titration is
done in a solution buffered to pH 5 with acetate-acetic acid buffer.
At very low concentration of lead (0.2%) this method is not reliable
enough, having an error of about 8% owing to the unsharp end-point of
the indicator.   For separation of the silver the authors proposed
either reduction of the silver with ascorbic acid or the precipitation

of lead and silver as the iodides, followed by decomposition of the
lead iodide by warming with potassium nitrite.

Dissolve 5 g of sample in 2M nitric acid, evaporate to dryness,
dissolve the residue in water and dilute to volume in a 100-ml
standard flask.   To a 50-ml aliquot add 3 drops of 0.5% aqueous
Variamine Blue solution and 40 ml of acetate buffer (pH 5).   Heat to
60°C and titrate the silver with 1M ascorbic acid.   Then add 4 - 5 ml
of ascorbic acid solution in excess, heat to 60°C and shake to
coagulate the precipitate.   Cool, add Xylenol Orange and titrate the
lead with 0.01M EDTA.   This method is suitable for a content of 0.2 -
2.0% of lead.

Sierra and Hernandez Canavate [21] first titrate the silver with
chloride or bromide, using the vanadate/o-dianisidine system as
oxidation-adsorption indicator.   They then titrate the lead with
EDTA, using Erio T as indicator.   These determinations can be used
for mixtures of silver and lead in ratios from 1:4000 to 50:1.

For the determination of silver in lead alloys Rzeszutko [22]
recommends its reduction with copper dust.   After filtration the
copper in solution is determined by EDTA titration, with murexide as
indicator.   According to the author the reduction of silver (up to
100 mg) proceeds quantitatively in dilute acetic acid within 15 min.

Multicomponent Pb-Ag alloys.   Silver solders, besides silver and a
small amount of lead, also contain copper, zinc and cadmium.   They
cannot be analysed by the methods described above.   According to our
own experience [23] we can perform the whole analysis as follows.

Determination of lead.   After dissolution of the sample in nitric
acid add to the solution an excess of 0.01M EDTA, make the solution
alkaline with concentrated ammonia solution, add sufficient potassium
cyanide to mask copper, zinc and cadmium, and titrate the excess of
EDTA with 0.01M calcium solution, using Thymolphthalexone as
indicator.   The end-point change to smoky grey or blue is very sharp.

Determination of lead and Cu + Zn (or Cd).   After dissolution of the
sample, precipitate silver with hydrochloric acid, filter off the
silver chloride, and wash it thoroughly with 2% nitric acid.   Make
up the filtrate to known volume.   To one fraction add an excess of
EDTA, partially neutralize with potassium hydroxide, then buffer with
hexamine to pH 5 - 5.5, add Xylenol Orange and back-titrate with 0.02M
EDTA.   The consumption of EDTA corresponds to the sum of all the
metals.   Now add to the solution 1,10-phenanthroline (see p. 281)
and titrate the liberated EDTA with lead nitrate.   The consumption
of lead corresponds to the sum Cu + Zn (or Cd).   In a further
aliquot mask copper with thiourea and titrate directly with EDTA the
sum Pb + Zn (or Cd) at pH 5, using Xylenol Orange as indicator.   From
the three titrations the amount of lead can be easily calculated.
It can be compared with the result obtained by the direct titration
of lead.

Determination of zinc and copper.   Mochalov and Bashkirova [24]
proposed a somewhat longer procedure, which might have its value for
the analysis of more complicated alloys.

Dissolve 1 g of sample in boiling nitric acid and dilute with water to
250 ml.   To 50 ml of this solution add 30 ml of ammonia (1 + 5) and

then small portions of 0.5M sodium borohydride solution in 0.5M
sodium hydroxide.   Filter off the reduced metallic silver and copper
and wash with water.   Destroy the excess of borohydride in the
filtrate with 2M hydrochloric acid and determine the zinc by EDTA
titration.

Dissolve the residue on the filter with nitric acid, precipitate
silver as AgCl, filter it off, and determine copper in the filtrate
with EDTA.

If the alloy also contains cadmium, in the reduction of copper and
silver with alkaline borohydride the cadmium is precipitated as the
hydroxide and collected with the reduced metals.   After dissolution
of this residue and precipitation of silver as AgCl, cadmium and
copper can be determined by complexometry (see p. 291).

9.3.10   Determination of Lead and Zinc in Brass and Bronzes

Small amounts of lead in these materials are usually determined
simultaneously with copper by electrolysis.   Amounts of lead above
1% may be determined complexometrically, as described by Böltz et al.
[25], who masked iron and aluminium with triethanolamine, and copper
and zinc with cyanide.   The zinc can be demasked with formaldehyde
and titrated, after determination of the lead.

To 1 g of sample add 0.25 g of sodium chloride (when tin is present)
and then 25 ml of nitric acid (1 + 1).   After dissolution dilute to
volume in a 100-ml standard flask.   Pipette 5 ml into a 300-ml
titration flask, add 0.5 g of hydroxylamine hydrochloride, 10 ml of
30% triethanolamine solution and 10 ml of ammonia buffer (pH 10).
Then add 10% potassium cyanide solution from a burette until the blue
colour disappears, and 10 ml in excess.   Dilute with 50 ml of water
and heat until the solution becomes slightly yellow or colourless.
Titrate hot with 0.01M EDTA, with Eriochrome Black T indicator, to
determine the lead.   Cool the solution, add 5 ml of formaldehyde
solution and titrate again immediately with 0.01M EDTA, to determine
the zinc.

If the sample also contains cadmium, only the sum Zn + Cd can be
determined by this method.

Igarashi [26] described a very similar method for the determination of
lead.   Before the titration, however, he adds to the solution a known
amount of lead nitrate (0.005M) to give a higher consumption of EDTA.
To sharpen the end-point he also adds Mg-EDTA complex, from which the
magnesium is displaced by lead.

Mihalcescu [27] proposed in principle the following method.   Metastannic and antimonic acids are separated by evaporation of the sample
with nitric acid.   After filtration and dilution to known volume,
part of the solution is evaporated to fumes with sulphuric acid to
separate $PbSO_4$, which is then dissolved in $7-8$ g of ammonium acetate,
and the lead determined in the diluted solution at pH 5 with 0.05M
EDTA, with Xylenol Orange as indicator.   Zinc can be determined in
the filtrate after separation of the lead sulphate.   Alternatively
the lead can be separated and determined by electro-deposition and
then zinc according to the method above.

REFERENCES

1.   K. Študlar and I. Janoušek, Hutn. Listy, 1958, 13, 805; Anal.
     Abstr., 1959, 6, 2097.
2.   C. Harzdorf, Z. Anal. Chem., 1964, 203, 101.
3.   J.L. Pinkston and C.T. Kenner, Anal. Chem., 1955, 27, 446.
4.   N.A. Filippova and T.F. Dubrovskaya, Zavod. Lab., 1956, 22, 907;
     Anal. Abstr., 1957, 4, 858.
5.   N.A. Filippova and T.F. Dubrovskaya, Zavod. Lab., 1960, 26, 711;
     Anal. Abstr., 1961, 8, 72.
6.   M. Dixon, P.J. Heinle, D.E. Humlicek and R.L. Miller, Chemist-
     Analyst, 1962, 51, 42.
7.   P.G. Shakhova and O.I. Korotkova, Zavod. Lab., 1965, 31, 295;
     Anal. Abstr., 1966, 13, 3508.
8.   M.J. Tschetter and R.Z. Bachman, Talanta, 1974, 21, 106.
9.   O. Ottendorfer, Metallurgia, 1958, 57, 105.
10. O. Ottendorfer, Chemist-Analyst, 1958, 47, 96.
11. J. Teperek, Chem. Analit. (Warsaw), 1973, 18, 625; Anal. Abstr.,
     1974, 26, 83.
12. R. Přibil and M. Kopanica, Chemist-Analyst, 1959, 48, 87.
13. R. Přibil, Chimia (Zurich), 1950, 4, 160.
14. E.A. Shteiman, Z.G. Dobrynina and E.A. Mordovskaya, Zavod. Lab.,
     1964, 30, 1200; Anal. Abstr., 1966, 13, 531.
15. P.G. Shakhova, Zavod. Lab., 1965, 31, 408; Anal. Abstr., 1966,
     13, 4032.
16. I.M. Yurist and O.I. Korotkova, Zavod. Lab., 1962, 28, 660;
     Anal. Abstr., 1961, 10, 31.
17. R. Přibil and V. Vesely, Chemist-Analyst, 1965, 54, 12.
18. I.M. Yurist, Zavod. Lab., 1964, 20, 805; Anal. Abstr., 1965, 12,
     31.
19. D.G. Biechler, Chemist-Analyst, 1962, 52, 48.
20. L. Erdey, Gy. Rády and O. Gimesi, Acta Chim. Acad. Sci. Hung.,
     1962, 12, 151; Anal. Abstr., 1962, 10, 1352.
21. F. Sierra and J. Hernandez Canavale, An. Soc. Esp. Fis. Quim B.,
     1962, 58, 219; Anal. Abstr., 1962, 9, 5072.
22. J. Rzeszutko, Chem. Analit. (Warsaw), 1974, 19, 81; Anal.
     Abstr., 1974, 27, 2445.
23. R. Přibil and V. Vesely, unpublished results.
24. K.N. Mochalov and T.T. Bashkirova, Zavod. Lab., 1969, 35, 795;
     Anal. Abstr., 1970, 19, 2080.
25. G. Böltz, H. Wiedman and W. Kurella, Metall, 1956, 10, 821.
26. S. Igarashi, Japan Analyst, 1961, 10, 193; Anal. Abstr., 1963,
     10, 2212.
27. M. Mihalcescu, Metallurgia (Bucharest), 1967, 19, 243; Anal.
     Abstr., 1969, 16, 49.

9.4   COPPER AND COPPER ALLOYS

Copper is a component of many alloys used in all fields of electronics
and also as a construction material in machine industries. Brass
and bronze are perhaps the best known. Many alloys of similar
composition are marketed under various names, for example the Cu-Ni-Zn
alloys called pakfong, argetan, alpaca etc.

Some metals are present in copper alloys in only small amounts, such
as in beryllium-aluminium-lead bronzes. Some alloys contain silicon
and/or phosphorus (silicon-phosphorus bronze).

From the analytical point of view all these alloys constitute a
diverse system of metals such as tin, lead, zinc, nickel, aluminium,
manganese, cobalt, silver, beryllium.   Even for pure (electrolytic)
copper there are some requirements as to its purity (ultratraces of
Bi, Pb, Fe).

With only a few exceptions complexometric analysis has considerable
value for these alloys, because most of them are composed of bivalent
metals, which can be easily determined by EDTA titration.   The
analytical procedures have changed with time and improved as complexometry itself has.   Many of the older methods which were based on
sound principles have been improved only by change to a more suitable indicator or masking agent.

Most of these alloys are soluble in nitric acid or hydrochloric acid
plus hydrogen peroxide.   Tin and antimony are separated by treatment
with nitric acid, and lead as the sulphate.   If copper is a major
component it can be separated by electrolysis.

Some authors recommend keeping all the components in solution by
adding hydrochloric acid to dissolve tin.

Many of the analyses are very simple, especially of the binary alloys
such as Cu-Sn or Cu-Zn, and the ternary alloys Cu-Ni-Al, Cu-Zn-Pb etc.
The principles of many procedures were worked out in the laboratories
of the firm Outokumpu Oy in Pori, Finland.

9.4.1   Determination of Tin in Bronze and Gun Metal [1]

Dissolve 0.25 g of sample in a 400-ml beaker with 3 ml of concentrated hydrochloric acid and 5 - 10 drops of 30% hydrogen peroxide.
Add 0.01M EDTA in excess (10 mg of tin will require 8.5 ml), then add
20 ml of saturated thiourea solution.   Adjust the pH to 2 by
addition of 2 - 5 g of ammonium acetate, add 3 drops of 0.2% Xylenol
Orange solution and titrate the excess of EDTA with 0.01M thorium
nitrate.

A similar method was described by Furuya and Tajiri [2].   For
determination both of Cu + Sn and Sn (after masking of copper) they
used 0.01M zinc chloride and Xylenol Orange for back-titration of
EDTA.

For large amounts of tin (40%) Kinnunen and Wennerstrand [1]
described a method based on the determination of the sum Cu + Sn by
back-titration of excess of EDTA with zinc at pH 5.   After masking
tin with ammonium fluoride they titrate the EDTA liberated, which
corresponds to the amount of tin.

The masking of tin with ammonium fluoride can also be used for other
metal combinations in alloys.   For example an alloy containing
copper, zinc and tin can be analysed as follows.   In one aliquot the
sum of all three is determined, then the tin is masked, and the
liberated EDTA titrated;   this gives the tin and the sum Cu + Zn.
In a second aliquot the Zn is determined after copper has been masked
with thiourea and tin with fluoride.

9.4.2   Determination of Lead in Brasses and Bronzes [3]

In a 400-ml beaker dissolve 1 g of alloy in 8 ml of nitric acid
(1 + 1).   If tin is present add 5 drops of hydrochloric acid.   After
dissolution boil the solution to expel brown gases, cool the solution
and transfer it to a separatory funnel.   Add 5 ml of 20% Rochelle
salt solution, 10 ml of concentrated ammonia solution, 30 ml of 20%
potassium cyanide solution and 10 ml of 2% sodium diethyldithiocarbamate solution.   Extract with 10 ml of chloroform and then with two
further 5-ml portions of chloroform.   Collect all the extracts in a
400-ml beaker, add 5 ml of concentrated nitric acid and heat gently
to expel chloroform.   Add 2 ml of 20% Rochelle salt solution,
neutralize with ammonia, add 10 ml of Schwarzenbach ammonia buffer
and dilute to 250 ml with water.   Add 3 ml of 20% potassium cyanide
solution and titrate with 0.01M EDTA, using Erio T as indicator.   If
manganese is present, the chloroform extracts are red.   The manganese
interference can be minimized by adding 4 g of potassium cyanide to
the solution to be titrated.

The authors described a further modification for the determination of
lead in lead concentrates.

Evaporation of the chloroform is not necessary if the lead is
stripped into water with mercuric chloride [4] or copper sulphate [3]
solution.

The authors extended their original method to cover the range 0.001-
20% lead [5].   For small amounts of lead they used conversion of the
colourless lead diethyldithiocarbamate complex into an equivalent
amount of the yellow-brown copper complex and measured this spectrophotometrically.

For large amounts of lead (> 20%) in bronzes, Caton et al. [6]
recommended direct titration in the presence of tartrate and
potassium cyanide.   A similar method was described by Igarashi [7],
who masked iron, manganese(III) and copper with a mixture of tartaric
acid, triethanolamine and potassium cyanide.   In these titrations
Erio T was used as indicator, but is less suitable than Methylthymol
Blue.

.4.3   Determination of Zinc and Copper in Cu-Zn Alloys

For these binary alloys several methods have been described.   For
zinc determination the most convenient methods are based on masking
copper with sodium thiosulphate [8,9] or thiourea [10,11], but suffer
interference from all other metals forming EDTA complexes.   Some of
these metals can be masked or separated (e.g. traces of Fe, Al, Pb).
Copper is usually determined by difference from the results for Cu +
Zn and for Zn (after masking of copper).   Only a few examples are
given here.

Determination according to Cheng [8]

Dissolve 0.25 g of sample in a 150-ml beaker with 3 ml of concentrated nitric acid, and evaporate gently to 1 - 2 ml.   Cool and dilute
to 50 ml with water.   Precipitate tervalent metals, tin, lead and
manganese by adjusting to pH 8 - 9 with ammonia (1 + 1).   Filter into a
250-ml standard flask, wash the precipitate and make up to volume.

If there is only a small amount of precipitate, it need not be
reprecipitated.   Take an aliquot of the solution containing not more
than 0.5 mmole of copper and 1 mmole of zinc, and adjust to pH 5 - 6.
Add sodium thiosulphate until the solution is colourless, and titrate
with 0.02M EDTA, using a few drops of PAN as indicator.   In a second
aliquot determine the sum of Cu + Zn in the same way but without
addition of the thiosulphate.

This is one of the oldest methods and its principle is used in other
methods except that Xylenol Orange is generally used instead of PAN.

Microdetermination according to Ashton [12]

Dissolve 10 - 15 mg of sample in the minimum of nitric acid $(1 + 1)$
and evaporate to a moist solid.   Add 2 ml of 1N sulphuric acid and
warm to dissolve the residue.   Cool and dilute to volume in a
standard flask (10 or 25 ml).   Transfer a fraction containing about
2 mg of zinc into a 50-ml titration flask and add 3 ml of water, 0.3 g
of hexamine and solid sodium thiosulphate until the solution is
colourless.   Then add a few small crystals of. thiosulphate and 3 or
4 drops of 0.1% Xylenol Orange solution and titrate with 0.004M EDTA
to a yellow end-point.   For determination of the sum Cu + Zn titrate
another fraction in very slightly ammoniacal solution with 0.004M
EDTA, using murexide as indicator.

Murexide is generally considered an inconvenient indicator for zinc
titration because of the competitive effect of ammonia on the zinc-
murexide complex.

Determination according to Borchert [13]

Dissolve 0.5 g of sample in 10 ml of concentrated nitric acid, boil
to expel nitrogen dioxide, dilute with 10 ml of water and add 5 ml of
hydrogen peroxide.   Warm to 70°C, make ammoniacal, cool, transfer to
a 100-ml standard flask and dilute to the mark.   Filter off the
precipitate (Pb, Sn, etc.) on a dry filter paper, but do not wash it.
Dilute a 10-ml aliquot of filtrate to 100 ml with water and acidify
it with nitric acid.   Add 5% thiourea solution dropwise until the
solution becomes colourless and add 2 ml in excess.   Add hexamine to
buffer the solution and titrate the zinc with 0.05M EDTA to the
yellow end-point of Xylenol Orange.   Then add 5 - 10 ml of hydrogen
peroxide and after 3 min 100 ml of water.   Titrate the copper with
0.05M EDTA from red-violet to blue and finally to green.

In the second titration the thiourea is destroyed by the peroxide
(demasking the copper) and the Xylenol Orange is partly decarboxylated (in the same way as with lead dioxide [14]) to give a product
which is a specific indicator for copper.

Determination according to Berák [15]

Dissolve the alloy in nitric acid and filter off any metastannic acid.
Mask iron (up to 0.5%) with tartaric acid, and the copper and zinc
with potassium cyanide in ammoniacal solution.   First titrate any
residual "impurities" with 0.01M EDTA (Erio T as indicator), then
demask zinc with formaldehyde and titrate it with EDTA.

Berák described a number of difficulties encountered during the work
(see p. 62).

9.4.4   Determination of Zinc and Lead in Brasses [16]

To 1 g of alloy in a conical flask add 0.25 g of sodium chloride if
tin is present, and dissolve the sample in 25 ml of nitric acid
(1 + 1).   Transfer the solution to a 100-ml volumetric flask.

Determination of lead.   Pipette a 5-ml aliquot (50 mg of alloy) into
a 300-ml titration flask, add 0.5 g of hydroxylamine, 10 ml of 30%
triethanolamine solution and 10 ml of Schwarzenbach buffer solution.
Add 10% potassium cyanide solution from a burette until the blue
colour disappears and then 10 ml more.   Dilute with 50 ml of water
and heat to boiling.   The resulting solution is colourless or
slightly yellow, and aluminium, iron, copper, nickel and zinc are all
masked.   Cool and titrate with 0.01M EDTA, using Erio T as indicator.

Determination of zinc.   To the solution just titrated add 5 ml of
10% formaldehyde solution to demask zinc and titrate immediately with
0.01M EDTA.

The authors also described determination of lead and zinc in lead
bronzes (10 - 30% Pb, 1.5% Zn) by the same method.   To improve the
colour change at the end-point they add 5 ml of 0.4% Mg-EDTA solution
before the titration.

Mihalcescu [17] developed a slightly longer method for the determination of zinc and lead in bronzes, but without the use of potassium
cyanide and its demasking.   A 5-g sample is dissolved in 60 ml of
nitric acid (1 + 1) and after evaporation and dilution the metastannic acid (and any Sb) is filtered off.   The filtrate is
evaporated with sulphuric acid to white fumes and lead sulphate
separated in the usual way.   Copper is precipitated as hydroxide
from the filtrate and filtered off.   In a suitable fraction of the
filtrate from the copper separation zinc is determined at pH $5 - 5.5$
with EDTA (Xylenol Orange as indicator).   The lead sulphate can be
dissolved with ammonium acetate and determined by EDTA titration.

9.4.5   Determination of Cadmium in Copper Alloys

Cadmium behaves similarly to zinc and in the absence of zinc can be
determined in copper alloys by all the methods described for zinc.
The extraction of cadmium as $Cd(NCS)_4^{2-}$ into a mixture of methyl ethyl
ketone and tributyl phosphate after masking of copper with thiourea,
proposed by Kinnunen and Wennerstrand [18] does not solve the problem
of the presence of zinc, which is co-extracted.   However, their
suggested preferential separation of the zinc thiocyanate complex by
extraction with an amyl alcohol/diethyl ether mixture (1 + 4 v/v)
seems satisfactory [18].   Boltz et al. [16] have proposed an interesting method based on the exchange reaction between $Cd(CN)_4^{2-}$ and the
Mg-EDTA complex:

$$Cd(CN)_4^{2-} + MgY^{2-} = CdY^{2-} + Mg^{2+} + 4CN^-$$

but zinc again interferes.

The analysis of Cu-Cd-Zn alloys by use of simple EDTA titrations has
not been completely solved.   One possibility is the method based on
indirect determination of cadmium with EGTA in sodium hydroxide

solution, where zinc is present as tetrahydroxozincate and does not
react with EGTA [19]. Lead, aluminium and iron do not interfere if
triethanolamine is also present, but the behaviour of copper in this
system has not been investigated. Cadmium could also be separated
from zinc by precipitation with 1,10-phenanthroline and iodide [20]
but copper would have to be removed first.

9.4.6   Determination of Manganese in Copper Alloys [21]

Dissolve 0.1 - 0.5 g of the alloy in 10 ml of nitric acid (1 + 1).
Destroy nitrogen oxides by addition of about 0.05 g of sulphamic acid
and boiling. Dilute to 350 - 400 ml, add solid ascorbic acid,
20% potassium cyanide solution (3 ml for each 0.1 g of sample), 10 ml
of concentrated ammonia solution and a few drops of Erio T indicator
and titrate with 0.05 - 0.005M EDTA. If the manganese content is
high, heat the solution before titration.

For aluminium bronzes, mask the aluminium by adding 2 - 3 g of
ammonium fluoride, boiling and then cooling to 50 - 60°C before the
addition of the ascorbic acid etc.

If lead is present it is precipitated by addition of a few ml of 10%
sodium diethyldithiocarbamate solution before addition of the
indicator. For determination of both the lead and manganese, their
sum is determined as described for manganese, then sodium diethyldithiocarbamate is added to precipitate the lead and the EDTA
liberated is titrated with manganese sulphate solution.

These methods were later improved by masking iron and aluminium with
triethanolamine and using Thymolphthalexone as indicator (it is
superior to Erio T when TEA complexes are present, see p. 185).

9.4.7   Determination of Aluminium in Copper Alloys

Freegarde and Allen [22], Mildner [23] and Holzbecher [24] all
recommend preliminary separation of the copper by electrolysis. The
Freegarde and Allen procedure is as follows.

Dissolve a 2-g sample in 60 ml of a mixture of sulphuric and nitric
acids (each 20% v/v), and heat until nitrous fumes have been expelled.
Add 10 ml of concentrated hydrochloric acid and evaporate until
fuming. Cool, add 100 ml of water and boil until all soluble matter
has dissolved. Set aside for 1 hr, filter off on paper-pulp, and
wash with 5% sulphuric acid until free from copper salts. Dilute
the filtrate to 300 ml and electrolyse the copper. Dilute the
solution to the mark in a 500-ml standard flask.

To a 50-ml aliquot add 25 ml of 0.05M EDTA, neutralize with ammonia
to Methyl Red and add 10 ml of buffer solution (15 g of sodium
acetate trihydrate and 5 ml of glacial acetic acid per litre). Boil
for 3 min, cool, add a few drops of alcoholic PAN solution and
titrate with 0.1M copper sulphate until the colour just changes to
violet. Add 30 ml of saturated sodium fluoride solution, boil for
3 min, cool, and titrate with 0.1M copper sulphate until the solution
is blue.

A similar method for low amounts of aluminium (0.1 - 0.4%) in copper

alloys was developed by Mildner [23]. Tin is separated as metastannic acid and copper by electrolysis, then excess of EDTA is added
and back-titrated with zinc sulphate (Xylenol Orange as indicator).
After addition of ammonium fluoride to mask aluminium, the EDTA
liberated is titrated with the zinc solution.

Holzbecher [24] has chosen another method. After dissolution of the
sample and removal of copper as above, he titrates aluminium under
ultraviolet light with 0.6M sodium fluoride, with salicylidene o-
aminophenol as fluorescent indicator, and then titrates zinc with
EDTA, with salicylaldehyde acetylhydrazone as fluorescent indicator.
Traces of copper and iron are masked with sodium thiosulphate.

9.4.8 Complete Analysis of Various Copper Alloys

Efforts have been devoted to develop general methods for fast
determination of not only the main components, but also accompanying
trace metals. As a result a few procedures have been described
which use complexometry in combination with other analytical methods
such as gravimetry, colorimetry or polarography. First the all-
complexometric methods will be described, and then two examples of
combined methods.

Analysis of a mixture of Cu, Pb, Zn, Ni and Mn [25]

Such a mixture is obtained after dissolution of the alloy in nitric
acid (tin being separated as metastannic acid). All five components
can be determined complexometrically provided that they are present
in at least milligram amounts (which can be partly regulated by
suitable choice of the aliquot volumes). Some determinations are
very selective and will be described here.

Determination of copper. To a slightly acidic sample aliquot add
sufficient 0.05M EDTA to complex all the metals present. Adjust to
pH 5 − 5.5 with hexamine, add some Xylenol Orange and back-titrate
with 0.05M lead nitrate.

Dilute a second aliquot with 150 - 200 ml of water and adjust to pH 2 -
3, add 5 - 20 ml of 20% thiourea solution and wait 1 - 2 min until the
solution is clear (it becomes brown during slow addition of the
thiourea). Add sufficient 0.05M EDTA to complex all the free metals
present (Cu is masked) and back-titrate as before with 0.05M lead
nitrate. The difference between the two titrations corresponds to
the amount of copper.

This determination is very selective for copper. If only one other
metal is present, e.g. zinc or nickel in Cu-Zn or Cu-Ni alloys, then
the second titration estimates its amount. Generally speaking, this
method is suitable for all binary alloys of copper.

The method described needs two aliquots. Budevsky and Simova [26]
recommend using only one aliquot as follows.

Dissolve 0.1 g of sample with 5 ml of nitric acid (1 + 1) (if tin is
present, boil the solution for 5 min). Cool the solution, transfer
it to a 100-ml standard flask and dilute to the mark. Transfer a
25-ml aliquot to a 150-ml beaker and add 15 ml of 0.05M EDTA, and
Xylenol Orange as indicator. Add ammonia solution dropwise until

the green colour of the solution acquires a blue tint, then add 2 ml
of acetic acid/ammonium acetate buffer solution (pH 5.5 - 6.0) and
back-titrate with 0.05M lead nitrate until the solution just becomes
blue. Then add 25 ml of 10% thiourea solution, followed by nitric
acid (1 + 1) dropwise until the solution becomes yellow. Stir for
10 min then add ammonia solution (1 + 1) until a violet colour
appears. Add 5 ml of the acetate buffer and titrate the liberated
EDTA with 0.025M lead nitrate from yellow to red-violet. This
second titration gives the copper content in the solution. The
first titration allows calculation of the sum of the two metals
present.

Attempts have also been made to isolate copper from other elements.
Banerjee and Dutta [27] precipitate copper in acid medium (8 - 10%
acid) with thioglycollic acid (25%) as a canary yellow compound, then
dissolve this in nitric acid and titrate for copper with 0.1M EDTA,
using PAN as indicator. Electrolytic separation of the copper is
more convenient, however.

Determination of nickel. To an aliquot of sample solution add 10 -
20 ml of ammonia solution (1 + 1) and mask copper, lead, zinc,
cadmium etc. by adding 16% thioglycollic acid (TGA) solution until
the solution becomes clear (if insufficient is added, the solution
turns black). Then add excess of 0.05M EDTA to complex nickel, add
Thymolphthalexone or Methylthymol Blue and back-titrate with 0.05M
calcium chloride.

This method is suitable for a large concentration of copper, because
the Cu-TGA complex is colourless in ammoniacal solution. If only
copper is present besides nickel we can determine the sum of Cu + Ni
by omitting the thioglycollic acid. After the back-titration,
addition of TGA will liberate EDTA from the Cu-EDTA complex and a
further back-titration will give the copper content.

Determination of lead. The method of Böltz et al. [16], p. 291, is
used, preferably with Thymolphthalexone as indicator.

The interference of manganese is problematical in this determination.
In the presence of hydroxylamine or ascorbic acid manganese is
present in bivalent form and is co-titrated with lead even in the
presence of potassium cyanide. Trace amounts can be masked with a
large concentration of potassium cyanide. If iron is simultaneously
present, the manganese is oxidized to form the very stable $Mn(CN)_6^{3-}$
complex, which does not react with EDTA.

Determination of zinc. No highly selective method for zinc is known.
Böltz et al. [16], after the determination of lead just described,
demask zinc from its cyanide complex by adding formaldehyde. If the
copper and lead are separated electrolytically the analysis for the
remaining metals is easier. The mixture of nickel and zinc can be
analysed by finding the sum of Ni + Zn by back-titration of excess of
EDTA in ammoniacal medium with magnesium chloride (Thymolphthalexone
as indicator) then masking zinc with thioglycollic acid and
titrating the liberated EDTA with magnesium chloride to obtain the
zinc content.

Analysis of aluminium bronzes [22].

The dissolution of the sample, electrolysis and complexometric

determination of aluminium were described on p. 292. The authors
proceed further as follows.

Determination of iron. After the dissolution, removal of lead
sulphate and silica, and electrolysis, they take a 10-ml aliquot from
the 500 ml of diluted sample solution, add 1 ml of 10% tartaric acid
solution, 10 ml of 0.05M EDTA, 10 ml of concentrated ammonia solution
and 4 ml of 30% hydrogen peroxide, dilute to 100 ml and measure the
absorbance at 520 nm in a 4-cm cell.

This method is that published by Schneider and Janko [29] for
determination of iron in metallic zirconium. Up to 7 mg in 100 ml
can be determined. Fluoride and colourless cations do not interfere,
but cyanide does.

Determination of nickel. A further 10-ml aliquot is used for
determination of nickel with nioxime.

Determination of manganese. In a 25-ml aliquot manganese is
determined by oxidation to permanganate with potassium periodate.

Determination of tin. A 5-ml aliquot is evaporated till white
fumes appear, then diluted with 5 ml of water and 2 ml of 10%
hydroxylamine hydrochloride solution and analysed polarographically.

Rapid complexometric analysis of brass [30].

For the complexometric determinations the authors used 0.01M DCTA.
They determine tin gravimetrically, and copper and lead by
electrolysis, the copper gravimetrically and the lead by dissolving
the PbO$_2$ off the anode and titrating with DCTA. Other elements such
as iron, nickel and zinc are also determined with DCTA. According
to the authors DCTA is much superior to EDTA when pH control is not
easy to maintain because of the presence of large amounts of ammonium
salts.

Determination of tin. Take a 1-g sample in a 150-ml beaker, cover
it with 5 ml of water and add 10 ml of concentrated nitric acid.
Cover the beaker with a watch-glass supported on glass hooks and heat
it on a steam-bath or hot-plate so that the solution will evaporate
to about 5 ml in approximately 1 hr. Cool, add some filter-paper
pulp, dilute with water to about 50 ml, and digest the precipitate on
a hot-plate for about 30 min. Filter off (Whatman No. 42 paper) and
wash thoroughly with hot dilute nitric acid (1 + 100). Transfer the
filter paper containing the precipitate to a weighed porcelain
crucible, ignite and weigh. Retain the filtrate for the further
analysis.

The SnO$_2$ obtained is not pure and entrains various metals. For high
accuracy, the authors use the method of Caley and Burford [31], based
on volatilization of tin(IV) iodide, as follows. To the weighed
impure stannic oxide add about 15 times its weight of pure ammonium
iodide and heat at 425–475°C in an electric furnace until all the
iodide has volatilized. Moisten the residue with concentrated
nitric acid, ignite and weigh. The loss in weight is equivalent to
the stannic oxide. Dissolve the residue in 2 ml of concentrated
hydrochloric acid by heating gently. Add 2 ml of 70–72% perchloric
acid and heat to fumes to remove the hydrochloric acid. Add the
solution to the filtrate and washings from the separation of tin.

Determination of copper and lead.   The authors electrolyse the
solution from the tin determination.   The principle is well known
and need not be described here.   They use a stationary gauze cathode
and rotating gauze anode (preferably sand-blasted to give good
adhesion of the lead dioxide.   The copper separated is weighed.

To determine the lead the lead dioxide is dissolved off the anode
with 25 ml of water, 0.5 ml of concentrated nitric acid and 0.25 g of
hydroxylamine hydrochloride.   Then 25 ml of 0.01M DCTA are added to
the solution (50 ml if the lead content is expected to exceed 0.5%)
and the surplus is back-titrated with 0.01M magnesium chloride (Erio
T as indicator).

Determination of iron.   After the electrolysis, iron is precipitated
as the hydroxide, which is dissolved in hydrochloric acid (and
preferably reprecipitated), and determined complexometrically by
addition of excess of DCTA and back-titration with magnesium solution.

Determination of nickel.   The filtrate from the iron separation is
used for precipitation of nickel with dimethylglyoxime, the
precipitate is dissolved in hot nitric acid and nickel determined by
back-titration as for iron.

Determination of zinc.   After separation of nickel part or all of
the filtrate is used for determination of zinc (by back-titration as
for iron etc.).

REFERENCES

1.   J. Kinnunen and B. Wennerstrand, Chemist-Analyst, 1957, 46, 34,
      92.
2.   M. Furuya and M. Tajiri, Japan Analyst, 1963, 12, 59, 388;
      Anal. Abstr., 1964, 11, 4735.
3.   J. Kinnunen and B. Wennerstrand, Chemist-Analyst, 1954, 43, 65.
4.   R. Přibil, Coll. Czech. Chem. Commun., 1953, 18, 783.
5.   J. Kinnunen and B. Wennerstrand, Metallurgia, 1970, 82, 81.
6.   L. Caton, E. Ciobanu and A. Hîrtopeanu, Revta. Chim. Bucharest,
      1968, 19, 552; Anal. Abstr., 1970, 18, 47.
7.   S. Igarashi, Japan Analyst, 1961, 10, 193; Anal. Abstr., 1963,
      10, 2212.
8.   K.L. Cheng, Anal. Chem., 1958, 30, 243.
9.   I.I. Kalinichenko, Zavod. Lab., 1958, 24, 266; Anal. Abstr.,
      1959, 6, 185.
10.   J.S. Fritz, W.J. Lane and A.S. Bystroff, Anal. Chem., 1957, 28,
      821.
11.   W.J. Ottendorfer, Chemist-Analyst, 1958, 47, 96.
12.   A.A. Ashton, Anal. Chim. Acta, 1963, 28, 296.
13.   O. Borchert, Neue Hütte, 1965, 10, 52.
14.   R. Přibil, Talanta, 1959, 3, 200.
15.   L. Berák, Hutn. Listy, 1957, 12, 817.
16.   G. Boltz, H. Wiedmann and W. Kurella, Metall, 1956, 10, 821.
17.   M. Mihalcescu, Metalurgia (Bucharest), 1967, 19, 443; Anal.
      Abstr., 1969, 16, 49.
18.   J. Kinnunen and B. Wennerstrand, Chemist-Analyst, 1954, 43, 34.
19.   R. Přibil and V. Veselý, Chemist-Analyst, 1966, 55, 4.
20.   R. Přibil and V. Veselý, Talanta, 1964, 11, 1613.
21.   J. Kinnunen and B. Merikanto, Chemist-Analyst, 1954, 43, 93.

22. M. Freegarde and B. Allen, Analyst, 1960, 85, 731.
23. L. Mildner, Hutn. Listy, 1974, 29, 129; Anal. Abstr., 1974, 27, 1834.
24. Z. Holzbecher, Chem. Listy, 1958, 52, 430; Anal. Abstr., 1959, 6, 77.
25. R. Přibil and V. Veselý, Chemist-Analyst, 1964, 53, 38.
26. O.B. Budevsky and L. Simova, Talanta, 1962, 9, 769.
27. S. Banerjee and R.K. Dutta, Indian J. Technol., 1978, 16, 42; Anal. Abstr., 1979, 36, 6B23.
28. R. Přibil and V. Veselý, Talanta, 1961, 8, 880.
29. P. Schneider and J. Janko, Chem. Listy, 1956, 50, 899.
30. C.A. Goetz and T.C. Loomis, Talanta, 1966, 13, 985.
31. E. Caley and M. Burford, Ind. Eng. Chem., Anal. Ed., 1936, 8, 114.

CHAPTER 10

Analysis of Rocks and Minerals

Rocks and minerals are an important part of the earth's resources. The production of building materials, porcelain, ceramics, fire-resistant materials etc., are wholly dependent on their output and processing. They are also of great importance in the chemical industry as well as the steel industry and metallurgy.

It is beyond the scope of this book to give a comprehensive survey of the use of complexometry in this field. For many of these products only certain specific physical and chemical properties are important and the complete chemical composition is not of interest. Therefore only selected analytical procedures for some of these materials will be outlined. Silicates, slags and cements will be dealt with in separate chapters.

10.1 ANALYSIS OF LIMESTONES, DOLOMITES AND MAGNESITES

Complete analysis of these important raw materials (determination of SiO_2, Fe_2O_3, Al_2O_3, CaO, MgO, CO_2, sometimes MnO, SO_3, TiO_2), developed in its classical gravimetric form in the 19th century, is not regarded as difficult, but is certainly very tedious, especially the gravimetric determination of calcium and magnesium. That is why the first communications on complexometric determination of those metals was welcomed in analytical laboratories and was soon applied to the analysis of these carbonates.

The literature on this topic is very ample, and the vast number of publications is not easy to survey. The first procedures suggested were simple, because only two indicators were available (murexide and Erio T). After the separation of silicic acid and precipitation of iron, aluminium and titanium hydroxides, calcium was determined in one portion of the filtrate, in sodium hydroxide medium. In a second portion of the filtrate the sum of Ca + Mg was determined in ammoniacal solution, with Erio T as indicator. This procedure was satisfactory enough for limestones and dolomites if the notorious drawback in the determination of calcium (adsorption of calcium on the magnesium hydroxide precipitate) was disregarded, but was, of course, not applicable for the determination of calcium in magnesites, because of the high Mg:Ca ratio. The procedure first published by

Jordan and Robinson [1] was later repeated many times by other workers, with slight modifications (see Section 6.3.1, p. 200).

In the earler procedures the authors limited their effort to the determination of calcium and magnesium and did not pay much attention to other metals, e.g. aluminium and manganese. Cheng [2] extended complexometry to iron determination in the same solution by EDTA titration, with salicylic acid as indicator at pH 2 - 3. Calcium could then be determined, without separation, in potassium hydroxide medium, with murexide as indicator. The iron hydroxide precipitated (up to 4 mg of Fe) did not interfere with the calcium determination.

A more advantageous (and nowadays generally used) procedure consists in the masking of iron, aluminium and traces of manganese with triethanolamine (TEA), which makes possible the reliable determination of the sum Ca + Mg (Thymolphthalexone or Methylthymol Blue as indicator). Owing to the low content of iron and aluminium, their precipitation with ammonia can be omitted in analysis of most of these raw materials.

The method for determination of iron and aluminium is governed by the requirements for the analytical results. According to some specifications the determination of the sum $Fe_2O_3 + Al_2O_3$ is required, in others the content of iron (as Fe_2O_3) must be given. The sum of both oxides is mostly determined gravimetrically, and the iron spectrophotometrically with 1,10-phenanthroline. The manganese content is usually very low, and is determined photometrically as permanganate. The titanium content is also low, and is determined spectrophotometrically with hydrogen peroxide.

Regardless of its oxidation state, iron can also be determined spectrophotometrically with ethylenediamine-N,N-bis(2-hydroxyphenyl-acetic) acid. The red 1:1 complex formed at pH 2 - 10 is extracted with a chloroform solution of methyltrioctylammonium chloride (Aliquat 336-S) and measured at 480 nm [3]. A fast method has also been worked out for the determination of manganese in natural carbonates, based on spectrophotometry of the green complex formed by manganese(III) with triethanolamine [4].

A very thorough piece of work on the analysis of magnesites and dolomites was published by Bennett and Reed [5]. Their method is included in a British Standard [6].

The commonly used determination of the sum of Ca + Mg in one portion of sample and of Ca in another is suitable when the Ca:Mg ratio is reasonable, but much more demanding is the determination of small amounts of calcium in magnesites or of magnesium (below 1%) in limestone. Several methods for this purpose were described in Section 6.3.1 (p. 200).

10.1.1   Determination of Calcium in Magnesites

All methods based on direct titration of calcium in the presence of precipitated magnesium hydroxide fail when large quantities of magnesium are present. Flaschka and Huditz [7] recommend precipitating calcium with naphthylhydroxylamine as the brick-red calcium salt, which on ignition gives a mixture of calcium oxide and hydroxide and only traces of co-precipitated magnesium. Calcium is then determined

complexometrically after dissolution in hydrochloric acid.   The
method was applied to routine analysis by Skalla [8] but has not
found general use.

A much simpler method was proposed by Flaschka and Jakobljevich [9],
based on the precipitation of magnesium hydroxide from a large volume
(500 ml) of strongly alkaline solution in the presence of triethanolamine and an excess of 0.05M EDTA.   According to the authors up to
850 mg of magnesium can be separated.   Calcium is then determined in
the filtrate by back-titration of the excess of EDTA with 0.05M
calcium chloride.   Murexide alone does not give a sharp enough end-
point, but is more satisfactory if mixed with Naphthol Green B in 2:5
ratio.

Determination of lime in magnesites

The Flaschka and Jakobljevich procedure [9] was reinvestigated by a
Sub-Committee of the British Ceramic Research Association [10].   The
method proved to be definitely superior to the others studied by the
group, but a sharper end-point was needed in the titration of calcium.
It was found that neither Calcein nor Calcon showed any improvement
over murexide, but a mixture of murexide with the Patton-Reeder
indicator [11] (Cal-Red) was found to be the best, giving a colour
change from violet to pink, and behaving much better in the presence
of precipitated magnesium hydroxide [11].   In this way the filtration of the magnesium hydroxide was eliminated.

Weigh 1 g of finely ground and dried (110°C) sample into a 250-ml
beaker.   Add 5 ml of water and 10 ml of concentrated hydrochloric
acid.   Cover the beaker with a watch-glass and boil to dissolve the
sample.   Rinse the watch-glass into the beaker and transfer the
contents into a 500-ml conical flask.   Cool, add 5 ml of triethanolamine solution (1 + 1) and (by pipette) 25 ml of 0.05M EDTA and then
dropwise with shaking 4M sodium hydroxide until a precipitate appears.
Dilute to about 250 ml, add 25 ml of 4M sodium hydroxide and 0.1 g of
the mixed indicator (per 100 ml of the solution), and titrate
immediately with 0.05M calcium chloride, until the last drop just
discharges the bluish-green colour and the mixture becomes grey with
a tinge of pink.   To confirm that the end-point has been reached,
add a further drop of calcium solution, shake, and let stand for
15 - 20 sec, when a definite colour change should develop.   Titrate
in good daylight, but not in sunlight.

The mixed indicator is made as follows.   Mix 23.75 g of sodium
chloride with 0.25 g of Patton-Reeder indicator.   Separately prepare
a mixture of 20 g of sodium chloride with 0.1 g of murexide.   Add to
the first mixture 1.25 g of the murexide mixture and thoroughly mix
by grinding.   Store in a well-stoppered tube protected from light.
These proportions may need to be varied to give a satisfactory end-
point, as different batches of both indicators may differ in colour
intensity.

The same method has been published as a collaborative work by
Thompson et al. [12].   All their attempts to improve the original
method by changing the indicator or using potentiometric titrations
were unsuccessful.

Though this procedure was acceptable at the time it was worked out
(1954 - 1959) it was not entirely satisfactory for routine work.

Improvements were subsequently made by the Refractories Working Group of the Analysis Committee of the British Ceramic Research Association. The flow-sheet for the complete analysis is shown in Fig. 10.1.

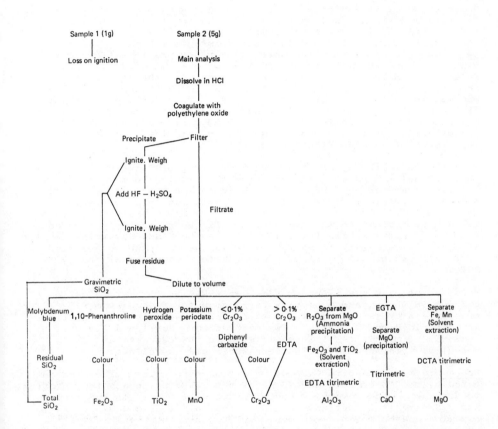

Fig. 10.1. Schematic diagram of the analysis of magnesites and dolomites [5] (by permission of the copyright holders, the Royal Society of Chemistry)

According to Bennett and Reed [5], the main features of this improved method are as follows.

(a) Magnesium is precipitated as the hydroxide in the presence of triethanolamine and EGTA. The substitution of EGTA for EDTA greatly improves the end-point for the calcium titration because of greater stability of the Ca-EGTA complex and the ease of precipitation of magnesium hydroxide from the relatively weak Mg-EGTA complex.

(b) Use of "Magflok" - a thick viscous liquid resin (Ridsdale and
Co. Ltd., Newham Hall, Newby, Middlesbrough, England) - as coagulant
for the magnesium hydroxide improves its collection by filtration
(which is included in this method in contrast to the previous ones).

The procedure for the lime determination is as follows.

Prepare 500 ml of a stock solution from a 2-g sample (see Fig. 10.1).
Transfer a 50-ml aliquot of the stock solution to a 250-ml standard
flask.   Add 5 ml of triethanolamine solution (1 + 1) and an
appropriate amount of 0.05M EGTA and dilute to 150 ml.   Add 25%
potassium hydroxide solution until no further precipitation takes
place and then 10 ml in excess, followed by 10 ml of 2% Magflok
solution.   Dilute to the mark, shake and allow the mixture to stand
for about 10 min to settle.   Filter the liquid through a 15 cm dry
Whatman No. 541 paper into a dry beaker.   Transfer 200 ml of the
filtrate into a 500-ml conical flask and add 15 ml of 25% potassium
hydroxide solution followed by about 0.03 g of screened Calcein
indicator, and titrate with 0.05M calcium chloride until the first
appearance of a green fluorescence.

Make the indicator by grinding together 0.2 g of Calcein, 0.12 g of
thymolphthalein and 20 g of potassium chloride.   To make the Magflok
solution transfer one drop of the resin to a beaker and weigh it,
then add sufficient water to give an approximately 2% w/v solution.

A blank determination must be done on all the reagents in accordance
with the general scheme (Fig. 10.1).   In the blank determination for
lime, it is necessary to add 50 ml of magnesium sulphate solution
containing about 0.06 g of magnesium, before addition of the 0.05M
EGTA.

More recently, a new approach has been taken to the determination of
calcium in the presence of very large amounts of magnesium [13].   The
principle of this method, the interferences, and optimal conditions
are described on p. 206ff.   Two methods were developed for the
analysis of magnesites;   they differed in their treatment of
potential interferents.   The procedures are as follows.

Procedure A.   Dissolve 0.2 - 0.5 g of the sample in 2 ml of concentrated hydrochloric acid and 10 ml of water and evaporate the solution
to dryness.   Add several drops of hydrochloric acid to the residue
and dissolve it in hot water.   Cool the solution, add 0.5 g of
hexamine and boil for 5 - 10 min.   Filter off the precipitated
hydroxides.   To the filtrate, in a titration flask, add 10 - 15 ml of
0.05M EGTA, 10 - 15 ml of saturated borax solution and a small amount
of solid Thymolphthalexone as indicator, and titrate the diluted
solution with 0.05M zinc from blue to colourless.

Procedure B.   Dissolve the sample as for procedure A.   Then adjust
the pH of the solution to 1, and extract the iron with three 25-ml
portions of a 50% solution of acetylacetone in chloroform.   Separate
the aqueous phase and boil off the dissolved acetylacetone (traces of
which decolorize the Mg-indicator complex and must not be present).
Neutralize the solution with sodium hydroxide and then continue as
described for the precipitation method.

10.1.2   Determination of Magnesium in Dolomites and Limestones

Determination of magnesium in dolomites is not a difficult problem
because the ratio CaO:MgO is about 2:1.   The Bennett and Reed [5]
method is satisfactory, and other methods are described in Section
6.3.1 (p. 200).   Determination of a very low content of magnesium
(0.5 - 2%) in the presence of calcium is rather difficult, however.
Various methods have been used for selective separation of either
calcium or magnesium before the titration.   Benewitz and Kenner [14]
precipitate calcium as oxalate and determine magnesium in the
filtrate by EDTA titration, with a very poor end-point (Erio T).
Their method has been applied by Sochevanova [15] and by Wengler and
Ausel [16].   Ghosh and Roy [17] precipitate calcium as molybdate.
On the other hand Muraca and Reitz [18] separate magnesium from
calcium as $Mg(OH)_2$ in the presence of mannitol, which prevents co-
precipitation of calcium, but double precipitation is still used.
These last two methods are practically unused by other authors.

Further approaches, such as determination of both elements by
combination of EGTA and DCTA titrations [19,20], with some modifications, were described in Section 6.3.1 (p. 200).

The problem of the determination of very low amounts of magnesium is
still open and it is doubtful whether it need be solved by complexometry at all, since so many highly sensitive instrumental methods are
available.

10.1.3   Determination of CaO in Burnt Limestone

Burnt limestones are often priced according to their content of
calcium oxide.   A series of extraction methods has been designed for
the separation of calcium oxide [and $Ca(OH)_2$] from calcium carbonate.
The final determination is performed by acid-base titration, which
yields somewhat low results [21].   One of the extraction methods was
adapted for use with complexometry by Verma et al. [22].

A suitable weight of sample was suspended in 500 ml of 2% sucrose
solution and this was shaken in a closed vessel for 2 hr.   After
titration, calcium was determined in a aliquot by EDTA titration,
with Erio T as indicator.   As a check, the authors determined
calcium in another portion of the solution by titration with hydrochloric acid.   According to the authors the complexometric method
gave "more probable" results, but there was about 8% relative
difference between the two sets of results.

Determination of free CaO + $Ca(OH)_2$ in calcium carbonate, silicate
and aluminate has been described by Splítek [23].   The method was
used for the analysis of the so-called white sludge that is formed in
the removal of silicate from tetrahydroxoaluminate solutions in the
production of Al_2O_3.   It is based on extraction of CaO + $Ca(OH)_2$
with a mixture of ethyl acetate and isobutyl alcohol by boiling under
reflux.   After re-extraction with hydrochloric acid, the total
calcium is determined complexometrically.

10.2   FLUORSPAR AND CRYOLITE

Fluorspar (fluorite, CaF_2) is a very important mineral product, huge
amounts of which are needed in the metallurgical industry (aluminium
and steel production). It is also added to cement, and is used in
the manufacture of special glasses and enamels. Fluorite is a major
raw material for the production of hydrofluoric acid and fluorine
(which plays an important role in the plastics industry).

Individual industries have various requirements for the chemical
composition of fluorite and its concentrates. Mostly the determination of CaF_2 and $CaCO_3$ is required. In the complete analysis the
loss by drying and by ignition is determined together with CaF_2,
SiO_2, CaO, MgO, S, P, Fe_2O_3 and $BaSO_4$. Complete analysis of
fluorite is very tedious and obviously seldom required. Complexometric analysis has found use only in the determination of iron,
total calcium and calcium carbonate.

10.2.1   Analysis of Fluorite According to Nielsch [24]

This method, published as early as 1955, is rather tedious, for it
uses opening of the sample by evaporation with sulphuric acid in a
platinum dish under an infrared lamp. The residue, containing
calcium sulphate, silicic acid and barium sulphate, is boiled in a
beaker with 400 ml of a mixture of sulphuric acid and hydrochloric
acid for 1 - 3 hr. Fractions of the filtrate are used for complexometric determination of iron and calcium.

The CaF_2 content which is of primary importance here is calculated
from the difference between the total calcium and the amount of $CaCO_3$
determined by the conventional method [25] based on the dissolution
of calcium carbonate in dilute acetic acid. Since this operation is
frequently encountered in other procedures its description is given
below.

Add to 1 g of the sample, in a 50-ml beaker, 10 ml of 10% acetic acid
and warm the solution for 1 hr on a water-bath. Use a watch-glass
to cover the beaker. Filter off on a membrane filter and wash the
residue with a small amount of water. The volume of filtrate must
not exceed 50 ml. Then add 10 ml of 15% triethanolamine solution
and a pinch of potassium cyanide. Adjust the pH to 11 - 12 and
titrate with 0.05M EDTA.

Because calcium fluoride is slightly soluble in acetic acid, Specht
subtracts 1 mg of calcium (for 1 g of the sample) from the results.
With this correction he, obtained very good results.

10.2.2   Analysis of Fluorite According to Povondra and Vébr [26]

This procedure differs substantially from the one just described, the
sample decomposition being much faster and the calcium carbonate
content calculated from the amount of bound carbon dioxide. The
method also makes it possible to determine SiO_2 and $BaSO_4$.

Dry the finely powdered sample at 100°C. To 1 g add 15 ml of
perchloric acid (1 + 3) saturated at 50°C with boric acid. Heat and
evaporate to dryness. Cool, add 5 ml of concentrated hydrochloric

acid and 80 ml of hot water and set aside for 10 min. Filter off
and ignite the silicic acid, weigh it, remove the silica in the
usual way with hydrofluoric and sulphuric acid, reweigh and determine
the SiO_2 by difference.   Heat the resiude with a few ml of hydrochloric acid $(1 + 4)$, dilute with boiling water, filter off on paper,
ignite the filter, convert any barium in the ash into $BaSO_4$, and
weigh.   If a high weight is obtained, fuse the $BaSO_4$ with sodium
carbonate and repeat the hydrochloric acid treatment.   Combine the
filtrates in a 250-ml standard flask and dilute to volume.   Take a
suitable aliquot, add 5 ml of 30% triethanolamine solution, make
alkaline with sodium hydroxide and titrate the total calcium with
EDTA, using murexide as indicator.   Larger amounts of iron and
aluminium must be removed by precipitation with hexamine.   The
precipitate can be ignited and weighed as R_2O_3.

For determination of the calcium Thymophthalexone is a more conven-
ent indicator.   To correct for the $CaCO_3$ content the authors
recommend volumetric determination of carbon dioxide (but do not say
how).   For large contents of $CaCO_3$ they recommend using the "acetic
acid" method, which has been criticized by Harzdorf and Steinhauser
[27] (see below).

Povondra and Vébr recommend this method for control of flotation
processes, because it takes only 30% of the time needed with the old
method.

A very similar method was worked out by Bhargava and Hines [28].
They decompose the sample by evaporating with 15% perchloric acid
(saturated with boric acid) until white fumes appear.   Calcium is
determined in the filtrate complexometrically, with Calcein (screened
with thymolphthalein) as fluorescent indicator.   The $CaCO_3$ is
extracted from a second portion of sample with acetic acid and
calcium determined as before.   The amount of CaF_2 is calculated by
difference.   Four samples can be analysed in $3\frac{1}{2}$ hr.   According to
the authors the method is precise with a standard deviation less than
0.13% for 90 - 93% CaF_2.

Determination of calcite in fluorite [29]

A modification of the conventional method was worked out by Popov
et al. [29].   To 0.5 g of sample add 5 ~ 8 ml of diethyl ether and
20 ml of 5% acetic acid solution, boil for 20 - 25 min and evaporate
to about 10 ml.   Add 10 ml of water, a little hydroxylamine and
20 ml of glycerol (to mask Sb, Sn, Cu and Zn), then 3 - 4 ml of 5M
sodium hydroxide, and titrate calcium with 0.05M EDTA (Acid Chrome
Dark Blue as indicator).   (The authors make a correction of 0.27% to
allow for the solubility of CaF_2).

It seems appropriate to mention here the study by Harzdorf and
Steinhauser [27].   They proved experimentally that the solubility of
calcium fluoride under the conditions of the conventional procedure,
e.g. boiling 1 g of fluorite sample with 10 ml of 10% acetic acid
solution, is much higher than generally assumed, amounting on average
to 3.7 mg of CaF_2.   The solubility of calcium fluoride is affected
by impurities present (Al, Fe, La) and is substantially lower in the
presence of yttrium.   This is explained by the fact that yttrium can
substitute to some extent for calcium in the crystal lattice, since
they have the same ionic radius.

AC - K

The correction for the solubility of CaF_2 also depends on other factors such as volume and boiling-time. This was shown as early as 1957 by Mashall and Geyer [30], who were able to determine 10 mg of calcium complexometrically in the presence of 20 mg of fluoride, by using a volume of at least 200 ml (the weight ratio Ca:F in CaF_2 is approximately 1:1).

10.2.3   Determination of CaF_2 in Fluorite

So far, all the methods have been based on the determination of total calcium and of the $CaCO_3$ soluble in acetic acid. The results obtained are corrected as mentioned above. The reverse procedure is also possible - to determine the content of CaF_2 after decomposition of the residue left after dissolution of the calcium carbonate.

Determination of CaF_2 after decomposition with acetic acid [31]

This method is derived from the older and less suitable method of Popov [32]. The calcium carbonate is dissolved in acetic acid and the insoluble CaF_2 is converted into soluble calcium chloride by boiling with a neutral solution of aluminium chloride. The calcium is then determined by EDTA titration, with Thymolphthalexone as indicator, after masking of the aluminium with triethanolamine.

To 0.1 - 0.2 g of sample add 8 ml of 10% acetic acid solution and heat (with stirring) for 30 min on a water-bath. Filter off the residue and wash it with three 30-ml portions of water. Transfer the filter and residue to the original beaker, add 20 ml of neutral 8% aluminium chloride solution, cover the beaker with a watch-glass and heat for 5 - 10 min. Then heat for $1\frac{1}{2}$ - 2 hr with stirring, maintaining the volume more or less constant by adding water during the heating. Filter, and wash the filter 4 or 5 times with water. Dilute the filtrate to 150 ml with water, add 10 ml of 33% triethanolamine solution, mix thoroughly, add 0.5 ml of 0.2% nitramine solution in ethanol and then 20% sodium hydroxide solution until the mixture is pale orange. Then add Thymolphthalexone and titrate with 0.05M EDTA from blue to orange or colourless.

Determination of calcite (and CaF_2) after decomposition with EDTA solution [33]

All the $CaCO_3$ is selectively dissolved by prolonged boiling with 0.05M EDTA. The calcium thus dissolved is determined by back-titration with magnesium solution (Thymolphthalexone as indicator). The total calcium is determined in a separate sample after fusion with sodium hydroxide in a silver crucible.

To 1 g of sample add 10 ml of 0.05M EDTA and 10 ml of water and heat for 30 min on a boiling water-bath with periodic shaking. Dilute the cooled mixture with 50 ml of water and filter. Make up the filtrate to volume in a suitable standard flask (e.g. 125 ml), dilute a 25-ml aliquot with 100 ml of water, add 15 ml of 50% v/v glycerol solution and 7.5 ml of 0.1M sodium hydroxide and titrate with 0.05M magnesium chloride, using Thymolphthalexone as indicator. This gives the $CaCO_3$ content.

(Remarks. It seems unusual to titrate with magnesium in the absence of ammonia and ammonium chloride; the efficiency of glycerol as a masking agent seems overestimated, compared with triethanolamine.)

To 0.1 g of sample in a silver crucible add 1 - 2 ml of water and 2 ml
of 50% sodium hydroxide solution.   Evaporate to dryness and fuse the
residue at 700 - 750°C for 3 - 6 min.   Leach the cooled melt with hot
water containing 1.5 g of boric acid, neutralize the solution to
Congo Red with 6M hydrochloric acid and heat until clear.   Dilute
the cooled solution to volume in a 250-ml standard flask and then
continue as above after addition of glycerol, sodium hydroxide and
indicator.   This gives the total calcium, and the CaF_2 is calculated
by difference.

(Remarks.   Experienced analysts will not be too enthusiastic about
the sample decomposition procedure;   that described by Povondra and
Vébr [26] is much simpler and faster.   Boiling with acidified
beryllium nitrate solution [34] or aluminium chloride solution
[31,35] has been proposed for the decomposition of CaF_2.   This
procedure has not so far been applied to the analysis of fluorites.)

10.2.4   Analysis of Fluorites According to Steinhauser et al. [36]

Steinhauser et al. designed a more complete analysis of fluorite in
which the use of complexometry is limited to determination of total
calcium.   The authors first determine the loss on drying at 110°C.
The dried sample is then partly decomposed by boiling with acetic
acid (determination of $CaCO_3$) as described earlier.   Silica is
determined gravimetrically by separation, weighing, and volatilization with hydrofluoric acid, or determined in a separate sample, as
silicomolybdic anhydride.   The residue from the hydrofluoric acid
treatment is fumed with sulphuric acid and then boiled with a mixture
of sulphuric acid and hydrochloric acid.   Barium sulphate - the only
undissolved component - is determined gravimetrically.   Calcium is
precipitated as the oxalate and determined either gravimetrically as
CaO or by photometric titration with EDTA solution.

The analysis is completed by the determination of sulphide e.g. by
releasing hydrogen sulphide from the sample, absorbing it in cadmium
acetate solution and determining it iodometrically.

10.2.5   Determination of Fluoride in Cryolite

This method, described in 1959 by Sakharova and Shishkina [37], is
based on the precipitation of fluoride as PbClF and complexometric
determination of lead in the precipitate.   The original method can
be improved in the light of present knowledge of complexometry.   The
authors' method of decomposition can be used, followed by a more
modern determination procedure.

Mix 0.25 g of the cryolite sample in a platinum crucible with 0.25 g
of ground fused silica and 6 g of a mixture of sodium and potassium
carbonates.   Fuse the mixture at 1000°C for 20 min.   Heat the
cooled melt with water and dilute to volume in a 250-ml standard
flask together with the insoluble residue.   After filtration through
a dry paper use 100 ml of the filtrate for determination of the
fluoride (p. 223).

10.3   ANALYSIS OF PHOSPHATES

Relatively little attention has been paid to complexometric analysis
of natural phosphates (chlorapatites, fluorapatites), because the
phosphomolybdate method is still considered satisfactory.

The determination of phosphate is also frequently encountered in the
analysis of superphosphates and other artificial fertilizers.

The principles will be given of three methods for the determination
of phosphorus in the presence of iron, aluminium and calcium.
Bakacs-Polgar [38] analyses such solutions by complexing the aluminium and iron in weakly acid medium with DCTA and then calcium with
an excess of EDTA, which is titrated with 0.01M magnesium chloride
(Erio T as indicator). After the addition of ethanol, titration
with the magnesium solution is continued until complete precipitation
of $MgNH_4PO_4.6H_2O$ is achieved; the Erio T serves as indicator for
this titration as well.

Szekeres et al. [39] determine calcium first by back-titration with
magnesium solution, using Thymolphthalexone as indicator, the iron
and aluminium being masked with triethanolamine. After addition of
1 g of ammonium chloride and 10 ml of ethanol the phosphate is again
precipitated by titration with 0.1M magnesium chloride. The volume
of the solution should not exceed 20 - 30 ml at the beginning of the
titration and 60 ml at the end-point, otherwise the results are not
correct (at high dilution the precipitation of $MgNH_4PO_4.6H_2O$ is too
slow).

Buss et al. [40] first precipitate the phosphate as phosphomolybdate,
collect this, dissolve it in ammonia solution, and finally transform
it by a tedious double precipitation into $ZnNH_4PO_4.6H_2O$. The zinc
is then determined complexometrically in a weakly acid medium. All
three procedures were applied to the analysis of natural phosphates,
and Buss et al. recommend their method for the determination of
phosphate in ores, slags, fertilizers and cast iron.

10.4   ANALYSIS OF MONAZITES

Analysis of monazites (monazite sands and concentrates) is not easy,
the difficulties beginning with the decomposition (because of the
presence of phosphate) and continuing to the end. The usual
decomposition procedure is protracted heating with sulphuric acid at
high temperature [41] or fusion with fluoride [42] or sodium peroxide
[43]. Pyrophosphoric acid has also been recommended for the
decomposition [44]. The separation of the metal phosphate from
strongly acid medium is another important step in the analysis of
monazites. Separation on various ion-exchangers has also proved
very useful. Most attention has been paid to the determination of
the trace metals in monazites, by known photometric methods. Chung
and Riley [41] give one such detailed scheme of complete analysis.

Complexometry has found use only in certain operations in monazite
analysis. A great deal of attention has been paid to thorium, which
can be separated from the lanthanides by ion-exchange. Mei-Cheng
Chen [45] separated thorium from 0.1M ammonium thiocyanate in 90%
methanol on Dowex C11-X8. The thorium was then eluted with 1M
hydrochloric acid and determined complexometrically (Alizarin Red as

indicator)   Another separation of thorium is based on its sorption
on KU-2 ion-exchanger from tartaric or trihydroxyglutaric acid
solution [43]. Thorium is then determined in the usual way with
Xylenol Orange as indicator, after elution with ammonium carbonate
solution.

Thorium can be separated from lanthanides by extraction with mesityl
oxide in the presence of a high concentration of aluminium nitrate as
salting-out agent [46].

Palei and Udal'tsova [47] recommend separating thorium from the rare
earths as the oxalate and determining it by direct titration with
0.1 - 0.005M EDTA, using amperometric end-point detection with two
polarizable platinum electrodes.

In some other procedures, the complexometric determination of
magnesium and calcium is mentioned but requires tedious separation
operations.

The sum of the rare earths present can also be determined complexometrically after separation of the phosphates. The only method that
makes it possible to determine the rare earths without separation of
phosphate is described on p. 196. It is based on the titration of
excess of DTPA with a zinc solution at pH 5 - 5.5, with Xylenol Orange
as indicator. The use of EDTA or DCTA is not possible in this case,
as the rare earths are displaced from the complexes and precipitated
as the phosphates [48].

This procedure has since been extended to complexometric determination of the phosphate as well as the rare earths [49], based on the
following principles. To a sufficiently acid solution (30 - 50 ml)
containing up to 50 mg of phosphate and 20 mg of lanthanides a
sufficient amount of 0.05M DTPA is added. After boiling for 1 - 2
min with 0.5 - 1.0 g of hydroxylamine or ascorbic acid at pH 4.5, the
solution is cooled, diluted to 100 - 150 ml and adjusted to pH 5 - 5.5
with hexamine. The excess of DTPA is titrated with 0.05M zinc
nitrate, with Xylenol Orange as indicator, from yellow to red-violet.
The consumption of DTPA corresponds to the sum of the lanthanides.

To the same solution an excess of 0.025M La-EDTA complex is added,
then the mixture is diluted to 200 - 300 ml, heated to 40 - 50°C and
titrated slowly with zinc solution to a red colour which must not
fade for at least 2 - 3 min. The consumption of zinc corresponds to
the amount of phosphate. In this titration the following reaction
takes place:

$$LaY^- + PO_4^{3-} + Zn^{2+} = ZnY^{2-} + LaPO_4$$

This displacement reaction is very fast at room temperature when a
large amount of phosphate is present, but for small amounts the
higher temperature of 40 - 50°C is more convenient.

In the first titration (determination of lanthanides) all elements
forming stable complexes with DTPA at the pH used are co-titrated.
The sum of the lanthanides can be determined in this case by their
masking with fluoride. Thorium is precipitated as phosphate even in
the presence of DTPA, but can be determined if we use TTHA instead of
DTPA for its complexation. The procedure is similar to that

described above. However, thorium, lanthanides and phosphates
cannot all be determined by this method, owing to the complicated
equilibria between the Th-TTHA complex and the lanthanide-DTPA
complexes [50].

Very recently, the determination of thorium in monazite has been
extended to the microscale by the use of very dilute titrant
solutions. Either HEDTA or EDTA is suitable for the purpose [51].

REFERENCES

1.   J.W. Jordan and K.L. Robinson, Chem. Ind. (London), 1953, 28,
      687.
2.   K.L. Cheng, T. Kurtz and R.H. Bray, Anal. Chem., 1952, 24, 1640.
3.   R. Přibil and J. Adam, Coll. Czech. Chem. Commun., 1972, 37,
      1277.
4.   J. Adam, J. Jírovec and R. Přibil, Hutn. Listy, 1969, 24, 739;
      Anal. Abstr., 1970, 19, 3896.
5.   H. Bennett and R.A. Reed, Analyst, 1971, 96, 640.
6.   B.S. 1902, Part 2E, 1970.
7.   H. Flaschka and F. Huditz, Radex-Rundschau, 1952, 181.
8.   N. Skalla, Radex-Rundschau, 1952, 186.
9.   H. Flaschka and H. Jakoblevich, Radex-Rundschau, 1954, 83;
      Anal. Abstr., 1955, 2, 2985.
10.   A.T. Green, A Report of the Analysis Sub-Committee of the
      British Ceramic Research Association, H.V. Thompson, (ed.),
      Special Publication No. 18, 1958.
11.   J. Patton and W. Reeder, Anal. Chem., 1956, 28, 1026.
12.   H.V. Thompson, A. Mayer, G. Padget, R.C. Chirnside and
      H. Bennett, Trans. Br. Ceram. Soc., 1959, 58, 353; Anal. Abstr.,
      1960, 7, 1674.
13.   R. Přibil and J. Adam, Talanta, 1977, 24, 177.
14.   J.J. Benewicz and C.T. Kenner, Anal. Chem., 1952, 24, 1186.
15.   M.M. Sochevanova, Zavod. Lab., 1955, 21, 530; Anal. Abstr.,
      1956, 3, 658.
16.   H.L. Wengler and W. Ausel, Arch. Eisenhüttenwesen, 1957, 28, 7.
17.   A.K. Ghosh and K.L. Roy, Anal. Chim. Acta, 1954, 14, 504.
18.   R.F. Muraca and M.T. Reitz, Chemist-Analyst, 1954, 43, 73.
19.   R. Přibil and V. Veselý, Talanta, 1966, 13, 233.
20.   R. Přibil and V. Veselý, Chemist-Analyst, 1966, 55, 82.
21.   M.R. Verma, V.M. Bhuchar, K.J. Therattil and S.S. Sharma,
      Analyst, 1958, 83, 160.
22.   M.R. Verma, V.M. Bhuchar, K.J. Therattil and S.S. Sharma, J. Sci.
      Ind. Res., 1955, 14B, 199.
23.   R. Splítek, Hutn. Listy, 1957, 12, 524.
24.   W. Nielsch, Glass Email Keramo Technik, 1955, 6, 73; Anal.
      Abstr., 1955, 2, 2985.
25.   F. Specht, Z. Anal. Chem., 1956, 149, 85.
26.   P. Povondra and J. Vébr, Coll. Czech. Chem. Commun., 1959, 24,
      637.
27.   C. Harzdorf and O. Steinhauser, Z. Anal. Chem., 1970, 251, 359.
28.   O.P. Bhargava and W.G. Hines, J. Iron Steel Inst., 1968, 206,
      1033; Anal. Abstr., 1970, 18, 793.
29.   M.A. Popov, N.M. Bershadskaya, K.A. Petaeva and V.A. Ostapenko,
      Uchen. Zap. Tsent. Nauchno-issled. Inst. Olovyan. Prom., 1965,
      46; Anal. Abstr., 1967, 14, 2447.
30.   J. Mashall and L. Geyer, Bull. Res. Council Israel, 1957, 6C, 74.

31. V.F. Lukyanov, E.M. Knyazeva and K.R. Orekhova, Zh. Analit.
     Khim., 1962, 17, 931; Anal. Abstr., 1963, 10, 4070.
32. M.A. Popov, Zavod. Lab., 1962, 26, 540; Anal. Abstr., 1961, 8,
     44.
33. A.P. Voronchikhina and L.S. Sokolkina, Zavod Lab., 1968, 34,
     1308; Anal. Abstr., 1970, 18, 1518.
34. F. Feigl and A. Schaeffer, Anal. Chem., 1951, 23, 251.
35. Soviet Standard GOST 7619-55.
36. O. Steinhauser, P. Fragstein and C. Harzdorf, Z. Anal. Chem.,
     1967, 232, 241.
37. Ya. G. Sakharova and N.I. Shishkina, Zavod. Lab., 1959, 25, 1442;
     Anal. Abstr., 1960, 7, 3239.
38. E. Bakacs-Polgar, Z. Anal. Chem., 1959, 167, 353.
39. L. Szekeres, E. Kardos and G.L. Szekeres, J. Prakt. Chem., 1965,
     28, 113.
40. H. Buss, H.W. Kohlschutter and M. Preiss, Z. Anal. Chem., 1962,
     193, 326.
41. K.S. Chung and J.P. Riley, Anal. Chim. Acta, 1963, 28, 1.
42. P.T. Joseph and C.N. Mony, Chem. Ind. (London), 1968, 1400.
43. Yu.A. Chernikov, V.F. Lukyanov and A.B. Kozlova, Zh. Analit.
     Khim., 1960, 15, 452; Anal. Abstr., 1961, 8, 1468.
44. B. Marchand, Atomkernenergie, 1962, 7, 371; Anal. Abstr., 1962,
     10, 4616.
45. Chen Mei-Cheng, Chemistry Taipei, 1968, 3, 118; Anal. Abstr.,
     1970, 18, 100.
46. Chun-Hwa Jen and Mei-Cheng Chen, Chemistry Taipei, 1968, 4, 148;
     Anal. Abstr., 1970, 18, 2303.
47. P.N. Palei and N.I. Udal'tsova, Tr. Komiss. po Analit. Khim.
     Akad. Nauk SSSR, 1960, 11, 299; Anal. Abstr., 1961, 8, 4572.
48. R. Přibil and V. Vesely, Chemist-Analyst, 1967, 56, 23.
49. R. Přibil, Talanta, 1975, 22, 688.
50. R. Přibil and V. Vesely, Talanta, 1963, 10, 899.
51. U.C. Maiwal and K. Srinivasulu, Mikrochim. Acta, in the press.

CHAPTER 11

Analysis of Silicates and Rocks

The complete classical analysis of silicates and rocks is a demanding and time-consuming task. The complexometric determination of calcium, magnesium and aluminium has contributed to the simplification and acceleration of the analysis. During the last thirty years the analysis has been made easier by the use of spectrophotometric and atomic-absorption or flame photometric methods for the determination of almost all the elements commonly encountered in silicate rocks, but it is doubtful whether the results are as accurate or precise as those obtainable by an expert using the classical methods.

One of the few classical methods still in use is gravimetric determination of silica, preferably as silicomolybdate, and another is the determination of iron(II), but in general the principles of the Shapiro and Brannock "rapid" method [1] or one of its variants are used, the whole analysis being done on aliquots from two "master" sample solutions, without separation by the classical scheme. The main requirement is complete sample decomposition, which is not always an easy matter, as will be discussed below. The scheme of Bennett and Reed, given on p. 301 in the previous chapter is applicable to silicate materials as well as to limestones and dolomites. For a full account of chemical methods of silicate analysis the reader is referred to the monograph by Jeffery and Hutchison [2].

Figure 11.1 shows another scheme, designed for Portland cement [3].

Narebski has described a scheme [4] for complexometric analysis of easily soluble silicates by use of the methods given in Chapter 6, and has applied it (slightly modified) to the analysis of zinc-magnesium spinel [5].

In the Narebski method, 1 g of finely powdered sample is fused with sodium carbonate, silica is separated in the usual way, by a single evaporation with hydrochloric acid, collected and purified. The residue from the silica purification is recovered by fusion and added to the original filtrate, which is then diluted to 250 ml, and fractions are taken for the rest of the analysis. The method requires only one precipitation of Fe + Al + Ti (+ phosphate).

312

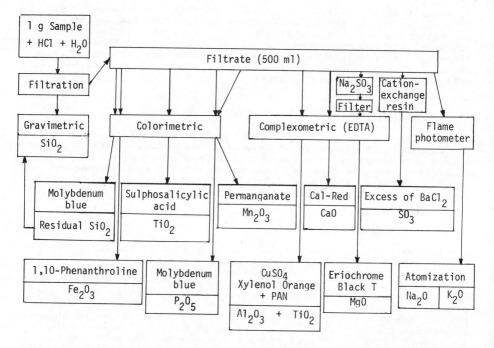

Fig. 11.1. Flow sheet for the rapid analysis of
Portland cement (after [3] by permission of the
copyright holders, the Royal Society of Chemistry)

For this Nabreski recommends using 20% pyridine solution, as the only
reagent which reliably allows separation of these hydroxides from
calcium, magnesium and manganese. According to his experience, in
spite of its unpleasant smell pyridine is preferable to hexamine,
since it is more stable and forms soluble complexes with bivalent
metal ions.

The hydroxide precipitate is dissolved and made up to known volume,
then a fraction is taken for determination of iron by EDTA titration
at pH 2 with salicylic acid as indicator, followed by determination
of the sum of Al + Ti by further addition of EDTA until an excess is
present, and back-titration at pH 5 - 6 with 0.05M zinc acetate (with
the ferricyanide redox system and dimethylnaphthidine as indicator).
In another fraction titanium is determined spectrophotometrically
with hydrogen peroxide.

The filtrate from the pyridine separation is diluted to standard
volume and calcium is determined in one aliquot by EDTA titration in
strongly alkaline medium (Patton-Reeder indicator). In a second
aliquot the sum of Ca + Mg (+ Mn) is determined at pH 10 (Schwarzenbach buffer) by direct titration with EDTA (Erio T as indicator).

A further entirely complexometric analysis of silicates, after removal

or separation of silica, was proposed by Přibil and Veselý [6].   In
aliquots of the master stock solution iron, aluminium and titanium
are determined in acidic medium by EDTA titrations.   In other
aliquots calcium, magnesium and manganese are also determined by EDTA
titration after masking of iron and aluminium with triethanolamine.
The method is also suitable for the analysis of slags, and details
will be given in the next chapter.

11.1   PARTIAL ANALYSIS OF SILICATES AND ROCKS

In this type of analysis attention has been focused mainly on the
determination of aluminium, calcium and magnesium.   Since complexometric determination of these metals is not possible without previous
separation steps or masking, fast decomposition procedures were
sought that would simultaneously yield a solution suitable for the
determination and separate interfering elements.

11.1.1   Aluminium

A relatively simple determination of aluminium in silicates and slags
was described by Chernikhov et al. [7].   The sample $(0.1 - 0.2$ g$)$ is
mixed with $3.5 - 5.0$ g of sodium hydroxide and 0.5 g of sodium
peroxide in a nickel crucible (plus $15 - 20$ mg of calcium oxide if
titanium is present), heated gently to melt the hydroxide and then
placed in a muffle furnace at $600 - 700°C$ for $15 - 20$ min.   The cooled
melt is extracted with 10 ml of 3.5% sodium hydroxide solution and
mixed with $0.2 - 0.6$ ml of 5% sodium sulphide solution (according to
the amount of Fe and Mn).   After a few minutes the solution is
filtered into a 250-ml standard flask and the precipitate is washed
6 or 7 times with 3.5% sodium hydroxide solution.   The filtrate is
then diluted to the mark and an aliquot containing $6 - 8$ mg of Al_2O_3
is mixed with a slight excess of $0.05M$ EDTA and after proper buffering with chloroacetate/chloroacetic acid mixture the excess of EDTA
is titrated with $0.05M$ thorium nitrate, with Alizarin Red S as
indicator.   According to the authors silica does not interfere in
amounts up to 50 mg.

Remarks.   The method reflects the state of knowledge and experience
at the time of its origin (1955).   The reliability of the separation
of aluminium from the hydroxide/sulphide precipitate is rather doubtful.   The relatively low consumption of EDTA $(2 - 4$ ml$)$ can also be
a source of significant error.

Bennett and co-workers have published several papers on the determination of aluminium in ceramic raw materials and in heat-resistant
ceramics.   They recommend back-titration of excess of EDTA with zinc
solution (dithizone as indicator), after prior separation of iron and
titanium by extraction with cupferron into chloroform.   Only manganese interferes, but does so quantitatively, so a correction can be
made [8].   This method has also been applied to chromium-containing
materials [9].   The sample is fused with sodium carbonate and boric
acid and the resulting chromate is extracted with a chloroform
solution of Amberlite LA-2, iron and manganese are extracted as their
diethyldithiocarbamates, and titanium is removed as the cupferronate.

Přibil and Veselý [10] found that rather unexpectedly the formation of
the DCTA complex of aluminium is instantaneous at room temperature

even in the presence of neutral salts.   This finding led to further
improvement of the determination of aluminium in the presence of
tervalent chromium and chromate [11].

Determination according to Bennett and Reed [12]

After separation of silica, and extraction of iron and titanium with
a chloroform solution of cupferron, aluminium is then determined
indirectly by titration of excess of DCTA with zinc solution,
dithizone being used as indicator.

Collect the aqueous solution from the cupferron-chloroform separation
in a 1000-ml conical flask.   Add a few drops of Bromophenol Blue and
just alkalize with concentrated ammonia solution.   Quickly just
acidify with concentrated hydrochloric acid and add 5 or 6 drops more.
Add enough 0.05M DCTA to provide a few ml excess.   Add ammonium
acetate buffer (120 ml of glacial acetic acid in 500 ml of water
mixed with 74 ml of concentrated ammonia solution, cooled, and
diluted to 1 litre) until the indicator turns blue, and 10 ml more.
Add an equal volume of ethanol, 20 ml of 10% hydroxylammonium
chloride solution and 1 - 2 ml of dithizone solution (0.025% in 95%
ethanol).   Titrate with 0.05M zinc chloride from green to the first
appearance of a permanent pink colour.

According to the authors, manganese is co-titrated quantitatively but
up to 0.5% of chromium(III) does not interfere and can be ignored.

Remarks.   The use of dithizone as indicator is not advantageous and
necessitates use of a 50% ethanol medium.   The authors themselves
admit that the colour change is sufficiently sharp only for a "pure
solution" of aluminium.   In the analysis of ceramic materials the
indicator proved to be rather unstable even during the titration.
According to the authors' opinion this might have been caused by a
decomposition product of cupferron which oxidizes the indicator.   A
high concentration of hydroxylamine removes this difficulty.   Most
indicators tried were reported to behave similarly to dithizone in
this system.

Determination in materials containing chromium [12]

The same procedure as above was used for the analysis of fire-
resistant bricks etc.   The authors changed their earlier statement
[13] that chromate was extracted quantitatively by Amberlite LA-2.
About 0.5% of the chromate escapes extraction, owing to partial
reduction.   In that case EDTA cannot be used for the determination
of aluminium because the Cr-EDTA complex is formed during the
necessary boiling of the solution.   This causes erroneously high
results for aluminium and the violet colour of the Cr-EDTA complex
makes a visual end-point impossible.   The use of DCTA overcomes
these obstacles completely.   Manganese is quantitatively co-titrated,
so that the results can be corrected accordingly.   Magnesium and
calcium do not interfere.

Determination according to Pritchard [14]

The separation of aluminium from iron, titanium, calcium and magnesium
by sodium hydroxide is improved by adding DCTA immediately after the
alkali and digesting this mixture at 100°C for an hour.   After
filtration, aluminium is determined by back-titration.

Mix 0.2 g of finely ground sample with 0.2 g of sodium hydroxide and
fuse the mixture in a platinum crucible at 650 -750°C for 10 min.
Add 25 ml of water to the cooled melt and let stand overnight.   Then
dissolve the melt with hot water and wash out the crucible with a few
drops of hydrochloric acid (1 + 1).   Transfer the solution, by means
of a polyethylene funnel with a long stem, into a 250-ml standard
flask containing 100 ml of hot 2M hydrochloric acid.   Cool, and make
up to the mark with water.   Pipette 50 ml of the solution into a
polypropylene beaker, add 60 ml of 5% sodium hydroxide solution, then
without delay 10 ml of 0.08M DCTA and a pinch of paper pulp.   Heat
the solution on a steam-bath for 1 hr.   After the solution has
cooled a little filter off on paper and wash the precipitate with six
6-ml portions of 2% sodium hydroxide solution.   Acidify the filtrate
with hydrochloric acid (1 + 1) (Bromocresol Green indicator), then
neutralize with 5M sodium hydroxide.   Finally add 25 ml of buffer
solution (pH 5.5) and a few drops of Xylenol Orange indicator, and
titrate the excess of DCTA with 0.025M zinc chloride, to a red-violet
colour。

To make the buffer solution dissolve 200 g of hexamine in 200 ml of
water, add 40 ml of concentrated hydrochloric acid and dilute to 1
litre with water.

Remark.   According to the author the use of a propylene beaker for
the digestion of the solution is essential.   Digestion in glass
vessels gives irregular results.   If present in amounts exceeding
1 mg, magnesium is precipitated, and some aluminium is adsorbed on
the precipitate during its aging.   Therefore the DCTA must be added
to the solution immediately after the sodium hydroxide.

The author considered that there was slight decomposition of DCTA
during boiling and recommended running a reagent blank.

A further method for aluminium is discussed on p. 177.

11.1.2   Calcium and Magnesium

When only determination of calcium and magnesium is required, for
example in refractories and the raw materials for making them, the
well known separation methods for calcium and magnesium are generally
used.   For example, Dippel [15] separates iron, aluminium, titanium
and manganese with ammonia and bromine water.   Obviously the
precipitation should be repeated.   In the filtrate he determines
calcium and magnesium by the common complexometric methods.   Trace
metals are masked with triethanolamine and potassium cyanide.

For the determination of calcium in high-alumina material Itakura and
Kodama [16] recommend the following procedure.   After fusion of
0.5 g of powdered sample with 1.5 g of boric acid in a platinum
crucible, the melt is boiled with 100 ml of water and 0.5 ml of
hydrochloric acid (1 + 1).   The crucible is removed and the mixture
is boiled for 5 min, then filtered through a filter paper containing
a small amount of filter-pulp.   The filtrate is diluted to volume in
a 250-ml standard flask with water.   In a 100-ml aliquot calcium is
determined after addition of a small amount of ascorbic acid,
triethanolamine, potassium cyanide and enough 20% potassium hydroxide
solution to give pH 12.8.   The authors use 0.01M EDTA, and a mixture
of Fluorexone and Methylthymol Blue (1 + 1) as indicator.   The colour

change at the end-point is from green to dark orange. According to
the authors the method has the advantage that the stock sample
solution can also be used for the determination of magnesium, sodium
and potassium.

Abdullah and Riley [17] describe an ion-exchange method for the
separation of calcium and magnesium from the major elements of
silicate rocks. It is based on the finding that practically all
interfering elements can be eluted from a column of the cation-
exchanger Zeo-Karb 222 (ammonium form) with a solution of the
diammonium salt of EDTA at pH 4.5.   Calcium and magnesium are eluted
successively, first magnesium with 0.5M ammonium chloride and then
calcium with 1M ammonium chloride.   Both metals are determined by
spectrophotometric EDTA titration, with Erio T as indicator.

11.1.3   Additional Methods

Methods for determination of iron and aluminium have been given by
Sajó [18], and for these elements and titanium by Přibil and Veselý
[19].

Silicon can be determined by extraction of silicomolybdic acid into
isoamyl alcohol and complexometric determination of the molybdenum
[20].

A complexometric method for sodium was described by Sinha and Roy [21].

Many of the methods described in the next chapter (on analysis of
slags) can be used for silicates which have compositions suitable for
use of titrimetric methods.

REFERENCES

1.   L. Shapiro and W.W. Brannock, U.S. Geol. Surv. Circ. No. 165,
      1952; Bull. U.S. Geol. Surv. No. 1036-C, 1956.
2.   P.G. Jeffery and D. Hutchison, Chemical Methods of Rock Analysis,
      3rd Ed., Pergamon Press, Oxford, 1981.
3.   A. Nestoridis, Analyst, 1970, 95, 51.
4.   W. Narebski, Bull. Acad. Pol. Sci., 1962, 10, 185.
5.   W. Narebski, Kwart. Geol., 1962, 6, 1; Anal. Abstr., 1963, 10,
      815.
6.   R. Přibil and V. Veselý, Chemist-Analyst, 1966, 55, 68.
7.   Yu.A. Chernikhov, B.M. Bobkina and I.M. Khersonskaya, Zavod.
      Lab., 1955, 21, 638; Anal. Abstr., 1956, 3, 948.
8.   H. Bennett, W.G. Hawley and R.P. Eardley, Trans. Brit. Ceram.
      Soc., 1962, 61, 201; Anal. Abstr., 1962, 9, 5090.
9.   H. Bennett and R.A. Reed, Res. Pap. Brit. Res. Assn., No. 595,
      1967; Anal. Abstr., 1968, 15, 3308.
10.   R. Přibil and V. Veselý, Talanta, 1961, 9, 23.
11.   R. Přibil and V. Veselý, Talanta, 1963, 10, 1287.
12.   H. Bennett and R.A. Reed, Analyst, 1970, 95, 541.
13.   H. Bennett and K. Marshal, Trans. Brit. Ceram. Soc., 1966, 65,
      681.
14.   D.T. Pritchard, Anal. Chim. Acta, 1965, 32, 184; Anal. Abstr.,
      1966, 13, 2911.
15.   P. Dippel, Silikattechnik, 1964, 15, 391.

16. M. Itakura and K. Kodama, Japan Analyst, 1978, 27, 531; Anal.
     Abstr., 1979, 37, 1B 288.
17. M.I. Abdullah and J.P. Riley, Anal. Chim. Acta, 1965, 33, 391.
18. I. Sajó, Acta Chim. Acad. Sci. Hung., 1955, 6, 251.
19. R. Přibil and V. Veselý, Talanta, 1963, 10, 233.
20. M.G. Vasil'eva and M.Z. Beshikdash'yan, Zh. Analit. Khim., 1970,
     25, 1952; Anal. Abstr., 1972, 22, 707.
21. B.C. Sinha and S.K. Roy, Talanta, 1979, 26, 596.

CHAPTER 12

Analysis of Slags

Slags are artificial silicates and their composition is of interest
to industrial chemists. Their chemical composition depends on the
raw materials used and the production conditions. The complete
analysis of slags usually comprises the determination of SiO_2, FeO,
MnO, Al_2O_3, TiO_2, CaO, MgO, P_2O_5, and S.

After the introduction of complexometry analysts departed from the
classical gravimetric methods for slag analysis in two ways. Some
took the step of trying to use complexometry for the whole analysis,
often disregarding the effect of the wide variation in composition.
Others took the more realistic approach of combining complexometry
with photometric methods. The procedures are generally similar to
those described for silicates in the preceding chapter.

Both approaches, however, try to avoid separation procedures wherever
possible, and to use a single sample solution, obtained after separation or removal of the silica. On the other hand, the determination
of individual components or groups of components invariably uses
separation procedures, and has led to development of special methods
for determination of calcium, or magnesium, or aluminium and
manganese, when they are present in amounts suitable for use of
titrimetric methods.

12.1 DETERMINATION OF INDIVIDUAL COMPONENTS IN SLAGS

12.1.1 Calcium

Calcium can easily be separated as oxalate even in the presence of
EDTA, which at pH 4 - 5 masks interfering elements [1]. This
procedure has been applied to the determination of calcium in lead-
containing slags [2]. Bieber and Večeřa [3] have thoroughly studied
the precipitation of calcium as oxalate, sulphite and tungstate. The
calcium in the precipitates was determined complexometrically, with
good results. Simple procedures for determination of calcium have
been described by Tanihara [4] and by Penkina et al. [5].

Procedure of Tanihara. Heat 0.1 g of open-hearth furnace sample with

319

5 ml of concentrated hydrochloric acid, 5 ml of 3% hydrogen peroxide
and 100 ml of water.   After addition of 10 ml of 4% magnesium
chloride solution, precipitate iron etc. together with $Mg(OH)_2$ with
10 ml of 20% sodium hydroxide solution.   Then add an excess of 0.05M
~~EDTA solution and back-titrate with 0.05M calcium chloride with Cal-~~
Red as indicator.   According to the author there is no interference
by aluminium (up to 7.5 mg), phosphate (up to 20 mg) and fluorite (up
to 50 mg).

Procedure of Penkina et al.   Dissolve 0.1 g of blast-furnace slag in
30 ml of 2.5% ammonium sulphate solution, boil, add 20 ml of hot
nitric acid (1 + 1), boil and cool, dilute to volume in a 250-ml
standard flask with water, and filter.   To 100 ml of filtrate add
5 ml of triethanolamine solution (1 + 3) and 30 - 35 ml (accurately
measured) of 0.015M EDTA (to prevent precipitation of Ca on addition
of the indicator) then 1.5 ml of mixed indicator and complete the
EDTA titration of the calcium.   Make the indicator by dissolving
0.25 g of Calcein and 2 mg of Thymolphthalein in 100 ml of 20%
potassium hydroxide solution.

According to the authors, iron, manganese and aluminium do not interfere, and their separation is unnecessary.   Up to 4% of BaO also
does not interfere.

12.1.2   Magnesium

Wakamatsu [6] and Endo and Hattori [7] described two practically
identical methods based on separation of interfering metals (Al, Fe,
Mn) and calcium by precipitation with ammonium acetate and ammonium
oxalate.   In the filtrate magnesium is determined by EDTA titration.

Procedure of Wakamatsu [6].   Dissolve 0.1 g of basic slag in 10 ml
of nitric acid (1 + 1) and dilute with 120 ml of water.   Add 20 ml
of 50% ammonium acetate solution and 20 ml of 50% ammonium oxalate
solution, boil for 2 min, and filter.   Dilute the filtrate to a
standard volume and to an aliquot add 50 ml of water, 10 ml of 5%
potassium cyanide solution (to mask iron), and ammonia to adjust the
pH to 10, then titrate with 0.01M EDTA, with Erio T as indicator.
According to the author the whole determination takes about 10 min.

12.1.3   Calcium and Magnesium

These methods are mostly based on separation of iron, aluminium,
titanium and manganese from the calcium and magnesium by precipitation
with ammonia, which is the simplest procedure but less reliable,
especially if large amounts of manganese are present.   In many cases
reprecipitation two or three times is necessary.   Several modifications of this principle are still described in a number of publications.   Only the Borissova-Pangarova and Mitropolitska method [8]
for the determination of calcium and magnesium in the presence of up
to 70% of manganese will be given here.

It is generally more reliable to use extraction for removing interfering elements, e.g. iron and titanium with cupferron and manganese
with sodium diethyldithiocarbamate [9].   Lapin et al. [10] extract
iron, titanium and manganese (without removal of silica) with oxine
from ammonium acetate buffer (pH 6.2) and then determine calcium and

magnesium by the usual EDTA methods.

Determination in the presence of manganese [8]. Fuse the sample
(containing up to 70% of MnO) with sodium carbonate and dissolve the
melt in nitric acid (1 + 1) with addition of a hot solution of
oxalic acid to dissolve hydrous MnO_2. Precipitate the manganese by
adding 5 g of potassium bromate and heating to 80°C (but not higher,
to avoid the separation of silicic acid). Make the filtrate up to
known volume, and use a fraction to determine the sum Ca + Mg after
addition of triethanolamine and Schwarzenbach buffer to give pH 9.5
(use Methylthymol Blue as indicator.) After this titration add
enough sodium or potassium hydroxide to give a clear violet colour
and titrate the liberated EDTA (corresponding to Mg) with a standard
calcium solution.

Remark. Determination of magnesium by this method has some limitations and can be variously modified (see Section 6.3.1).

12.1.4   Aluminium

For the individual determination of aluminium oxide in slag (or
silicates) the formation of soluble tetrahydroxoaluminate by alkaline
fusion is used for its separation. Only the method of Fedorov and
Ozerskaya [11] is given here.

Fuse 0.25 - 0.50 g of sample in an iron crucible with 10 g of sodium
peroxide at 800 - 850°C. Extract the cooled melt with 80 - 100 ml of
water, dilute to 500 ml and set aside for 12 hr. Filter and wash,
dilute the filtrate to standard volume and boil an appropriate
aliquot for 8 - 10 min, cool, add 25 ml of 0.05M EDTA, neutralize to
phenolphthalein with concentrated hydrochloric acid, and add 15 ml of
acetate buffer solution (pH 5 - 6). Boil the solution for 4 - 5 min,
cool, add Xylenol Orange and titrate with 0.05M zinc solution to a
colour change from yellow to crimson.

12.1.5   Titanium

Traces of titanium in slag are usually determined spectrophotometrically with hydrogen peroxide. For complexometric determination of
large amounts of titanium, any iron and aluminium must be separated
by precipitation of the titanium with sodium hydroxide in the presence
of triethanolamine. The method is suitable for titanium contents
above 0.5% [12].

After separating silicic acid, precipitate iron, aluminium and
titanium and filter off. Dissolve the precipitate with 20 - 30 ml of
hot hydrochloric acid (1 + 1), cool, add 20 - 30 ml of 20% triethanolamine solution, make alkaline with 20% potassium hydroxide solution
and boil. Allow the titanium precipitate to settle, filter off and
wash it with hot water, then dissolve it in hot concentrated nitric
acid containing 2 - 5 ml of 3% hydrogen peroxide solution. Cool to
10°C, add an excess of 0.05M EDTA, a few drops of Xylenol Orange and
neutralize with concentrated ammonia solution to a red colour. Then
adjust to pH 2 (test paper) with 1M nitric acid and titrate with 0.05M
bismuth nitrate from yellow to red.

Remarks. The titration must be performed on a sufficiently dilute

solution because of the intense yellow colour of the Ti-EDTA-peroxo
complex.   The limiting amount of titanium is 25 mg.   Amounts
greater than 40 mg of titanium in 400 ml cannot be determined because
the yellow colour obscures the end-point.   If the material is very
rich in titanium the determination can be shortened by direct
precipitation with potassium hydroxide and triethanolamine (which
masks Fe and Al).   Co-precipitated magnesium does not interfere in
the final titration of titanium.

12.1.6   Manganese

Large amounts of manganese in slags and ores can be determined
complexometrically in the presence of calcium if this is masked with
ammonium fluoride [13].   After removal of silicic acid, the
hydroxides of Fe, Al and Ti are separated by double precipitation
with hexamine.   The combined filtrates are diluted to a suitable
standard volume (e.g. 500 or 1000 ml).   Then 50 or 100 ml of the
filtrate is mixed with an excess of 0.05M EDTA and 0.5 g of hydroxylamine hydrochloride and warmed to 70°C.   After addition of ammonia
buffer (pH 9), 1 ml of 10% triethanolamine solution and a few
crystals of potassium cyanide, the solution is stirred thoroughly and
calcium is precipitated by dropwise addition of 10 ml of 10% ammonium
fluoride solution.   Then the excess of EDTA is titrated with 0.05M
manganese sulphate (containing 1% of hydroxylamine chloride), with
Erio T as indicator.

Results obtained by this method for samples containing 25 - 55% of MnO
were in good agreement with those from classical methods, but in
general about 1% lower.

A simpler method for manganese (4 - 50%) in slags (concentrates and
other metallurgical products) was described by Chekhlarova et al. [14].

Dissolve 0.5 g of sample in hydrochloric acid and perchloric acid or
fuse it with sodium carbonate.   Dissolve the residue in hydrochloric
acid, heat to 60 - 70°C and adjust the pH to 1.5 - 2.0.   Mask the iron
with DCTA (titration with salicylic acid as indicator).   To the
cooled solution add hydroxylamine hydrochloride, potassium tartrate
and ammonium fluoride, and adjust to pH 9 - 9.5 with ammonia.   Titrate
with 0.05M EDTA with Thymolphthalexone as indicator.

12.1.7   Metallic Iron

According to Wakamatsu [15] the metallic iron is dissolved quantitatively in an ethanolic solution of mercuric chloride.   After
separation of the insoluble residue, iron in the filtrate is oxidized
with ammonium persulphate and determined complexometrically.

Mix the powdered sample (0.5 g, 100-mesh) with solid mercuric chloride
(10 times the expected weight of metallic iron) and 50 ml of ethanol.
Stir the mixture at 60°C for 20 min in an atmosphere of dry carbon
dioxide.   Filter off the insoluble matter and wash it three times
with ethanol.   To the combined filtrate and washings add 2 ml of 6M
hydrochloric acid and 10 ml of 20% ammonium persulphate solution,
dilute with 100 ml of water and adjust to pH 2.0 - 2.5 with 50%
ammonium acetate solution.   Titrate at 40°C with 0.01M EDTA, using
1 ml of 1% tiron solution as indicator.

12.1.8   Magnesium and Magnesium Oxide in Uranium Slags

This is a special problem connected with uranium production by
reduction of uranium tetrafluoride with metallic magnesium.   The
slag formed during this process contains mainly magnesium fluoride,
magnesium oxide and metallic uranium, uranium dioxide and tetrafluoride, aluminium and iron.   McKend [16] worked out a method for
the complexometric determination of both metallic magnesium and the
sum of Mg + MgO.   The method is based on selective dissolution of
magnesium and magnesium oxide in a buffered solution of EDTA, and
then selective dissolution of magnesium, activated by dichromate, in
acetic acid.   The procedure requires the several conditions to be
observed and therefore the reader is referred to the original source
for further details.

12.2   COMPLETE ANALYSIS OF SLAGS

As said in the introduction to this chapter, several papers describe
complete analysis of slags by complexometry alone or in combination
with photometric methods for determination of the trace components.
Some examples are introduced here.

12.2.1   Analysis of Basic Slags According to Wakamatsu [17]

Dissolve 0.5 g of sample in 5 ml of hydrochloric acid (1 + 1) and
determine silica as usual.   Dilute the filtrate and washings to
250 ml in a standard flask with water.   Dilute 50 ml of this stock
solution with 40 ml of water, adjust to pH 2 with ammonium acetate
and titrate iron with 0.0125M EDTA, using salicylic acid as indicator.

To the same solution add 0.1 ml of 0.0125M copper sulphate, adjust to
pH 3 with ammonium acetate, boil, add PAN as indicator and then
0.0125M EDTA in slight excess, and back-titrate the excess with
0.0125M copper sulphate.   This gives the aluminium content.   Heat
the same solution to 60°C, bring to pH 4.5, add another excess of
EDTA and back-titrate with copper solution as before, to obtain the
manganese content.

Dilute the solution from the manganese titration to 250 ml in a
standard flask.   Take a 50 or 100 ml aliquot, adjust to pH 10, add
EDTA to the red solution until it becomes yellow, then 2 ml in excess.
Add 3 drops of PAN indicator and titrate with copper solution.   This
gives the sum of Ca + Mg.

Take another 100-ml portion of the diluted solution from the manganese
titration, add saturated ammonium oxalate solution (20 ml), filter,
and titrate magnesium in the filtrate with EDTA by the usual method.
The whole analysis takes about 3 hours.

Remark.   A similar method of rapid analysis of magnesia clinker for
basic refractories was published by Honjo et al. [18].

12.2.2   Analysis of Slags According to Endo [19]

Dissolve the sample in 15 ml of hydrochloric acid (1 + 1) and a few
drops of 30% hydrogen peroxide.   Dilute to 150 ml, add 10 ml of 25%

ammonium acetate solution and adjust the pH to 1.5.   Heat to 30°C
and titrate iron with 0.05M DCTA (salicylic acid as indicator).   To
the solution add 1 ml of 0.05M copper sulphate, adjust the pH to 2 -
2.2 with 5% ammonium acetate solution, add PAN indicator and titrate
with 0.05M DCTA to obtain the amount of aluminium.   Dilute the
solution after titration to 250 ml in a standard flask.   To a 100-ml
portion add 5 - 10 ml of 0.1M EDTA, and ammonia to make the pH 3.5 -
4.4.   Boil the solution with 20 ml of 5% ammonium oxalate solution,
filter off the calcium oxalate and determine the calcium with
permanganate.

To another 100-ml portion add ammonia to make the pH 10, then 20 ml
of 20% ammonium chloride solution, 5 ml of triethanolamine and 15 ml
of 20% potassium cyanide solution and titrate with 0.05M EDTA, using
Erio T as indicator, to find the sum of Ca + Mg.

Remarks.   This method was published in 1958 when not much was known
about the behaviour of DCTA, which is now known to react with
aluminium at room temperature.   According to our unpublished work
the exchange reactions (for example between the DCTA-Fe complex and
TEA or DCTA-Mn and cyanide etc.) proceed slowly.

12.2.3   Analysis of Basic Slags [20]

Endo and Takagi considerably improved the previous method.   The slag
sample (0.5 g) is dissolved in a mixture of nitric acid and hydrochloric acid, and diluted accurately with water to 250 ml.   Four
50-ml portions are used for titration.   In the first portion iron
and aluminium are titrated with EDTA (at pH 2) with salicylic acid
and the copper-PAN complex (at pH 3) as indicators respectively.   In
the second portion calcium is precipitated as oxalate (at pH 4) and
subsequently determined, interferences being masked with EDTA.   In
the third portion the sum of Mg + Ca is found by EDTA titration at
pH 10, with Thymolphthalexone as indicator, after masking of Fe and
Al with triethanolamine, and of manganese with potassium cyanide.
In the fourth portion the sum of Ca + Mg + Mn is titrated with EDTA
at pH 10 in the presence of triethanolamine and hydroxylamine hydrochloride, with Thymolphthalexone as indicator.   The contents of
manganese and magnesium are calculated from the difference between
the three titration results (for Ca, Ca + Mg and Ca + Mg + Mn).

Remarks.   The authors were able to avoid the determination of calcium
in the presence of precipitated Mg(OH) , which is known to lack
precision, especially at high magnesium levels.   It is also known,
that high concentrations of aluminium cause positive errors in the
results for iron.   The masking of iron and manganese (with TEA + KCN)
also has limitations.   After addition of cyanide the Fe-TEA complex
solution becomes yellow, which can spoil the colour change of the
indicator.   The amount of manganese present should not exceed a few
milligrams either, since otherwise a large amount of potassium
cyanide must be used (see pp. 180, 181).

12.2.4   Analysis of Slags According to Přibil and Veselý [12]

Dissolve 1 g of powdered sample in 20 - 25 ml of hydrochloric acid
(1 + 1), heating on a sand-bath.   During the dissolution add 0.1 g
of potassium chlorate and remove silicic acid in the usual way.

Dilute the filtrate to 250 ml in a standard flask.

Transfer an aliquot containing 0.1 - 0.2 g of sample to a 400-ml
beaker, dilute to 100 - 150 ml with water and add sufficient 0.05M
DCTA to complex the iron and aluminium present.   Adjust to pH 5.0 -
5.5 with hexamine, add some Methylthymol Blue and titrate with 0.05M
zinc chloride to an intense blue colour.   Then add 1 - 2 g of
ammonium fluoride, adjust to pH 2 - 4 with hydrochloric acid (1 + 1),
cover the beaker and boil the solution for 3 min.   Cool, adjust the
pH to 5.0 - 5.5 with hexamine and titrate the liberated EDTA with zinc
solution to an intense blue.   The first titration gives the sum
Fe + Al, the second the amount of aluminium.

Dilute a second aliquot with 100 ml of water, add 20 ml of 20%
triethanolamine solution and then 1M potassium hydroxide until the
solution is colourless.   Add some Fluorexone and titrate slowly with
0.05M EGTA to the disappearance of the green fluorescence.   This
titration corresponds to calcium.

To a third aliquot add 20 ml of 20% triethanolamine solution, 20 -
50 ml of ammonia buffer (pH 10) and wait for 5 min.   Add the amount
of 0.05M EGTA required by the calcium and 1 - 2 ml in excess, then add
some Methylthymol Blue and titrate with 0.05M DCTA from intense blue
to smoky grey or yellow.   The result corresponds to magnesium. (If
more than 20 mg of magnesium is present $Mg(OH)_2$ precipitates but can
be redissolved by heating, and the titration can be continued dropwise.

Remarks.   Traces of manganese do not interfere, since they will be
present as $Mn(III)$-TEA complex.   Up to 2 mg of manganese can be
masked if 1 ml of 5% hydrogen peroxide is added and the solution is
let stand for 5 min before the titration.   Further modifications can
be made for samples containing larger amounts of manganese, and/or
titanium, both of which can also be determined complexometrically.

12.3   COMBINED METHODS FOR SLAG ANALYSIS

These methods are based on complexometric determination of the main
components (mostly only Ca and Mg).   Other components are determined
photometrically (more recently by flame or atomic-absorption spectrometry).

A relatively fast method for such analysis was described as early as
1955 by Williams [21].   After fusion of the sample with sodium
carbonate, silicic acid is determined in the usual way.   Aliquots of
the filtrate are used for the photometric determination of iron (with
1,10-phenanthroline), manganese (with periodate), and titanium (with
hydrogen peroxide).   Another part of the filtrate is used for the
precipitation of iron, aluminium, titanium and manganese with ammonia
and ammonium persulphate.   The precipitate is ignited, the mixed
oxides are weighed and the aluminium content is calculated by
difference.   The filtrate from the hydroxide separation serves for
the EDTA titration of calcium (murexide as indicator) and the sum of
Ca + Mg (Erio T as indicator).   Similarly procedures are described
by Basilevskaya [22] and by Szarvas et al. [23].

12.3.1   Analysis of Cupola Slags

The composition of cupola slags varies considerably.   Bieber and
Večeřa [24] quote the ranges 22 - 55% SiO_2, 1 - 10% Fe_2O_3, 8 - 30% Al_2O_3,
0 - 2% TiO_2, 1 - 5% MnO, 20 - 45% CaO, 1 - 25% MgO, 0.2% S and 0.5% P_2O_5.
Only their work and that of Clarke [25,26] will be dealt with here.
Both publications give many valuable technical hints that are
neglected by other authors, whose findings are usually based on the
study of synthetic solutions or of slag solutions after the separation
of silicic acid.

Preparation of slag samples

Bieber and Večeřa consider that pouring the liquid slag sample into
water is not a satisfactory procedure, as it does not ensure
representative sample composition.   With basic slags some decomposition may occur in the water and some soluble components escape
determination.   It is preferable to take a liquid sample with an
iron spoon and pour it over a smooth steel plate, where it will
quickly solidify.   The sample can then be easily broken up with a
hammer and ground to a powder in a steel mortar.   About 3 - 6 g of
this powdered sample is separated by coning and quartering and further
ground in an agate mortar to a fine powder that passes a 100-mesh
sieve.   Any iron grains mechanically trapped in the slag should be
removed with a strong magnet before the fine grinding.   The whole
preparation of the sample takes 10 - 15 min.

Dissolution of the sample

Most cupola slags are soluble in acids.   Clarke recommends
dissolution in the following way.

Place 0.5 g of sample in a beaker, add 25 ml of water and heat to
boiling.   Add concentrated hydrochloric acid dropwise with continuous
stirring.   Ensure that the solid does not sediment on the bottom of
the beaker, and that the precipitated silica gel does not form a
coating on any undissolved sample and hinder its otherwise almost
instantaneous dissolution.   Even slags containing up to 50% SiO_2 can
be dissolved rapidly by this procedure if the CaO content is
sufficiently high (20 - 25%).   The solution obtained by this procedure
is used by Clarke exclusively for the determination of SiO_2 after
evaporation with a few drops of nitric acid and 40 ml of concentrated
perchloric acid.

A separate 1-g sample is used for the determination of aluminium etc.
This sample is dissolved in a mixture of 15 ml of concentrated
perchloric acid, 15 ml of concentrated nitric acid and 5 ml of
concentrated hydrofluoric acid, in a platinum dish.   The solution is
carefully evaporated to fumes of perchloric acid and allowed to fume
gently for 5 min but not taken to dryness, since doing so forms a
poorly soluble precipitate from slags that are rich in aluminium and
titanium.   The residue is taken up in dilute acid in the usual way
and diluted to a standard volume.

For determination of silica in slags that are insoluble in hydrochloric acid Clarke uses a fusion with sodium carbonate in a platinum
dish, followed by evaporation with perchloric acid.   The silica is
determined gravimetrically in the usual way.

According to the procedure of Bieber and Večeřa, 0.2 g of sample is placed in a platinum dish, wetted with 2 - 3 ml of water, mixed with a platinum wire, and then boiled with 5 ml of concentrated nitric acid until evolution of brown fumes ceases. The solution is cooled and after addition of 10 ml of concentrated hydrofluoric acid and 1 ml of concentrated sulphuric acid is heated on a sand-bath till fuming. After cooling and addition of 10 ml of concentrated hydrochloric, the solution is transferred to a 250-ml standard flask and diluted to volume with water.

The analysis of the stock sample solutions by the procedures of Bieber and Večeřa [24] and of Clarke [25,26] is outlined below in schematic form.

Analysis of Cupola Slags

Bieber and Večeřa	Clarke
Sample: 0.2 g in 250 ml	Sample: 1.0 g in 200 ml
FeO: from 10 ml, as Fe^{3+} $KSCN + H_2O_2$	FeO: up to 2% from 10 ml, as Fe^{3+} $NaSCN + (NH_4)_2S_2O_8$ FeO: over 2% with acetylacetone
TiO$_2$: from 50 ml $H_2O_2 + H_3PO_4$	TiO$_2$: from 50 ml $H_3PO_4 + H_2O_2$
MnO: separation of R_2O_3 and $MnO(OH)_2$ with peroxodisulphate, then photometrically as $KMnO_4$ or volumetrically with arsenite	MnO: from 25 ml $H_3PO_4 + (NH_4)_2S_2O_8 + Ag^+$ photometrically as $KMnO_4$ of volumetrically with arsenite.
Al$_2$O$_3$: from 50 ml Precipitation of $R(OH)_3$ with gaseous ammonia, EDTA titration of Al + Fe + Ti and correction for Fe and Ti	Al$_2$O$_3$: from 50 ml Precipitation with DDCNa, extraction with $CHCl_3$, addition of NH_4F, and titration with 0.1Ṁ HCl
CaO: separation of $MnO(OH)_2$, then EDTA titration at pH 12	CaO: from 20 ml Extraction of other metals with acetylacetone and DDCNa, then EDTA titration with murexide as indicator
CaO + MgO: separation of $MnO(OH)_2$, then EDTA titration with Erio T as indicator	CaO + MgO: from 20 ml Other metals extracted as for CaO Titration with EDTA (Erio T as indicator)

At first glance the procedures are not very different.   Clarke
prefers extraction to precipitation, to avoid the danger of co-
precipitation.   He prefers to make each determination on a separate
sample or aliquot, so that no results need be obtained by difference
(except for Mg).   For higher iron contents (around 10%) Clarke
recommends permanganate titration after reduction with stannous
chloride.   The complexometric titrations recommended are of the
oldest type and could advantageously be replaced by more suitable
ones (use of new indicators, EGTA titration for calcium etc.).

12.3.2   Analysis of Slags with High Manganese and Phosphorus Contents

Staats [27], in a brief communication, discusses the obstacles in the
determination of calcium in the presence of high amounts of manganese
and phosphate.   Masking with triethanolamine or direct titration of
calcium cannot be used for such slags.   He recommends precipitation
of manganese by boiling with ammonium persulphate in weakly acidic
solution, which is almost quantitative.   For example 91 mg of
manganese can be precipitated by boiling for 15 min with 2.5 g of
ammonium persulphate, leaving only 0.1 - 0.2 mg of Mn in the filtrate.
This remaining manganese can easily be masked with triethanolamine
and calcium determined by back-titration of excess of EDTA, with
fluorexone as indicator.   The end-point is sufficiently sharp if
good illumination is available.

REFERENCES

1.   R. Přibil and L. Fiala, Chem. Listy, 1952, 46, 331.
2.   A.N. Karolev and R.A. Karanov, Zavod. Lab., 1959, 23, 413.
3.   B. Bieber and Z. Večeřa, Coll. Czech. Chem. Commun., 1961, 26,
      59.
4.   H. Tanihara, Japan Analyst, 1960, 9, 572;   Anal. Abstr., 1962,
      9, 3250.
5.   G.B. Penkina, M.G. Khlevnaya and I.M. Zamyshlyaeva, Zavod. Lab.,
      1978, 44, 1402;   Anal. Abstr., 1979, 36, 6B 125.
6.   S. Wakamatsu, Japan Analyst, 1956, 5, 264;   Anal. Abstr., 1957,
      4, 29.
7.   Y. Endo and K. Hattori, Japan Analyst, 1957, 6, 243;   Anal.
      Abstr., 1958, 5, 383.
8.   R. Borissova-Tangarova and E. Mitropolitska, Dokl. Bolg. Acad.
      Nauk, 1977, 30, 385;   Anal. Abstr., 1977, 33, 4B 143.
9.   V. Rett, Hutn. Listy, 1965, 20, 134;   Anal. Abstr., 1966, 13,
      2964.
10.   N.N. Lapin, A.T. Slyusarev and N.S. Prilutskaya, Sb. Nauch.
      Trud. Zhdanovsk. Metallurg. Inst., 1960, 393;   Anal. Abstr.,
      1961, 8, 1859.
11.   A.A. Fedorov and F.A. Ozerskaya, Sb. Trud. Tsentr. Nauch.-Issled.,
      Inst. Chern. Metallurg., 1963, 195;   Anal. Abstr., 1964, 11,
      4298.
12.   R. Přibil and V. Veselý, Chemist-Analyst, 1966, 55, 68.
13.   P. Povondra and R. Přibil, Coll. Czech. Chem. Commun., 1961, 26,
      2164.
14.   I. Chekhlarova, I. Gabrovski and Ts. Tsolov, Metalurgyia (Sofia),
      1975, 30, 13;   Anal. Abstr., 1976, 30, 2B 146.
15.   S. Wakamatsu, J. Iron Steel Inst. Japan, 1961, 47, 612;   Anal.
      Abstr., 1962, 9, 4718.

16. J. McKend, Anal. Chem., 1960, 32, 1193.
17. S. Wakamatsu, J. Iron Steel Inst. Japan, 1959, 45, 717; Anal. Abstr., 1960, 7, 2747.
18. T. Jonjo, H. Goto and Y. Watabe, J. Ceram. Assoc. Japan, 1959, 67, 275; Anal. Abstr., 1960, 7, 2743.
19. Y. Endo, Japan Analyst, 1958, 7, 611; Anal. Abstr., 1959, 6, 3390.
20. Y. Endo and H. Takagi, Japan Analyst, 1960, 9, 998; Anal. Abstr., 1962, 9, 3737.
21. V. Williams, Iron Steel (London), 1955, 28, 525.
22. I.N. Basilevskaya, Zavod. Lab., 1956, 22, 166.
23. P. Szarvas, J. Korondán and I. Raisz, Magy. Kem. Lap., 1967, 3, 149.
24. B. Bieber and Z. Večeřa, Prace Csl. Vyzkumu Slevarenskeho, 1957, 4, 73; Anal. Abstr., 1959, 6, 2184.
25. W.E. Clarke, J. Res. Brit. Cast Iron Assoc., 1958, 7, 249; Anal. Abstr., 1959, 6, 1210.
26. W.E. Clarke, J. Res. Brit. Cast Iron Assoc., 1959, 7, 867; Anal. Abstr., 1960, 7, 3778.
27. O. Staats, Z. Anal. Chem., 1966, 223, 185.

CHAPTER 13

Analysis of Cements

Various sorts of cements, raw materials and intermediates in the cement industry constitute a suitable field for the practical use of complexometry. Earlier methods limited themselves to the determination of calcium and magnesium and were based on the use of the classic indicators Erio T and murexide [1 - 3]. Huppertsberg [3], for instance, determined calcium without the separation of iron and aluminium. The iron was precipitated as the hydroxide and aluminium masked as tetrahydroxoaluminate, which did not interfere with the determination. Langmyhr and Saether [4] recommended for fast analysis of clinker that silica should be determined gravimetrically, and iron and aluminium spectrophotometrically with ferron (7-iodo-8-hydroxy-quinoline-5-sulphonic acid). Wallraf [5] used triethanolamine for masking iron and aluminium in the spectrophotometric EDTA titration of calcium, with murexide as indicator, and of the sum of Ca + Mg (Erio T). According to the author the whole determination takes 20 min and its precision is comparable with that of the classical methods. Deur–Siftar and Bauman [6] determined silica after coagulation with gelatine, separated R_2O_3 with ammonia and bromine water (or used TEA for masking) and determined only calcium and magnesium complexometrically. Lahovský and Michálek [7] extended the application of complexometry by EDTA titration of iron (with salicylic acid as indicator) followed by determination of aluminium at pH 4.1 - 4.2 with Cu-PAN as indicator. These authors rightly noted that when higher amounts of aluminium are present the results for iron are somewhat high but still within acceptable limits. Aliquots of the filtrate from the separation of silicic acid were used for the EDTA determination of calcium (murexide) and the sum of Ca + Mg (Methylthymol Blue). When titanium is present in higher quantities, photometric determination of iron after masking of titanium with fluoride is preferred. The effect of the presence of titanium and manganese was not investigated in detail in this work.

Several methods for the determination of calcium and magnesium have been developed. Some examples are given below.

13.1 DETERMINATION OF CaO AND MgO

Maier and Wust method [8]

This is based on the method given by Přibil and Veselý [9], based on
direct titration of calcium with EGTA and separate determination of
magnesium with DCTA, the calcium being masked with EGTA and iron and
aluminium with triethanolamine.

Mix 1 g of sample with 1 g of ammonium chloride and carefully add
10 ml of concentrated hydrochloric acid.   Crush any small clumps of
sample with a glass rod and heat the mixture on a steam-bath for
30 min, with occasional stirring.   Add 50 ml of hydrochloric acid
(1 + 20) and filter into a 500-ml standard flask.   Wash the residue
on the filter first with 5 ml of hot hydrochloric acid (1 + 20) and
then with hot water till the washings are free from chloride.   Cool
the solution in the flask and make up to the mark with water.

Determination of CaO.   Pipette 20 ml of the sample solution (40 mg
of sample) into a 600-ml beaker, add 10 ml of 20% TEA solution and
one drop of saturated auramine solution.   Then add 2M sodium hydroxide dropwise till the yellow colour disappears.   Dilute to 350 ml,
add another 10 ml of 2M sodium hydroxide and titrate photometrically
with 0.02M EGTA, using murexide as indicator.

Determination of MgO.   Take 20 ml of sample solution in a 600-ml
beaker, add as much 0.02M EGTA solution as was consumed in the
calcium determination plus 2 ml in excess.   Then add 10 ml of 20%
TEA solution and dilute to 350 ml with water.   Adjust to pH 10.0 -
10.5 with 40 ml of buffer [30 g of NH_4Cl in one litre of ammonia
solution (1 + 1)].   Titrate photometrically (620 nm) with 0.01M DCTA,
using Methylthymol Blue as indicator.   If the magnesium content is
below 1%, addition of a known amount of magnesium (e.g. equivalent to
1%) is recommended.

The authors give some other procedures for determination of CaO and
MgO in cement materials, but these differ only in the preparation of
the sample solution.

Armannssen and Magnusson method [10]

The sample is decomposed with hydrochloric acid (1 + 2) and silica
and iron are removed in the usual manner.   To an aliquot of the
filtrate borax buffer and a small amount of Zn-EGTA complex are added
and the calcium (through its displacement of zinc) is titrated
photometrically with 0.01M EGTA, with zincon as indicator.   The
buffer contains 25 g of $Na_2B_2O_7.10\ H_2O$, 3.5 g of NH_4Cl and 5.7 g of
NaOH per litre.

According to the authors this method is more precise than are the
gravimetric (oxalate precipitation) and atomic-absorption methods.

More recently a comprehensive paper was published by Hitchen and
Zechanowitsch [11] dealing with the determination of calcium and
magnesium in various materials (basic slags, iron ores, clays,
limestone, dolomite etc.).   Their procedures are applicable over the
range 0.1 - 85% CaO and/or MgO.   The principle is not new.   After
the separation of silica the sum of CaO + MgO is determined by EDTA
titration after masking of interfering elements with triethanolamine

and potassium cyanide, a mixture of Cresolphthalein complexone
(Phthaleincomplexone) and Naphthol Green B being used as indicator.
Calcium alone is determined indirectly by titration of excess of EDTA
with calcium chloride, with Calcein - thymolphthalein mixture as
indicator.

Remarks. The method avoids any separation steps except for removal
of silicic acid, but does not avoid precipitation of magnesium
hydroxide (if present in large amounts) in the separate determination
of calcium. The authors consider their method as generally
applicable to a great number of materials. Lopez Gomez [12] has
analysed 100 samples of cement for calcium (Calcein or Calcon as
indicator) and magnesium (Methylthymol Blue and Erio T as indicators)
by EDTA titration. His statistical analysis of the results showed
that for calcium the scatter was greater with Calcein than with
Calcon as indicator, and with either indicator the method can be
considered only as a rapid (non-referee) method. With magnesium the
results were rather worse, only those with use of Erio T (and
titration of a hot solution) being acceptable.

A similar method has been re-examined by Tovares Cravo [13], who
concluded that the determination of calcium in cement cannot be taken
as accurate, as some calcium is lost by adsorption on the initial
precipitation of aluminium and iron with ammonia (a fact long known
to classical analysts, but either forgotten or not known by many of
the newer generations). The possibility of the adsorption of calcium
on the precipitated magnesium hydroxide is not mentioned in the
abstract of the paper. From all this it can be concluded that EGTA
is the preferred titrant for calcium, especially when a large amount
of magnesium is present [14]. It is also preferred in the potentiometric titration.

Gillard et al. [15] have used a J-type mercury-drop electrode as
indicator electrode for potentiometric EDTA determination of CaO and
MgO in cements. Silica is removed as usual, and aliquots of the
resulting solution are used for the analysis. For calcium determination a few drops of 0.01M Hg-EGTA (to stabilize the electrode
potential) and enough 20% triethanolamine solution to mask Al and Fe
are added and the solution is titrated at pH 8.5 with 0.05M EGTA.

Another aliquot is buffered to pH 10 (Schwarzenbach buffer) and
titrated potentiometrically with addition of TEA. The results are
claimed to be very accurate.

Remarks. Halides interfere if present in more than μM concentrations, so nitric acid must be used in the decomposition step. The
authors claim that the results for slags, clay, limestone and
dolomite agree well with those obtained gravimetrically, and that the
method should be adaptable for automation.

An amperometric method for the stepwise titration of calcium with
EGTA and magnesium with EDTA in ratios from 1:10 to 10:1 was
described by Roueche and Monnier [16], who used it for the determination of Ca and Mg in water, cement and soil. They used 0.1M
ethanolamine (pH 10.6) as buffer, and ammonium citrate and potassium
sodium tartrate for masking interferences.

13.2   DETERMINATION OF ALUMINIUM

Burglen and Longuet [17] recommend that the solution from separation
of silica should be heated, and then treated with 5M sodium hydroxide
and sodium carbonate solution.   Iron, titanium, manganese and
magnesium are precipitated as hydroxides, calcium as carbonate,
phosphate as ferric or calcium phosphate.   The filtrate contains
sodium tetrahydroxoaluminate.   These authors determined the aluminium by the outdated method of back-titration of excess of EDTA in hot
solution with aluminium chloride (acetate buffer), with haematoxylin
as indicator.   They recommend this method for the fast determination
of aluminium not only in cements but also in clinkers, clays and iron
ores.

13.3   TOTAL ANALYSIS OF CEMENTS

In the course of time more detailed and sometimes critical studies
on the analysis of cements appeared.   Some of them were published in
journals not primarily concerned with analytical chemistry.   Besides
giving the findings from the investigation, these papers often refer
extensively to previous works and to some extent have the nature of
reviews.   We may mention in particular those published by Burglen
and Longuet [18] (125 references), Voinovitch et al. [19] (61 references) and Levert [20] (33 references).   These authors tried to
combine common complexometric determinations with other methods such
as spectrophotometry.

Burglen and Longuet [18], for instance, chose spectrophotometric
titration of calcium and magnesium, with Methylthymol Blue as
indicator, and derived the corresponding stability constants from
detailed study of the spectra of the indicator complexes.   They
determined calcium in strongly alkaline solution with Calcon as
indicator.   Voinovitch et al. [19] determined calcium and magnesium
with EGTA and DCTA in the same way as in analysis of silicates [21].

On the whole there is little difference between the total analysis of
cements and that of silicates described in detail in the previous
chapter (for example, see p. 313 for an outline of analysis of
Portland cement).

Some procedures will be outlined below, but only briefly since the
individual determinations are well known.

13.3.1   Analysis of Cements (Jugovic [22])

Principle.   After fusion of 1 g of sample with sodium hydroxide the
cooled melt is dissolved in hydrochloric acid and the solution diluted
to 250 ml.   Suitable aliquots are used for the following determinations.

    SiO_2:   spectrophotometrically as silicomolybdic acid
    Fe and Al: spectrophotometrically with ferron at 365 and 600 nm
        respectively
    Ti: spectrophotometrically with tiron
    Ca and Mg: by EDTA titration
    P_2O_5:   spectrophotometrically as phosphomolybdic acid
SO_3 is determined gravimetrically as $BaSO_4$, in a separate sample.

Remark. The author claims that the determination of all five
components takes about 75 minutes. Simultaneous analyses of six
samples can be performed in $3\frac{1}{2}$ hours.

13.3.2   Rapid Analysis of Cements (Stiglitz and Cornet [23])

Principle. A 1-g sample is fused with 10 g of sodium carbonate.
The cooled melt is dissolved in warm hydrochloric acid (1 + 4) and
the solution diluted to 500 ml, from which aliquots for each determination are taken.

SiO_2:   spectrophotometrically as silicomolybdic acid at 430 nm
Fe and Al: by EDTA titration at pH 1.2 and 3.2 respectively
Ca and Mg: by EDTA titration at pH 10.0 (Ca + Mg) and calcium
             at pH 12.5
Free CaO: by EDTA titration after extraction with ethanediol.

Remarks. The method is suitable for the analysis of alumina,
cements, bauxite, limestone and clays.

13.3.3   Analysis of Aluminosilicates (Debras-Guédon [24])

Principle. After separation of silica the following determinations
are done.

Fe_2O_3:   spectrophotometrically with sulphosalicylic acid
TiO_2:   spectrophotometrically with chromotropic acid or H_2O_2
Al_2O_3:   with EDTA after separation of Fe and Ti with cupferron
R_2O_3:   (Al, Fe, Ti) gravimetrically as basic benzoates.
CaO and MgO: with EDTA after masking of Ti, Fe, Al with
                tartaric acid.
Na_2O, K_2O, Li_2O: by flame photometry

Remark. According to the author more than 1000 samples have been
analysed in the laboratories of the French Ceramic Society with very
accurate results.

13.3.4   Analysis of Cements (Naidu and Sastry [25])

Principle. Iron and aluminium are precipitated as hydroxides with
ammonia and thoroughly washed. The following determinations are
then done.

Al and Fe: the hydroxides are dissolved in 1M hydrochloric acid;
iron is determined with EDTA at pH 1.5 - 2.5 with Variamine Blue as
indicator, then excess of EDTA is added and aluminium determined by
back-titration with zinc solution at pH 5 - 6 (Xylenol Orange as
indicator).

Ca and Mg: the sum of Ca + Mg is determined with EDTA and Erio T,
then magnesium is precipitated with sodium hydroxide and the EDTA
freed is titrated with calcium chloride (Cal-Red as indicator).

Remark. Although this method would be expected to suffer from
adsorption errors, the authors report surprisingly good results, with
an average error of about 0.3%. Twenty samples can be analysed in

10 hours.

The four examples above were selected to illustrate various modifications of analytical procedures used in cement analysis. Complexometric titrations are fast enough, but they can, of course, be replaced by more expensive methods. For example, Fifield and Blezard [26] used an Auto-Analyzer in the automated analysis of cements, using spectrophotometric determinations of silica, iron, aluminium, and calcium, and turbidimetric determination of sulphate. They emphasize the precision and reproducibility of the method, which is based on commonly known reactions. The authors claim that "by using five simultaneous systems an analysis of one sample can be completed within 15 minutes, and one operator can process at least 24 samples during a 7-hours period."

Recently increasing attention has been paid to the use of atomic-absorption spectrophotometry in cement analysis (e.g. [27,28]), but the necessity for high dilution of the samples may be regarded by some as a drawback.

13.3.5 Complexometric Determination of Sulphate in Cements

Relatively little attention has been paid to the possibility of complexometric determination of sulphate in cements. Calleja and Fernandez-Paris [29] recommend that the sample should be digested with hot nitric acid and the sulphate (after filtration) precipitated as lead sulphate, which is collected and then dissolved in potassium sodium tartrate solution and titrated at pH 10 with EDTA (Erio T as indicator).

In another study [30] the same authors dissolved the lead sulphate in excess of EDTA in acetic acid medium (pH 4.5) and back-titrated with lead solution, using either dithizone or potassium chromate as indicator.

Remarks. The use of Xylenol Orange as indicator would be more suitable, as well as a different buffer for the pH adjustment in view of the formation of lead acetate complexes (see p. 153).

REFERENCES

1.   F. Becker, Rev. Mater. Constr. (Paris), 1951, Nos. 431/432, 1.
2.   H. Habock, TIZ-Zbl., 1954, 78, 171.
3.   R. Huppertsberg, Zement-Kalk-Gibs, 1956, 45, 249.
4.   F. Langmyhr and B. Saether, Zement-Kalk-Gibs, 1956, 45, 429.
5.   M. Wallraf, TIZ-Zbl., 1957, 81, 41.
6.   D. Deur-Siftar and E. Bauman, Kem. i Ind. (Zagreb), 1959, 8, 1;
     Anal. Abstr., 1959, 6, 3414.
7.   J. Lahovský and Z. Michálek, Stavivo, 1958, 2.
8.   U. Meier and H. Wüst, Zement-Kalk-Gibs, 1968, 21, 403.
9.   R. Přibil and V. Veselý, Talanta, 1966, 13, 233.
10. H. Armannsson and H.F. Magnusson, Anal. Chim. Acta, 1975, 74,
     208.
11. A. Hitchen and G. Zachanowitsch, Talanta, 1980, 27, 269.
12. R. Lopez Gomez, Cemento-Horm., 1975, 46, 749; Anal. Abstr.,
     1976, 31, 3B 190.

13. M. do Rosário Tovares Cravo, Tecnica (Liston), 1964, 38, 121; Anal. Abstr., 1966, 13, 1296.
14. R. Přibil and J. Adam, Talanta, 1977, 24, 177.
15. M. Gillard, J. Daube, S. Genot and H. Drobbel, Chim. Anal. (Paris), 1970, 52, 747; Anal. Abstr., 1967, 20, 3861.
16. A. Rouèche and D. Monnier, Anal. Chim. Acta, 1964, 31, 426; Anal. Abstr., 1966, 13, 1550,
17. L. Burglen and P. Longuet, Rev. Mater. Constr., C,1960, 327; Anal. Abstr., 1961, 8, 4095.
18. L. Burglen and D. Longuet, Rev. Mater. Constr., 1966, No. 604, 1; No. 605, 49; No. 606, 107: Anal. Abstr., 1967, 14, 3225.
19. I.A. Voinovitch, R. Barbaras, G Cohort, G. Koelbel, G. Legrand and J. Louvrier, Chim. Anal. (Paris), 1968, 50, 334; Anal. Abstr., 1969, 17, 2143.
20. J.M. Levert, Ind. Chim. Belge, 1969, 34, 287; Anal. Abstr., 1970, 19, 144.
21. R. Přibil and V. Veselý, Chemist-Analyst, 1966, 55, 82.
22. Z.T. Joguvic, A.S.T.M. Stand. Spec. Tech. Publ., 1965, 395; Anal. Abstr., 1967, 14, 1998.
23. P. Stiglitz and J. Cornet, Rev. Mater. Constr., 1963, 271; Anal. Abstr., 1965, 12, 138.
24. J. Debras-Guédon, Ind. Ceram., 1963, 345; Anal. Abstr., 1965, 12, 136.
25. P.P. Naidu and C.A. Sastry, Z. Anal. Chem., 1971, 253, 206.
26. J.A. Fifield and R.G. Blezard, Chem. Ind. (London), 1969, 1286.
27. J.T.H. Roos and W.J. Price, Analyst, 1969, 94, 89.
28. T.O. Lees, Chem. Ind. (London), 1969, 1249.
29. J. Callega and J.M. Fernandez-Paris, Rev. Cienc. Apl., 1961, 15, 120; Anal. Abstr., 1962, 6, 162.
30. J. Calleja and J.M. Fernandez-Paris, Rev. Cienc. Apl., 1962, 16, 312; Anal. Abstr., 1963, 10, 1448.

CHAPTER 14

Analysis of Glasses

Complexometric analysis of various sorts of glasses has developed in a similar manner to analysis of silicates. A survey of the oldest methods was compiled by Flaschka [1]. In the early days most attention was paid to determination of calcium and magnesium. Cluley [2] volatilized silica with hydrofluoric acid and extracted the heavy metals with oxine before determining calcium and magnesium by known methods. Flaschka [1] considered the possibility of complexometric determination of iron and aluminium and gave a detailed procedure for complete analysis of glasses, including the determination of potassium. A procedure for the analysis of bottle glass was worked out by Kříž [3]. He determined iron with thiocyanate, and the sum of iron and aluminium by EDTA titration (indirectly by back-titration with copper sulphate, using Pyrocatechol Violet as indicator). Manganese was determined either colorimetrically or by EDTA titration after masking of calcium and magnesium with ammonium fluoride.

Generally any method described in the previous chapters on silicates and cements may be used for the complete analysis of glasses. In the analysis of special glasses some further determinations may be required that are not included in the usual analytical procedures. Some examples of these will be given below.

14.1 DETERMINATION OF CALCIUM, MAGNESIUM AND MANGANESE IN GLASSES

In a critical study Sinha and Dasgupta [4] have focused attention on the masking of manganese with potassium cyanide in the determination of calcium and magnesium. They have found the optimal conditions for the oxidation of manganese(II) to the tervalent state by aerial oxygen in solutions containing triethanolamine and potassium cyanide. The resulting $Mn(CN)_6^{3-}$, as is well known, is inert towards EDTA.

Procedure. The dried sample $(0.5\ g)$ is moistened with water in a platinum basin and treated with 10 ml of hydrofluoric acid and 1 ml of 9M sulphuric acid. The mixture is heated on a sand-bath until fumes of sulphur trioxide are evolved. The treatment is repeated with a further 10 ml of hydrofluoric acid, and the solution evaporated to dryness. The residue is fused with $2-3\ g$ of potassium hydrogen sulphate, cooled, and dissolved in hydrochloric acid $(1 + 20)$. Iron, titanium or zirconium (if more than 5 mg is present) are removed as hydroxides by precipitation with 20% hexamine solution and filtration. The filtrate is diluted to volume in a 250-ml standard flask.

Determination of Ca + Mg.   Dilute a suitable aliquot to 100 ml with water, add 10 ml of 30% triethanolamine solution and mix thoroughly. Then add slowly, with swirling, a mixture of 30 ml of Schwarzenbach buffer (pH 10) and 30 ml of 10% potassium cyanide solution.   Dilute to 200 ml and bubble air through the solution for 10 min at 30-40°C. Titrate with EDTA, using Cresolphthalexone - Naphthol Green mixture as indicator.   The colour change is from pink to green.

Determination of manganese.   After the titration of Ca and Mg add 2-3 g of ascorbic acid to the solution.   Titrate the manganese with EDTA, using as indicator a mixture of Erio T, Titan Yellow and Naphthol Green.   Titrate slowly to the first change of colour from pink to blue.   Then warm the solution to about 40°C and if the pink colour reappears continue with the titration to a permanent blue.

Determination of calcium.   Dilute another portion of the solution to 150 ml.   Add 2 ml of thymolphthalein (0.1% in 30% TEA solution) and slowly, with swirling, 10 ml of TEA.   Then add dropwise 1M potassium hydroxide till the solution turns blue or green.   After addition of 5 ml of 10% potassium cyanide solution and 15 ml of 5M potassium hydroxide titrate with EDTA in the presence of 0.2 g of indicator (a 1% solid mixture of Calcein with potassium chloride).   The end-point is shown by a colour change to red, with disappearance of the green fluorescence.

Remarks.   If we compare the authors' procedures for the determination of Ca + Mg and of only Ca one point is not clear.   In the first case the $Mn(CN)_6^{2-}$ is prepared by bubbling air through the solution as described.   In the determination of calcium the aerial oxidation is apparently not used even though the amount of manganese is the same as in previous titration.   Presumably the very high alkalinity used accelerates the oxidation, or the aeration step was accidentally omitted in writing the paper.

14.2   DETERMINATION OF ZIRCONIUM

Probably the oldest method is that described by Panasyuk and Mirovskaya [5].

Procedure.   Fuse 0.5 g of sample in a silver crucible with 5 g of potassium hydroxide and 2 g of borax.   Leach the cooled clear melt with water and transfer it with the insoluble residue to a 250-ml standard flask.   Add 50 ml of concentrated hydrochloric acid and dilute to the mark.   Boil an aliquot (25 ml) together with 25 ml of 2M hydrochloric acid and 50 ml of water.   Titrate the solution with 0.05M EDTA after addition of Erio T.   Heat again just before the end-point is reached.

Remark.   This determination is not so simple to perform and often leads to erratic results (see Section 6.1.1).

Determination of zirconium according to Su [6]

Decompose sufficient finely ground sample containing 5 - 50 mg of ZrO_2,

in a 150-ml platinum dish on a steam-bath with about 3 ml of water,
10 ml of 9M sulphuric acid and 10 ml of concentrated hydrofluoric
acid.   Evaporate the solution on a hot-plate to complete dryness and
fuse the residue with sodium hydrogen sulphate.   Dissolve the cooled
melt in 30 - 40 ml of 6M hydrochloric acid and transfer the solution
to a 200-ml tall beaker.   Add 50 ml of 5% mandelic acid solution and
digest at 80 - 85°C for 20 min.   Filter off the precipitate on a
medium porosity sintered-glass crucible and wash with 2% hydrochloric
acid containing 5% mandelic acid.   Place the crucible back in the
original beaker and add excess of 0.05M EDTA and about 100 ml of
water.   Cover the crucible, add 1.5 ml of 6M hydrochloric acid and
boil for 20 - 25 min to form the Zr-EDTA complex.   During the boiling
take care to wash down any precipitate adhering to the wall of the
beaker.   Cool the solution, adjust to pH 5 - 8 with 7M ammonia
solution, add enough ammonium acetate buffer (pH 5) and titrate the
excess of EDTA to the colour change from greenish-yellow to orange.
Su et al. [7] later simplified this method by avoiding separation of
the zirconium.

Fuse the sample, containing less than 25 mg of aluminium, with a
mixture of sodium carbonate and borax.   Dissolve the melt in 2M
perchloric acid and dilute with an equal volume of water, then
titrate the zirconium at 90 - 95°C with EDTA, using Xylenol Orange as
indicator.   Aluminium can be determined in the same solution by
complexation with EDTA at pH 4.5 - 5 and back-titration.

Sugawara et al. [8], in view of the results obtained by these two
procedures [6,7] claimed that for the accurate determination of
zirconium in low concentration in glasses (1 - 5%) it is necessary to
separate it from the matrix elements.   For this they used precipitation of zirconium with p-bromomandelic acid.   For the determination
of zirconium contents as low as 1% dissolution of the sample with
hydrofluoric acid is not reliable owing to the retention of fluoride,
which is difficult to remove by evaporation with perchloric acid [9].
For this reason they fused the sample with sodium carbonate and borax.
They also used spectrophotometric detection of the Xylenol Orange end-
point.

The procedure seems a little tedious but according to the authors
gives better sensitivity, precision and accuracy.   The details can
be found in the original publication [8].

14.3   DETERMINATION OF THORIUM [10]

To 0.2 - 0.3 g of finely ground sample in a 400-ml beaker add 10 - 20 ml
of water and 2.5 ml of concentrated perchloric acid.   Warm the
solution for some minutes, with swirling, but avoid boiling.   Transfer
the solution to a 500-ml beaker, dilute to about 250 ml with water,
adjust the pH to 3.5 with dilute ammonia solution, add 5 - 8 drops of
0.1% Pyrocatechol Violet solution and titrate with 0.1M EDTA to lemon-
yellow.

Remarks.   The authors analysed optical glasses containing 12 - 14% of
ThO_2.   Other components (Ba, La, Al, As, Sb) do not interfere.   As
1 ml of 0.1M EDTA corresponds to 26.41 mg of ThO_2, it is better to
titrate the thorium at pH 2.5 with 0.02 - 0.05M EDTA, with Xylenol
Orange as indicator.

14.4   DETERMINATION OF THORIUM, LANTHANUM, BARIUM AND BORON [11]

For the analysis of optical glass (non-silicate) containing 20% of
ThO_2, 20% of La_2O_3, 20% of BaO and 40% of B_2O_3, Malát and Mučka [11]
~~developed a relatively rapid complexometric method.   In principle~~
they proceed as follows.   Thorium is determined as described in the
previous paragraph.   In a further sample solution thorium is
separated from lanthanum etc. by precipitation with hexamine at pH 4.
After the separation of $Th(OH)_4$ the filtrate is divided into two
parts.   In one the lanthanum is determined by EDTA titration at pH
5.5 with Xylenol Orange as indicator.   In the second the sum of
La + Ba is determined in ammoniacal medium by direct EDTA titration
(after addition of a small amount of Mg-EDTA complex) with Erio T as
indicator.   In a separate portion boron is determined after the
metals have been masked with EDTA, according to the method described
by Přibil and Wünsch [12].

Remark.   Xylenol Orange cannot be used for stepwise titration of
thorium and lanthanum because of the formation of the coloured
ternary complex Th-EDTA-XO at about pH 5.   Therefore the thorium
must be separated in the Malát and Mučka method.   The separation of
thorium is not necessary in the method described in the next
paragraph.

Determination of thorium, lanthanum and barium with TTHA and EDTA [13]

Dissolve a weighed sample (2 g) of the glass (ground to 100-mesh
particle size) in 10 ml of nitric acid (1 + 1), heating on a sand-bath
for 15 - 20 min.   Add 10 ml of hydrochloric acid (1 + 1) and evaporate
to dryness.   Evaporate again with 10 ml of hydrochloric acid (1 + 1).
To the residue, add a few drops of concentrated hydrochloric acid and
50 - 100 ml of hot water.   Filter off on a fine-pore paper, collecting the filtrate in a 250 or 500-ml standard flask, washing the
filter thoroughly with hot water to free the silica from adsorbed
boric acid.   Retain the filter paper and contents if silica is to be
determined in a conventional manner.   Cool the filtrate and dilute
to the mark with water.

Determination of Th.   Place a portion of the sample solution in a
400-ml beaker and adjust to pH 2.5 - 4.0 with sodium hydroxide or
nitric acid and dilute to 150 - 250 ml with water.   Add some Methylthymol Blue and titrate with 0.05M TTHA to a pure yellow.   The
titration must be conducted to the "full" end-point, otherwise minute
traces of thorium will block the indicator after addition of pyridine.

Determination of La.   After titration of the thorium gradually add
20% aqueous pyridine solution until the blue indicator colour returns
and then 10 - 15 ml in excess (pH 5 - 6).   Titrate with 0.05M TTHA to
a grey-yellow colour.   If the blue colour fades during the titration
add a few more drops of pyridine solution.   The result corresponds
to lanthanum.   The lanthanum must be titrated as soon as possible
after the thorium, since after 10 - 15 min the lanthanum will start to
displace thorium from its complex.

Determination of Ba.   After titration of the lanthanum, if the amount
of barium does not exceed 40 mg, add 20 ml of 20% of triethanolamine
solution and 5 - 10 ml of 1M sodium hydroxide.   Titrate with 0.05M
EDTA to a colour change from intense blue to smoky-grey.   The
addition of TEA is necessary for obtaining a sharp end-point.   If

iron is present it must be removed after separation of the silica;
this can be done by forming ferroin and extracting its ion-association
complex with iodide into chloroform, or by extraction with diethyl-
dithocarbamate.

Some experimental optical glasses (low-silica or silica-free) have
very complicated composition, and their analysis, even if EDTA
titrations are used, is a little tedious.   For example we have
worked out such an analysis for a sample containing zirconium,
thorium, lanthanum, aluminium, calcium, barium, silicon and boron
oxides.   The details can be found in the original communication [14].

14.5   PARTIAL ANALYSES OF GLASSES

For some special glasses only one or two components need be determined and a complete analysis is not necessary.   Besides spectrophotometric determination of trace elements (Fe, Ti etc.) complexometry can be used, for instance in the determination of scandium,
aluminium, zinc or lead.   Some examples will be given here.

Determination of scandium [15]

The determination is relatively simple but less selective.   The
finely ground sample is dissolved by heating in a mixture of hydrofluoric acid and sulphuric acid, and the solution is evaporated.
The residue is dissolved in 25 ml of hydrochloric acid (1 + 4) and
diluted to 250 ml with water.   The pH is adjusted to 2 - 3 and
scandium titrated with 0.05M EDTA, with Xylenol Orange as indicator.

Determination of aluminium in glass containing phosphorus [16]

Dissolve 0.1 g of sample in 5 - 8 ml of 25% sodium hydroxide solution
and then add 3M hydrochloric acid until the precipitate formed during
the neutralization redissolves.   Dilute to 100 ml and heat the
solution for 2 hr (with addition of hydrochloric acid if a precipitate is formed) to form orthophosphate.   Pass the solution through a
column of Dowex 50-X4 (0.25 - 0.5 mm particle size).   Wash the column
free from phosphate with 250 ml of 0.1M hydrochloric acid.   Then
elute aluminium with 80 ml of 5M perchloric acid.   Add excess of
EDTA to this eluate and back-titrate with lead nitrate, using Methylthymol Blue as indicator, to obtain the aluminium content.

Determination of aluminium (and zirconium) [7]

Treat the sample with sulphuric acid and hydrofluoric acid and
evaporate the solution to dryness.   Repeat this treatment.   Ignite
the residue and fuse it with a sodium carbonate and borax mixture.
Dissolve the metl in 2M perchloric acid, and dilute with an equal
volume of water.   Titrate the zirconium at 90°C with EDTA solution,
using Xylenol Orange as indicator.   Then add excess of EDTA, boil
for 30 min, adjust to pH 4.5 - 5.0 (acetate buffer);   add a little
Hg-EDTA and back-titrate the excess of EDTA potentiometrically with
zinc solution, using a mercury-ring electrode and a saturated
mercurous sulphate reference electrode.

Determination of zinc [17]

Weigh 0.1 - 0.3 g of finely ground sample into a platinum dish, moisten

with water and add 10 - 20 ml of 40% hydrofluoric acid.   Mix thoroughly with a platinum wire and slowly evaporate nearly to dryness.
Transfer the moist residue with warm water into a 150-ml beaker.
Add Methylthymol Blue, adjust to pH 5 - 5.5 with solid hexamine (blue-
violet colour) and titrate with 0.05M EDTA to lemon-yellow.

Remarks.   If lead is present in the glass, dissolution of the sample
with hydrofluoric acid and sulphuric acid creates difficulties,
because the precipitated lead sulphate cannot be filtered off even
with a very close filter and a small amount passes into the filtrate.
The results for zinc are then about 0.5% too high.   A modified
procedure is used if lead is present.

Moisten the sample in a platinum dish with water, add 3 g of sulphamic
acid and 10 - 20 ml of 40% hydrofluoric acid.   Evaporate slowly as
described above.   To the residue add 100 ml of water, filter off,
wash with water and determine zinc in the clear filtrate as above.

Note.   Such samples could probably be dissolved in hydrofluoric acid
and nitric acid.   The solution would then contain zinc and lead,
which can be easily determined (zinc being masked with 1,10-phenan-
throline, see p. 72).   Aluminium is masked by the fluoride present.
Calcium and barium do not interfere.

Determination of lead and barium

Sinha and Roy [18] studied in detail the possibility of determination
of lead and barium, which often occur in various sorts of optical
glass, low-melting frits etc.   In fact they did not use a complexometric determination but employed EDTA only for the separation of
$PbSO_4$ and $BaSO_4$.   They decomposed the sample with sulphuric acid and
hydrofluoric acid by evaporation.   They extracted lead sulphate from
the residue selectively with slightly acid EDTA (pH around 4.3).   The
barium sulphate was filtered off and determined gravimetrically.
From the filtrate containing the Pb-EDTA complex, lead sulphate was
precipitated by acidification and also determined gravimetrically.
The authors in fact made indirect use of a method published long ago
[19] for the separation of lead and barium in the presence of EDTA.

Remarks.   Complexometry can be easily adopted for analysis of a
mixture of $PbSO_4$ and $BaSO_4$.   After the extraction of $PbSO_4$ with a
known volume of EDTA, the excess of complexone can be determined by
back-titration at pH 5 - 5.5 with zinc solution (Xylenol Orange as
indicator).   Both $BaSO_4$ and $PbSO_4$ are soluble in an alkaline solution
of EDTA, the excess of which can be back-titrated with calcium
solution after addition of triethanolamine, and Thymolphthalexone as
indicator.

REFERENCES

1.   H. Flaschka, Sprechsaal Keram-Glas-Email, 1955, 88, 188.
2.   H.J. Cluley, Analyst, 1954, 54, 567.
3.   M. Kříž, Chemické rozbory obalového skla. Výzkumný ústav
     strojního skla, Teplice, 1956.
4.   B.C. Sinha and S. Dasgupta, Talanta, 1978, 25, 693.
5.   V.I. Panasyuk and N.A. Miroevskaya,   Zavod. Lab., 1959, 25, 147;
     Anal. Abstr., 1959, 6, 4718.

6.   Y.S. 'Su, Anal. Chem., 1965, 37, 1067.
7.   Y.S. Su, W.R. Strzegowski, A.R. Kacyon and I.E. Lichtenstein,
      Anal. Chim. Acta, 1976, 81, 167.
8.   K.F. Sugawara, Y.S. Su and W.R. Strzegowski, Talanta, 1978, 25,
      669.
9.   Y.S. Su and D.E. Campbell, Anal. Chim. Acta, 1971, 55, 265.
10.   M. Malát, J. Pelikán and V. Suk, Chemist-Analyst, 1956, 45, 61.
11.   M. Malát and V. Múčka, Chemist-Analyst, 1961, 50, 110.
12.   R. Přibil and L. Wünsch, Chem. Listy, 1952, 46, 337.
13.   R. Přibil and V. Veselý, Chemist-Analyst, 1964, 53, 12.
14.   R. Přibil and V. Veselý, Chemist-Analyst, 1964, 53, 43.
15.   E.P. Bil'tyukova, V.F. Prokopets and A.M. Eserchko, Steklo,
      1973, 2, 33; Anal. Abstr., 1975, 29, 4B 234.
16.   V.A. Luginin and I.A. Tserkovnitskaya, Zavod. Lab., 1971, 37,
      287; Anal. Abstr., 1972, 22, 166.
17.   M. Paleček and R. Přibil, Silikáty, 1962, 6, 296.
18.   B.C. Sinha and S.K. Roy, Talanta, 1975, 22, 763.
19.   R. Přibil and D. Maričová, Chem. Listy, 1952, 46, 542.

CHAPTER 15

Analysis of Ores and Concentrates

It might be expected that this chapter would be more extensive and varied. Very often only one component is determined in control analyses used to check ore enrichment processes such as flotation etc. Ore quality is often subject to certain specifications stipulating the permitted concentration range of a particular component. Thus only special procedures for partial analysis of industrial ores are usually found in the literature on complexometric analysis. Some industrial procedures have been developed for "in house" use and have not been published in the usual journals or have remained in "technical reports".

Many of the methods given below are of older origin but still remain applicable or can easily be modified according to present knowledge of complexometry (choice of indicators, use of spectrophotometric or potentiometric titrations etc.).

The classical determination of calcium and magnesium has never attained wide popularity among analytical chemists. This is reflected in the existence of complexometric procedures for the determination of both elements in all kinds of ores. These determinations will be discussed in a separate section at the end of this chapter.

Total analysis of ores is not often done, but if the need arises it is possible to combine complexometry with polarography, spectrophotometry and other methods.

15.1   LEAD ORES AND CONCENTRATES

Most of the methods are based on the precipitation of lead sulphate during decomposition of the sample. The lead sulphate is dissolved in sodium tartrate, ammonium acetate or EDTA solution and determined complexometrically. The procedures given by individual authors differ from each other only in the method of titration.

15.1.1 Lead in Galena [1]

Dissolve 50 mg of sample in a mixture of 5 ml of concentrated hydrochloric acid and 1 ml of concentrated nitric acid, then add 10 ml of
concentrated sulphuric acid and evaporate to dryness. After cooling
add a little water and evaporate again. After cooling digest the
residue with 80 - 100 ml of water, boil briefly, cool and filter off
(e.g. with a Schleicher and Schüll 5895 paper). Place the filter
and precipitate in a 300-ml conical beaker, add 80 ml of water, 20 -
25 ml of 0.01M EDTA, 5 ml of ammonia buffer (pH 10), a few crystals
of potassium cyanide (to mask zinc), a small amount of Mg–EDTA
complex and a small amount of tiron (to mask aluminium). Warm the
solution to 30 - 40°C and disperse the filter paper by vigorous
shaking. Titrate the excess of EDTA with 0.01M magnesium chloride,
using Erio T as indicator.

Several methods for the determination of lead have been described in
the literature. Zaichikova [2] dissolves lead concentrates containing less than 2% of $BaSO_4$ in a hydrochloric and nitric acid mixture,
and evaporates the solution with sulphuric acid. The lead sulphate
is collected and then dissolved in ammonium acetate solution and
titrated with EDTA (Erio T as indicator). Karolev and Koichev [3]
determine lead by direct titration with EDTA, using Xylenol Orange
(pH 5.4 - 5.9) or Methylthymol Blue (5.7 - 7.6), which give a very
sharp end-point.

Budevsky [4] described a similar method for the determination of lead
in ores and concentrates. The isolated lead sulphate was dissolved
in a mixture of tartaric acid and sodium hydroxide, then the solution
was boiled for 10 min (in the absence of barium) or for up to 45 min
in the presence of barium sulphate. After adjustment of the pH to
6 - 7 the solution was titrated with 0.025M EDTA, with Methylthymol
Blue as indicator. According to the author precipitation of lead
may occur in some cases and therefore the titration should be performed very slowly near the end-point.

15.1.2 Determination of Lead and Sulphur in Galena [5]

Dissolve 1 g of sample in a mixture of concentrated nitric acid and
bromine water. Oxidize the solution with potassium chlorate and
evaporate it twice to dryness, with addition of hydrochloric acid
after the first evaporation. Dissolve the residue in water and make
up to volume with water in a 200-ml standard flask.

Determination of lead. Dilute 25 ml of the sample solution with
water, add ammonia buffer (pH 10), and potassium cyanide to mask trace
elements, and titrate with 0.02M EDTA, using Erio T as indicator.
TEA should also be present to prevent precipitation of lead.

Determination of sulphate. To another 25-ml aliquot add the amount
of EDTA consumed in the lead titration, heat to boiling, precipitate
sulphate with 0.02M barium chloride and boil for 2 min. Then cool,
add 10 ml of 0.02M magnesium chloride solution, a few ml of 10%
potassium cyanide solution, 10 ml of ammonia buffer (pH 10) and
titrate again with 0.02M EDTA (Erio T as indicator).

15.1.3   Determination of Lead in Ores [6]

Donaldson, in a detailed study, examined the results for lead determination from 25 Canadian laboratories.   Two samples of ore, containing 1.93 and 6.98% of lead, were tested by atomic-absorption spectrometry (AAS), X-ray fluorescence, polarography, gravimetry (as $PbSO_4$),
volumetric methods (with molybdate or chromate) and the ASTM EDTA
method.   She came to the conclusion that whenever a precipitate of
lead ($PbCrO_4$, $PbSO_4$) has to be filtered off, a small amount of lead
is lost during washing and is found in the filtrate, and there is also
some left in the filter paper after dissolution of precipitates.
For these reasons she recommends as the best method the separation of
lead from a solution containing tartaric acid and potassium cyanide
at pH $\pm$ 0.1, with a chloroform solution of sodium diethyldithiocarbamate.   After mineralization of the evaporated extract, lead is
determined with 0.01M EDTA, with Xylenol Orange as indicator.   A
method similar in principle was published long ago (1954) by Kinnunen
and Wennerstrand [7] and applied to the determination of lead in
various copper alloys [8].   Donaldson claims that her method, after
prior separation of indium (extraction of the bromide with isopropyl
alcohol) and bismuth (extraction with potassium ethyl xanthate and
chloroform) is nearly specific;   only thallium interferes, and is
co-titrated with the lead.

15.1.4   Determination of Lead in Ores after Ion-Exchange Separation
          [9]

Dissolve 0.1 - 0.5 g of sample in 25 ml of concentrated hydrochloric
acid, with addition of 1 - 2 ml of hydrogen peroxide.   Evaporate the
solution almost to dryness and dissolve the residue by boiling with
50 ml of 1.5M hydrochloric acid containing 2 g of sodium chloride.
Filter the solution into a column of 1 g of AN-2F anion-exchange
resin (grain size 0.1 - 0.5 mm) conditioned by washing with 40 - 50 ml
of 1.5M hydrochloric acid).   Wash the filter paper and anion-exchange
column with 150 ml of 1.5M hydrochloric acid, then elute lead with
100 ml of 0.01M hydrochloric acid, adjust the pH to 5 - 5.5 with solid
hexamine, add Xylenol Orange and titrate the red-violet solution with
0.01M EDTA solution to a lemon-yellow colour.

Remark.   According to the authors the method can also be used for
the determination of lead in the presence of zinc and cadmium.

15.2   ZINC ORES AND CONCENTRATES

For the determination of zinc, lead and cadmium in zinc ores, poly-
metallic ores, ore-dressing products etc., several methods have been
proposed.   They are mostly based on separation of lead as lead
sulphate and of interfering metals by precipitation as hydroxide with
sodium hydroxide or ammonia.   After this treatment zinc is usually
determined by EDTA titration in slightly acidic medium, with Xylenol
Orange or Methylthymol Blue as indicators.

Probably the oldest method is the one described in 1957 by Gundlach
[1].   The treatment of the sample is the same as in the analysis of
galena (see p. 345).   After separation of lead sulphate and the
hydroxides (precipitated with ammonia) zinc is determined in ammonia
buffer by EDTA titration, with Erio T as indicator.   Aluminium must

be masked with ammonium fluoride and traces of copper reduced with
hydroxylamine. Cadmium is quantitatively co-titrated but can be
masked with sodium diethyldithiocarbamate [10].

Similar methods were published by Budevsky [11] and Karolev [12].

More recently the same principle was applied by Rao et al. [13] and
by Ray et al. [14]. As an example only the method of Budevsky [11]
will be given.

Place a suitable amount of sample in a 300-ml conical flask, partly
dissolve it in 10 ml of warm hydrochloric acid, then add 5 ml of
concentrated nitric acid and heat till dissolution is complete.
Evaporate the solution to 1 - 2 ml. If the sample contains large
quantities of lead add 8 ml of sulphuric acid (1 + 1) and evaporate
to appearance of white fumes (when decomposing lead slags add 0.2 -
0.5 g of NH_4F). If the precipitate of lead sulphate is not very
large it need not be filtered off. Add 0.5 g of ammonium persulphate
to the solution and precipitate hydroxides by boiling and addition of
concentrated ammonia solution. Boil the solution for a few more
minutes and then filter. Dissolve the hydroxides in 10 ml of hydrochloric acid (1 + 1) and precipitate again in the same way. Boil
the combined filtrates to remove ammonia, add 50 mg of ammonium
fluoride and some Xylenol Orange and then make the solution acid by
dropwise addition of hydrochloric acid till yellow. After final
adjustment of pH (5 - 5.5) by addition of solid hexamine, titrate the
zinc with 0.035M - 0.040M EDTA to a pure yellow colour.

Remarks. Other procedures differ from this one only in some details.
Rao et al. [13] precipitate the hydroxides with 25% sodium hydroxide
solution. Zinc is then determined at pH 5 - 5.5 (Xylenol Orange)
after masking of copper with thiosulphate and of aluminium with
ammonium fluoride. Ray et al. [14], recommend separation of lead as
the sulphate, then reduction of copper with ascorbic acid and
thiourea, and masking of cadmium with brucine iodide.

15.2.1   Determination of Zinc in Concentrates [15]

Dissolve 2.5 g of sample in bromine (2 - 3 ml) and 50 ml of concentrated nitric acid, and treat the insoluble residue in the usual way
(removal of silica, fusion, dissolution). Transfer the combined
filtrate, washings and recovered material in a 250-ml standard flask
and dilute to volume. Pipette 25 ml into a separatory funnel and
neutralize with ammonia solution. Add 1 - 2 g of ammonium bifluoride,
2 - 5 ml of saturated thiourea solution, and 50 ml of 50% ammonium
thiocyanate solution, then extract the zinc with 40 ml of methyl
isobutyl ketone. Transfer the organic layer to an 800-ml beaker,
add 30 ml of ammonia buffer (pH 10) and 50 - 100 ml of acetone, and
dilute to 400 ml with water. Add 2.5 ml of 20% potassium cyanide
solution, a small amount of Erio T and titrate with 0.05M EDTA
solution to a pure blue. Add 4% formalin solution, in small portions,
and titrate again until further addition of formalin has no effect on
the end-point (i.e. the blue colour does not revert to red-violet).

This extraction method is very effective. According to the authors
up to 125 mg of zinc can be extracted from 50 ml of solution. The
authors determined zinc not only in concentrates (50% Zn) but also in
lead concentrates (12%), slags and various alloys, and even in nickel

sulphate (0.040% Zn).   Iron is masked as the fluoride complex, and
copper with thiourea.   Any co-extracted elements are masked with
cyanide or titrated with EDTA before the zinc is demasked with
formaldehyde and titrated.

Remarks.   Twenty years later essentially the same method was
published by Kraft and Dosch [16].   It also is based on extraction
of the zinc thiocyanato complex into methyl isobutyl ketone.   The
extract is diluted with ethanol, mixed with buffer of pH 5.5 and
directly titrated with EDTA, Xylenol Orange being used as indicator.
Cadmium (if over 0.5%) is masked with potassium iodide,   Copper,
if present, is extracted as its 2-nitroso-1-naphthol complex into
chloroform.   The method is suitable for the range 5 - 60% of zinc.
The extraction and titration take 30 minutes.

15.2.2   Determination of Zinc in Ores [17]

Dissolve the sample (0.5 - 1.0 g, containing not more than 50 mg of
cadmium) in 20 ml of nitric acid (1 + 1).   Add 10 ml of sulphuric
acid (1 + 1) and heat to white fumes.   Cool, dissolve the residue in
50 ml of water, filter and evaporate to dryness.   Dissolve the
residue in 50 ml of 10% solution of sodium chloride in 0.12M hydrochloric acid.   Pass the solution through a column (2.5 cm diameter)
containing 15 g of Amberlite IRA-140 (40 - 60 mesh, conditioned with
the same solution).   Wash the column with the same hydrochloric acid/
sodium chloride solution and elute zinc and cadmium with a 2% solution
of ammonium chloride in 1M ammonia solution.   Add 0.5 g of sodium
diethyldithiocarbamate to the eluate (to precipitate cadmium), heat
to 60°C and titrate with EDTA to the blue colour of Erio T.

15.2.3   Determination of Sulphur in Zinc Concentrates [18]

Fuse 0.15 g of sample with a mixture of 1.5 g of sodium peroxide and
0.3 g of sodium carbonate for 10 min at 700°C.   Extract the cooled
melt with water, add nitric acid until the pH is 7 ± 0.5, then heat
on a steam-bath for 1 hr.   Filter off any precipitated Fe(OH)$_3$.
Dilute the filtrate to volume in a 100-ml standard flask.   To a 10 ml
aliquot add 5 ml of 0.05M lead nitrate (prepared in 10% ethanol
solution), filter off the precipitated lead sulphate and determine
lead in the filtrate with 0.01M EDTA, using Xylenol Orange as
indicator.

Remark.   Liteanu and Dulamita [19] proposed a more laborious method.
The sample is fused in a crucible with zinc or cadmium powder and
Eschka mixture at 920 - 950°C (Zn) or 760 - 780°C (Cd).   The sulphide
formed is then decomposed in a Drechsel bottle with hydrochloric acid
and the hydrogen sulphide evolved is absorbed in a known volume of
standard copper acetate solution.   The excess of copper is then
determined by EDTA titration, with murexide as indicator.   According
to the authors the method is suitable for 3 - 5 mg of sulphur.   The
whole procedure takes 90 minutes.

15.3   ZINC-LEAD ORES

An example of nearly complete complexometric analysis of these ores
was described in 1955 by Amin and Farah [20].   Their method needs a

number of separation steps such as ion-exchange separation, and
precipitation, for example iron with cupferron, lead and manganese
with thioacetamide, aluminium with ammonia.   Their analysis scheme
is reproduced in Fig. 15.1.

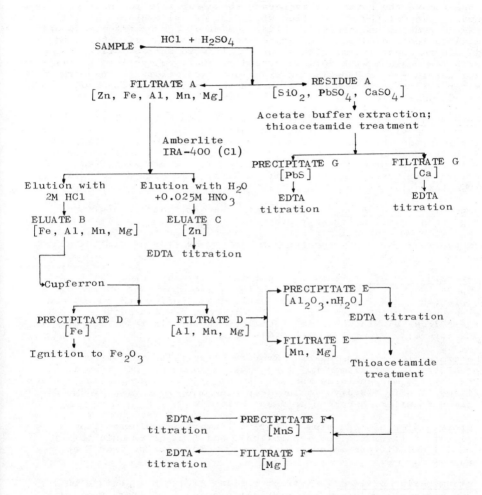

Fig. 15.1 Scheme for analysis of zinc-lead ores
           [20] (by permission of the copyright
           holders, J.T. Baker Chemical Co.)

Procedure.   Place in a 150-ml tall beaker 1 g of the finely ground
sample (200-mesh), moisten with about 5 ml of water and, with stirring
to prevent particles adhering to the bottom, add 15 ml of concentrated
hydrochloric acid and 1 ml of concentrated nitric acid to ensure
dissolution of the sulphides.   Cover the beaker immediately with a
watch-glass and heat on the water-bath until the reaction subsides,
then remove the cover and evaporate the clear solution nearly to dryness.   Repeat the evaporation with 5 ml of hydrochloric acid if some
sample remains undissolved.   Add 5 - 8 ml of concentrated sulphuric
acid to the cooled mass and evaporate to white fumes.   Cool and
dilute to about 150 ml with water.   Allow to settle, filter, and
wash the residue with cold water and alcohol, collecting the filtrate
and washings in a 250-ml standard flask, and dilute to the mark with
water (filtrate A).   The residue on the filter contains SiO_2, $PbSO_4$
and $CaSO_4$ (residue A).

Determination of zinc.   To 25 ml of filtrate A (containing not more
than 50 mg of Zn) add 3 ml of concentrated hydrochloric acid and pass
the solution through a column of Amberlite IRA-400 (chloride form).
Wash the column with 50 ml of 2M hydrochloric acid and keep this
solution for further treatment (eluate B).   Elute zinc from the
column with water and 0.025M nitric acid.   Determine zinc with 0.10M
EDTA (ammonia buffer and Erio T).

Determination of iron.   From eluate B precipitate iron with 6%
cupferron solution (freshly prepared).   After 15 min filter off, and
wash with 2M hydrochloric acid and water.   Ignite (with the usual
precautions for cupferronates) and determine iron gravimetrically as
Fe_2O_3.

Determination of aluminium.   Heat the filtrate D from the iron
separation with ammonium persulphate to decompose the excess of
cupferron.   To the colourless solution add buffer (pH 8 - 10) and 2 g
of ascorbic acid, then heat for 15 min on the water-bath to precipit-
ate⸱ hydrous aluminium oxide.   Collect the precipitate, dissolve it,
and determine aluminium with 0.01M EDTA.

Determination of manganese.   Evaporate the filtrate E (from
precipitation of aluminium) to a small volume in a platinum dish.
Precipitate manganese as MnS with 20 ml of 2% thioacetamide solution.
Filter it off, dissolve it and determine manganese with 0.01M EDTA
(ammonia buffer, ascorbic acid and Erio T).

Determination of magnesium.   Boil off hydrogen sulphide from
filtrate F (from the MnS separation), add 5 ml of ammonia buffer
(pH 10) and titrate magnesium with 0.01M EDTA, using Erio T as
indicator.

Determination of lead.   Extract residue A with hot acetate buffer
(20% ammonium acetate, 2.4% v/v acetic acid).   Filter, wash with
water, and dilute the filtrate and washings to volume in a 250-ml
standard flask.   To 100 ml of this solution add 10 ml of 2% thioacetamide solution and 5 ml of buffer solution.   Boil for 5 min and
filter off the precipitated PbS.   Dissolve it in nitric acid, add
pH 10 buffer and 1 ml of 10% potassium cyanide solution, and titrate
with 0.01M EDTA (Erio T as indicator).

Determination of calcium.   Boil filtrate G from separation of the
lead, make alkaline with aqueous ammonia and titrate with 0.01M EDTA

(murexide as indicator).

Remarks.   Some operations in the scheme could be improved in the
light of current knowledge of complexometry.   For example filtrate D,
containing aluminium, manganese and magnesium, can be analysed without separation, and residue A could also be analysed more simply for
lead and calcium.

15.4   CHROMIUM ORES

In the analysis of chromium ores or chromium-magnesium ores complexometry has not been much applied, and mostly only for the determination
of calcium and magnesium.   For chromium itself the redox methods are
more suitable.   Richards and Boyman [21], in their procedure for the
analysis of chromium ores, determine most of the components spectrophotometrically, calcium by flame photometry, and only the sum of
calcium and magnesium by DCTA titration.   Gricius [22] described a
method for the determination of magnesium in small amounts after
separation from interfering elements by precipitation with hexamine
and sodium diethylthiocarbamate.   Calcium is masked with EGTA and
the magnesium is titrated with EDTA, with Calmagite as indicator.

Banerjee and Vizzini [23] have criticised some of the ASTM methods
for the determination of calcium in chromium ores and in chromium-
magnesite refractories for being laborious and time-consuming.   The
procedure they suggest instead depends on the solubility in acids of
the magnesium and calcium silicates present (merwinite, monticellite
or forsterite).   They treat the finely ground sample (0.5 g) with
20 ml of hydrochloric acid (1 + 1).   In the analysis of chromium
ores the insoluble residue is filtered off and thoroughly washed with
hot water.   Calcium is determined in a fraction of the filtrate,
iron and aluminium being masked with triethanolamine.   Adsorption of
calcium on the $Mg(OH)_2$ precipitate is prevented by the addition of
5 ml of 0.1% poly(vinyl alcohol) solution.   Back-titration of excess
of 0.025M EDTA with 0.025M calcium chloride, with Hydroxynaphthol
Blue as indicator, gave the best results.

The authors claimed that six samples could be analysed in 3 - 7 hours
(depending on sample composition) whereas the ASTM procedure would
take 25 hours.

15.4.1   Complete Complexometric Analysis of Chromium Ores [24]

Determination of Fe, Al and Cr.   Fuse 0.2 g of sample with a mixture
of sodium carbonate and borax and a little potassium chlorate.
Dissolve the cooled melt in water and reduce chromium(VI) with
hydrogen peroxide.   Titrate iron with EDTA at pH 2, using sulphosalicylic acid as indicator.   Add an excess of EDTA and boil the
solution to form the aluminium and chromium EDTA complexes.   Titrate
the excess of EDTA potentiometrically with zinc acetate solution at
pH 5, using the ferrocyanide/ferricyanide redox system.   Demask
aluminium from its EDTA complex by boiling with ammonium fluoride and
titrate the freed EDTA with zinc acetate.

Determination of Ca and Mg.   Dissolve 0.5 g of sample in a mixture
of sulphuric acid and perchloric acid, filter off silica and determine
it gravimetrically.   Precipitate the hydroxides of iron and aluminium

with hexamine.   In one fraction of the filtrate determine the sum of calcium and magnesium by EDTA titration (Erio T).   In a second fraction determine the calcium in potassium hydroxide medium (Calcein).

15.4.2   Analysis of Chromium Ores According to Král [25]

The method is based on known principles, as follows.   After decomposition of the sample by fusion with potassium hydrogen sulphate, silica is determined gravimetrically.   Chromium is determined by titration with iron(II) after oxidation to chromium(VI) with ammonium persulphate (and silver as catalyst).

In a separate sample iron and aluminium hydroxides are precipitated with ammonia.   Aluminium is separated from iron with sodium hydroxide and determined by EDTA titration.   Similarly iron is determined with EDTA, with salicylic acid as indicator.   Titanium is determined photometrically with hydrogen peroxide.   Calcium and the sum of calcium and magnesium are determined complexometrically (Fluorexone and Erio T, respectively, as indicators).

15.5   MANGANESE ORES

In the complexometric analysis of manganese ores or concentrates most attention has been paid to the determination of manganese, calcium and magnesium.   Because of the relatively low stability of the Mn-EDTA complex, manganese is mostly determined in ammoniacal medium, where calcium and magnesium interfere and so must be masked with ammonium fluoride;   likewise iron and aluminium must be masked, with triethanolamine.   Some authors prefer previous separation of iron and aluminium by boiling with hexamine.   Various ion-exchange separations have also been successfully applied.

15.5.1   Determination of Manganese (Povondra and Přibil [26])

Dissolve the sample (0.5 - 1.0 g) in hydrochloric acid (1 + 1) and oxidize with potassium chlorate.   Filter off the insoluble residue and dilute the filtrate to about 300 ml.   Add a few drops of hydrogen peroxide and neutralize with 5% sodium hydroxide solution to pH 2.   Heat the solution to 60°C, add 20 ml of 20% hexamine solution and boil for 5 min.   Filter off the precipitated hydroxide, wash with hot water, dissolve it in hydrochloric acid, and repeat the precipitation.   Dilute the combined filtrates to 1 litre in a standard flask. To 50 or 100 ml of this solution add a measured excess of 0.05M EDTA and 0.5 g of hydroxylamine hydrochloride, heat the solution to 70°C, then add ammonia buffer (pH 9), 1 ml of 10% triethanolamine solution and few mg of potassium cyanide.   Add slowly, with vigorous stirring, 20 ml of 10% ammonium fluoride solution.   After 1 min titrate the excess of EDTA with 0.05M manganese solution (containing 1% of hydroxylamine hydrochloride) to the red-violet colour of Erio T. Results for manganese (25 - 50% of MnO) are in fair agreement with those of classical methods, but in general about 1% lower.

15.5.2   Determination of Manganese (Endo and Koroki [27])

Fuse 0.5 g of sample with a mixture of sodium and potassium carbonates

and dissolve the cooled melt with water and 30 ml of hydrochloric
acid (1 + 1).   Add a few drops of 3% hydrogen peroxide to assist in
dissolving manganese, and boil the solution.   After cooling, dilute
to 250 ml in a standard flask.   To 50 ml of this solution add 20 ml
of 20% triethanolamine solution, dilute to 200 ml with water, make
ammoniacal, add 35 ml of 25% ammonium fluoride solution and stir for
2 - 3 min.   After addition of 20 ml of 10% hydroxylamine hydrochloride
solution and 2 - 3 drops of 5% potassium cyanide solution adjust the
pH to 10 - 10.5 with ammonia and titrate with 0.05M EDTA to the
disappearance of the blue colour of Thymolphthalexone.

Remarks.   This method seems quicker than the previous one.   The
authors used this method for the analysis of ores rich in manganese
(40 - 50%) and with a small content of iron (2 - 10%).   Large amounts
of iron cannot be masked properly with TEA in ammoniacal medium, so
this method has certain limitations.   In sodium hydroxide solution
the Fe-TEA complex is absolutely colourless, but this medium is
suitable only for the indirect determination of calcium and manganese
(in the absence of fluoride).

Nearly the same method for high content of manganese in ores was
published some years later by Vetrova [28].   Crişan and Taloescu [29]
modified the Povondra and Přibil method as follows.   Iron and
manganese are precipitated as the hydroxides with sodium hydroxide.
After dissolution of the precipitate in hydrochloric acid, iron is
determined at pH 1 with EDTA (sulphosalicylic acid as indicator).
An excess of EDTA is added to the titrated solution, the pH adjusted
to 5 - 6 (to form the Mn-EDTA complex) and the excess is back-titrated
with ferric chloride (the sulphosalicylic acid already present serves
as indicator).   The filtrate from the hydroxide separation is used
for the usual determination of calcium and the sum of Ca + Mg.   It
may be remarked that it might be more reliable to determine iron as
described, and then manganese in the same solution, in ammoniacal
medium, by direct titration with EDTA (Thymolphthalexone as indicator).
If only Erio T is available for both titrations, DCTA must be used
[30].

15.5.3   Determination of Manganese after Ion-Exchange Separation [31]

Decompose the sample and remove SiO_2 and WO_3 or other insoluble
residue.   Evaporate a fraction of the filtrate to dryness on a sand-
bath.   Take up in water, add 3 g of citric acid and dilute to about
100 ml.   Adjust the pH to 3.3 - 4.0 (pH-meter) and dilute to 150 ml.
Pass the solution through a column (at 2 - 3 ml/min) of Amberlite
IR-120 (sodium form) and wash with three 13-ml portions and one 50-ml
portion of water.   Elute the manganese from the column with 15 ml of
pH-4.6 acetate buffer containing a known volume of 0.05M EDTA.
Neutralize the eluate, add ammonia buffer (pH 10) and titrate the
excess of EDTA with zinc solution, using Erio T as indicator.

The calcium and magnesium retained on the column can subsequently be
eluted with an EDTA/borax solution at pH 11, and the sum Ca + Mg
determined by back-titration with zinc.

15.6   ANALYSIS OF COPPER ORES

The older methods for analysis of copper ores mainly use well-known

separation steps for removing interfering metals (Pb as $PbSO_4$, Fe and
Al as hydroxides) or use selective separation of copper from the
sample. Fainberg and Filatova [32] precipitate copper as the
sulphide by boiling with sodium thiosulphate. After ignition, the
residue is dissolved in nitric acid and copper determined with EDTA.
Szelag and Kozlicka [33] separate lead as sulphate, and precipitate
copper with potassium ethyl xanthate. After mineralization of the
precipitate they determine copper with EDTA (murexide or Glycinethymol
Blue as indicator). Bhaduri and Jayaprakash [34] proposed separation
of copper by precipitation with a pyridine solution of fumaric acid,
iron being masked with ammonium fluoride.

The following method [35] is very selective and suitable for routine
work. It is based on cementation of copper with lead.

Preparation of the column. In a Jones reductor make a column of
lead granules (99.99% pure, grain-size 0.25 - 1.0 mm) 10 - 15 cm long.
Wash the column several times with hydrochloric acid (1 + 4) and then
with hot water till the effluent is free from iron. After reduction
of a sample, wash the column with hot 10% ammonium acetate solution
and then hot water, to remove the lead sulphate formed during the
removal of copper from the column. For storage fill the column with
hydrochloric acid (1 + 20).

Sample separation. Place 0.25 - 1.0 g of sample in an Erlenmeyer
flask, add 15 - 20 ml of concentrated hydrochloric acid, and heat on a
sand-bath. Add 5 - 10 ml of concentrated nitric acid, evaporate to
dryness and repeat the evaporation after addition of 10 ml of concentrated hydrochloric acid. Dissolve the residue in 5 ml of concentrated hydrochloric acid and 50 ml of distilled water. (The evaporation step is important, because the presence of traces of nitric acid
or nitrogen oxides interferes with reduction of the copper). Pass
the solution through the column at 25 - 75 ml/min. Wash the column
with water and then twice with 0.5% sulphuric acid. Dissolve the
separated copper from the column with 200 ml of 2% solution of
sulphuric acid (containing 12 - 18 g of H_2O_2 per litre), added in
several portions, each being allowed to drain completely before the
next is added, since the considerable amount of oxygen evolved hinders
the flow of liquid through the column.

Determination of copper. Add 20% sodium hydroxide solution to the
eluate from the column until the solution is turbid, then clarify
with 0.5% sulphuric acid. Boil the solution for 15 min to decompose
the peroxide, add 5 ml of 1% ammonium molybdate solution (to precipitate $PbMoO_4$) and boil for a further 2 min. Cool the solution, adjust
the pH to 8 with ammonia and titrate the copper with 0.04 - 0.05M EDTA,
using murexide as indicator.

Remarks. This method is highly selective. There is no interference
from the presence of large amounts of iron, aluminium, manganese,
zinc, cadmium, nickel, calcium and magnesium. Only silver and
bismuth are reduced. Silver does not interfere with the EDTA
titration and bismuth hydrolyses during the elution. The method can
be simplified as follows. The eluate is collected in a 250-ml
standard flask and diluted to the mark. In 100 ml of this solution
the sum of Pb + Cu is determined by EDTA titration (Xylenol Orange).
In another aliquot, after addition of sodium thiosulphate or thiourea
to mask the copper, lead is determined as above.

15.7   DETERMINATION OF Ca AND Mg IN ORES AND CONCENTRATES

The choice of analytical procedures for the determination of these
two elements is entirely dependent on the chemical composition of the
sample.   The main aim is to remove all interfering components, to
obtain at the end a "pure solution" of calcium and magnesium.

Jankovský [36] examined the determination of calcium and magnesium in
lead, zinc, copper and antimony concentrates, and worked out methods
for their decomposition.

Procedure.   One g of sample is decomposed with hydrochloric acid and
nitric acid, silica is removed by evaporation with sulphuric acid and
hydrofluoric acid, and antimony by evaporation with hydrobromic acid
and sulphuric acid.   Heavy metals are separated by precipitation
with sodium diethyldithiocarbamate from acetate medium (pH 5.5) and
extracted into chloroform.   Large amounts of titanium are separated
by chloroform extraction of its cupferronate.   Aluminium is masked
with triethanolamine.   In the resulting solution calcium and
magnesium are determined complexometrically.   According to the author
the results agree with those obtained after mercury-cathode electrolysis, which he had used [37] in the determination of calcium and
magnesium in iron ores.

For the determination of calcium and magnesium in manganese ores
Jurczyk [38] recommends separation of iron, aluminium, chromium and
titanium as basic benzoates.   Manganese (and also Co, Ni, Cu, Zn) is
precipitated with sodium diethyldithiocarbamate.   In the resulting
solution calcium is determined at pH 12 - 13 (Fluorexone-thymolphthal-
ein mixture) and magnesium at pH 10 (Erio T).

For the determination of calcium and magnesium Jurczyk and Szarowicz
[39] published an interesting method based on the separation of iron
as the EDTA complex (formed by titration with EDTA, with salicylic
acid as indicator) with the ion-exchanger Wofatit KPS.   Calcium and
magnesium are eluted with hydrochloric acid/sodium chloride.   Interfering elements (Mn, Pb, Zn, Cu, Ni) are precipitated with sodium
diethyldithiocarbamate and extracted with chloroform.   Calcium and
magnesium are then determined as above.   Other details can be found
in the literature.

Pánek et al. [40] applied the well-known zinc oxide method for the
separation of easily hydrolysed metals.   After separation of the
zinc oxide suspension, manganese is precipitated by heating with
potassium persulphate in ammoniacal medium, with addition of zinc
chloride to prevent adsorption of calcium and magnesium.   In the
resulting solution the calcium and magnesium are determined by EDTA
titration after addition of potassium cyanide to mask the zinc
present.

REFERENCES

1.   M. Gundlach, Erzmetall, 1957, 10, 177.
2.   L.B. Zaichikova, I.N. Lutchenko and V.P. Ioffe, Zavod. Lab.,
     1957, 24, 910; Anal. Abstr., 1958, 5, 1806.
3.   A.N. Karolev and M.K. Koichev, Zavod. Lab., 1959, 25, 546;
     Anal. Abstr., 1960, 7, 928.

4.   O.B. Budevsky, Zavod. Lab., 1960, 26, 50; Anal. Abstr., 1960, 7, 3689.
5.   K. Pannami, Z. Anal. Chem., 1955, 207, 22.
6.   E.M. Donaldson, Talanta, 1976, 23, 161.
7.   J. Kinnunen and B. Wennerstrand, Chemist-Analyst, 1954, 43, 65.
8.   J. Kinnunen and B. Wennerstrand, Metallurgia, 1971, 82, 81.
9.   V.A. Khalizova, I.G. Krasyukova, V.A. Donchenko, A.Ya. Alekseeva and E.P. Smirnova, Zavod. Lab., 1967, 33, 1064; Anal. Abstr., 1968, 15, 7220.
10. R. Přibil, Coll. Czech. Chem. Commun., 1954, 19, 58.
11. O.B. Budevsky, A.N. Karolev, R.A. Karanov and L. Simona-Filipova, Zavod. Lab., 1959, 25, 1439; Anal. Abstr., 1960, 7, 3152.
12. A.N. Karolev, Zavod. Lab., 1967, 33, 811; Anal. Abstr., 1968, 15, 5882.
13. K.S. Rao, A.M. Barve, C.S.R. Rao, S.D. Ved and O.P. Sharma, Res. Ind., New Delhi, 1974, 19, 70; Anal. Abstr., 1975, 29, 3B46.
14. A.K. Ray, R.K. Ghonge and D.J. Tahalramani, Res. Ind., New Delhi, 1975, 20, 25; Anal. Abstr., 1976, 31, 2B52.
15. J. Kinnunen and B. Wennerstrand, Chemist-Analyst, 1953, 42, 80.
16. G. Kraft and H. Dosch, Erzmetal, 1975, 28, 506; Anal. Abstr., 1977, 32, 2B61.
17. M. Hisada and K. Kashiwara, Japan Analyst, 1959, 8, 235; Anal. Abstr., 1960, 7, 1680.
18. O.B. Budevsky and L. Pencheva, Z. Anal. Chem., 1964, 203, 14; Anal. Abstr., 1965, 12, 4506.
19. C. Liteanu and N. Dulamita, Revta. Chim. Bucharest, 1969, 19, Anal. Abstr., 1965, 18, 1523.
20. A.M. Amin and M.Y. Farah, Chemist-Analyst, 1955, 44, 62.
21. C.S. Richards and E.C. Boyman, Anal. Chem., 1964, 36, 1790.
22. A.J. Gricius, South Afric. Chem. Inst., 1965, 18, 100; Anal. Abstr., 1966, 13, 4809.
23. S. Banerjee and J.B. Vizzini, Anal. Chim. Acta, 1976, 82, 439.
24. Z. Sosin and I. Strzeszewska, Chem. Analit. (Warsaw), 1964, 8, 425; Anal. Abstr., 1965, 12, 4512.
25. S. Král, Hutn. Listy, 1966, 21, 496.
26. P. Povondra and R. Přibil, Coll. Czech. Chem. Commun., 1961, 26, 2164.
27. Y. Endo and C. Koroki, Japan Analyst, 1960, 9, 992; Anal. Abstr., 1962, 9, 3695.
28. N.I. Vetrova, Trudy Vses. Nauch. Issled. Geol. Inst., 1964, 117, 17; Anal. Abstr., 1965, 12, 5794.
29. I.A. Crişan and S. Taloescu, Revta. Chim. Bucharest, 1966, 17, 430; Anal. Abstr., 1967, 14, 6109.
30. R. Přibil, Coll. Czech. Chem. Commun., 1955, 20, 162.
31. P. Povondra and T. Šulcek, Coll. Czech. Chem. Commun., 1959, 24, 2398.
32. S.Yu. Fainberg and K.N. Filatova, Zavod. Lab., 1958, 24, 534; Anal. Abstr., 1959, 6, 845.
33. M. Szelag and M. Kozlicka, Chem. Analit. (Warsaw), 1962, 7, 813; Anal. Abstr., 1963, 10, 1711.
34. B.P. Bhaduri and K.C. Jayaprakash, Indian J. Chem., 1970, 8, 1141.
35. R. Karanow, A. Karolew and D. Toschewa, Talanta, 1962, 9, 409.
36. J. Jankovský, Z. Anal. Chem., 1966, 220, 26.
37. J. Jankovský, Hutn. Listy, 1961, 16, 506.
38. J. Jurczyk, Z. Anal. Chem., 1968, 240, 236.
39. J. Jurczyk and I. Szarowicz, Hutnik, 1969, 36, 295; Anal. Abstr., 1970, 19, 1264.
40. Z. Pánek, D. Zemanová and J. Novák, Silikaty, 1967, 11, 67; Anal. Abstr., 1968, 15, 3190.

CHAPTER 16

Analysis of Ferrites

Ferrites, formed by sintering of iron oxide with one or more other metal oxides, have considerable importance due to their magnetic properties. Ferrites differ not only in their chemical properties but also in crystal structure. Of greatest importance are the cubic ferrites with spinel structure and general formula MFe_2O_4 where M is a bivalent metal ion such as iron, manganese, zinc, cobalt, nickel, copper, magnesium, cadmium. Orthoferrites have the formula $M'FeO_3$ where M' is a tervalent metal ion such as aluminium or a lanthanide and garnets have the formula $M_3'Fe_5O_{12}$. Hexaganol ferrites are derived from magnetoplumbite, $PbFe_{12}O_{19}$, with various bivalent metal ions replacing lead. The properties of ferrites are altered considerably by even slight changes in chemical composition and therefore their analysis is very important for quality assessment. Complexometry has found wide application in this field because of its rapidity and accuracy. The structure of ferrites depends on the oxidation state of the metals present and on the amount of bound oxygen and the deviation from oxygen stoichiometry.

The main task in the analysis of ferrites is the separation of iron (as matrix element), followed by determination of the other components, usually by complexometry or polarography.

This chapter will give a brief survey of the analysis of ferrites. Some attention will also be given to the preparation of samples for analysis. The analysis of ferrites has been reviewed by McCrory-Joy and Joy [1], and the use of complexometry for the purpose has been surveyed by Funke [2].

16.1 DISSOLUTION OF FERRITES

Most ferrites can be dissolved in hot hydrochloric acid. Difficulties are met mainly with nickel ferrites, depending on their composition and thermal processing. Funke [2] recommends dissolution in hydrochloric acid under pressure. The decomposition is performed in a sealed glass ampoule by heating to 150°C (or 200°C if the ferrite contains nickel). Decomposition under pressure is advantageous in that it is not necessary to grind the sample finely, a task which is

difficult with strongly fused ferrites and can lead to contamination
of the sample even if done in an agate mortar.   Only a small excess
of acid is needed for the decomposition, so time-consuming evaporations are avoided, and so is fusion of any residue, which usually
increases the risk of contamination with other metals.

Funke recommends use of 250-mm long tubes with a wall thickness of
1.5 mm, and samples with a grain size up to 3 mm.   Decomposition
with hydrochloric acid is usually complete after heating for two
hours.

16.2   SEPARATION OF IRON

Most metals in ferrites are determined complexometrically, and therefore the ferrite solution must be free from excess of iron.   Many
authors agree that masking of iron with triethanolamine cannot be
used in ferrite analysis because of its "low masking capacity".
Nevertheless we have demonstrated that up to several tens of milligrams of iron can be masked in strongly alkaline sodium hydroxide
medium, forming a completely colourless complex.   However, this
medium is not suitable for the determination of zinc (formation of
tetrahydroxozincate) or copper (partial formation of the Cu-TEA
complex).   On the other hand nickel can be determined in the presence
of iron in strongly ammoniacal solution, with murexide as indicator
[3].   This approach has been used in the analysis of Fe-Ni-Li-ferrite
[4].

Separation on ion-exchangers

Funke [2] describes the use of a strongly basic anion-exchanger with
iminodiacetate functional groups (Wofatit L 150) for the separation
of iron and nickel from zinc, or zinc from lead etc.   Separation of
cations on ion-exchangers in complexometric analysis of ferrites has
also been used by other authors [5,6].

Separation by extraction

Several methods have been proposed for the separation, most of them
specially designed for a particular determination.   Funke [2]
re-examined the extraction of iron into ether from a hydrochloric acid
medium.   In a specially constructed extractor, up to 300 mg of iron
can be extracted in 30 minutes.   Reliable separation is achieved
from a mixture of iron (15 - 150 mg) with nickel (30 - 60 mg), manganese
(30 - 60 mg) and magnesium (15 - 40 mg), with a relative error of less
than 1%.   However, zinc and cobalt are partially extracted (5 - 10%)
into the organic phase.

Fano and Licci [7] proposed extraction of iron with acetylacetone for
microanalysis of ferrites (75 - 1000 µg of sample).   The iron can then
be spectrophotometrically determined directly in the organic phase,
at 440 nm.   These authors used this extraction in the analysis of
the special ferrites $Ba_2Zn_2Fe_{12}O_{22}$ and $Ba_2Co_2Fe_{12}O_{22}$.   Zinc and
cobalt were determined with dithizone, barium by microtitration with
EDTA.   The whole analysis took two hours.   These authors also
recommended extraction of iron from 8M hydrochloric acid with
di-isopropyl ether [8].

Extraction with cupferron

Funke [2] used precipitation of iron with cupferron as proposed by
Fritz et al. [9], which provides reliable separation of iron from
many other metals.   The iron is precipitated from hydrochloric acid
medium (pH 0.3) with a small excess of cold freshly prepared 5%
cupferron solution.   The complex is extracted with a 1:1 v/v mixture
of isoamyl alcohol and benzene by shaking in a separatory funnel for
60 sec.   After separation of the aqueous phase the organic phase is
washed with hydrochloric acid (pH 0.3) and the aqueous phases are
combined.   This solution is not suitable for further separation by
ion-exchange, owing to the presence of ammonium salts.   There is
good separation from manganese, magnesium, cobalt and zinc, but
copper will be considerably co-extracted [9].

Extraction with sodium diethyldithiocarbamate

Extraction with this reagent is very frequently used for the
extraction of manganese in the determination of calcium and magnesium
in silicates etc.   Its application for the extraction of iron was
devised by Pedan et al. [10,11], and has the advantage of giving
simultaneous extraction of copper and manganese as well.   The
extraction must be repeated at least thrice because of the low
capacity of ethyl acetate, when this is used as the organic solvent.

16.3   ANALYSIS OF INDIVIDUAL FERRITES

The analysis of two-component ferrites (Fe + M) is a relatively easy
task if the iron is properly separated.   Traces of iron remaining
can often be masked with triethanolamine.   Zinc is usually determined
in ammoniacal solution by EDTA titration with Erio T as indicator.
Similarly magnesium is titrated with EDTA, with Erio T or Thymolphthalexone as indicator.   A strongly ammoniacal solution is used
for the EDTA titration of nickel with murexide as indicator.   The
best procedure for the determination of cobalt seems to be back-
titration of excess of EDTA with 0.05M magnesium chloride in ammonia
solution (Thymolphthalexone as indicator).

For the determination of iron various redox titrations have been
proposed, besides simple EDTA titrations at pH 1 - 2.5 (usually with
sulphosalicylic acid as indicator).   Such determinations can often be
done without separation of the iron, either a separate sample or a
fraction of sample solution being used.   Though determination of the
metals in "pure solutions" is very simple, a few examples of the
analysis of more complex ferrites will be given below.

16.3.1   Rapid Determination of Magnesium in Ferrites [10,11]

Dissolve 0.1 g of sample by boiling with hydrochloric acid for 2 - 3 hr
(keep the acid volume constant).   Dilute the solution to 250 ml in a
standard flask.   Neutralize 100 ml of this solution with 20 - 30%
sodium hydroxide solution, add 5 g of sodium diethylthiocarbamate,
and extract the precipitate by shaking with 100 ml of ethyl acetate
for 5 - 10 min.   Repeat the extraction if some precipitate remains
undissolved.   To the aqueous phase add 20 ml of ammonia buffer
(pH 10) and titrate with 0.1M EDTA to the colour change of Chromogene
Black ET00 (C.I. Mordant Black II; C.I. 14645) from cherry-red to blue.

Remark.   The authors claimed that the presence of aluminium oxide in such ferrites did not affect the results.

16.3.2   Determination of Strontium in Sr-Ferrites [12]

Two methods for the determination of strontium in these ferrites by EDTA titration were described by Vetéssy et al. [12].   The first is based on precipitation of iron with hexamine.   In the filtrate strontium is determined in ammonia buffer with EDTA and Erio T after addition of Zn-EDTA complex.   In the second, the residue after evaporation of the sample solution is treated with 80% (fuming) nitric acid, which dissolves out the iron salts.   The residual strontium nitrate is washed with concentrated nitric acid, then dissolved in water, and the strontium is determined as before.

16.3.3   Determination of Zinc in Zn-Mn Ferrites [13]

Dissolve the sample in hydrochloric acid, dilute the solution to suitable volume (to give a hydrochloric acid concentration of $\sim 2M$), make up to known volume, and take a fraction containing 10 - 25 mg of zinc.   Reduce the iron by adding ascorbic acid until the solution is colourless, and add 200 mg in excess.   Extract zinc from the solution with 5% dioctylmethylamine solution in trichloroethylene, and repeat the extraction.   Shake the combined extracts for 1 min with dilute hydrochloric acid to remove traces of co-extracted copper, cobalt and nickel.   Mix the organic extract with 100 ml of alcohol and 25 ml of dilute ammonia solution and titrate the zinc with 0.01M EDTA (Erio T as indicator).

16.3.4   Analysis of Mn-Zn-Mg Ferrites [14]

This method is entirely based on well-known complexometric titrations. From the dissolved sample iron is extracted as the cupferronate into a 1:1 mixture of benzene and isoamyl alcohol.   In the aqueous phase all three elements are determined complexometrically.

(i)   The sum of Mn + Zn + Mg is determined by direct titration with EDTA and Erio T.

(ii)   In the same solution magnesium is displaced from its EDTA-complex by the addition of ammonium fluoride.   The released EDTA is titrated with manganese chloride solution (containing ascorbic acid).

(iii)   Zinc is then determined by adding potassium cyanide and titrating the released EDTA with manganese solution.   The amount of manganese is calculated by difference.

Remarks.   A somewhat different method was proposed by Chiaki et al. [15].   A 1-g sample is dissolved in hydrochloric acid (followed by fusion of the residue with $K_2S_2O_7$, leaching and combination with the main solution), then the solution is diluted to volume in a 100-ml standard flask.   To a 20-ml aliquot 20 ml of concentrated hydrochloric and 2 - 5 ml of concentrated nitric acid are added and iron is extracted with 20 ml of isobutyl methyl ketone.   The aqueous phase is then evaporated with 10 ml of perchloric acid to white fumes.   The residue is dissolved in 20 ml of 20% ammonium chloride solution, the pH adjusted to 6.5 - 8.5, and zinc and manganese are precipitated with 25 ml of 4% sodium diethyldithiocarbamate solution and extracted with 20 ml of trichloroethylene. In the aqueous phase magnesium is determined by direct EDTA titration, with Erio T indicator. Both zinc and manganese are stripped from the organic phase with sodium hydroxide solution. After the addition of ammonia buffer, ascorbic acid and potassium cyanide, manganese is determined by EDTA titration, and then zinc after demasking of $Zn(CN)_4^{2-}$ with formaldehyde.

16.3.5   Analysis of Mn-Zn Ferrites [16]

Diantipyrinylmethane (DAPM), a favourite extraction reagent of the Russian school, is used in the rapid complexometric analysis of these ferrites.

Dissolve 0.6 g of powdered sample by boiling with 2 ml of concentrated hydrochloric acid. After cooling, dilute the solution to volume in a 200-ml standard flask. To a 10-ml aliquot add ascorbic acid to discharge the yellow colour. Extract the zinc with 0.1M DAPM in dichloroethane. Separate the organic phase, add 35 ml of ethanol and 10 ml of ammonia buffer (pH 9.5 - 10 0), then titrate zinc with EDTA (Erio T as indicator).

Extract another 10-ml aliquot directly with 10 ml of DAPM without addition of ascorbic acid. The organic phase contains the iron and zinc, the aqueous phase only the manganese. Wash the aqueous phase by shaking it with dichloroethane, add several crystals of hydroxylamine hydrochloride, dilute to 100 ml, adjust the pH to 7 with ammonia or ammonia buffer, and titrate the manganese with EDTA (Erio T indicator).

Remarks. This method seems to be quicker than the Rózycki method [14] and was modified by Kuchkina et al. [17] for the analysis of Mn-Zn ferrites containing a small amount of nickel. These authors proceed as follows. They extract zinc with a chloroform solution of DAPM in the presence of ascorbic acid and determine zinc in the extract as in the Shevchuk method above. In separate aliquots they determine iron (by dichromate titration) and nickel spectrophotometrically with dimethylglyoxime. Manganese is determined in a separate sample (0.3 g) by potentiometric titration with permanganate in pyrophosphate medium.

16.3.6   Analysis of Co-Mn-Mg Ferrites [18]

Dissolve 0.5 - 1.0 g of sample in aqua regia and evaporate to dryness. Dissolve the residue in 5 ml of concentrated hydrochloric acid and 100 ml of hot water. Filter off silica and evaporate the filtrate to 50 ml. Add 60 - 70 ml of concentrated hydrochloric acid and extract iron into ether in a suitable extractor. After the extraction remove ether from the aqueous phase by evaporation to a small volume, then dilute to volume in a 250-ml standard flask.

Determination of Co + Mn. To a portion of the solution, containing 2 - 50 mg of each metal, add 0.05M EDTA in excess, adjust the pH to 5 - 5.5 with hexamine, warm the solution to 90 - 95°C and titrate with 0.05M lead nitrate, using Xylenol Orange as indicator.

Remark.   In this medium at room temperature, manganese is displaced
by lead at the end-point, forming the robust Mn - XO complex, and the
colour change of XO becomes irreversible.   In hot solution cobalt
reacts with EDTA very quickly and the end-point is reversible unless
the Mn:Co ratio exceeds 18:1.

Determination of Co + Mn + Mg.   To an aliquot containing 2 - 20 mg of
each metal add an excess of 0.05M EDTA, a small amount of ascorbic
acid, and then ammonia, and titrate the excess of EDTA with 0.05M
magnesium chloride, using Thymolphthalexone as indicator, to an
intense blue colour.

Determination of cobalt.   To the solution from the Co + Mn + Mg
determination add 0.2 - 1.0 g of potassium cyanide, warm the solution
to 40°C (to mask Co) and after 5 min dilute with water and continue
the titration with magnesium chloride to the colour change from
yellow to blue or blue-green.

Remarks.   The sum Co + Mn + Mg can be determined without difficulty
over a wide range of metal concentrations, because the Co-EDTA complex
does not affect the colour change. After addition of potassium
cyanide the intensely yellow $Co(CN)_4^{2-}$ is formed and obscures the end-
point when 15 mg of Co are present in 150 ml.   Up to 40 mg of Co can
be present if the solution is diluted to 500 ml or more.

Calculation.   Because the sum of Co + Mn is known from the first
titration (result A) and also the sum of Co + Mn + Mg (result B) and
the amount of cobalt (result C) the amounts of all the elements can
easily be calculated: C = Co;   A - C = Mn;   B - A = Mg.

16.3.7   Analysis of Co-Ba Ferrites [19]

After extraction of iron with a 1:1 mixture of tributyl phosphate and
benzene, the aqueous solution is evaporated to dryness.   The residue
is dissolved in pH-5.5 buffer solution and cobalt is titrated with
0.02M EDTA (Methylthymol Blue as indicator).   In the same solution
barium is determined by EDTA titration after adjustment of the pH to
9.5.   In a separate sample iron is determined by the usual dichromate
titration.

16.3.8   Analysis of Pb-Ni-Mg Ferrites [20]

After separation of iron the sum of lead and nickel is determined in
slightly acidic medium (by back-titration of excess of EDTA with lead
nitrate, Methylthymol Blue being used as indicator).   To the same
solution a further excess of EDTA is added, the solution is made
ammoniacal, and magnesium is determined by back-titration with
calcium chloride.   Potassium cyanide is then added to mask nickel
and the EDTA liberated is determined by titration with calcium
chloride.

16.4   POTENTIOMETRIC ANALYSIS OF FERRITES

Fano and Licci [8] describe, in a detailed study, the advantages of
automated potentiometric titration with EDTA.   It is demonstrated
that these titrations are more precise than the visual titrations and

can be performed with samples weighing only a few mg.   The analysis
of a particular ferrite $(Ba_2Zn_2Fe_{12}O_{22})$ will be given here as an
example.

Determination of iron

Dissolve the sample in concentrated hydrochloric acid at $50 - 60°C$,
add a few drops of hydrogen peroxide, and dilute until the acid
concentration is 8M.   Extract the iron with di-isopropyl ether
(preferably by four extractions each with $80 - 100$ ml of the ether).
Strip the iron from the organic phase with three 20-ml portions of
water, add an excess of EDTA $(0.01 - 0.05M)$ to the combined strippings,
adjust the pH to 5 with ammonium acetate and titrate the unconsumed
EDTA with 0.01M ferric chloride, using a platinum-wire indicator
electrode and saturated calomel electrode (SCE) as reference [21].

Determination of zinc

Evaporate the aqueous phase to remove hydrochloric acid (three times,
with addition of water between evaporations), adjust the pH to $1 - 1.5$,
add an excess of EDTA, heat almost to boiling and precipitate barium
sulphate with hot 0.5M sulphuric acid.   Filter off and wash the
precipitate with hot water.   To the cooled filtrate and washings add
ammonium acetate to bring the pH to 5 and titrate the unconsumed EDTA
potentiometrically with ferric chloride as before.

Remark.   The precipitation of barium sulphate in the presence of
EDTA prevents the adsorption of zinc on the precipitate.

Determination of barium

Place the filter paper and precipitate of barium sulphate in the
original beaker, add sufficient EDTA solution and 2 ml of concentrated
ammonia solution.   Boil the solution to dissolve the barium sulphate
completely.   After addition of 20 ml of pH-10 buffer (16 g of
ammonium chloride and 28 ml of concentrated ammonia solution per
litre), and 2 or 3 drops of 0.0025M Hg-EDTA, deaerate the solution
with nitrogen and titrate the excess of EDTA with 0.01M zinc nitrate,
using a mercury-cup indicator electrode and SCE reference for end-
point detection.

Remarks.   The authors deliberately avoided stepwise titration of
iron (at pH 2) and zinc (at pH 5.5), because the results obtained
were erroneous and irreproducible.   The reader is referred to the
original paper for further details.

Further methods will be found in the review by McCrory-Joy and Joy [1],
or can be worked out by the reader from the information given in
Chapter 6.

REFERENCES

1.   C. McCrory-Joy and D.A. Joy, Talanta, in the press.
2.   A. Funke, Z. Anal. Chem., 1969, 244, 105.
3.   R. Přibil, Coll. Czech. Chem. Commun., 1954, 19, 58.
4.   F. Tanos, Tavkozl. Kut. Inst. Kozl., 1966, 11, 113; Ref. Zh.
      Khim., 1967, 4G 111.

5.   V. Bărcănescu, E. Potamian and S. Călugăreanu, Revta. Chim.
      Bucharest, 1964, 15, 561;   Anal. Abstr., 1966, 13, 3572.
6.   V.P. Vasyutin, N.M. Morozova and A.S. Konisheva, Novya Metody
      Khim. Anal. Mater., 1971, 2, 46.
7.   V. Fano and F. Licci, Mikrochim. Acta, 1975 II, 561.
8.   V. Fano and F. Licci, Analyst, 1975, 100, 507.
9.   J.S. Fritz, J.M. Richard and A.A. Bystroff, Anal. Chem., 1957,
      29, 577.
10.   G.P. Pedan, Yu.T. Karavanskaya and G.V. Kukhtenkova, Zavod. Lab.,
      1964, 30, 1448;   Anal. Abstr., 1966, 13, 1768.
11.   G.P. Pedan, Yu.T. Karavanskaya and G.V. Kukhtenkova, Ukr. Khim.
      Zh., 1965, 31, 722;   Anal. Abstr., 1966, 13, 6213.
12.   Z. Vetéssy, S. Veres and L. Csányi, Z. Anal. Chem., 1979, 297,
      159.
13.   T.R. Andrew and P.N.R. Nichols, Analyst, 1961, 86, 676.
14.   C. Rózycki, Chim. Analit. (Warsaw), 1967, 12, 573;   Anal. Abstr.,
      1968, 15, 4703.
15.   E. Chiaki, Y. Tomita and M. Ezawa, Japan Analyst, 1966, 15, 1047;
      Anal. Abstr., 1968, 15, 2630.
16.   I.A. Shevchuk, N.N. Nikol'skaya and T.N. Simonova, Zavod. Lab.,
      1965, 31, 545;   Anal. Abstr., 1966, 13, 4821.
17.   E.D. Kuchkina, G.V. Litvinova and N.M. Panina, Metody Anal.
      Kontrolya Proizvod. Khim. Prom-sti, 1977, No. 3, 39;   Anal.
      Abstr., 1979, 36, 3B182.
18.   R. Přibil and V. Veselý, Chemist-Analyst, 1961, 50, 108.
19.   E.D. Kuchkina, L.T. Sorochenko, N.M. Panina and E.V. Lotar,
      Metody Anal. Kontrolya Proizvod Khim. Prom-sti, 1977, No. 3, 42;
      Anal. Abstr., 1979, 36, 3B181.
20.   R. Přibil and V. Veselý, Chemist-Analyst, 1961, 50, 73.
21.   R. Přibil, Z. Koudela and B. Matyska, Coll. Czech. Chem. Commun.,
      1951, 16, 80.

CHAPTER 17

Analysis of Semiconductors

Polarography and spectrophotometry, among other instrumental methods, have been very widely used in the analysis of semiconductors and their parent materials. Most of the methods were reviewed several years ago [1]. Methods suitable for determination of trace levels of less common metals in semiconductors have been surveyed by Goryushina [2]. Problems in the analysis of high-purity samples of semiconductor materials were discussed by Karpov and Alimarin [3].

Complexometry has been applied to the determination of indium, and of phosphorus and arsenic in some phosphides and arsenides.

Determination of indium in In-Sb-Te alloy [4]

Fuse 30 mg of the sample in a micro-crucible with 150 - 300 mg of sodium hydrogen sulphate. Dissolve the cooled melt with 3 ml of a mixture of sulphuric acid and hydrochloric acid (25 ml of H_2SO_4 + 45 ml of HCl + 180 ml of H_2O). Dilute to volume in a 10-ml standard flask with the acid mixture. Take 1 ml of this solution, dilute with 1 ml of 10% potassium sodium tartrate solution, add ammonia buffer (pH 8 - 10), warm the solution to 75°C and titrate with 0.01M EDTA (Erio T as indicator).

Remarks. Xylenol Orange is mostly used as indicator for the determination of indium, in acidic medium, for example by Shafran et al. [5] for the determination of indium in indium-antimony-arsenic solid solutions.

Determination of indium in indium phosphide and arsenide [6]

Dissolve 0.4 - 0.5 g of the sample by heating under reflux in nitric acid or in a mixture of nitric acid and hydrochloric acid, then evaporate to dryness. Dissolve the residue in sulphuric acid and dilute to volume in a 100-ml standard flask. To 5 or 10 ml of this solution add 15 - 20 ml of ammonium citrate solution to give pH 3.5. Heat the solution to 55°C and titrate the indium with 0.2M EDTA, with Xylenol Orange as indicator. Precipitate phosphate from the same solution as $MgNH_4PO_4.6H_2O$ (preferably twice) and determine it gravimetrically as $Mg_2P_2O_7$. Determine arsenic(V) iodometrically.

AC - M

Remarks.   Alimarin et al. [7] used coulometric EDTA titrations for
the determination of nickel (3%) and indium (2 - 5%) in lead-based
semiconductors.   Mirovich [8], for the analysis of Zn-In-S semiconductors recommended spectrophotometric determination of zinc with
dithizone, and of indium (in a separate solution) with Xylenol Orange.
He also recommends polarographic determination of zinc and indium in
acetate buffer (pH 4.75).   The half-wave potentials are -0.78 V for
indium and -1.15 V for zinc.

Analysis of boron phosphide and boron arsenide [9]

Phosphorus in BP.   Fuse the sample with carbonate mixture.   Dissolve
the cooled melt and precipitate phosphate with bismuth nitrate
solution.   Dissolve the precipitated $BiPO_4$ in nitric acid and
titrate the bismuth at pH 1 - 1.2 with EDTA, using Xylenol Orange as
indicator.   For further details see Section 7.1.3.

Analysis of solid solution of BP + BAs.   Fuse the sample with a
mixture of K_2CO_3 + Na_2CO_3 + $NaNO_3$ (40 - 50-fold excess) and extract
the cooled melt with hot water.   Precipitate phosphate and arsenate
with a standard solution of bismuth nitrate.   Filter off the
precipitate and in the filtrate determine the excess of bismuth by
EDTA titration at pH 1 - 1.2, with Xylenol Orange as indicator.   In
another portion of sample solution determine arsenic with potassium
bromate after reduction to As(III) with hydrazine and distillation
with hydrochloric acid.

Determination of boron.   Convert the boron into boric acid and
titrate it by the mannitol method.

Remark.   A fast method for the determination of phosphate has been
based on the following principle [10].   The bivalent and tervalent
elements present in the solution are determined indirectly by
addition of excess of EDTA and back-titration with zinc chloride
solution at pH 5.5, with Xylenol Orange as indicator.   Then an
excess of La-EDTA complex is added and after heating to 30 - 40°C the
solution is very slowly titrated with 0.05M zinc chloride from yellow
to red violet.   During the titration lanthanum is displaced from its
EDTA-complex and immediately precipitated as phosphate:

$$LaY^- + Zn^{2+} + PO_4^{3-} = ZnY^{2-} + LaPO_4$$

The first drop of excess of zinc changes the indicator colour.

REFERENCES

1.   P.F. Kane, Anal. Chem., 1966, 38, No. 3, 29A.
2.   V.G. Goryushina, Zavod. Lab., 1971, 37, 513; Anal. Abstr., 1972,
      22. 1420.
3.   Yu.A. Karpov and I.P. Alimarin, Zh. Analit. Khim., 1979, 34,
      1402; Anal. Abstr., 1980, 38, 2B207.
4.   Yu.S. Lyalikov and L.S. Kopanskaya, Izv. Akad. Nauk Mold. SSR,
      1961, 12, 47; Anal. Abstr., 1963, 10, 4547.
5.   I.G. Shafran, M.Z. Partashnikova and T.I. Pletneva, Trudy Vses.
      Nauchno-Issled. Inst. Khim. React., 1966, No. 28, 92; Anal.
      Abstr., 1967, 15, 706.

6.   Yu. S. Lyalikov and K.E. Kolchina, Zavod. Lab., 1966, 32, 1319;
      Anal. Abstr., 1968, 15, 707.
7.   I.P. Alimarin, M.N. Petrikova and T.A. Kokina, Zh. Analit. Khim.,
      1970, 21, 1008;   Anal. Abstr., 1971, 22, 2505.
8.   L.V. Mirovich, Trud. Kompo. Analit. Khim., 1968, 16, 141;   Anal.
      Abstr., 1971, 16, 2865.
9.   A.A. Reshchikova, Z.S. Medvedova and G.F. Dimitrieva, Trud.
      Kompo. Analit. Khim., 1968, 16, 216;   Anal. Abstr., 1969, 16,
      2909.
10.   R. Přibil, Talanta, 1975, 22, 688.

CHAPTER 18

Analysis of Catalysts

Complexometric analysis of catalysts does not differ much from the analysis of alloys. The only difference is that the catalysts are simpler in composition, usually containing only two or three components - metals or metal oxides. The chemical nature of catalysts is determined by their function in chemical reactions (hydrogenation, hydrosulphuration, hydrorefining catalysts etc.).

The complexometric methods used in this field do not extend the framework of common procedures; indeed some recent works even disregard the latest findings in complexometry (concerning choice of indicators, masking agents, pH adjustment etc.). Hence most methods will be described briefly below, with fuller details only when necessary.

Analysis of Raney nickel [1]

The sample is dissolved in sulphuric acid (1 + 1) and the solution diluted to known volume. The sum of Al + Ni is determined by boiling the sample with excess of EDTA for 2 - 3 min and then back-titrating with zinc solution at pH 5 - 6. After demasking of the Al-EDTA complex with ammonium fluoride, the liberated EDTA (equivalent to the aluminium) is determined by further titration with zinc solution.

Remark. The authors reported relative errors of 0.3% for Ni and 0.6% for Al, but stated that owing to the fire risk, the sample could not be weighed exactly, and the sum of Ni and Al was assumed to be 100% (50% of each).

Determination of Al and Ni in hydrosulphuration catalysts [2]

After preparation of the sample solution molybdenum and other metals are separated on Dowex 1-X8. The effluent is evaporated to dryness and the residue dissolved in water. In a suitable part of the solution the sum of Al + Ni is determined by back-titration of excess of EDTA with zinc solution, with dithizone as indicator. Aluminium is then determined as for Raney nickel, by demasking it from the Al-EDTA complex by addition of ammonium fluoride.

Determination of Al in hydrorefining catalyst [3]

The sample is dissolved in hydrochloric acid (1 + 1).   To an aliquot
of the sample solution containing 1 - 12 μmole of aluminium an excess
of DCTA is added, and the surplus is back-titrated at pH 3.5 oscillometrically with 0.0075M lead acetate in anhydrous methanol.   Nickel
and cobalt are masked with 1,10-phenanthroline.

This method has been simplified by the same authors [4], by use of
direct DCTA titration of aluminium.   Aluminium (10 - 60 μmole) in
aqueous isopropyl alcohol (1 + 3) medium is titrated oscillometrically
at pH 3.3 - 3.5 with 0.005M DCTA.   Cobalt and nickel are again masked
with 1,10-phenanthroline.

Determination of metallic nickel in Ni-Si-Al catalyst [5]

Metallic nickel is selectively dissolved in a 1:9 v/v mixture of
bromine and methanol at 20°C.   After removal of bromine and methanol
by evaporation with sulphuric acid, nickel is determined in strongly
ammoniacal solution with 0.05M EDTA, with murexide as indicator.

The method is claimed to be superior to those involving magnetic
determination or extraction of $Ni(CO)_4$.

Analysis of Al-Cu catalyst [6]

A 0.3-g sample containing 5 - 7% of copper and 36 - 40% of aluminium is
dissolved in sulphuric acid (1 + 1) and the solution is diluted to
suitable volume.   An aliquot is boiled at pH 2 with EDTA and the sum
of Al + Cu is determined by back-titration at pH 5 - 5.5 with 0.02M
zinc chloride (Xylenol Orange as indicator).   Then the Al-EDTA
complex is decomposed by boiling with ammonium fluoride and the
liberated EDTA is titrated with zinc chloride.

Analysis of Al-Ni-Cu catalyst [7]

The sample (0.5 - 1.0 g) is dissolved in nitric acid and the solution
is evaporated to dryness.   The residue is dissolved in water and
made up to 250 ml in a standard flask.   One aliquot is used for
determination of the sum Al + Ni + Cu and of Al as above.

In a second aliquot the sum of Cu + Ni is determined in ammoniacal
triethanolamine medium by direct titration of the hot solution with
EDTA (murexide as indicator).   A third aliquot serves for the determination of nickel in ammoniacal medium, after reduction of the
copper with sodium thiosulphate.

Remarks.   The whole analysis could be done in weakly acidic medium.
The sum of Al + Ni + Cu and of Al alone could be determined as for Al-
Cu catalysts.   Nickel can be determined reliably in another aliquot
by back-titration with zinc chloride (Xylenol Orange as indicator)
after masking of aluminium with fluoride and of copper with thiourea.
This principle can also be used for the combinations Al-Zn-Cu, Al-Co-
Cu or Al-Co-Mg.   Somewhat less suitable methods for analysis of the
first and third combinations have been published by Abdurakhimova and
Shamsiev [8] and Levkovich [9], respectively.

Analysis of Fe-Cu-Zn catalysts [10]

These catalysts for carbon monoxide oxidation consist essentially of
a CuO-ZnO mixture (approx. 1:2) and sometimes contain 1 - 2% of iron
oxide.   An analysis based on ion-exchange separation followed by
EDTA titrations was proposed by Chaudhuri and Sen [10].

The sample is dissolved in 9M hydrochloric acid and the solution is
passed through a Dowex 1-X4 column.   Copper is eluted with 3M hydrochloric acid and determined with EDTA at pH 10, with murexide as
indicator.   Iron is eluted with 0.5M hydrochloric acid and determined
with EDTA at pH 2 - 3 and 40 - 50°C, in the presence of sulphosalicylic
acid as indicator.   Zinc is eluted with 0.005M hydrochloric acid and
titrated at pH 10 with EDTA (Erio T as indicator).   The authors
claim that their method is quicker than the conventional methods and
that the results are satisfactory.

Analysis of Fe-Zn-Cr catalyst [11]

The sample (0.2 g) is decomposed by successive addition of 40 ml of
perchloric acid, 37 ml of hydrochloric acid and 37 ml of sulphuric
acid.   The solution is then made up to 500 ml.   In suitable
portions of the solution iron and zinc are determined complexometrically.   Chromium is determined iodometrically after oxidation with
hydrogen peroxide.

Remark.   In the presence of iron and copper, chromate can only be
determined iodometrically if DCTA is added to mask these two metals.

Analysis of Zn-Cu-Cr catalyst [12]

This catalyst for methanol synthesis, containing approximately 30%
each of zinc and chromium and about 5% of copper, was analysed by
Karolev [12] by the following procedure.

A 0.5-g sample is dissolved with sulphuric acid and the solution is
heated till fuming;   graphite is oxidized with nitric acid.   The
solution obtained is made up to volume in a 50-ml standard flask.
Fractions are taken for individual determinations.   Zinc is determined by masking aluminium with ammonium fluoride and copper with
thiourea, adding excess of EDTA and back-titrating at pH 5.3 - 5.8
with 0.05M zinc sulphate, and Xylenol Orange as indicator.   The sum
of Zn + Cu is determined in the same manner, without the addition of
thiourea.   The sum of Cr + Cu + Zn is obtained by boiling a portion
of the solution with excess of EDTA for 15 - 20 min.   After cooling,
the surplus EDTA is titrated at pH 5.9 - 6.3 with 0.05M zinc sulphate
(Methylthymol Blue as indicator).   The chromium is then calculated
by difference.

Analysis of Fe-Cr catalyst [13]

Fuse 2 g of finely ground sample with sodium carbonate.   Dissolve
the cooled melt in hydrochloric acid (1 + 1) and make up the solution
to 500 ml.   Heat an aliquot with hydrogen peroxide for 10 min to
reduce chromium(VI) to chromium(III).   Cool the solution and add
excess of 0.05M EDTA (about three times as much as is needed for
complexation of both metals).   Then add 1 - 2 g of sodium bicarbonate,
and some acetate buffer (pH 5).   Titrate the unconsumed EDTA with
0.05M copper sulphate, with Catechol Violet as indicator.   This gives

the sum of Fe + Cr.   In another aliquot determine iron as usual with
0.05M EDTA and salicylic acid after reduction of chromate with
peroxide.

Remark.   Formation of the violet Cr-EDTA complex is very slow at
room temperature but is catalysed by addition of bicarbonate and
acetate [14], so boiling of the solution is avoided, but this
catalytic reaction needs at least 10 min at pH 5.3 - 6.0 for
quantitative formation of the Cr-EDTA complex [15] (see p. 129).

Determination of cobalt in zeolite catalyst [16]

The sample (0.2 - 0.5 g) of the catalyst (used for toluene disproportionation) is decomposed with hydrofluoric acid and sulphuric acid.
The residue from the evaporation to remove silica is dissolved in
hydrochloric acid.   The solution is neutralized with ammonia and
aluminium is separated by precipitation with pyridine.   In a
fraction of the filtrate, cobalt is determined by back-titration of
excess of EDTA at pH 10 with zinc chloride solution, with Erio T as
indicator.

Analysis of Pd-Pb catalyst [17]

The analysis of this catalyst, containing calcium carbonate as
carrier and used for the synthesis of linalyl acetate, was proposed
by Kurakina et al. [17].

A 0.5-g sample is dissolved in dilute hydrochloric and nitric acids.
In one portion of the solution the sum of Pb + Pd is determined by
back-titration of excess of EDTA with 0.05M zinc acetate at pH 6,
with Xylenol Orange as indicator.   In another portion lead is
precipitated at 80°C with ammonia and ammonium chloride, as $Pb(OH)Cl$;
in the filtrate calcium is determined by EDTA titration, with Acid
Chrome Dark Blue (C.I. Mordant Blue 13) as indicator.   The precipitated $Pb(OH)Cl$ is dissolved in acetate buffer (pH 6), excess of EDTA
is added, and the surplus titrated with zinc chloride as above.

Remarks.   More recently, Raoot and Raoot [18] have published a
highly selective method for the complexometric determination of
palladium, which could very easily be used for this analysis.   The
method is based on masking palladium with thiourea.   In principle, a
solution containing palladium and other cations is mixed with an
excess of EDTA and back-titrated with lead nitrate at pH 5 - 5.5, with
Xylenol Orange as indicator.   Thiourea is then added and the EDTA
liberated by formation of the palladium-thiourea complex is
titrated in the same way.

Determination of molybdenum in catalysts

Not much attention has been paid to complexometric determination of
molybdenum.   Only two methods are of interest: (a) determination
of molybdenum(V), which forms a 2:1 complex with EDTA;   (b) determination of molybdenum(VI) as the ternary 1:1:1 complex with EDTA and
hydroxylamine.   Both methods were described in detail in Section 6.2.

Only the first method has been applied in the analysis of catalysts
(containing aluminium and cobalt or nickel), by Uvarova and Rik [19].
The molybdenum is reduced in the presence of 0.005M EDTA by boiling
with 3 g of hydrazine sulphate for 5 min in slightly acid medium.

Aluminium is masked with saturated ammonium fluoride solution.   The
excess of EDTA is then titrated at pH 4 - 5 with 0.005M copper
sulphate, PAN being used as indicator.

Before the determination of molybdenum any cobalt present must be
separated as $Co(OH)_3$ in alkaline solution in the presence of hydrogen
peroxide.   Nickel must also be separated beforehand, as $Ni(OH)_2$.
The hydroxide precipitate is dissolved in acid and cobalt (or nickel)
determined with EDTA (back-titration of excess with magnesium
sulphate, with Erio T as indicator).

Remarks.   For the determination of molybdenum in catalysts containing iron and cobalt, Cotiga [20] recommends the precipitation of
molybdenum as the oxinate in the presence of EDTA, which masks iron
and cobalt.   The precipitate is dried at 120°C and weighed.   In the
filtrate from the oxinate precipitation, cobalt is displaced from its
EDTA complex by addition of calcium nitrate solution, and simultaneously precipitated as cobalt oxinate, which is collected, dried and
weighed.

The individual methods on which this analysis was based were originally published nearly twenty years earlier [21 - 23].

There are other analyses where complexometry is used only as an
alternative method.   For example, Perchik and Khotsyanivs'kii [24]
in the analysis of tungsten catalysts containing WS_2, NiS and Al_2O_3
determined all components gravimetrically and recommended EDTA
titration only for nickel (in absence of aluminium!).   Habesberger
[25], in the analysis of vanadium catalyst (for H_2SO_4 production)
which is stabilized with phosphoric acid, determines phosphate by
EDTA titration after precipitation as $MgNH_4PO_4 \cdot 6H_2O$.

REFERENCES

1.   V.E. Ermolaeva, N.I. Frolova and G.E. Burtseva, Khim. Farm. Zh.,
      1976, 10, 134; Anal. Abstr., 1977, 33, 6B62.
2.   T. Fernandez, C. Sanchez-Pedreno and J.M. Rocha, Ion, Madr.,
      1975, 35, 408; Anal. Abstr., 1976, 31, 2B253.
3.   C. Sanchez-Pedreno, T. Fernandez and J.J. Arias, An. Quim.,
      1975, 71, 50; Anal. Abstr., 1975, 29, 3C48.
4.   T. Fernandez, C. Sanchez-Pedreno, L. Martinez-Conerejo and J.J.
      Arias, An. Quim., 1975, 71, 71; Anal. Abstr., 1975, 29, 3C48 (II).
5.   H. Urbain, R. Bacaud, H. Charcosset and L. Tournayan, Chim. Anal.
      (Paris), 1968, 50, 242; Anal. Abstr., 1969, 17, 2126.
6.   F.I. Bondarevskaya and V.V. Malikhov, Metody Analit. Kontrolya
      Kach. Prod. Khim. Prom-sti, 1978, No. 5, 45; Anal. Abstr., 1979,
      37, 2B226.
7.   M.K. Abdurakhimova and S.M. Shamsiev, Dokl. Nauk. Uzbek. SSR,
      1972, No. 4, 36; Anal. Abstr., 1973, 24, 2694.
8.   M.K. Abdurakhimova and S.M. Shamsiev, Uzbek. Khim. Zh., 1971,
      No. 4, 31; Anal. Abstr., 1973, 23, 2288.
9.   A.N. Levkovich, Zavod. Lab., 1972, 38, 1416; Anal. Abstr., 1973,
      24, 3358.
10.   S. Chaudhuri and S.P. Sen, Fert. Technol., 1977, 14, 5; Anal.
      Abstr., 1978, 35, 5B191.
11.   A. Turuta, T. Crisan and A. Cernenco, Revta. Chim. Bucharest,
      1971, 22, 307; Anal. Abstr., 1972, 22, 1443.

12. A. Karolev, Chim. Ind. (Sofia), 1970, 42, 212; Anal. Abstr.,
     1971, 20, 3602.
13. P.C. Gupta, D.N. Salai, D. Dutta and A. Sarkar, Technol. Sindri.,
     1974, 11, 252; Anal. Abstr., 1975, 29, 6B146.
14. V.K. Rao, D.S. Sundar and M.N. Sastri, Chemist-Analyst, 1965,
     54, 86.
15. V.K. Rao, D.S. Sundar and N.M. Sastri, Z. Anal. Chem., 1966,
     218, 92.
16. A.M. Garnish and S.B. Fazul'zhanova, Metody Analit. Kontrolya
     Kach. Prod. Khim. Prom-sti, 1978, No. 5, 47; Anal. Abstr., 1979,
     37, 2B229.
17. T.V. Kurakina, A.G. Sokolova and N.M. Popova, Maslob. Zhir.
     Prom-sti, 1976, No. 11, 30; Anal. Abstr., 1978, 34, 1B240.
18. K.N. Raoot and S. Raoot, Talanta, 1981, 28, 327.
19. E.I. Uvarova and V.M. Rik, Khim. i Tekhnol., Topliv. i Masel,
     1964, No. 5, 67; Anal. Abstr., 1965, 12, 6559.
20. M. Cotiga, Revta Chim. Bucharest, 1968, 19, 615; Anal. Abstr.,
     1970, 18, 1476.
21. R. Přibil and M. Malát, Coll. Czech. Chem. Commun., 1950, 15,
     120.
22. R. Přibil and V. Sedlář, Coll. Czech. Chem. Commun., 1951, 15,
     69.
23. R. Přibil, Analytical Applications of EDTA and Related Compounds,
     pp. 138-143. Pergamon Press, Oxford, 1973.
24. F.I. Perchik and O.I. Khotsyanivs'kii, Vis. Kiivsk. Politekh.
     Inst. Ser. Khim. Mashinobuluv. Tekhnol., 1966, No. 3, 150;
     Anal. Abstr., 1968, 15, 3314.
25. K. Habesberger, Chem. Prum., 1962, 12, 547.

CHAPTER 19

Analysis of Pigments

Complexometry has become a very useful tool in both the macro and micro analysis of pigments. The main reason for this is that pigments are relatively simple inorganic materials of natural or synthetic nature. The complexometric procedures for the analysis of lithopone, zinc white, cadmium yellow, for example, are rather old, and reflect the state of complexometry at the time of their origin.

Since the analyses involve only simple titrations, individual procedures will be given in abbreviated form.

Analysis of zinc oxides [1]

Zinc oxide contains traces of lead as impurity, and therefore both zinc and lead are determined.

Depending on the expected lead content, 50 - 100 mg of sample are dissolved in the necessary amount of hydrochloric acid. The sum Pb + Zn is then determined with 0.01M EDTA in alkaline medium containing tartaric acid. Erio T is used as indicator.

In a separate sample the lead is similarly determined after masking of zinc with potassium cyanide. If calcium or magnesium is present, lead and zinc must be separated by precipitation with thioacetamide before their titration.

Remark. In slightly acidic medium the sum Pb + Zn can be determined by EDTA titration at pH 5 - 5.5 (Xylenol Orange as indicator). In the same solution zinc is determined by masking it with 1,10-phenanthroline and titrating the liberated EDTA with 0.01M lead nitrate.

Analysis of pigments containing cadmium

These materials contain barium sulphate as well as cadmium sulphide and zinc sulphide, the relative amounts influencing the colour shade of the pigments.

Two methods have been described for the determination of zinc and cadmium after separation of the barium sulphate. The older of the

two, described on p. 166, is based on the determination of the sum
Zn + Cd by direct titration with EDTA in ammoniacal solution, with
Erio T as indicator.   Cadmium is then precipitated from its EDTA
complex with sodium diethyldithiocarbamate and the liberated EDTA is
titrated with magnesium chloride solution [2].   The second method
(see p. 170) is based on the titration of cadmium with EGTA [3].

Dissolve 1 g of pigment in 15 ml of nitric acid (1 + 3) and boil to
remove oxides of nitrogen.   Dilute with water and filter (paper).
Wash the barium sulphate with hot water.   Transfer the filtrate to
a 250-ml standard flask and dilute to the mark with water.   Pipette
an aliquot containing 0.1 or 0.2 g of the sample to a beaker, dilute
with water and add a measured volume of 0.05M EGTA sufficient to
complex all the cadmium.   Add 1 - 2M sodium hydroxide until the zinc
hydroxide precipitate redissolves.   If only a small amount of zinc
is present no precipitate appears, so in this case add 10 ml of 1 - 2M
sodium hydroxide for every 100 ml of slightly acid solution taken.
Allow the solution to stand for at least 15 min, to allow complete
conversion of zinc into tetrahydroxozincate.   Add 1% aqueous metalphthalein solution until a faint pink colour is observed, and titrate
with 0.05M calcium chloride solution to a sharp colour change from
pink to intense violet.

Remark.   Up to 300 mg of zinc can be tolerated in this procedure.
The tetrahydroxozincate anion forms an almost colourless complex with
the indicator.   If no zinc (or lead) is present the colour of the
free indicator in strongly alkaline medium is also intensely violet
and the end-point indication fails.   In such a case the procedure
can be salvaged by deliberate addition of zinc to a second sample.
Thymolphthalexone or Methylthymol Blue cannot be used, because the
masking conditions make the solution too alkaline.

The method for cadmium is very selective, because iron and aluminium
can be masked with triethanolamine.   Lead is also masked by the
sodium hydroxide present.

If it is of interest to determine the zinc, another fraction of the
solution is used.   The sum of Cd + Zn is determined by direct
titration with 0.05M EDTA at pH 5 - 5.5 (hexamine buffer) with Xylenol
Orange as indicator.   Zinc can be found by difference, from the
results of the two titrations.

The method has been applied successfully to mixtures of cadmium red
and lithopone in various ratios, containing from 0.05 to 69% cadmium.

Analysis of lithopone

The oldest methods for the determination of zinc (in ZnO or ZnS) are
based on the EDTA titration of zinc with Erio T as indicator [4,5].
The determination can also be done on the micro-scale [1].   Veldwijk
[6] recommended Xylenol Orange for titration of zinc in slightly
acidic medium.

The sample (0.3 - 0.6 g) is dissolved in 15 ml of 4M hydrochloric acid,
and the solution is boiled to remove hydrogen sulphide and oxidized
by boiling with a few drops of nitric acid for several minutes.   Then
the solution is diluted to 100 ml, buffered with solid hexamine to
pH 5 - 5.5, and titrated with 0.1M EDTA.

Determination of lead in lead chromate [7]

The sample (0.5 g) is dissolved in 50 ml of 10% sodium hydroxide solution and diluted to the mark in a 250-ml standard flask.   A 10-ml aliquot is diluted with 50 - 60 ml of water and neutralized with hydrochloric acid to pH 5 - 6.   After addition of 20 ml of 0.01M EDTA and ammonia buffer (pH 10), the excess of EDTA is titrated with 0.01M magnesium chloride, Erio T being as indicator.

Analysis of anticorrosion basic lead silicochromate [8]

The sample (0.5 - 0.6 g) is fused with 4 g of a mixture of potassium and sodium carbonates at 900 - 950°C.   The cooled melt is treated with hydrochloric acid and the silica determined gravimetrically in the usual way.   The lead in the filtrate is determined by EDTA titration after masking of aluminium (which would otherwise interfere) with ammonium fluoride.   Chromium [present in the filtrate as Cr(III)] is oxidized to chromium(VI) and determined iodometrically.

Remark.   Complexometry has also been used for evaluation of various lead chromates, dissolved in 0.25% hydrochloric acid.   In such solutions of chrome yellows, reds and oranges, as well as in molybdate pigments, lead is determined by EDTA titration [9].

Analysis of zinc yellow ($ZnO + PbCrO_4$) [7]

A 0.5-g sample is dissolved in 25 ml of dilute hydrochloric acid in a 250-ml standard flask and diluted to the mark.   In 50 ml of this solution the sum of lead and zinc is determined by titration with 0.05M EDTA at pH 10, with Erio T as indicator.   Because of the chromate present the colour change is from orange to green.

Remark.   If only a small amount of lead is present Borchert recommends its extraction from cyanide solution with a chloroform solution of dithizone.   The organic phase is diluted with ethanol and the lead determined with 0.001M EDTA, according to Kotrlý [10]. Zinc can be determined in the aqueous phase, after demasking with formaldehyde from the cyanide complex, by the usual titration with EDTA and Erio T.

REFERENCES

1.   H. Flaschka, Fette Seifen, 1952, 54, 267.
2.   R. Přibil and V. Veselý, Coll. Czech. Chem. Commun., 1953, 18, 783.
3.   R. Přibil and V. Veselý, Chemist-Analyst, 1966, 55, 4.
4.   M.R. Verma and V.M. Bhuchar, Paint Mnfc., 1957, 27, 384; Anal. Abstr., 1958, 5, 1462.
5.   A. Schaller and E. Mihalovics, Oestr. Chem. Ztg., 1959, 60, 338.
6.   G. Veldwijk, Verfkroniek, 1967, 40, 427; Anal. Abstr., 1969, 16, 1979.
7.   O. Borchert, Plaste Kautschuk, 1959, 6, 562.
8.   N.E. Dunyushevskaya, O.V. Alekseeva, B.M. Wasserman and V.I. Kashmarova, Lakokras Mater. Primenenie, 1975, No. 4, 52; Anal. Abstr., 1976, 31, 5C95.
9.   W. Kondratchoff and J.M. Pujade-Renaud, Double Liaison, 1974, 21, 593; Anal. Abstr., 1975, 29, 1C86.
10.   S. Kotrlý, Coll. Czech. Chem. Commun., 1957, 22, 1765.

CHAPTER 20

Analysis of Varnish Driers

To the group of varnish driers (siccatives) belong various metal salts of resinous or naphthenic acids $(C_nH_{2n-1}COOH)$, the latter type being most often used because of its higher content of catalytically effective metal (lead, cobalt, manganese, zinc or calcium). Although determination of the individual metals is simple, preparation of the sample solution may present problems. The earliest methods used extraction with acid. After the separation of the aqueous phase the individual elements were determined complexometrically [1,2]. Lucchesi and Hirn [3] dissolved the driers in methanol-benzene mixture, with addition of acid. The individual elements were determined (after pH adjustment) by EDTA titration. Veldwijk [4] does not regard this mixture of solvents as very advantageous, since benzene dispersed in the solution considerably decreases the sharpness of the indicator colour change. He also does not stick to a uniform dissolution method as other authors do, but recommends for individual driers slightly modified procedures, which require only the use of simple reagents. Only two indicators are required for all the complexometric determinations - Erio T and Xylenol Orange. His procedures are given below.

Lead siccative. Weigh 1 g of sample (300 - 500 mg of Pb) into a 300-ml glass-stoppered Erlenmeyer flask. Then add successively 20 ml of carbon tetrachloride and 20 ml of acetone, 5 ml of 4M nitric acid and 10 ml of distilled water. Shake the flask vigorously for 1 min. Add a little Xylenol Orange (1% w/w mixture with potassium chloride) and solid hexamine till a red colour appears, and then 5 g of hexamine and 5 ml of water. Titrate with 0.1M EDTA till yellow. Shake for 1 min more and continue slowly with the titration. Repeat the shaking and titration till the yellow colour remains unchanged for 3 min.

Cobalt siccative. Dissolve 1 g of sample (about 50 mg of Co) in 20 ml each of carbon tetrachloride and benzene as above, and add 3 ml of 4M hydrochloric acid and 10 ml of water. After shaking, add 10 g of hexamine and 25 ml of 0.1M EDTA and shake again for 1 min. Add 1 ml of 0.001M 1,10-phenanthroline and some Xylenol Orange, and titrate with 0.1M zinc sulphate from yellow to red-violet. Before the end-point shake the solution thoroughly. The phenanthroline is

377

said to prevent blocking of the Xylenol Orange with displaced traces
of cobalt.

Manganese siccative. Dissolve 1 g of the drier (about 65 mg of Mn)
in 20 ml each of acetone and benzene, and add 5 ml of 4M hydrochloric
acid and 0.5 g of ascorbic acid.   Shake for 1 min, add 25 ml of 0.1M
EDTA, adjust to pH 10 and titrate excess of EDTA with 0.1M zinc
sulphate, using Erio T as indicator.

Zinc siccative.   Dissolve 1 g of the sample (about 100 mg of Zn) as
above.   Add 5 g of hexamine and titrate with 0.1M EDTA to the yellow
colour of Xylenol Orange.

Calcium siccative.   Prepare the sample as above.   Determine calcium
in the usual way with EDTA after addition of a small amount of Mg-
EDTA complex, with Erio T as indicator.   Thymolphthalexone can be
used instead of Erio T.

Analysis of combined siccatives

In the analysis of combined driers containing two or more metals,
various masking agents can be used to make the individual titrations
more selective.   Graske [5] has given a procedure for analysis of
siccatives containing lead, cobalt, zinc and manganese.   The sample
is dissolved in benzene-ethanol mixture, the solution is divided into
four parts, and each is treated with an excess of EDTA and made
ammoniacal with pH-10 buffer.   For all the back-titrations a
manganese sulphate solution is used.   The sum of Pb + Co + Zn + Mn
is determined in the first part of the solution (without any masking).
In the second, Pb + Mn is determined after masking of Co and Zn with
potassium cyanide.   In the third part zinc is determined after
demasking from its cyanide complex by addition of formaldehyde.   In
the fourth part lead is masked with sodium diethyldithiocarbamate.

REFERENCES

1.   J. Polorný and J. Přibyl, Chem. Zvesti, 1955, 9, 20.
2.   G. Leggieri, Chimica (Milan), 1955, 10, 287.
3.   C.A. Lucchesi and C.F. Hirn, Anal. Chem., 1958, 30, 1877.
4.   G. Veldvijk, Verfkroniek, 1968, 41, 362; Anal. Abstr., 1970, 18,
      1059.
5.   A. Graske, Dig. Fed. Soc. Paint Technol., 1961, 33, 855; Anal.
      Abstr., 1962, 9, 3348.

CHAPTER 21

Analysis of Electroplating Solutions

Complexometry has found one of its widest fields of application in the analysis of electroplating baths. This is a consequence of the process-control nature of these analyses, for which titration methods are very suitable. Before the introduction of complexometry, titration methods were used in plating-bath analysis only for the determination of acids, bases, boric acid, nickel, chromium, cyanide, chloride and formate. Needless to say, several instrumental methods such as polarography, spectrophotometry and atomic-absorption spectrometry have been described in the literature for the determination of metallic components of plating baths, but all these methods require equipment that is rather expensive and not available in small narrowly specialized laboratories. On the other hand complexometric analysis can be performed with very simple equipment such as common glassware (standard flasks, pipettes, burettes) and a minimum number of chemicals. A technician can acquire sufficient training within one or two days.

It is therefore not surprising that complexometry penetrated this field even in the early stages of its development [1 - 3]. Many papers were published on this topic and can be found in both analytical and non-analytical journals (Metal Finishing, Plating, Metalloberflache, Elektrie, Electroplating Metal, etc.)

The continuous advances in electroplating technology have made it necessary to develop new analytical procedures. A brief description of the analytical procedure is not sufficient in some of the more complicated analyses since it does not give an explanation of difficulties and failures that might be encountered. Therefore a more detailed discussion of this useful analytical field will be given below. As an aid to understanding, the baths may be divided into three groups: acid baths, alkaline baths containing cyanide and metal cyanide complexes, and special baths (containing for example gold, palladium, indium, gallium etc.). Chromium-plating baths can be considered separately.

21.1 ACID BATHS

According to Borchert [4] such baths usually have the following compositions (concentrations in g/l.).

> Copper plating solution: Cu 50 - 60, H_2SO_4 30
> Nickel plating solution: Ni 15 - 70, H_3BO_3 0 - 40
> Zinc plating solution: Zn 60 - 70, Al 1.5, H_2SO_4, 1
> Tin plating solution: Sn 20 - 30, H_2SO_4 30, phenol up to 64
> Lead plating solution: Pb 100 - 120, HBF_4 up to 125

Hence plating baths are rather concentrated metal solutions, so only a small volume of sample needs to be taken for the analysis, usually 1 - 2 ml. This produces a drawback often not expected by an inexperienced worker. For example, if the difference between two titrations of 1 ml of a copper bath is 0.1 ml of 0.05M EDTA, the difference in the calculated copper content will be 318 mg/l. This error will increase with increasing equivalent weight of the metal titrated. In the determination of lead a difference of 0.1 ml of 0.05M EDTA represents 1.03 g of Pb per litre. Such errors can be decreased by using finely graduated burettes (subdivided in 0.02-ml divisions) and calibrated pipettes, but it is preferable to dilute the sample accurately (with calibrated pipettes and standard flasks) and to use large volumes of diluted sample and of titrant. The precision also depends on the properties of the indicator used, the colour change of which should be not only sharp but also lie as close as possible to the end-point (see Section 2.1.4, p. 18).

The determination of individual metals in these baths is in fact very simple, but detailed descriptions will be given, for the reasons already stated.

21.1.1   Analysis of Copper Plating Baths

Copper is usually determined by titration with 0.05 - 0.1M EDTA. Indicators used include murexide, Glycinethymol Blue, and Xylenol Orange "activated" with a trace of 1,10-phenanthroline or as its oxidized form (see pp. 36, 49 and 290).

Determination with murexide. Dilute 1 ml of the bath with 50 ml of water and add ammonia till the solution smells of it slightly. Titrate with 0.05M EDTA from yellow-green to pure violet.

Remark. An insufficient amount of ammonia causes an unsharp end-point. In that case add one or two drops more ammonia solution and continue with the titration (one drop of EDTA may be sufficient).

Determination with Glycinethymol Blue. Dilute 2 ml of the bath with 50 ml of water. After addition of the indicator adjust the pH to 5 - 5.5 with 3 ml of 20% hexamine solution. Titrate with 0.1M EDTA to the colour change from blue to emerald-green.

Remark. This titration gives more accurate results because it is not disturbed by any calcium and magnesium present in the electrolyte.

Determination with Xylenol Orange. The indicator cannot be used directly, because it is blocked by cupric ions (the metal-indicator complex reacts too slowly with EDTA). Its oxidation product, or strictly speaking, its decarboxylation product, see p. 290) reacts specifically with copper, and it behaves similarly to Glycinethymol Blue.

Preparation of the indicator [4].   Dissolve 0.1 g in 100 ml of water,
add 3 ml of 30% hydrogen peroxide, and boil for 10 min.   Cool, and
dilute to the original volume.   This solution is stable.

21.1.2   Analysis of Nickel Plating Baths

Dilute 2 ml of the bath with water and add concentrated ammonia to
make the solution strongly alkaline.   The colour change at the end-
point is from yellow, through orange, to bright violet.

Determination of nickel and zinc

An excess of EDTA is added, followed by Schwarzenbach buffer, and the
surplus EDTA is titrated with magnesium sulphate solution (Erio T as
indicator).   This gives Ni + Zn.   2,3-Dimercaptopropanol is added
to mask zinc, and the EDTA liberated is titrated with the magnesium
solution [5].

21.1.3   Analysis of Zinc Plating Baths

Determination of zinc with Erio T.

Dilute 1 ml of the bath with 50 - 100 ml of water, add 10 ml of
ammonia buffer (pH 10) and a little Erio T, and titrate with 0.05M
EDTA from wine-red to steel-blue.

Determination of zinc with Xylenol Orange

Dilute 1 ml of the bath with water as above, add a few drops of 0.5%
aqueous solution of Xylenol Orange, then solid hexamine in small
portions until the solution becomes red-violet, and then a little
extra.   Titrate with 0.05M EDTA until pure lemon yellow.   If the
colour fades before the end-point add a small amount of hexamine.

21.1.4   Analysis of Tin Plating Baths [3]

Mix 2 ml of the bath with exactly 20 ml of 0.1M EDTA.   Neutralize
the solution with ammonia, add 5 ml of ammonia buffer (pH 10) and
some Erio T and titrate with 0.1M magnesium sulphate solution from
blue to wine-red.

21.1.5   Analysis of Lead Plating Baths

Because of the very high concentration of lead, the bath must be
sufficiently diluted.   Dilute 25 ml to 500 ml in a standard flask
and mix thoroughly.   Take 25 ml of this solution, dilute with 100 ml
of water, add Xylenol Orange and neutralize with ammonia solution
(1 + 1) till violet.   Add dilute nitric acid until the solution is
yellow, then hexamine to restore the violet colour.   Titrate with
EDTA to the sharp colour change to lemon yellow.

The procedures described above are convenient for solutions of the
"pure" electrolytes.   Traces of other metals interfere according to
their chemical character.   Some of them are co-titrated (the results
are not much influenced) but some block the indicator, and the colour

changes of the indicators then do not correspond to those expected,
making the end-points of the titrations uncertain.   Individual masking agents usually help to improve the end-point.

21.2   ALKALINE BATHS

These plating solutions contain an excess of cyanide, complex metal
cyanides, sodium hydroxide or carbonate and auxiliary compounds
(tartaric acid etc.).   According to Borchert [4] such baths have the
following compositions (quantities in g/l.).

> Silver plating solution: Ag 3 - 30, CN⁻ 20 - 30, K_2CO_3 0 - 100
> Copper plating solution: Cu 14 - 20, CN⁻ 22 - 27, NaOH 40 - 90
> Brass plating solution: Cu 10 - 30, Zn 5 - 30, CN⁻ 25 - 80,
>   NaOH 0 - 15
> Zinc plating solution: Zn 20 - 35, CN⁻ 30 - 50, NaOH 40 - 90
> Cadmium plating solution: Cd 10 - 20, CN⁻ 40 - 50
> Gold plating solution: Au 0.5 - 4, CN⁻ 2 - 10, Na_2HPO_4 20 g

Several papers have been published describing the complexometric
analysis of these baths.   The experiences of the authors can differ
in details.   For this reason several such methods are described for
comparison.

21.2.1   Analysis of Silver Plating Baths

Determination of silver, sodium cyanide, sodium carbonate and sometimes iron as an impurity, is usually required in the analysis of
these baths.

Direct titration of silver with EDTA is not suitable since the Ag-
EDTA complex has too low a stability.   Nevertheless a method has
been published based on its EDTA titration with Thymolphthalexone as
indicator (see p. 218).

For the indirect determination of silver the well-known displacement
reaction is used:

$$Ni(CN)_4^{2-} + 2Ag^+ = 2Ag(CN)_2^- + Ni^{2+}$$

Nickel is then determined in ammoniacal medium by EDTA titration,
with murexide as indicator.   These methods require complete decomposition of the bath.   A certain disadvantage lies in the high
equivalent for silver $\left(2Ag \equiv Ni;\right.$ 1 ml of 0.05M EDTA = 10.79 mg of Ag$\left.\right)$.

Procedure (Konishi [6]).   Add 2 g of potassium persulphate to 5 ml
of the bath and evaporate the solution almost to dryness.   Moisten
the residue with a few drops of nitric acid (1 + 1) and heat again.
Cool, then add 150 ml of water, 15 ml of concentrated ammonia
solution and 5 ml of 5% $K_2Ni(CN)_4$ solution.   Titrate the liberated
nickel with 0.05M EDTA, using murexide as indicator.

Olivero and Villata [7] use a similar approach.   They prepare a
solution of the $Ni(CN)_4^{2-}$ complex ion by titration of 25 ml of 0.1M
nickel sulphate (diluted to 150 ml) with 0.1M potassium cyanide in
ammoniacal medium, using murexide as indicator.   A sample of silver
electroplating solution (5 ml) containing 5 - 25 g of silver per litre

is decomposed by evaporation with sulphuric acid and nitric acid.
The residue is dissolved in 100 ml of water and made slightly
ammoniacal.   To this solution is added the previously prepared
solution of $Ni(CN)_4^{2-}$, and the nickel released is titrated with 0.1M
EDTA as above.

Lalitha and Natarajan [8] decompose the bath sample (containing 10 -
30 mg of Ag) by addition of bromine until a yellow colour appears and
then boiling for 10 min.   This step is repeated with a small
addition of bromine.   Then 10 - 20 ml of dilute ammonia solution and
5 ml of $K_2Ni(CN)_4$ solution (about 2% of Ni) are added and the
liberated nickel is titrated as previously.

Malát and Holeček [9] have paid attention to the determination of
nickel in cyanide silvering baths.   Such electroplating solutions,
containing $K_2Ni(CN)_4$, KCN, $KAg(CN)_2$ and K_2CO_3, are used for silvering
steel, copper, brass and nickel parts.   The presence of $K_2Ni(CN)_4$
increases the hardness of the plate and its wear resistance.   The
authors used for the determination of nickel the displacement reaction
with silver (see equation above).   However in this case all the free
cyanide must also be bound as $Ag(CN)_2^-$ or it will interfere in the
subsequent EDTA titration of nickel with murexide as indicator.   The
nickel is displaced from its cyanide complex by using an ammoniacal
solution of silver chloride, and a silver nitrate solution.   The
solution obtained is completely clear and the nickel can be determined
either by direct EDTA titration (murexide) or by back-titration of
excess of EDTA with magnesium sulphate solution (Erio T).

Procedure.   Dilute a 5 ml sample of the silvering bath with about
100 ml of water.   Add 1 - 2 ml of 5% ammonia solution, 4.5 ml of
ammoniacal silver chloride solution (71.67 g of AgCl + 600 ml of
concentrated ammonia solution per litre) and then, dropwise, just
sufficient 1M silver nitrate to cause dissolution of any precipitate
formed.   If a slight turbidity persists, clear the solution by
addition of a few drops of 5% ammonia solution.   Add a pinch of 1%
solid mixture of murexide with sodium chloride and titrate with 0.05M
EDTA to the colour change from yellow to red-violet.   For the
indirect determination of nickel prepare the sample in the same way
but add a known and excessive volume of 0.05M EDTA.   If a slight
turbidity develops, clear the solution by gentle warming.   Add
Erio T (1% solid mixture with sodium chloride) and titrate the excess
of EDTA with 0.05M zinc solution from blue to red-violet.

21.2.2   Analysis of Copper Plating Solutions

For the complexometric determination of copper in cyanide baths it is
necessary to remove cyanide completely either by evaporation with
strong acid or by oxidation with hydrogen peroxide [10,11].   The
evaporation with acid is not pleasant because of the HCN liberated.
The method described by Dubský [12] seems very convenient.

Procedure.   Dilute 10 ml of the bath to 100 ml in a standard flask.
To 10 ml of this diluted solution add sufficient water, 1 - 2 g of
glycine and 3 ml of 30% hydrogen peroxide solution.   Warm the
solution slowly and carefully.   During this warming the solution
becomes blue (Cu-glycine complex).   After 10 min of boiling, cool
the solution to 80°C, add 20 ml of pyridine-acetate buffer (pH 5.5)
and some Xylenol Orange (1% solid mixture with sodium chloride) until

the solution is violet.   Titrate with 0.05M EDTA to light green-yellow.

Remarks.   If the boiling is too prolonged, any hexacyanoferrate present will decompose, (as will tetracyanozincate in brass analysis), and the solution will become cloudy owing to hydrolysis.

21.2.3   Analysis of Zinc Plating Solutions

The baths contain an excess of cyanide, $Zn(CN)_4^{2-}$, sodium hydroxide and carbonate.   For the determination of zinc it is not necessary to remove the cyanide by decomposition with acids.   Most methods are based on the usual demasking of the zinc-cyanide complex with formaldehyde.   Zinc is then determined by direct titration with EDTA.

Procedure [3].   To 2 ml of the bath add formaldehyde solution drop-wise till the appearance of the first turbidity.   Then dilute with 20 ml of water, add 5 ml of ammonia buffer (pH 10) and titrate with 0.01M EDTA, using Erio T as indicator.

Heyer [13] recommends determining zinc in slightly acid solution after demasking from its cyanide complex with formaldehyde.

Procedure [13].   Dilute 5 ml of the bath to 50 ml in a standard flask with water.   Dilute 10 ml of this solution with 70 ml of water, add 20 ml of 4% formaldehyde solution, let stand for at least 1 min and then neutralize the solution with 6M hydrochloric acid to the yellow colour of Phenol Red.   Add a few drops of acid in excess (pH 2 - 3).   In the presence of copper the solution is slightly turbid but clears after addition of 1 ml of 1M thiourea.   Then add 8 ml of 1M hexamine (to give pH 6 - test with indicator paper) and Methylthymol Blue to give an intensely blue colour.   Titrate with 0.05M EDTA to pure yellow.

21.2.4   Analysis of Cadmium Plating Solutions

Determination of cadmium in cyanide baths is practically identical with the determination of zinc.   Kubišta [3] uses Erio T and ammoniacal medium.

Procedure [3].   To 5 ml of the bath add 5 ml of formaldehyde and let stand until the solution is turbid.   Add 5 ml of ammonia buffer (pH 10), dilute with 50 ml of water and titrate with 0.1M EDTA, using Erio T as indicator.

Heyer [13] recommends Methylthymol Blue as more convenient.

Procedure [13].   Dilute 5 ml of the bath to 50 ml in a standard flask.   Dilute 10 ml of this solution with 70 ml of water and add 5 - 10 ml of concentrated ammonia solution.   After addition of formaldehyde let stand for 1 min.   Then add Methylthymol Blue, to obtain an intensely blue colour, and titrate with 0.05M EDTA.   The colour change from blue to yellow-grey is very sharp.

Determination of cyanide and zinc (or cadmium)

The main problem is how to determine the amount of free or total

cyanide in cadmium or zinc cyanide baths. Two methods are based on
indirect determination of free cyanide by binding it with nickel, and
a third is based on the titration of cyanide with silver nitrate and
subsequent complexometric determination of zinc (or cadmium) in the
same solution. All three will be given here.

Determination according to Olivero and Villata [14]

Dilute 1 ml of the bath with 200 of water, add 10 ml (accurately
measured) of 0.1M nickel sulphate, enough ammonia solution (1 + 1) to
dissolve the precipitate, and 5 or 6 drops of saturated aqueous
murexide solution.   Dilute to 500 ml and titrate with 0.1M EDTA.
Under these conditions any zinc present is co-titrated with the
excess of nickel (not bound to cyanide) and so must be determined
separately. A second sample is therefore titrated with EDTA at
pH 10 (Erio T as indicator), then zinc is demasked with formaldehyde
and titrated.   The cyanide content is calculated from the results of
both titrations.

According to the authors sodium carbonate (up to 60 g/l.) and
sulphide (up to 1 g/l.) do not interfere.

Determination according to Heyer [13]

Dilute 5 ml of the bath to 100 ml in a standard flask.   Take an
aliquot containing 10 - 25 mg of zinc (or 15 - 30 mg of cadmium),
dilute with 70 ml of water and add 10 ml of concentrated ammonia
solution.   Add slowly, with stirring, 10 ml (accurately measured) of
0.1M nickel sulphate.   After addition of murexide the solution must
be yellow.   If not, add a further 2 - 3 ml of nickel sulphate (again
measured accurately).   Titrate the excess of nickel and all the zinc
(or cadmium) with 0.05M EDTA.   Add, with stirring, 10 ml of 0.1M
silver nitrate.   Titrate the displaced nickel (corresponding to CN^-)
with 0.05M EDTA.

Remark.   In the second titration an excess of silver nitrate must be
used, otherwise the end-point is sluggish.

Calculation.   If the volume of $NiSO_4$ used is A ml, consumption of
0.05M EDTA in the first titration B ml, and in the second titration
C ml, then with use of a 0.5-ml sample

$$[NaCN] = 19.6 \, C \, g/l.$$

$$[Zn] = 6.54 \cdot (B - A + C) \, g/l.$$

Determination according to Salka [15]

This author considers the Liebig-Denigès argentometric determination
of cyanide as not satisfactory. The end-point given by the turbidity
of silver iodide is dependent on the amount of potassium iodide added
and comes too soon or too late.   He recommends determining cyanide
in sodium hydroxide medium, with diphenylcarbazide as indicator.
Zinc can be determined in the same solution by EDTA titration.   This
method seems to have fallen into oblivion, but because it is simpler
than the previous ones it is given here in detail.

Dilute the sample (10 ml) to 100 ml in a standard flask.   Dilute a
10-ml aliquot to 100 ml with water.   Add 5 ml of 20% sodium hydroxide

solution and 5 - 8 drops of indicator (1% diphenylcarbazide solution in ethanol). Titrate with 0.1M silver nitrate to the sharp colour change from red to yellow. The consumption of silver nitrate corresponds to the total amount of cyanide. Zinc is present in the solution as tetrahydroxozincate.

To the same solution add 10 ml of 1M ammonium chloride and 5 ml of concentrated ammonia solution. Adjust the alkalinity to pH 8 - 8.8 with dilute hydrochloric acid. Titrate the zinc with 0.1M EDTA, using Erio T as indicator. From the two titrations the $Zn:CN^-$ ratio as well as the total cyanide content can be calculated.

21.2.5   Analysis of Brass Plating Solutions

The baths contain NaCN, Na_2CO_3, and NaOH in addition to $Cu(CN)_3^{2-}$ and $Zn(CN)_4^{2-}$. The methods are mostly based on demasking of zinc from its cyanide complex. Kubišta [3] uses two samples. In one (2 ml) he determines the sum of Zn + Cu after decomposition of the cyanides with sulphuric acid. Both metals are determined in ammoniacal solution by EDTA titration, with murexide as indicator. In another 2-ml sample zinc is determined after demasking with formaldehyde.

Procedure for zinc. To 2 ml of sample add formaldehyde dropwise till the appearance of a slight turbidity. Dilute the mixture with 20 ml of water, add 5 ml of ammonia buffer (pH 10) and titrate with 0.1M EDTA, using Erio T as indicator.

Much attention has been paid to the displacement of zinc from its cyanide complex. Besides formaldehyde, chloral hydrate (see p. 61) and some other compounds such as acetone, ethyl methyl ketone, benzaldehyde and cyclohexanone [16] can be used for the demasking. Acetone has proved to be very satisfactory, especially in the analysis of baths containing copper. In Dubský's procedure [17] 30 - 50 ml of acetone and the usual ammonia buffer are added to the bath sample and zinc is titrated directly with EDTA (Erio T as indicator).

The method can also be applied to the analysis of cadmium plating solutions, except that excess of EDTA is added and back-titrated with magnesium sulphate solution, with Erio T as indicator.

There have been some attempts to avoid demasking the zinc and decomposing cyanide baths with acids. For example, zinc can be precipitated from cyanide solution as ZnS with 30% sodium sulphide solution and the precipitate dissolved in hydrochloric acid and the zinc determined by EDTA titration [17]. In non-cyanide solutions any copper must first be removed by precipitation with sodium hydroxide.

Analysis of brass plating baths according to Borchert [18]

Determination of zinc. After decomposition of the cyanide (by boiling with 10 ml of 3% hydrogen peroxide for 3 min carefully add 10% thiourea solution until the solution is decolorized and then 2 ml in excess. Add Xylenol Orange and 10% hexamine solution until the solution is red-violet (formation of XO-Zn complex). Titrate with 0.05M EDTA from red-violet to lemon-yellow.

Remark. Thiourea and hexamine must not be added in great excess, because at pH > 6 copper tends to precipitate as sulphide if present in large concentration.

Determination of copper.   After the zinc titration add to the solution 10 ml of 30% hydrogen peroxide.   The yellow solution immediately turns red-violet.   Continue with the EDTA titration from red-violet, through blue, to light green.

Remark.   Under normal conditions Xylenol Orange is blocked by copper, but is decarboxylated by the hydrogen peroxide added (see pp. 36, 290) and is then a specific indicator for copper [19].

21.2.6   Analysis of Nickel Plating Solutions

Besides simple nickel baths (containing magnesium) there are various alkaline baths containing the cyanide complexes of nickel and zinc for plating out Ni-Zn alloys.

In the analysis of Ni-Mg baths, Yurist and Shakhova [20] determine nickel indirectly by back-titration of 0.1M EDTA with 0.1M ferric chloride (sulphosalicylic acid as indicator).   They determine the sum of Ni + Mg in ammoniacal solution with Erio T as indicator.

Analysis of nickel-zinc plating solutions

These baths are used for the deposition of zinc-nickel alloys. Amsheeva [21] described a slightly tedious method for the determination of zinc and nickel in these baths.   It is based on successive demasking of zinc and then nickel with formaldehyde.   Her method will be given only briefly.

Determination of zinc.   Dilute 1 ml of sample with water to 100 ml and mix it with 2 ml of 40% formaldehyde solution.   After adjusting the pH, titrate the zinc with EDTA.

Determination of nickel.   Dilute a 20 or 25 ml sample with 200 ml of water and mix it with 30 ml of 40% formaldehyde solution.   After 15 min neutralize the solution with hydrochloric acid (1 + 1) to Congo Red paper, warm to 70 - 80°C and precipitate nickel in ammoniacal medium with dimethylglyoxime.   Dissolve the precipitate in hydrochloric acid, evaporate to 2 - 3 ml, and determine the nickel by adding excess of EDTA and back-titrating at pH 10 with 0.05M zinc sulphate.

Remarks.   Amsheeva analysed baths containing 25 - 35 g of zinc, 0.2 - 0.5 g of nickel, 65 - 75 g of sodium cyanide, 70 - 80 g of sodium hydroxide and 2 - 3 g of sodium sulphide per litre.   The Zn:Ni ratio is very inconvenient for the determination of nickel without its separation.   With a more convenient ratio we can use a simpler method: after decomposition of the sample with acid the sum of Ni + Zn is determined indirectly by addition of excess of EDTA and back-titration with magnesium sulphate (Erio T), and then direct titration of zinc after its demasking with thioglycollic acid.

21.2.7   Determination of Impurities in Brass and Nickel Baths

Two complexometric procedures have been published..   The first, for

the determination of iron in copper solutions, was described by
Borchert [18] and is based on the masking of copper with thiourea in
the presence of ammonium fluoride which was worked out in our
laboratory [22] (see p. 116).

The second was published by Ernest [23], and describes determination
of copper, zinc and cadmium in nickel plating solutions. This author
used chromatographic ion-exchange separation of all three metals as
follows. A 50-ml sample is mixed with 45 ml of concentrated hydrochloric acid and passed at 15 ml/min through a column of 25 ml of
Dowex 1 (in chloride form). Nickel is eluted with 6M hydrochloric
acid, copper with 120 - 150 ml of 3.5M hydrochloric acid and cadmium +
zinc with 0.005M hydrochloric acid. The individual metals are
determined by EDTA titration. The author described further possibilities for separation. The results were accurate to 1%. Saccharin
and 2-furaldehyde (added as brighteners) do not interfere in the ion-
exchange separations.

21.3   ANALYSIS OF CHROMIUM PLATING SOLUTIONS

For the analysis of chromium baths the following determinations are
required: tervalent and sexivalent chromium, iron, sulphate, and in
some cases fluoride, zinc, nickel. The proposed methods are
relatively simple, such as iodometric determination of chromate, or
sulphate as $BaSO_4$, etc. Nevertheless complexometry can speed up
some determinations as will be shown below.

Determination of sulphate [24]

To 3 ml of the plating solution add 25 ml of 16% hydroxylamine hydrochloride solution and heat to 75°C. Precipitate the sulphate with
10% barium chloride solution. After 5 min cool the mixture and
centrifuge, wash the precipitate and dissolve it in 20 ml of 0.02M
EDTA. Add 3 ml of concentrated ammonia solution and titrate the
excess of EDTA with magnesium sulphate solution, using Erio T as
indicator. Trace amounts of copper, iron, nickel and manganese do
not interfere.

Remarks. A critical study of determination of sulphate in chromium
baths was published by Uhlig and Schäfer [25]. After extraction of
chromate with tributyl phosphate they determined sulphate by EDTA
titration and compared the results with those obtained by iodometric,
nephelometric and volumetric (rhodizonate) determinations. Goldstein
[5] has given a method similar to that of White and Henry [24], but
precipitates the barium sulphate in presence of picric acid.

Determination of iron

The classical determination of iron as Fe_2O_3 has been replaced by
simple EDTA titration [26].

Dilute 5 ml of the sample to 150 - 200 ml with water, adjust the pH to
2.5, add 10% tiron solution as indicator and titrate with 0.02M EDTA
until the blue-green colour is just discharged.

Remark. Salka [27] recommends oxidizing the chromium(III) and
precipitating the iron as $FePO_4$, dissolving this precipitate and
titrating the iron with EDTA, using ferron (7-iodo-8-hydroxyquinoline-

5-sulphonic acid) as indicator.   The colour change from green to yellow at the end-point is very sharp.   The method is suitable for iron concentrations up to 15 - 20 g/l.

Determination of zinc in chromium baths

Tajima and Kurobe [28] recommend the extraction of zinc from chromium baths as the $Zn(SCN)_4^{2-}$ complex, into isobutyl methyl ketone.   After stripping (together with copper) with ammonia solution, the zinc is determined by EDTA titration as usual.   If copper is present, both must be masked with potassium cyanide and then zinc demasked with formaldehyde.

Determination of chromium(III)·in chromium plating solutions

Two complexometric methods for chromium(III) have been published. Both require prior separation of chromate.

Procedure according to Goldstein [29].   Chromate is separated (together with most of the iron) by two successive extractions from hydrochloric acid medium into isobutyl methyl ketone at 0°C.   The aqueous phase, after addition of excess of 0.0125M EDTA, is adjusted to pH 2.8 - 3.5 and boiled for 10 min (to form the violet Cr-EDTA complex), then back-titrated with 0.005-0.01M thorium nitrate, Xylenol Orange being used as indicator.

Procedure according to Arsen'ev [30].   A sample containing up to 15 g of iron and chromium is diluted to 100 ml in a standard flask. A 10-ml aliquot is diluted with 150 ml of water, and 25 ml of barium chloride solution (containing 25 g of $BaCl_2.2H_2O$, 50 ml of glacial acetic acid and 25 ml of 25% ammonia solution per 500 ml) are added. The mixture is diluted to volume in a 200-ml standard flask, let stand for 5 min, and filtered.   In 100 ml of the filtrate iron is titrated with 0.025M EDTA, with sulphosalicylic acid as indicator. After addition of a further amount of EDTA and boiling for 10 min, 15 ml of 1M sodium hydroxide are added and the unconsumed EDTA is titrated with 0.025M zinc acetate, with Xylenol Orange as indicator.

Remark.   Only one method which allows determination of aluminium and tervalent chromium in the presence of chromate has been worked out, in our laboratory [31].   It is based on the use of DCTA, which forms a complex with aluminium at room temperature and is also stable to boiling even in the presence of chromate [32].   The procedure is as follows.   To the sample solution containing aluminium and not more than 25 mg of chromium(III) and 75 mg of chromium(VI) add sufficient 0.05M DCTA to complex all the aluminium, adjust the pH to 5.5 with hexamine, dilute to 250 ml, add Xylenol Orange, and titrate with 0.05M zinc sulphate.   Add to the titrated solution a further excess of DCTA, boil for 10 min, dilute to 600 ml and titrate again with 0.05M zinc sulphate.   As far as is known, this procedure has not been applied to chromium bath analysis.

21.4   ANALYSIS OF SPECIAL PLATING SOLUTIONS

These electrolytes contain gallium, indium or gold in combination with other metals which improve the quality of the deposits.   During the electrolysis alloy deposits are formed, which have certain desired qualities.   They have great value in electronics.

21.4.1   Analysis of Gallium-Indium Electrolytes [33]

These glycerol baths contain gallium and indium in about 1:25 ratio, and the analysis is based on extraction of gallium with ether.   The electrolyte sample, containing 13-15 mg of gallium, is diluted to 50 ml with water, and made 6M in hydrochloric acid.   The gallium is extracted twice with diethyl ether saturated with hydrogen chloride. The combined ether phase is washed 3 or 4 times with ether-saturated hydrochloric acid.   The ether phase and aqueous phase (plus washings) are evaporated to drive off the ether.   The residue from the organic phase contains all the gallium, that from the aqueous phase contains all the indium.   Both metals are determined by usual EDTA titration.

Remark.   Determination of gallium and indium without previous separation is described on p. 125.

21.4.2   Analysis of Indium-Tin Electrolytes [34]

Dilute the sample, containing about 31 mg of indium, to 50 ml with 2% hydrochloric acid.   Add 0.7 g of sodium fluoride to mask tin, and 1 ml of saturated sodium acetate solution, and adjust the pH to 2.5 with acetic acid.   Heat the solution, add a few drops of Cu-EDTA solution and some PAN solution and titrate with 0.02M EDTA to a clear yellow end-point.   The tin is determined iodometrically (with iodate and iodide) after precipitation as metastannic acid, dissolution and reduction with iron powder.

21.4.3   Determination of Gallium in Gold Baths [35]

Mix 2 ml of the electrolyte with 50 ml of pH-2.2 buffer (50 ml of 0.2M potassium phthalate + 64.6 ml of 0.2M hydrochloric acid, diluted to 200 ml with water).   Add 30 ml of 0.01M EDTA and some Bromopyrogallol Red as indicator, boil the solution (slow formation of Ga-EDTA complex) and back-titrate with 0.01M bismuth nitrate.   The colour change is from orange-red to violet-red.

According to the author, tartrates, $Au(CN)_2^-$, glucose, and mg amounts of copper, zinc, cadmium and lead do not interfere.   Iron is co-titrated, and nickel and cobalt cause errors of up to +10%.

21.4.4   Determination of Indium in Gold Baths [36]

Adjust the pH of 20 ml of the bath to pH 3 with 0.1N sulphuric acid, dilute to 50 ml, heat to 70°C and titrate with 0.02M EDTA, using Xylenol Orange as indicator.   Iron, aluminium, copper, lead, nickel and zinc do not interfere if their total amount does not exceed 1% of the indium content.

21.4.5   Determination of Indium, Nickel and Zinc in Gold Baths [36]

These baths are suitable for gold-alloy plating, because the thin coatings obtained are mechanically resistant compared with pure gold, which cannot be used for such purposes.   One type of bath contains these elements as their citrates.   The citric acid must first be decomposed.

Sample preparation.   To 25 - 50 ml of the plating solution add 25 ml
of sulphuric acid (1 + 1) and 10 ml of concentrated nitric acid.
Evaporate to white fumes, add 10 ml of concentrated sulphuric acid
and evaporate again.   Cool, carefully add 100 ml of water, boil for
2 - 5 min, and allow to stand for 2 hr.   Filter off the gold on paper
and wash with hot water.   Transfer the filtrate to a 250-ml standard
flask and dilute to the mark.

Determination of gold.   Ignite the filter paper and weigh the
residue of gold.

Determination of nickel.   To an aliquot of the solution, containing
10 - 80 mg of nickel, add sufficient 0.05M EDTA to complex all the
nickel, 25 - 30 ml of concentrated ammonia solution and 10 ml of 20%
mercaptoacetic acid solution, dilute to 150 - 200 ml with water and
allow to stand for 3 - 5 min.   Add some Methylthymol Blue and titrate
with 0.05M calcium chloride to an intense blue.

Determination of indium.   To another aliquot of the solution,
containing 25 - 125 mg of indium, add 5 - 10 ml of nitric acid (1 + 1)
and 20 ml of 20% triethanolamine solution.   Mix, add 25 - 30 ml of
concentrated ammonia solution and 0.5 - 1.0 g of potassium cyanide,
and allow to stand for 5 - 6 min.   Then add sufficient 0.05M DCTA to
complex all the indium and allow to stand for 15 min.   Heat the
solution to 40°C and titrate with 0.05M calcium chloride, using
Methylthymol Blue as indicator.

Determination of In + Ni + Zn.   Dilute an aliquot of the solution,
containing 10 - 20 mg of Ni, 25 - 125 mg of In and 3 - 40 mg of Zn, with
water and add sufficient 0.05M EDTA to complex all three metals
present.   Add 35 - 50 ml of concentrated ammonia solution and either
allow the solution to stand for 15 min or heat it slightly and titrate
as above with 0.05M calcium chloride (Methylthymol Blue as indicator)
to an intense blue colour.

Remark.   Alternatively determine all three metals in acidic medium
by adding an excess of 0.05M EDTA and back-titrating with zinc or
lead solution at pH 5.5 (hexamine buffer), using Xylenol Orange as
indicator.

In both cases the value for zinc is obtained from the difference
between this sum and the sum of Ni + In found from previous
titrations.

21.4.6   Determination of Indium, Nickel and Aluminium in Gold Baths
         [37]

The method is very similar to the one just described.   After
separation of the gold, nickel is determined in one portion of the
filtrate after masking of aluminium with triethanolamine and indium
with thioglycollic acid.   In a second portion the sum of Al + Ni + In
is determined in slightly acidic medium by back-titration of excess
of EDTA with lead nitrate solution, with Xylenol Orange as indicator.
In the same solution, after the usual demasking of aluminium with
ammonium fluoride, the liberated EDTA is titrated with the lead
solution to obtain the aluminium content.

21.4.7 Analysis of Cyanide Gold Baths

Borchert [38] described as an example the analysis of a bath having
the average composition (g/l.): Au 2.5, Cu 0.5, Ni 0.6, Ag 0.2,
Na_2HPO_4 70 and NaCN 10.

Principle. A 10-ml sample is decomposed with 20 ml of concentrated
hydrochloric acid and evaporated to 2 - 3 ml. After addition of
2 - 3 ml of aqua regia the solution is heated to boiling, diluted to
100 ml, and boiled again. The precipitated AgCl is filtered off and
determined complexometrically by the displacement reaction with
$Ni(CN)_4^{2-}$. In the filtrate Cu + Ni is determined by EDTA titration,
with murexide as indicator. After the titration the solution is
made slightly acid and gold is determined similarly to silver, with
$Ni(CN)_4^{2-}$. In a separate portion of sample nickel and copper are
determined successively (masking of Cu with mercaptoacetic acid)
both determinations being done by back-titration of excess of EDTA
with calcium chloride (Thymolphthalexone as indicator).

21.4.8 Determination of Antimony in Gold-Antimony Baths [39]

Bem and Rubel studied several methods for the determination of
antimony in fresh and used baths. They considered the amperometric
titration of antimony with DCTA (dropping mercury electrode) as more
accurate than the usual bromate titration. Details can be found in
the original literature [39,40].

21.4.9 Determination of NTA and Cadmium in Cadmium Baths

Many attempts have been made to use various complexone-type compounds
as a component of electroplating solutions, but only one paper seems
to have been published on analysis of such solutions. This was by
Gorshenina [41] and described determination of cadmium and free NTA
(nitrilotriacetic acid). For the determination of cadmium 1 ml of
the solution is treated with hydrochloric acid and potassium bromate
to decompose the NTA. Cadmium is then determined complexometrically.
The free NTA is determined in a separate portion of the bath by
titration with cadmium sulphate.

21.4.10 Determination of Rochelle (Segnette) Salt

Sodium potassium tartrate is often present in various plating
solutions. Its determination is based on its complex formation with
bismuth [42]. Basic bismuth nitrate is added in excess to the sample
solution in alkaline medium, and the $Bi(OH)_3$ precipitated is filtered
off, dissolved in nitric acid and titrated with EDTA, with Pyrocatechol Violet as indicator.

REFERENCES

1.   K.E. Langford, Analysis of Electroplating and Related Solutions,
     3rd Ed., Draper, Teddington, England, 1962.
2.   N. Hall, Metal Finishing Guide Book, Metal and Plastics Pubs.,
     Westwood, New York, 1966.

3.   Z. Kubišta, Rozbory lazní pro povrchovou úpravu kovu, SNTL, Prague, 1966.

4.   O. Borchert, Abwässer aus der chemischen Oberflächebehandlung, Verlag für Bauwesen, Berlin, DDR, 1966.

5.   M. Goldstein, Chim. Anal. (Paris), 1966, 48, 214; Anal. Abstr., 1967, 14, 4535.

6.   S. Konishi, Metal Finish., 1965, 63, 77; Anal. Abstr., 1967, 14, 1888.

7.   A. Olivero and E. Villata, Galvanotecnica, 1966, 17, 95; Anal. Abstr., 1967, 16, 1779.

8.   K.S. Lalitha and S.R. Natarajan, Metal Finish., 1968, 66, 82; Anal. Abstr., 1969, 16, 1779.

9.   M. Malát and K. Holeček, Chemist-Analyst, 1961, 50, 115.

10. F.W. Gutman, Plating, 1956, 45, 345; Anal. Abstr., 1957, 4, 3855.

11. K.E. Langford, Electroplat. Metal, 1958, 11, 439; Anal. Abstr., 1959, 6, 3537.

12. I. Dubský, Metalloberfläche, 1961, 15, 13; Anal. Abstr., 1961, 8, 3172.

13. F.C. Heyer, Chemist-Analyst, 1967, 56, 86.

14. A. Olivero and E. Villata, Plating, 1966, 53, 770; Galvanotechnica, 1966, 17, 55; Anal. Abstr., 1967, 14, 6014.

15. A. Salka, Metal Finish., 1960, 58, 59; Anal. Abstr., 1961, 8, 48.

16. I. Dubský, Chem. Prum., 1959, 9, 80; Anal. Abstr., 1959, 6, 4694.

17. Ts.I. Dobrobovskaya and P.V. Gorokhovskaya, Kauchuk i Rezina 1957, 12, 31; Anal. Abstr., 1959, 6, 476.

18. O. Borchert, Elektrie, 1962, 16, 254.

19. O. Borchert, Massanalytische Schnellmethoden für Beiten und galvanische Elektrolyte, Deutscher Verlag für Grundstoffindustrie, Leipzig, 1966.

20. I.M. Yurist and P.G. Shakhova, Zavod. Lab., 1959, 25, 1298; Anal. Abstr., 1960, 7, 3272.

21. A.A. Amsheeva, Zavod. Lab., 1962, 28, 278; Anal. Abstr., 1962, 9, 5080.

22. R. Přibil and V. Veselý, Talanta, 1961, 8, 743.

23. G. Ernest, Metalurgia (Bucharest), 1967, 19, 208; Anal. Abstr., 1969, 16, 167.

24. W.W. White and M.C. Henry, Plating, 1972, 59, 429; Anal. Abstr., 1973, 24, 777.

25. F. Uhlig and M. Schäfer, Metalloberfläche, 1966, 20, 269.

26. H.W. Detther, Metal Finish, 1957, 55, 67; Anal. Abstr., 1957, 4, 3323; Metalloberfläche, 1957, 11, 12; Anal. Abstr., 1958, 5, 2634.

27. A. Salka, Metal Finish., 1958, 56, 64; Anal. Abstr., 1959, 6, 3482.

28. N. Tajima and M. Kurobe, Japan Analyst, 1961, 10, 528; Anal. Abstr., 1963, 10, 3240.

29. M. Goldstein, Chim. Anal. (Paris), 1971, 53, 227; Anal. Abstr., 1972, 22, 765.

30. V.I. Arsen'ev, Zavod. Lab., 1971, 37, 416; Anal. Abstr., 1972, 22, 766.

31. R. Přibil and V. Veselý, Talanta, 1962, 9, 23.

32. R. Přibil and V. Veselý, Talanta, 1963, 10, 1287.

33. E.P. Cocozza, Chemist-Analyst, 1960, 49, 46.

34. E.P. Cocozza, Chemist-Analyst, 1960, 49, 124.

35. S. Rubel, K. Brajter and I. Szynkarszuk, Chem. Analit. (Warsaw), 1972, 17, 955; Anal. Abstr., 1973, 24, 3351.

36. S. Rubel and J. Bem, Chem. Analit. (Warsaw), 1971, 16, 595; Anal. Abstr., 1973, 24, 3351.

37.   R. Přibil and V. Veselý, Chemist-Analyst, 1966, 55, 38.
38.   O. Borchert, Elektrie, 1962, 16, 253.
39.   J. Bem and S. Rubel, Chem. Analit. (Warsaw), 1972, 17, 1259;
      Anal. Abstr., 1973, 24, 3352.
40.   S. Rubel and J. Bem, Chem. Analit. (Warsaw), 1972, 17, 279;
      Anal. Abstr., 1972, 23, 3824.
41.   A.P. Gorshenina, Zavod. Lab., 1975, 41, 940; Anal. Abstr., 1976,
      30, 5B77.
42.   S. Sriveeraraghaven and S.R. Natarajan, Metal Finish., 1975, 73,
      37; Anal. Abstr., 1976, 31, 1B34.

CHAPTER 22

Further Applications of Complexometry

The account of practical examples where complexometry can be used
instead of older analytical methods seems to be endless. In some
fields, however, complexometry is used only for the determination of
calcium or magnesium. In others, despite its advantages, it is only
slowly making its way into practical use; this is especially the case
when standard methods have been established for analysis of the
material in question. The preparation of these standards involves a
great deal of effort by a number of specialized and expert laboratories
before they are accepted by the appropriate commission, and put into
print. The same is true for changes and supplements to the standards.
For example, complexometric determination of magnesium in fertilizers
was not included in the AOAC standard methods until 1965 [1], almost
20 years after its first publication. Individual national
pharmacopoeias can also be considered as a special set of standards
for assessing content and purity of pharmaceutical products. One of
the main reasons hindering the application of complexometry to some
fields is that preparation of a new standard is an expensive as well
as a tedious process, which perhaps explains why standards are not
updated too often.

Some areas in which complexometry is sometimes applied are discussed
briefly below.

22.1   PHARMACEUTICAL ANALYSIS

Pharmaceutical compounds containing metals other than alkali metals
are comparatively few in number, and are usually very simple compounds
of calcium, magnesium, zinc, aluminium, bismuth or mercury. The
methods for their determination are generally straightforward applications of the methods developed in the early stages of complexometry,
and have been surveyed [2].

Complexometry can be exploited in another way in this field, however.
Many drugs give insoluble metal compounds, the metal content of which
can be determined by EDTA titration and used for assay of the drugs.
This topic has been dealt with systematically by Budĕšínský, who used
already known reagents, such as Dragendorff's reagent, for the
determination of caffeine, methylcaffeine, codeine, antipyrine etc.

395

Similarly mercuric chloride solution was used for the determination of amidopyrine, theobromine, theophylline, etc.   Among other precipitants studied by Budešínský may be mentioned Marmé's reagent K (K_2CdI_4) and the similar complex $K_2Cd(SCN)_4$.   Copper compounds have also been used for quantitative precipitation of some drugs, for example quinine with an acetone solution of cupric chloride or theophylline with an ethanolic solution of copper acetate.   Silver salts have been used for precipitation of sulphamides and indirect determination of 6-mercaptopurine.

Most methods require a modified working technique, and many compounds must be determined in non-aqueous solution.   A main drawback of the methods, however, is the relatively high equivalence factor, which for many drugs is over 20 mg per ml of 0.05M EDTA.

Further details of these procedures and many other methods can be found in the monograph referred to [2].

22.2   CLINICAL ANALYSIS

Determination of calcium in blood and blood serum is frequently needed in clinical examination.   Much attention has been paid to microdetermination of calcium, as large samples of blood cannot be taken off, especially from babies.   Holasek and Flaschka [3] have described several methods for determination of calcium in blood, serum and urine.

22.3   ANALYSIS OF FOODS

Complexometry occupies only a minor place in the analysis of foods. Most attention in the period 1950 - 1962 was paid to the determination of calcium, magnesium, phosphate and reducing sugars.   Since then little has appeared in the literature, presumably because all problems were regarded as sufficiently solved.   For the information of the reader, some references to the earlier work are given here. At first the applications were confined to determination of calcium in milk [4 - 9], fruit juices [10,11] and other drinks (beer, wine), then attention was paid to the determination of calcium (magnesium) in juices, sugar, syrup and molasses [12 - 18].

In 1954 - 1957 some work was done on determination of reducing sugars. Potterat and Eschmann [19] replaced Fehling's solution by alcoholic Cu-EDTA solution.   After the reaction with glucose and other reducing sugars, the precipitated Cu_2O was filtered off and the copper determined by EDTA titration.   It is well known that the Fehling method is purely empirical and requires careful observation of conditions such as boiling-time, concentration of solution, cooling, washing of the Cu_2O, etc.   These authors described the procedures in detail and gave calibration curves for the determination of glucose, fructose, lactose and maltose.   Their methods were applied by other authors for the determination of sugar in dietetical preparations [20] fermented drinks [21], and so on, and were variously modified [22].

REFERENCES

1.   C.V. Gehrke and J.P. Ussary, Anal. Chem., 1965, 37, No. 5, 67R.
2.   R. Přibil, Komplexometrie, Vol. IV, Pharmaceutisch und organische
     Analyse, Deutscher Verlag für Grundstoffindustrie, Leipzig, 1966.
3.   A. Holasek and H. Flaschka, Komplexometrische und andere
     titrimetrische Methoden des klinisches Laboratoriums, Springer-
     Verlag, Vienna, 1961.
4.   E.R. Ling, Analyst, 1958, 83, 179.
5.   C.W. Raadsveld, Ned. Melk. Zuiveltijdschr., 1956, 10, 114.
6.   G. Christianson, R. Jennes and S.T. Coulter, Anal. Chem., 1954,
     26, 1923.
7.   R. Jennes, Anal. Chem., 1953, 25, 966.
8.   J. Mašek, Prum. Potravin, 1961, 12, 384.
9.   H. Konrad, Z. Lebensm. Unters. Forsch., 1962, 118, 35.
10. M.C. Bennet, Intern. Sugar J., 1958, 60, 225.
11. C.C. Strachan, Can. Food. Ind., 1951, 22, 25.
12. W. Fivian and M. Moser, Zucker-Beihefte, 1951, 4, 317.
13. F. Schneider and A. Emmerich, Zucker-Beihefte, 1951, 4, 53.
14. R. Saunier and A. Lemaitre, Sucrerie Franc., 1951, 92, 231.
15. A. Gee, L.P. Domingues and V.R. Deitz, Anal. Chem., 1954, 26,
     1487.
16. G.S. Benin and S.L. Gutina, Tr. Vses. Tsentr. Nauchn. Issled.
     Inst. Sakharn. Prom. 1955, 3, 206.
17. S. Zagrodski and H. Zaorska, Gaz. Cukrow., 1956, 59, 282.
18. L. Kohanka, Listy Cukrovar, 1955, 71, 191.
19. M. Potterat and H. Eschmann, Mitt. Lebensm. Hyg., 1954, 45, 312.
20. H. Hadorn and H. Suter, Mitt. Lebensm. Hyg., 1955, 46, 341;
     J. Agric. Food. Chem., 1955, 3, 862.
21. H. Rentschler, H. Tanner and P. Dejimk, Mitt. Lebensm. Hyg.,
     1957, 48, 238.
22. H. Eschmann, Chemist-Analyst, 1956, 45, 5.

Index